Analytical Chemistry: Quantitative and Qualitative Analysis

Analytical Chemistry: Quantitative and Qualitative Analysis

Editor: Bernard Wilde

NY RESEARCH
P R E S S

New York

Published by NY Research Press
118-35 Queens Blvd., Suite 400,
Forest Hills, NY 11375, USA
www.nyresearchpress.com

Analytical Chemistry: Quantitative and Qualitative Analysis
Edited by Bernard Wilde

© 2018 NY Research Press

International Standard Book Number: 978-1-63238-564-2 (Hardback)

Cataloging-in-Publication Data

Analytical chemistry : quantitative and qualitative analysis / edited by Bernard Wilde.
    p. cm.
Includes bibliographical references and index.
ISBN 978-1-63238-564-2
1. Chemistry, Analytic. 2. Chemistry, Analytic--Quantitative.
3. Chemistry, Analytic--Qualitative. I. Wilde, Bernard.
QD75.22 .A53 2018
543--dc23

Contents

Permissions

List of Contributors

Index

Preface

Analytical chemistry is the branch of chemistry which separates, identifies and measures matter. The methods used in analytical chemistry can be classified into classical, wet chemical and instrumental methods. It can be applied in a number of fields such as medicine, forensic science, environmental science, etc. This book contains some path-breaking studies in the field of analytical chemistry. A number of latest researches have been included to keep the readers up-to-date with the global concepts in this area of study. This book is an essential guide for both academicians and those who wish to pursue this discipline further.

All of the data presented henceforth, was collaborated in the wake of recent advancements in the field. The aim of this book is to present the diversified developments from across the globe in a comprehensible manner. The opinions expressed in each chapter belong solely to the contributing authors. Their interpretations of the topics are the integral part of this book, which I have carefully compiled for a better understanding of the readers.

At the end, I would like to thank all those who dedicated their time and efforts for the successful completion of this book. I also wish to convey my gratitude towards my friends and family who supported me at every step.

Editor

Determination of nanoparticle size distribution together with density or molecular weight by 2D analytical ultracentrifugation

Randy P. Carney[1,2], Jin Young Kim[1], Huifeng Qian[3], Rongchao Jin[3], Hakim Mehenni[4,5], Francesco Stellacci[1,2] & Osman M. Bakr[4,5]

Nanoparticles are finding many research and industrial applications, yet their characterization remains a challenge. Their cores are often polydisperse and coated by a stabilizing shell that varies in size and composition. No single technique can characterize both the size distribution and the nature of the shell. Advances in analytical ultracentrifugation allow for the extraction of the sedimentation (s) and diffusion coefficients (D). Here we report an approach to transform the s and D distributions of nanoparticles in solution into precise molecular weight (M), density (ρ_p) and particle diameter (d_p) distributions. M for mixtures of discrete nanocrystals is found within 4% of the known quantities. The accuracy and the density information we achieve on nanoparticles are unparalleled. A single experimental run is sufficient for full nanoparticle characterization, without the need for standards or other auxiliary measurements. We believe that our method is of general applicability and we discuss its limitations.

[1] Department of Materials Science and Engineering, Massachusetts Institute of Technology, Cambridge, Massachusetts 02139, USA. [2] Institute of Materials, École Polytechnique Fédérale de Lausanne, Station 12, 1015 Lausanne, Switzerland. [3] Department of Chemistry, Carnegie Melon University, Pittsburgh, Pennsylvania 15213, USA. [4] Center for Solar and Alternative Energy Science and Engineering, King Abdullah University of Science and Technology, Thuwal 23955-6900, Saudi Arabia. [5] Physical Sciences and Engineering Division, King Abdullah University of Science and Technology, Thuwal 23955-6900, Saudi Arabia. Correspondence and requests for materials should be addressed to F.S. (email: francesco.stellacci@epfl.ch) (or) to O.M.B. (email: osman.bakr@kaust.edu.sa).

Hybrid inorganic–organic core–shell nanoparticles (NPs) are finding a wide range of applications in solar cells[1], opto-electronics[2,3], nanophotonics/plasmonics[4], catalysis[5–7], drug delivery[8,9] and biomedical imaging agents[10,11]. Their chemical[12], electronic[13], optical[2,12,14], magnetic[15] and catalytic[13,16] properties, and self-assembly[17,18] inherently depend on their size and composition. Hence, as industrial and research needs grow more complex, it becomes imperative to find a versatile, reliable and scalable method for the full characterization of these particles. The total evaluation of a NP entails a global analysis that pieces together measurements taken from multiple techniques (Fig. 1). However, such an approach usually presents an incomplete picture of the sample in question since NPs are rarely perfectly monodisperse. Methods that characterize the organic shell give only macroscopic averages of the whole sample distribution, and size analysis techniques provide distribution data only of the total NP (see Fig. 1). Combined analyses are possible, but become challenging as the complexity (for example, polydispersity) of the sample increases. For example, the recent advances, where fractionation and size analysis are used in series, provide a wealth of information, but at the expense of numerous assumptions and laborious approaches[19]. Arguably, NP characterization has become a rate-limiting step, hindering the development and prospective uses of these promising materials.

NP research lacks a single platform that quickly, easily and completely characterizes the size, density (an indicator of composition) and molecular weight of each unique particle species among a heterogeneous mixture, with a single experimental measurement. However, recent 2D mathematical and computational modelling advancements in sedimentation-velocity analytical ultracentrifugation (SV-AUC)[20] allow for the mapping of sedimentation coefficient and diffusion coefficient distributions of species present in solution[21,22].

AUC is performed using an ultracentrifuge fitted with one or more optical detection systems, allowing the observation of the fractionation process of a species dissolved in solution. The sedimentation process is monitored by a scanning UV/VIS optical detection system that records the concentration profile, $c(r, t)$, with respect to radial distance from the rotor (r) and time (t). The $c(r, t)$ is subsequently numerically modelled and transformed into a sedimentation coefficient (s) and diffusion coefficient (D) distribution, $c(s, D)$[21].

SV-AUC is a tool particularly suited for the study of NPs[23–26]. First, the technique characterizes the sample in solution; hence it provides an opportunity to observe the properties in conditions similar to conditions present in most applications (for example, self-assembly, solution casting, physiological … etc.). Second, because the data of an AUC experiment is analysed in the framework of fundamental thermodynamic principles[23], it does not require any standard or calibration. Finally, AUC requires little sample (< 1 mg) and minimal preparation and it encompasses virtually any particles soluble in a liquid phase.

The presumed requirement for *a priori* knowledge of the density for the target species has always prevented wider implementation of AUC. This becomes an even greater hurdle in core–shell NPs, because their density depends on the ratio between the size of the core and that of the shell. Despite numerous attempts to circumvent the issue of direct NP density measurement to obtain quantitative AUC characterization[27,28], the problem still exists, particularly for samples for which density depends on size (for example, particles with a fixed length shell but variable sized cores).

Here we demonstrate a simple scheme to fully characterize NPs with 2D SV-AUC that not only overcomes the limitation of *a priori* density measurement, but also allows us to obtain the density distribution of a species, in addition to its size and molecular weight distributions. Our method is enabled by the simultaneous extraction of both the sedimentation and diffusion coefficient distributions from the sedimentation process of the NP species present in the sample. Our approach differs from other SV-AUC studies on NPs—including Svedberg's original experiments[25,26]—

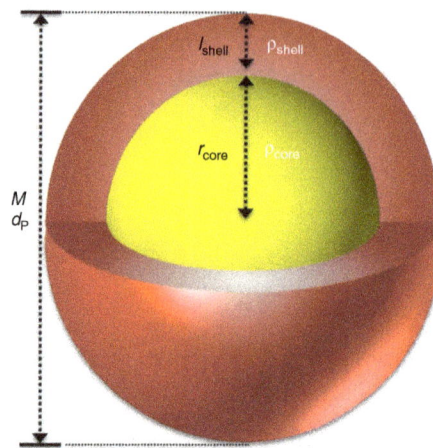

Figure 1 | Typical characterization schemes for core–shell nanoparticles. The core density (ρ_c) is normally taken as the bulk density[47] or quantified by x-ray diffraction assuming a conformation to a particular model. TEM, scanning transmission electron microscopy, and x-ray diffraction are the most commonly used methods to investigate core radius (r_c) and composition; yet, these techniques provide little information on the organic ligand shell (due to the low contrast of organic material). Furthermore, core size distributions extracted from TEM images are generally skewed by the choice of the selected area of the sample, small sampling size (a few thousand particles at most) and the undercounting of the smallest particles (due to their low TEM contrast). To measure total particle diameter (d_p), dynamic light scattering (DLS) is not particularly sensitive for small particles and atomic force microscopy (AFM) or scanning tunnelling microscopy (STM) are slow and share the same sampling selection limitations of TEM. Both size exclusion chromatography and gel electrophoresis require a standard. For small particles (< 40 kDa), mass spectroscopy (MS), in particular electrospray ionization MS (ESI-MS), is the preferred method to measure particle molecular weight (M); but, problems in stability and complexity limit this technique[33]. To characterize the organic ligand shell density (ρ_L), thermogravimetric analysis can be utilized, but it is accurate only for a monodisperse species[48]. Nuclear magnetic resonance (NMR) can also provide information on the composition and density of the ligand shell, but counterions present a major problem, and again, the technique is limited only to large sizes[49]. No single technique can simultaneously measure all six parameters with a single experiment.

in that we require no prior measurement nor make any assumptions regarding the density of the NPs and that we utilize both the diffusion data (ignored or unavailable in most studies) and the sedimentation data obtained from 2D SV-AUC to determine NP density. It should be further noted that we assume a 1:1 correspondence between the hydrodynamic Stokes' diameter and the actual diameter of the particle, which we prove to be an accurate description for a wide range of NPs. As our methodology is simple, rapid, accurate and scalable, we expect the findings in this research article to be useful to anyone interested in the properties and applications of NPs. This method could become useful for those applications that are especially sensitive to NP size and overall variability.

Results

Theory. The Lamm equation describes the evolution of a solute concentration distribution under centrifugation[23,29]:

$$\frac{\partial c}{\partial t} = D\left(\frac{\partial^2 c}{\partial r^2} + \frac{1}{r}\frac{\partial c}{\partial t}\right) - \omega^2 s\left(r\frac{\partial c}{\partial r} + 2c\right) \quad (1)$$

A solution to the Lamm equation is a spatially and temporally resolved concentration function, $c(r, t)$, sigmoidal in shape, real-valued and differentiable. The analytical ultracentrifuge records

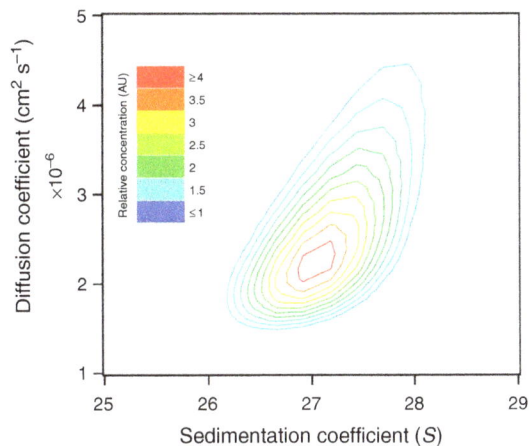

Figure 2 | 2D AUC distribution for $Au_{144}(SR)_{60}$ nanoclusters. The sedimentation and diffusion coefficient distributions for the $Au_{144}(SR)_{60}$ magic-sized nanocluster in toluene (T = 20 °C).

experimental concentration profiles that satisfy the Lamm equation. Although the general Lamm equation is impossible to solve with a closed analytical solution[30], modern computational software (such as SEDFIT[31] or Ultrascan[22]) can directly fit approximate solutions of the Lamm equation to a set of experimental data points for an entire sedimentation process (for example, Supplementary Fig. S1). SEDFIT, used for this work, employs a numerical finite element method with an adaptive grid-size algorithm and moving frame of reference, which drastically improves diffusion modelling compared with other computational models and numerical solutions[20]. The numerical method analysis results in a high-resolution differential distribution of sedimentation coefficients $c(s)$, a model that deconvolutes diffusion from sedimentation and reduces peak broadening. The $c(s)$ model can be easily extended to include two-dimensional size-and-shape information, $c(s, D)$[21]. Finally, all peaks can be tested to ensure authenticity by Monte-Carlo analysis and F-statistics[32].

The basis for AUC's theoretical treatment is in thermodynamic and hydrodynamic principles[23]. During sedimentation velocity experiments, three forces act on a sedimenting solute. A centrifugal force induced by the acceleration of the rotor, $\omega^2 r$, is proportional to particle mass, so that larger particles sediment faster than smaller ones (provided the mass of the particle is greater than the displaced mass of the solvent). The buoyant force (governed by the Archimedes' principle) and frictional force act in opposite direction to the centrifugal force, impeding sedimentation. The frictional force is generated by movement of the solute through the solvent according to the hydrodynamic treatment of viscous drag and is proportional to a frictional coefficient and the solute terminal velocity (u). The three forces come into balance very quickly (within $\sim 10^{-6}$ s[23]) and the particle achieves terminal velocity. Rearrangement of the force balance yields the well-known Svedberg equation[23]:

$$\frac{DM(1-\bar{v}_p\rho_s)}{RT} = \frac{u}{\omega^2 r} \equiv s \qquad (2)$$

where the sedimentation coefficient, s, is defined as the solute terminal velocity per unit centrifugal field. Here ρ_s is solvent density and $\bar{v}_p$ is the partial specific volume, which can generally be equated with the inverse density of the particle (that is, $\bar{v}_p^{-1} = \rho_p$). For a smooth, compact sphere under the limit of low Reynolds number (that is, small size), Stokes' law and the Stokes-Einstein[23] relation combine to give D in terms of the hydrodynamic Stokes' diameter d_H:

$$d_H = \frac{1}{D}\frac{k_b T}{3\pi\eta_s} \qquad (3)$$

where η_s is the viscosity of the liquid and k_b is Boltzmann's constant ($k_b = R/N_A = 1.38065 \times 10^{-23} J K^{-1}$). Equations 2 and 3 are merged and rearranged to form the well-known expression providing the basis for measuring Stokes-equivalent spherical diameters with AUC[23,25,26]:

$$d_H = \sqrt{\frac{18\eta_s s}{(\bar{v}_p^{-1} - \rho_s)}} \qquad (4)$$

In our treatment, we assume that the Stokes' diameter and the particle diameter are approximately equal (that is, $d_H = d_p$), and combine Equations 3 and 4—and re-express Equation 2 in terms ρ_p—to obtain formulas that allow us to determine the particle parameters ρ_p, M, and d_p in terms of the experimentally measured s and D value distributions:

$$\rho_p = \rho_s + 18\eta_s s\left(\frac{1}{D}\frac{k_b T}{3\pi\eta_s}\right)^{-2} \qquad (5a)$$

$$M = \frac{sRT}{D}\left(1 - \frac{\rho_s}{\rho_p}\right)^{-1} \qquad (5b)$$

$$d_p = \sqrt{\frac{18\eta_s s}{(\rho_p - \rho_s)}} \qquad (5c)$$

Equations 5a–c represent the theoretical basis underlying our 2D SV-AUC measurements. In contrast to most SV-AUC approaches where only Equation 5c is used by assuming a particle density (ρ_p), we find that the ρ_p can instead be calculated from s and D through Equation 5a. Our approach's accuracy at predicting the density (ρ_p), molecular weight (M) and diameter of NPs (d_p) is demonstrated experimentally in the next section.

2D AUC of single monodisperse standard nanoclusters. As an illustrative test case, we first measured the 'magic-sized' NP $Au_{144}(SR)_{60}$ ($R = -CH_2CH_2Ph$) (synthesis[33] and ESI-MS characterization detailed in Methods). 2D SV-AUC was performed on a sample of $Au_{144}(SR)_{60}$ nanoparticles (Experimental details provided in Methods). Figure 2 plots the analysis obtained from the data collected for the $Au_{144}(SR)_{60}$ cluster.

The particle density was first calculated by Equation 5a using the integrated, weight-averaged diffusion and sedimentation coefficients, $\rho_p = 4.51 \, g\,cm^{-3}$. With the particle density in hand, Equation 5b is utilized to calculate molecular weight, $M = 35,260 \pm 180$ Da. Remarkably, the molecular weight predicted by theory for $Au_{144}(SR)_{60}$ is $M = 36,597$ Da, giving our method < 4% error. The molecular weight found by ESI-MS[33] for the sample studied was $M = 36,598$ Da. Finally, the ρ_p and sedimentation coefficient are transformed into particle diameter (d_p) by Equation 5c. The distribution of the diameters is weight-averaged to a single value after normalizing by the concentration distribution, $d_p = 2.83 \pm 0.01$ nm.

2D AUC of a mixture of monodisperse standard nanoclusters. As a next step, we tested the validity of our method on a mixture of three atomically discrete thiolated gold NPs with the same type of organic ligand, also of exactly known molecular formula[33–38]. Again we obtain an excellent match (~2–3% error) between the actual molecular weights measured by ESI-MS and the ones we obtained with our method. Three Au nanocluster standards were mixed together: $Au_{25}(SR)_{18}TOA$, $Au_{38}(SR)_{24}$ and $Au_{144}(SR)_{60}$ ($R = -CH_2CH_2Ph$). The $Au_{25}(SR)_{18}TOA$ nanocluster was bound to a tetraoctylammonium ion (TOA^+) due to its negative monovalency[39]. After AUC, sedimentation coefficient distributions were fit to the experimental data with SEDFIT ($\sigma = 0.683$), and the s- and D-distributions

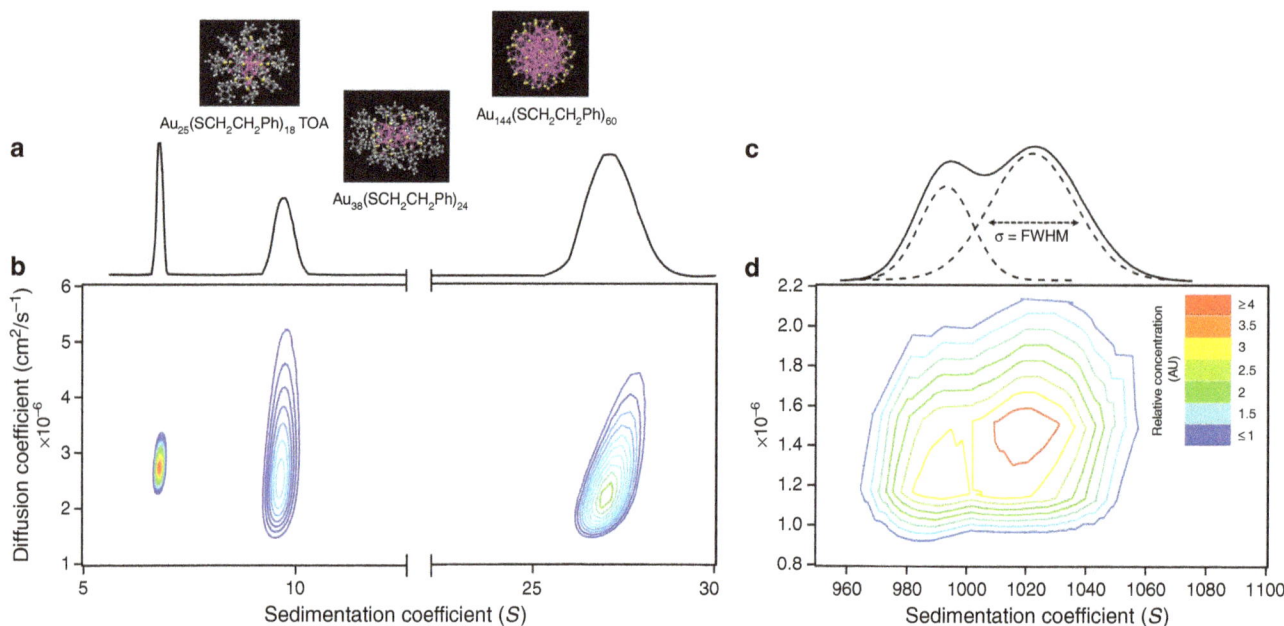

Figure 3 | 2D AUC distributions for both nanoclusters and PDT-capped nanoparticles. The three 'magic-sized' clusters were mixed in toluene (T = 20 °C) and sedimented simultaneously, to illustrate the accuracy and ease of measuring sedimentation and diffusion coefficients for a distribution of species. (**a**) The integrated 1D sedimentation coefficient distribution over the respective values of diffusion coefficients. (**b**) 2D sedimentation and diffusion coefficient distributions for three thiolated gold nanoclusters. (**c**) The integrated one-dimensional distribution of sedimentation coefficients taken over all diffusion coefficients and the multi-peak Gaussian fit illustrating the resolution of two PDT-NP peaks. (**d**) The sedimentation and diffusion coefficient distributions for PDT-NPs.

Table 2 | Data from the 2D AUC distributions for both nanoclusters and PDT-capped nanoparticles.

Clusters	s (S)	D (cm^2 s^{-1})	d_p (nm)	ρ_p (g cm^{-3})	M by AUC (Da)	M by ESI-MS	MW error (%)
Au$_{25}$(SR)$_{18}^-$ TOA$^+$	6.8±0.1	2.8×10^{-6}	2.41±0.02	1.95	8,000±90	7,862	1.75
Au$_{38}$(SR)$_{24}$	9.6±0.1	2.7×10^{-6}	2.58±0.01	2.43	11,030±50	10,778	2.33
Au$_{144}$(SR)$_{60}$	27.1±0.4	2.5×10^{-6}	2.83±0.01	4.51	35,260±180	36,598	3.65

PDT-NPs	s (S)	D (cm^2 s^{-1})	d_p (nm)	ρ_p (g cm^{-3})	M by AUC (Da)	d_{core} (nm)	d_{core} from TEM (nm)
Peak A	985±13	1.2×10^{-6}	11.54±0.15	4.65	2,256,790±25,300	6.82±0.15	6.7±0.07
Peak B	1022±21	1.4×10^{-6}	10.37±0.22	5.80	2,037,690±34,820	6.68±0.22	

Experimental data for Figure 3. (top) Three Au nanocluster standards (R= − CH$_2$CH$_2$Ph). Presented are the measured weight-averaged sedimentation coefficients (toluene, 20 °C), diffusion coefficients, hydrodynamic diameter, density, molecular weights by both AUC and ESI-MS, and average molecular weight percent errors from molecular formula as determined by ESI-MS. The standard deviation in s was taken as the full width at half maximum, which was then propagated through the calculations. (bottom) For both peaks of the PDT-NPs, the 2D distributions were transformed into density, molecular weight, and particle and core diameter. Also included is the average core diameter from TEM, in order to compare with the core diameters measured for both peaks by AUC.

(Fig. 3b) were transformed into diameters, densities and molecular weights, by the same procedure used for the single Au$_{144}$(SR)$_{60}$ nanoclusters. The 2D distributions can be integrated over all diffusion coefficients and reduced to a one-dimensional plot (Fig. 3a). Table 2 summarizes the 2D experimental data for the three NPs sedimented together in solution.

Determination of particle composition. In the very common case of a NP composed of two phases (for example, core–shell NPs and monolayer protected NPs) we can obtain a reasonable estimate of the composition of the particle by combining the measured M and ρ_p with the bulk densities (usually tabulated or easily measured) of the two phases. For example, the values for the core and ligand densities[27,28] are: bulk gold: 19.32 g cm^{-3}; bulk ligand: 1.03 g cm^{-3}. Combining mass conservation

$$M_p = N_{core} M_{Au} + M_{shell} M_{ligand} \quad (6)$$

where N_{core} and N_{shell} represent the number of atoms in the core and the number of molecules in the shell, respectively, with an additive volume consideration:

$$\rho_p^{-1} M_p = \rho_{core}^{-1} N_{core} M_{Au} + \rho_{shell}^{-1} N_{shell} M_{ligand} \quad (7)$$

a set of equations is yielded that can be solved for N_{shell} and N_{core}, and thus molecular formula:

$$N_{shell} = \frac{M_p \rho_{core}^{-1} - M_p \rho_p^{-1}}{M_{ligand} \left(\rho_{core}^{-1} - \rho_{shell}^{-1} \right)} \quad (8a)$$

$$N_{core} = \frac{M_p \rho_p^{-1} - M_p \rho_{shell}^{-1}}{M_{Au} \left(\rho_{core}^{-1} - \rho_{shell}^{-1} \right)} \quad (8b)$$

For the Au$_{144}$(SR)$_{60}$ nanoparticle, the estimated molecular formula by AUC was calculated with Equation 8 to be Au$_{137}$(SR)$_{60}$, a decent estimate of the actual molecular formula (Table 1). Given instead the actual molecular formula, it is possible to back calculate (by Equation 8) the density of the ligand shell. For the Au$_{144}$(SR)$_{60}$ nanocluster, the ligand shell density was found to be $\rho_L = 1.2$ g cm^{-3}, slightly higher than the bulk ligand density, as expected for these types of curved systems[28].

2D AUC of PDT-coated nanoparticles. In order to prove the validity of our technique for large NPs, we applied the method to larger, pentadecanethiol gold nanoparticles (PDT-NPs) synthesized using the Stucky method (see Methods for details)[40]. These particles are illustrative of the often-encountered case of so-called 'nearly' monodisperse particles. The 2D sedimentation and diffusion plot for this sample is shown in Figure 3d. From this data we can extract the size, density and molecular weight of the sample, as outlined in previous paragraphs. Furthermore, we can use the bulk densities of the core ($\rho_{Au} = 19.32 \, g \, cm^{-3}$) and shell ($\rho_{PDT} = 0.85 \, g \, cm^{-3}$), to estimate the average molecular formula, the core diameter and the shell thickness. The data is summarized in Table 2. Unlike the previous NP standards (Fig. 3b), the PDT-NPs, although narrowly dispersed, are not discrete species. This signifies that the standard deviation in core diameter is indicative of sample properties rather than the intrinsic instrumental broadening of the peaks. By transmission electron microscopy (TEM) (Supplementary Fig. S2), the average core diameter (d_{core}) and standard deviation of the PDT-NPs was 6.7 ± 0.07 nm—essentially a perfect match with our AUC analysis. Yet our results reveal that

AUC provides a more complete analysis as our results show that the core diameter is actually composed of a bi-modal distribution of nanoparticles, irresolvable with TEM, 6.68 ± 0.10 nm and 6.82 ± 0.06 nm.

2D AUC of polydisperse gold nanoparticles. After proving the effectiveness of our technique for NPs of various sizes, we now demonstrate the power of our technique by characterizing truly polydisperse particles, the most general case encountered in unrefined nanoparticle systems. To this end, we tested a polydisperse sample of sodium 11-mercaptoundecanesulfonate-capped AuNPs (MUS-NPs). The core diameter distribution of these particles ranges from ~4–9 nm, as suggested by TEM (Supplementary Fig. S2). The NPs were dissolved in 0.15 M NaCl and sedimented with AUC at 20 °C at a speed of 5,000 r.p.m. The 2D sedimentation and diffusion plot for this sample is shown in Figure 4.

Discussion

The molecular weights, densities and hydrodynamic diameters obtained by 2D SV-AUC are in extremely good agreement with the theoretical and experimental data in the literature for those clusters, and were all obtained in solution with a single experiment. Parameters are reported to three significant digits, because the radial positions and solvent density are known accurately to the same significance. A comparison between the cluster's sedimentation behavior when run alone compared to when it was run together with the other clusters revealed no statistically significant difference (*t*-test, σ = 0.95) in the measured *s* and *D* values of the nanoclusters between both situations. We notice that the errors for the parameters determined for the three clusters vary between the three types. Even though they slightly increase with the cluster size at this point, we believe that the error is related to the non-ideality in the shape of the nanocrystals. Indeed, according to crystallographic data, $Au_{25}(SR)_{18}TOA$ is quasi-spherical with an aspect ratio very close to 1, whereas $A_{38}(SR)_{24}$ is rod-like with an aspect ratio of ~1.3 (refs 35,39). It is, therefore, not surprising that the error in molecular weight for $Au_{38}(SR)_{24}$ is slightly higher compared with $Au_{25}(SR)_{18}TOA$, because

Table 1 | Estimated molecular formula and ligand shell density for standard nanoclusters.

Actual formula	Estimated formula*	Estimated ligand shell density (g cm⁻³)†
$Au_{25}(SR)_{18}^-TOA^+$	$Au_{23}(SR)_{21}$, $Au_{27}(SR)_{18}^-$ $TOA^{+‡}$	1.1
$Au_{38}(SR)_{24}$	$Au_{38}(SR)_{24}$	1.0
$Au_{144}(SR)_{60}$	$Au_{137}(SR)_{60}$	1.2

Actual number of core and ligand molecules compared with estimate numbers obtained from AUC and bulk density. The last column lists the estimated ligand shell density (g cm⁻³).
*Estimated from the data in Figure 3a, assuming bulk densities of thiol ligand and gold core.
†Given first the molecular formula of each cluster and plugging into Equation 8.
‡For a sample with unknown components or counterions, it would be impossible to predict the composition based on the molecular weight.

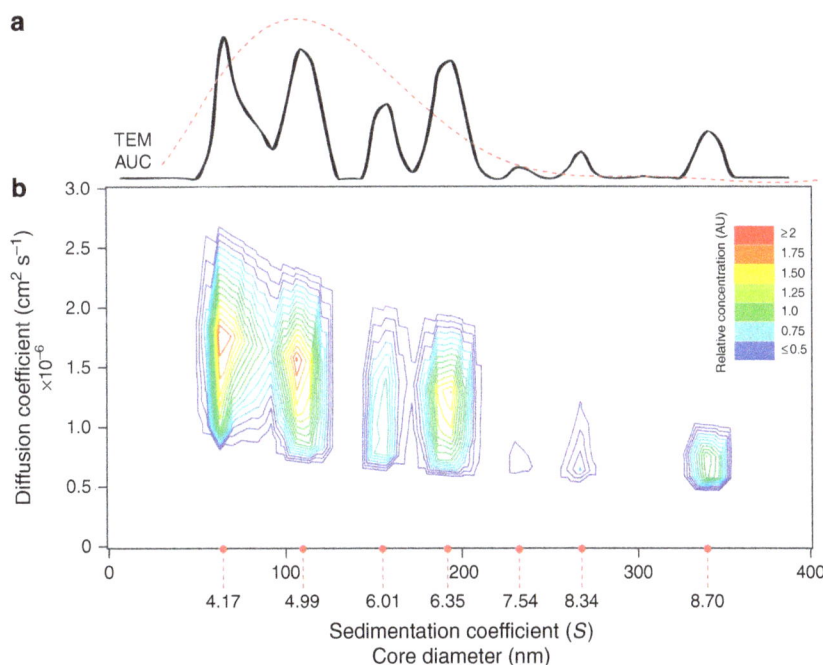

Figure 4 | 2D AUC distribution of polydisperse gold nanoparticles. (a) The integrated 1D distribution of sedimentation coefficients taken over all diffusion coefficients and the overlaid particle size distribution by TEM, to illustrate the increase in resolution by AUC. (**b**) The sedimentation and diffusion coefficient distributions for the polydisperse NP sample.

the basis of our analysis is the validity of Stoke's Law for describing the diffusion of the particles.

The encouraging results obtained from the nanoparticle standards open the possibility of applying the 2D SV-AUC method outlined above to naturally polydisperse samples, which exhibit dispersity in size, density and molecular weight. In this way, we are able to examine a particular population in a distribution, without requiring homogeneity in density (typically required in AUC experiments). These results also confirm that our methodology is robust enough to handle NPs that deviate from the spherical-shape requirement in Stokes Law (Equation 3); as the NPs in Figure 3b are known to be highly faceted or slightly prolate[35,39]. Yet, one should not push this concept too far. In samples with particles with widely varying aspect ratios, it would not be possible to apply our method without a priori knowledge on the aspect ratio of the specific (fractionated) populations. For example, the application of our method to analyse nanorods requires the knowledge of their aspect ratio.

Equation 8 was used to approximate the molecular formulas and ligand shell densities (Table 1), given the bulk core/ligand densities and theoretical molecular formulas. The estimated molecular formulas for the three NP standards confirm that the bulk ligand density could be used to reasonably approximate the actual ligand shell density. On the other hand, if prior knowledge of the molecular formula is available, one may be interested in using our methodology to measure the density of the ligand shell, as illustrated in Table 1.

It should be noted that nanoparticles in the size range of the PDT-NP sample (Fig. 3d), or larger, have small diffusion coefficients and typically sediment quickly, hence care must be taken in the choice of rotor speed. The general approach is to decrease the speed to increase the influence of diffusion during sedimentation while still maintaining sedimentation as the dominant transport mechanism. This yields a more accurate measurement of the diffusion coefficient, without compromising precision in the measurement of the sedimentation coefficient. In other words, a slower speed is necessary to accurately observe the diffusion of larger nanoparticles. The time scale of the experiment, however, remains only a few hours and still just a single run is required. Further work will be needed to determine a precise criterion to determine the optimal speed for a given sample.

For the sample of polydisperse MUS-coated NPs, the core diameter was calculated from the 2D sedimentation and diffusion plot and presented in Figure 4b. The core diameter distribution obtained by TEM was qualitatively overlaid onto the 1D AUC plot in Figure 4a to illustrate the level of resolution unobtainable by TEM, even when combined with other techniques. Even though the core size distributions show a good overall agreement in size range, 2D SV-AUC reveals the presence of distinct populations in the sample that are obscured by the comparatively tiny sampling capability of TEM. Previously, Colfen et. al. have demonstrated the angstrom resolution of AUC analysis and its potential power in replacing other lower-resolution techniques[41]. SV-AUC distributions sample millions to billions of particles[23], wheras TEM histograms can accommodate a few thousand particles at most. Given this level of detail, it should be immediately obvious to the reader that SV-AUC could be a powerful tool to understand nanoparticle growth[24]. Also, the core diameter distributions are the only parameters easily obtained by TEM. Even though insight into the ligand shell is in theory possible with advanced high-angle annular dark-field scanning transmission electron microscopy[42], the technique is not practical for statistically assessing its properties, especially in an industrial setting. In a single experiment, our method provides the core size, overall particle diameter, and density for each population of NPs present in our sample without the need for a global analysis.

As well as the shape variations described in previous sections, charges on nanoparticles can also lead to non-ideality. The MUS-coated particles described above were prepared in salt solutions

to screen the highly charged surface (zeta potential (ζ) = $-38\pm 5.30\,mV$)[8]. The classical way to measure whether electrostatic gradients induced by centrifugation affect the sedimentation coefficient of a sample is to run the solute in a series of solutions differing in ionic strength[23]. Provided that one can accurately measure the density and viscosity of the ionic buffer solution, the application of our method can be applied to highly charged solutes.

We believe that our method has general applicability, even beyond nanoparticles; yet, we do acknowledge limitations for our approach. First, some non-ideal samples would require assumptions. For example, the analysis of rods with varying shape should be coupled with another sizing technique to predetermine the aspect ratio. For a solute which is relatively monodisperse in axial ratio, our analysis could be applied with little adaptation. Recent developments in multi-wavelength AUC by Colfen et al. could prove excellent for this type of sample by providing a third axis of information that could be used to eliminate the need for prior axial ratio determination[43]. Alternatively, preparative fractionation prior to AUC could significantly reduce the complexity of the solute. Future work by our group will explore this possibility. Other types of non-ideality could be samples with widely varying charge density, or with density varying with size, but in unpredictable ways. Moreover, the sample shown in Figure 4 is actually composed of nicely discrete population of particles; this helps to reduce the error in our analysis. Samples with continuously varying sedimentation coefficient distribution, such as polymers or polymer-coated nanoparticles, would require an artificial binning of the data that would introduce another source of error. For example, we tried to determine the iron content of ferritin type-I and found that we needed a set of additional assumptions to achieve a qualitative description of the complex[44,45].

The 2D SV-AUC NP characterization approach described herein provides unprecedented access to the size, density and molecular weight distributions of NPs, from a single experiment and without the use of standards or auxiliary measurements—as is usually required for quantitative characterization with AUC. We proved the accuracy and the generality of our approach by testing NPs of various size distribution modes (for example, mixtures of atomically discrete particles, narrowly dispersed particles and polydisperse particles of a wide size distribution). We find that our approach delivers precise insight into the density, molecular weight and size distributions. Furthermore, no burdensome or speculative global analysis is needed to correlate these property distributions to one another—as is required when one studies different sample properties with multiple characterization methods.

Even though our methodology relies on Stokes' law to derive the various property distributions of NPs, we find that it is robust enough to handle NPs that slightly deviate (for example, oblate and faceted) from the spherical shape assumption of Stokes' law (formulated for a hard-sphere in a continuum liquid). Furthermore, we were able to accurately measure the properties of some of the smallest NPs that can be presently synthesized ($d_p \sim 2\,nm$) and thus exist in the regime of dominant intermolecular forces where Stokes' law is expected to break down. This discovery is of great importance to anyone interested in techniques that rely on Stokes' law for particle sizing (for example, dynamic light scattering (DLS), FFF and electrophoresis) and nanorheolgy in general.

NP research has seen an upsurge in size fractionation techniques in recognition of the need to minimize heterogeneity in NP samples that are intended for various applications. The methodology demonstrated in this work is ideally suited to work hand-in-hand with this trend, because it simultaneously maps a particle's density, size and molecular weight to its s and D values. Therefore, it provides not only the properties of the NPs present in a given sample, but also a solution allowing researchers to select (that is, fractionate) species in the sample based on pre-desired properties: a technique we are currently developing for future work. We believe this advantage, in addition to the accuracy, ease and general nature of our 2D SV-AUC treatment, will render this technique indispensible for the future studies and applications for all types of core–shell NPs.

Methods

Analytical ultracentrifugation. AUC was performed using a Beckman Optima XL-A, An-60 Ti rotor, scanning absorbance optics, with 12-mm path length double sector centerpieces with sapphire windows. Epon centerpieces were used for water (MUS-NPs) and hexanes (PDT-NPs), and aluminum centerpieces for samples in toluene (NCs). All measurements were made at 520 nm, 20 °C, at speeds ranging from 3,000 r.p.m. to 40,000 r.p.m. and data ranges from over 50–100 scans were chosen to be representative of the whole run (radial step size of 0.003–0.008 cm). SV runs typically required 0.05 to 0.5 mg of material in 400 μL solution. Each sample was prepared at varying concentrations to ensure that the sedimentation and diffusion coefficients were not concentration dependent. An inhomogeneous solvent model was applied to account for solvent compressibility caused by high pressure build-ups at the centrifugal fields obtained at high rotor speed[23]. The sedimentation and diffusion coefficients are provided under normal conditions (20 °C, water) and must be converted to the appropriate solvent[46].

Measurement of AUC experimental error. To obtain a measure of the experimental standard deviation of AUC, the $Au_{144}(SR)_{60}$ NP was repeatedly measured a total of 10 times at 20 °C in toluene at 8,000 r.p.m. The results were extraordinarily reproducible in both s and D, measured to be: 27.1 ± 0.1 S and $(2.5 \pm 0.1) \times 10^{-6}$ cm^2 s^{-1}, respectively. Hence, for both cases the standard deviation is limited by the line width we measure for a single experiment (~4% of the average). To compare the difference between AUC analysis of the standard nanoclusters sedimented separately and together, a t-test was used ($\sigma = 0.95$), assuming the data was normally distributed (visual test). For a grid of 10 f/f_0 values and 30 s values, the degrees of freedom $n = 300$.

Synthesis of $Au_{25}(SC_2H_4Ph)_{18}$TOA nanoclusters. HAuCl$_4$·3H$_2$O (0.4 mmol) was dissolved in 5 ml Nanopure water, and tetraoctylammonium bromide (TOAB, 0.47 mmol) was dissolved in 10 ml toluene; the two solutions were combined and vigorously stirred to facilitate phase transfer of Au(III) salt into the toluene phase. After phase transfer was completed, the aqueous phase was removed using a glass pipette. The toluene solution of Au(III) was cooled to 0 °C in an ice bath over a period of 30 min under constant magnetic stirring. PhC$_2$H$_4$SH (~3 equivalents per mole of gold) was added and stirring was reduced to a very low speed. The deep red solution turned faint yellow over a period of ~5 min, and finally clear over ~1 h. After that, the stirring speed was changed to fast stirring (~1100 r.p.m.) and, immediately, an aqueous solution of NaBH$_4$ (4 mmol, 10 equivalents per mole of gold, freshly made in 7 ml ice-cold nanopure water) was rapidly added all at once. The reaction was allowed to proceed overnight. Then the aqueous layer at the bottom of the flask was removed and the toluene solution was dried on a rotary evaporator. Ethanol or methanol was added to separate Au$_{25}$ clusters from TOAB and side products (for example, disulfide), and so on. Pure Au$_{25}$ clusters were collected by extracting with acetonitrile[38].

Synthesis of $Au_{38}(SC_2H_4Ph)_{24}$ nanoclusters. In a typical experiment, 0.5 mmol HAuCl$_4$·3H$_2$O and 2.0 mmol glutathione powders were mixed in 20 ml acetone at room temperature under vigorous stirring for ~20 min. The mixture (yellowish cloudy suspension) was then cooled to ~0 °C in an ice bath. After ~20 min, a solution of NaBH$_4$ (5 mmol, dissolved in 6 ml ice-cold nanopure water) was rapidly added to the suspension under vigorous stirring. The solution colour immediately turned black after addition of NaBH$_4$, indicating the formation of Au nanoclusters. After ~20 min, black Au$_n$(SG)$_m$ nanoclusters precipitated out of the solution and deposited onto the inner wall of the flask, leaving a clear acetone solution. The clear solution was then decanted and 6 ml water was added to dissolve the Au$_n$(SG)$_m$ clusters. A solution of Au$_n$(SG)$_m$ (around 200–300 mg, dissolved in 6 ml nanopure water) was mixed with 0.3 ml ethanol, 2 ml toluene and 2 ml PhC$_2$H$_4$SH. Note that ethanol is added to prompt phase transfer of Au$_n$(SG)$_m$ from water to organic phase. The diphase solution was heated to 80 °C and maintained at this temperature, under air. The Au$_n$(SG)$_m$ clusters were found to transfer from the water phase to the organic phase in less than 10 min. The thermal process was allowed to continue for ~40 h at 80 °C. Over the long etching process, the initial polydisperse Au$_n$ nanoclusters were finally converted to monodisperse Au$_{38}$(SC$_2$H$_4$Ph)$_{24}$ clusters. The Au clusters were precipitated and washed thoroughly with ethanol (or methanol) to remove excess thiol. Then, the Au$_{38}$(SC$_2$H$_4$Ph)$_{24}$ nanoclusters were simply separated from Au(I)-SG (poorly soluble in almost all solvents) by extraction with dichloromethane or toluene[36].

Synthesis of $Au_{144}(SC_2H_4Ph)_{60}$ nanoclusters. HAuCl$_4$·3H$_2$O (0.45 mmol) was dissolved in 5 ml nanopure water, and TOAB (0.52 mmol) was dissolved in 10 ml toluene. After mixing, the solution was vigorously stirred until phase transfer of Au(III) was completed (the toluene phase became deep red, whereas the initial yellow aqueous phase became clear). The clear aqueous phase was removed using

a 10 ml syringe. The toluene phase containing Au(III) were cooled to 0 °C in an ice bath for ~30 min. Then, PhC$_2$H$_4$SH (~3 equivalents of Au) was added under fast magnetic stirring. The deep red solution turned yellow gradually and finally almost clear in ~1 h. NaBH$_4$ solution (4.5 mmol, dissolved in 5 ml ice-cold nanopure water) was rapidly added to solution all at once. The solution colour immediately changed to black. The reaction was allowed to proceed for ~24 h. After ~24 h, the aqueous phase was discarded and the black toluene phase was dried by rotary evaporation. Ethanol was used to separate the Au nanoclusters from TOAB and other side products. To obtain truly monodisperse Au$_{144}$ nanoclusters, excess PhCH$_2$CH$_2$SH was used to etch the as-prepared Au nanoclusters from the first step. Typically, 20 mg Au nanoclusters was dissolved in 1 ml toluene, and 0.5–0.8 ml neat PhC$_2$H$_4$SH was then added to the Au nanoclusters solution. The solution was heated to 80 °C and maintained at 80 °C for about 24 h under constant magnetic stirring. After that, 20 ml methanol was added to the solution to precipitate Au nanoclusters. Only Au$_{144}$ nanoclusters and Au(I)-SCH$_2$CH$_2$Ph exist in the black precipitation. Au$_{144}$ nanoclusters were extracted with CH$_2$Cl$_2$, and Au(I)-SCH$_2$CH$_2$Ph residuals (poorly soluble) were discarded[33].

Molecular weight determination of nanoclusters. Electrospray ionization (ESI) mass spectra were used to determine the molecular weight of Au$_{25}$(SC$_2$H$_4$Ph)$_{18}$$^-$ TOA$^+$, Au$_{38}$(SC$_2$H$_4$Ph)$_{24}$ and Au$_{144}$(SC$_2$H$_4$Ph)$_{60}$ nanoclusters (the latter two clusters are charge neutral) (see Supplementary Fig. S3). ESI mass spectra were recorded using a Waters Q-TOF mass spectrometer equipped with Z-spray source. The source temperature was kept at 70 °C. The sample was directly infused into the chamber at 5 μl min^{-1}. The spray voltage was kept at 2.20 kV and the cone voltage at 60 V. The ESI sample was dissolved in toluene (1 mg ml^{-1}) and diluted (1:2 vol) by dry methanol (containing 50 mM CsOAc to enhance cluster ionization in ESI). The experimental error of formula weight determination is typically < 0.3 Da (within the range of < 10,000 Da).

Synthesis of pentadecanethiol gold nanoparticles. 0.25 mmol chlorotriphyenylphosphine gold (AuPPh$_3$Cl) was mixed with 0.75 mmol PDT in 20 ml of benzene to form a clear solution, to which 2.5 mmol of tert-butyl amine borane complex was then added. The mixture was stirred at 55 °C for 1 h, and then cooled to room temperature. 100 ml of ethanol was then added to precipitate the NPs. The precipitated NPs were collected by centrifugation and washed three times with a mixture of benzene and ethanol. The highly reproducible monodisperse NPs were obtained without any further treatment[40].

References

1. Ravirajan, P. et al. Hybrid polymer/zinc oxide photovoltaic devices with vertically oriented ZnO nanorods and an amphiphilic molecular interface layer. J. Phys. Chem. B. **110**, 7635–7639 (2006).
2. Daniel, M. C. & Astruc, D. Gold nanoparticles: assembly, supramolecular chemistry, quantum-size-related properties and applications toward biology, catalysis and nanotechnology. Chem. Rev. **104**, 293–346 (2004).
3. Weller, H. et al. Photochemistry of semiconductor colloids .14. Photochemistry of colloidal semiconductors - onset of light-absorption as a function of size of extremely small CdS particles. Chem. Phys. Lett. **124**, 557–560 (1986).
4. Lal, S., Link, S. & Halas, N. J. Nano-optics from sensing to waveguiding. Nat. Photonics **1**, 641–648 (2007).
5. Haruta, M. & Date, M. Advances in the catalysis of Au nanoparticles. Appl. Catal. **222**, 427–437 (2001).
6. Zhu, Y., Qian, H. F., Drake, B. A. & Jin, R. C. Atomically precise Au-25(SR)(18) nanoparticles as catalysts for the selective hydrogenation of alpha,beta-unsaturated ketones and aldehydes. Angew. Chem. Int. Edit. **49**, 1295–1298 (2010).
7. Zhu, Y. et al. Exploring stereoselectivity of Au-25 nanoparticle catalyst for hydrogenation of cyclic ketone. J. Catal. **271**, 155–160 (2010).
8. Verma, A. et al. Surface-structure-regulated cell-membrane penetration by monolayer-protected nanoparticles. Nat. Mater. **7**, 588–595 (2008).
9. Ghosh, P., Han, G., De, M., Kim, C. K. & Rotello, V. M. Gold nanoparticles in delivery applications. Adv. Drug Deliv. Rev. **60**, 1307–1315 (2008).
10. Huang, X. H., Jain, P. K., El-Sayed, I. H. & El-Sayed, M. A. Gold nanoparticles: interesting optical properties and recent applications in cancer diagnostic and therapy. Nanomedicine **2**, 681–693 (2007).
11. Schopf, B. et al. Methodology description for detection of cellular uptake of PVA coated superparamagnetic iron oxide nanoparticles (SPION) in synovial cells of sheep. J. Magn. Magn. Mater. **293**, 411–418 (2005).
12. El-Sayed, M. A. Some interesting properties of metals confined in time and nanometer space of different shapes. Accounts Chem. Res. **34**, 257–264 (2001).
13. Trindade, T., O'Brien, P. & Pickett, N. L. Nanocrystalline semiconductors: synthesis, properties and perspectives. Chem. Mater. **13**, 3843–3858 (2001).
14. Kelly, K. L., Coronado, E., Zhao, L. L. & Schatz, G. C. The optical properties of metal nanoparticles: the influence of size, shape, and dielectric environment. J. Phys. Chem. B. **107**, 668–677 (2003).
15. Seo, W. S. et al. Size-dependent magnetic properties of colloidal Mn3O4 and MnO nanoparticles. Angew. Chem. Int. Ed. **43**, 1115–1117 (2004).

16. Mohanty, A., Garg, N. & Jin, R. A universal approach to the synthesis of noble metal nanodendrites and their catalytic properties. *Angew. Chem. Int. Ed.* **49,** 4962–4966 (2010).

17. Rao, C. N. R., Kulkarni, G. U., Thomas, P. J. & Edwards, P. P. Size-dependent chemistry: properties of nanocrystals. *Chem. Eur. J.* **8,** 29–35 (2002).

18. Boal, A. K. *et al.* Self-assembly of nanoparticles into structured spherical and network aggregates. *Nature* **404,** 746–748 (2000).

19. Borger, L. & Colfen, H. Investigation of the efficiencies of stabilizers for nanoparticles by synthetic boundary crystallization ultracentrifugation. *Analyt. Ultracentrifugation V* **113,** 23–28 (1999).

20. Brown, P. H. & Schuck, P. A new adaptive grid-size algorithm for the simulation of sedimentation velocity profiles in analytical ultracentrifugation. *Comput. Phys. Commun.* **178,** 105–120 (2008).

21. Brown, P. H. & Schuck, P. Macromolecular size-and-shape distributions by sedimentation velocity analytical ultracentrifugation. *Biophys. J.* **90,** 4651–4661 (2006).

22. Brookes, E., Cao, W. M. & Demeler, B. A two-dimensional spectrum analysis for sedimentation velocity experiments of mixtures with heterogeneity in molecular weight and shape. *Eur. Biophys. J.* **39,** 405–414 (2010).

23. Mächtle, W. & Börger, L. *Analytical Ultracentrifugation of Polymers and Nanoparticles* (Springer, 2006).

24. Planken, K. L. & Colfen, H. Analytical ultracentrifugation of colloids. *Nanoscale* **2,** 1849–1869 (2010).

25. Svedberg, T. & Rinde, H. The determination of the distribution of size of particles in disperse systems. *J. Am. Chem. Soc.* **45,** 943–954 (1923).

26. Svedberg, T. & Rinde, H. The ultra-centrifuge, a new instrument for the determinition of size and distribution of size of particle in amicroscopic colloids. *J. Am. Chem. Soc.* **46,** 2677–2693 (1924).

27. Jamison, J. A. *et al.* Size-dependent sedimentation properties of nanocrystals. *ACS Nano.* **2,** 311–319 (2008).

28. Lees, E. E. *et al.* Experimental determination of quantum dot size distributions, ligand packing densities and bioconjugation using analytical ultracentrifugation. *Nano. Lett.* **8,** 2883–2890 (2008).

29. Lamm, O. The theory and method of ultra centrifuging. *Z. Phys. Chem. A-Chem. T.* **143,** 177–190 (1929).

30. Hansen, P. C. *Rank-Deficient and Discrete Ill-Posed Problems: Numerical Aspects of Linear Inversion* (SIAM, 1998).

31. Dam, J. & Schuck, P. Calculating sedimentation coefficient distributions by direct modeling of sedimentation velocity concentration profiles. *Methods. Enzymol.* **384,** 185–212 (2004).

32. Schuck, P. & Demeler, B. Direct sedimentation analysis of interference optical data in analytical ultracentrifugation. *Biophys. J.* **76,** 2288–2296 (1999).

33. Qian, H. & Jin, R. Controlling nanoparticles with atomic precision: the case of Au144(SCH2CH2Ph)60. *Nano. Lett.* **9,** 4083–4087 (2009).

34. Jin, R. C. *et al.* Size focusing: a methodology for synthesizing atomically precise gold nanoclusters. *J. Phys. Chem. Lett.* **1,** 2903–2910 (2010).

35. Qian, H. F., Eckenhoff, W. T., Zhu, Y., Pintauer, T. & Jin, R. C. Total structure determination of thiolate-protected Au-38 nanoparticles. *J. Am. Chem. Soc.* **132,** 8280–8281 (2010).

36. Qian, H. F., Zhu, Y. & Jin, R. C. Size-focusing synthesis, optical and electrochemical properties of monodisperse Au-38(SC2H4Ph)(24) nanoclusters. *ACS Nano.* **3,** 3795–3803 (2009).

37. Wu, Z., Suhan, J. & Jin, R. C. One-pot synthesis of atomically monodisperse, thiol-functionalized Au-25 nanoclusters. *J. Mater. Chem.* **19,** 622–626 (2009).

38. Zhu, M., Lanni, E., Garg, N., Bier, M. E. & Jin, R. Kinetically controlled, high-yield synthesis of Au25 clusters. *J. Am. Chem. Soc.* **130,** 1138–1139 (2008).

39. Zhu, M., Aikens, C. M., Hollander, F. J., Schatz, G. C. & Jin, R. Correlating the crystal structure of A thiol-protected Au-25 cluster and optical properties. *J. Am. Chem. Soc.* **130,** 5883–5885 (2008).

40. Zheng, N., Fan, J. & Stucky, G. D. One-step one-phase synthesis of monodisperse noble-metallic nanoparticles and their colloidal crystals. *J. Am. Chem. Soc.* **128,** 6550–6551 (2006).

41. Pauck, T. & Colfen, H. Hydrodynamic analysis of macromolecular conformation. A comparative study of flow field flow fractionation and analytical ultracentrifugation. *Anal. Chem.* **70,** 3886–3891 (1998).

42. Abad, J. M., Gass, M., Bleloch, A. & Schiffrin, D. J. Direct electron transfer to a metalloenzyme redox center coordinated to a monolayer-protected cluster. *J. Am. Chem. Soc.* **131,** 10229–10236 (2009).

43. Strauss, H. M. *et al.* Performance of a fast fiber based UV/Vis multiwavelength detector for the analytical ultracentrifuge. *Colloid Polym. Sci.* **286,** 121–128 (2008).

44. Colfen, H. & Volkel, A. Hybrid colloid analysis combining analytical ultracentrifugation and flow-field flow fractionation. *Eur. Biophys. J.* **32,** 432–436 (2003).

45. May, C. A. *et al.* The sedimentation properties of ferritins. New insights and analysis of methods of nanoparticle preparation. *Biochim. Biophys. ACTA* **1800,** 858–870 (2010).

46. Schuck, P. AUC Direct boundary modeling with SEDFIT, www. analyticalultracentrifugation.com (2010).

47. Schaaf, P., Senger, B. & Reiss, H. Defining physical clusters in nucleation theory from the N-particle distribution function. *J. Phys. Chem. B.* **101,** 8740–8747 (1997).

48. Balasubramanian, R., Guo, R., Mills, A. J. & Murray, R. W. Reaction of Au-55(PPh3)(12)Cl-6 with thiols yields thiolate monolayer protected Au-75 clusters. *J. Am. Chem. Soc.* **127,** 8126–8132 (2005).

49. Kohlmann, O. *et al.* NMR diffusion, relaxation and spectroscopic studies of water soluble, monolayer-protected gold nanoclusters. *J. Phys. Chem. B.* **105,** 8801–8809 (2001).

Acknowledgments

We thank P. Schuck and P. Brown for helpful discussions and for the continuous development of SEDFIT. This research was supported by the National Science Foundation Graduate Research Fellowship Program. We thank D. Pheasant and the Massachusetts Institute of Technology Bioinstrumentation Facility for technical assistance. H.M. and O.M.B. acknowledge the use of the King Abdullah University of Science and Technology Analytical Chemistry Core Lab.

Author contributions

R.P.C. and O.M.B. conceived and designed the experiments. R.P.C. and H.M. performed the AUC experiments. H.Q., J.Y.K. and R.P.C. synthesized the nanoparticles, and J.Y.K., H.M. and R.P.C. performed the TEM characterization. R.P.C. analysed the data and co-wrote the paper with O.M.B. and F.S. All authors discussed the results and commented on the manuscript. Correspondence and requests for materials should be addressed to F.S. or O.M.B.

Additional information

A graphene field-effect transistor as a molecule-specific probe of DNA nucleobases

Nikolai Dontschuk[1], Alastair Stacey[1], Anton Tadich[2,3], Kevin J. Rietwyk[3,†], Alex Schenk[3], Mark T. Edmonds[3,†], Olga Shimoni[1,†], Chris I. Pakes[3], Steven Prawer[1] & Jiri Cervenka[1]

Fast and reliable DNA sequencing is a long-standing target in biomedical research. Recent advances in graphene-based electrical sensors have demonstrated their unprecedented sensitivity to adsorbed molecules, which holds great promise for label-free DNA sequencing technology. To date, the proposed sequencing approaches rely on the ability of graphene electric devices to probe molecular-specific interactions with a graphene surface. Here we experimentally demonstrate the use of graphene field-effect transistors (GFETs) as probes of the presence of a layer of individual DNA nucleobases adsorbed on the graphene surface. We show that GFETs are able to measure distinct coverage-dependent conductance signatures upon adsorption of the four different DNA nucleobases; a result that can be attributed to the formation of an interface dipole field. Comparison between experimental GFET results and synchrotron-based material analysis allowed prediction of the ultimate device sensitivity, and assessment of the feasibility of single nucleobase sensing with graphene.

[1] The School of Physics, The University of Melbourne, Melbourne, Victoria 3010, Australia. [2] Australian Synchrotron, 800 Blackburn Road, Clayton, Victoria 3168, Australia. [3] Department of Physics, La Trobe University, Bundoora, Victoria, Australia. † Present addresses: Center for Nanotechnology & Advanced Materials, Bar Ilan University, Ramat Gan, Israel (K.J.R.); School of Physics, Monash University, Clayton, Victoria, Australia (M.T.E.); School of Physics and Advanced Materials, University of Technology, Sydney, New South Wales, Australia (O.S.). Correspondence and requests for materials should be addressed to N.D. (email: ndo@student.unimelb.edu.au) or to S.P. (email: s.prawer@unimelb.edu.au) or to J.C. (email: jiri.cervenka@gmail.com).

Efforts to determine the order of the four nucleobases in a strand of DNA have been advancing at an extremely rapid pace in recent years[1,2]. Next-generation DNA sequencing technologies promise to provide cheap and high-throughput analysis of a complete DNA strand and achieve reduced sequencing artefacts. Electrical sequencing using graphene and nanopore technologies has recently attracted great attention due to the possibility to provide real-time sequencing of a whole single DNA molecule[3-11]. These methods are based on the use of graphene as an electrical readout-based chemical sensor while a strand of DNA is fed through a nanopore[2-4,10]. Most of the graphene-based sequencing technologies are fundamentally reliant on detecting molecular-specific interactions of individual nucleobases with a graphene surface[4] or its defects[5]. Ultimately, these interactions need to induce electronic modifications that are detectable by graphene electrical devices. Although the interaction of DNA nucleobases with graphene have been studied theoretically using molecular dynamics simulation and density functional theory (DFT) methods[12-16], the experimental evidence of such specific interaction in electrical transport within the graphene has not yet been demonstrated.

Graphene-based electrical sensors have already demonstrated exceptional sensor characteristics by being capable of detection of adsorption and desorption of individual gas molecules from the graphene surface[17]. This sensitivity is a direct consequence of the two-dimensional crystal structure, unique electronic properties[17,18] and exceptionally low-noise intrinsic characteristics of graphene based devices[19,20]. So far, however, single-molecule sensitivity has been reported only for nitrogen dioxide[17], and it is not clear whether graphene can also detect other molecules, which do not induce strong charge transfer doping to graphene, down to the single-molecule level. Despite the high sensitivity of graphene devices, a key practical aspect of any chemical sensor is also their ability to distinguish between target analytes, in this case the nucleobases. Recent research has used graphene chemically functionalized with specific chemical binding groups to achieve analyte specificity[21,22]. Chemical functionalization, however, comes at the cost of decreased graphene mobility[23] and increased distance between the sensor and analytes, potentially hampering sensitivity. For these reasons a direct analyte–sensor interaction may be favourable for chemical sensing of individual DNA nucleobases. This leads to an important and fundamental question—can graphene detect and identify each of the nucleobases on its surface based on their specific graphene–molecule interaction? Understanding how these nucleobase–graphene interactions manifest in electric transport will be of key importance for the next-generation graphene-based DNA sequencing technologies.

Here we investigate the effects of molecular adsorption of submonolayer coverages of individual DNA nucleobases on the graphene electronic structure using electric transport measurements. We demonstrate that graphene field-effect transistors (GFETs) are capable of detecting distinct coverage-dependent conductance signatures upon adsorption of the four different DNA nucleobases: adenine, guanine, cytosine and thymine. We examine the correlation between adsorption-induced nucleobase dipoles and electron transport as a function of molecular coverage and analyse the implications for single-molecule detection and identification. Furthermore, using simultaneous synchrotron-based X-ray photoelectron spectroscopy (XPS) measurements we explore the molecule-specific limit of adsorbed dipole-induced doping in a range of graphene materials. We analyse the sensitivity of our GFET devices and provide an estimate for the resolution limit in realistic scaled-down devices.

Results

Device fabrication and measurements. GFET sensors, shown schematically and optically in Fig. 1, were constructed from single-layer graphene (SLG) channels 50 μm wide and 50–200 μm long sitting on top of a SiO_2 on doped Si transistor gate. The channels were designed to be significantly larger than the common chemical vapour deposition (CVD) graphene grain size ($5{-}10\,\mu m^2$) to ensure our results reflect the bulk properties of these polycrystalline graphene sheets, averaging out the effects of a range of defects present in the layers. XPS measurements were made on large area graphene samples ($25\,mm^2$) fabricated identically to GFET devices but without evaporated electrical contacts. The GFET channels are comparable in size to the X-ray spot area of $50 \times 100\,\mu m^2$, allowing for reasonable comparison between the two techniques. All measurements were conducted at the Australian Synchrotron on the Soft X-ray beam line in an ultra-high vacuum (UHV) setup with base pressure $<1 \times 10^{-9}$ mbar to suppress the influence of contaminating molecules adsorbed on the surface. Since water[24] and other airborne contaminants[25] have been shown to interact strongly with graphene on SiO_2 substrates, samples were thoroughly annealed after insertion into UHV. Deposition of the molecules on graphene utilized *in situ* low-temperature effusion cells loaded with a pure powders of the DNA nucleobases (Methods).

The interaction of nucleobases with graphene was monitored with two independent measurements, via electric transport and with core level XPS. Electric transport measurements were obtained by setting a constant voltage between the source and drain contacts (V_{SD}) and monitoring the source drain current (I_{SD}), while the gate voltage (V_G) was swept from -12 to 12 V. Carbon and nitrogen 1s core level (C 1s and N 1s) spectra were taken with 330 and 450 eV X-ray photons to attain high surface sensitivity and large photoionization cross section. Molecular coverage was controlled by incremental deposition and determined by comparing the ratio between the molecular N 1s peak area to the SLG C 1s peak area (Methods). Finally, we conducted angular resolved C and N K-edge near edge X-ray adsorption fine structure (NEXAFS) measurements to investigate the molecular orientation with respect to the graphene surface. Extensive beam damage studies revealed that under repeated X-ray exposure the graphene C 1s peak would shift. To ensure this effect did not interfere with any molecule-induced shifts each new XPS measurement was done on a previously unmeasured spot of the

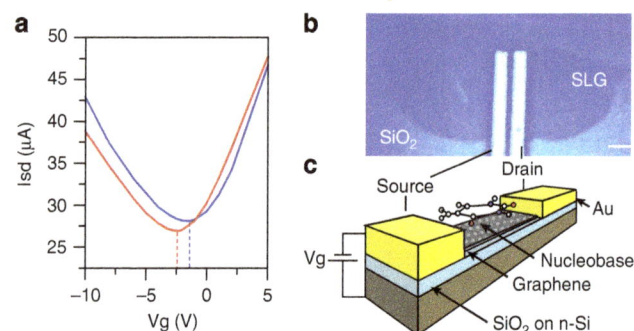

Figure 1 | Two terminal graphene devices used to monitor molecule-induced graphene properties. (**a**) Source drain current at a constant source drain voltage (10 mV) with varying gate voltage for clean (blue) graphene and 0.1 ML guanine (G) covered (red) graphene. (**b**) Optical and (**c**) schematic images of a GFET device used for electrical experiments and adsorption of each of four DNA nucleobases on graphene. Positive gate voltage is defined as a positive potential at the doped Si contact. The scale bar shown in **b** is 200 μm.

graphene sample. This also precluded direct XPS measurement of the GFET devices.

Probing molecule-specific interactions. Figure 1 shows a typical result of electric transport of a cleaned graphene GFET and after exposure to one of the nucleobases. The GFET spectra of all four DNA nucleobases can be found in Supplementary Information (Supplementary Fig. 1). Figure 1a clearly demonstrates the high sensitivity of GFETs to adsorption of a small number of molecules, in this case ~ 0.1 monolayer ($\approx 10^9$ molecules) of guanine. The conductivity spectrum of graphene shifts to the left after guanine adsorption, indicating an effective n-type doping of graphene in response to adsorption of the molecules. The voltage shift of the spectra can be determined by the position of the current minima (marked by the dashed lines). The gate voltage corresponding to the minimum in I_{SD}, known as the charge neutrality point (CNP), represents the Dirac point in the graphene band structure. The number of charge carriers in graphene is calculated using a parallel plate capacitor approximation of the doped Si/SiO$_2$/graphene stack. This gives $n = C_{BG}V/e = 2.46 \times 10^{11}$ cm^{-2}V^{-1} V_G, where $C_{BG} = \epsilon/d$ for a SiO$_2$ dielectric layer with $\epsilon = 3.54 \times 10^{-11}$ Fm^{-1} and a thickness of $d = 90$ nm. For the case of G in Fig. 1, a 0.7 V shift indicates a relative increase of 1.7×10^{11} electrons cm^{-2} in the conduction band of graphene.

In Fig. 2, we compare the induced charge carrier density in the graphene (determined from CNP shift) by adsorption of different DNA nucleobases as a function of molecular coverage. The observed coverage dependence is clearly molecule specific; however, for each molecule it can be divided into two similar regimes. In the very low coverage regime the induced charge scales approximately linearly with coverage (linear fits are displayed as dashed lines in Fig. 2). As the coverage increases the induced charge per molecule tails off to saturation, where adding additional molecules induces no further charge carriers. This coverage dependence is fitted by solid lines using an electrostatic depolarization model (equation 2), which attributes the observed shifts to the interface electric field formation from molecular dipoles. The depolarization model predicts the observed linearity at low coverages, and this has been experimentally confirmed for coverage-dependent work function shifts, which are also described by this model[26,27].

To assess the potential of bulk graphene devices for DNA sequencing, we have used the observed linear behaviour in the low coverage regime to extrapolate the response of GFET devices to the adsorption of single DNA nucleobase molecules, when scaling down the active sensor area. Figure 2c shows the expected change in carrier density from a single nucleobase on a GFET with 100×100 nm^2 of exposed graphene calculated from the linear fit to the low coverage dependence. Transport properties of GFETs of this size are still expected to be dominated by the bulk properties of graphene. The predicted induced charge carrier density in the scaled-down GFETs by the individual nucleobases is of the order of $\Delta n = 10^6$–10^8 e, with the largest signal for adsorbed guanine showing an increase of $\Delta n = 3.8 \times 10^8$ e. Even though carrier density changes in graphene of this magnitude (10^8 e) have previously been experimentally detected on larger GFETs with a device area of 1 μm^2 (ref. 17), the predicted signals are well below the observed noise level of our simple devices ($\Delta n = 10^{10}$ e). It is important to point out that these calculated signals apply to our sensors operating in a UHV environment. In ambient conditions or in a buffered ion solution where actual DNA sensors operate this sensitivity may not be achievable.

An option to improve GFET sensitivity to single nucleobases is to further scale down to the graphene nanoribbon regime. Shrinking the channel area, however, increases the expected low-frequency device noise, making nucleobase signal detection more challenging. For nanoribbon devices, with devices areas of 2,000–4,000 nm^2, noise amplitudes of the order of 10^{-6} A have been reported[28,29]. Moreover, the electron transport in graphene nanoribbons has been shown to be dominated by edge and quantum confinement effects[30]. Another way to achieve single-molecule resolution is to improve the sensitivity of GFET devices by using a different detection method or mechanism. For instance, monitoring changes in I_{SD} at a fixed V_G is expected to significantly improve the device sensitivity in high-mobility graphene devices, because the measured signal becomes a function of the graphene charge carrier mobility (μ) with $\Delta I_{SD} = \mu_G e \Delta n$ (ref. 31). The recent advances in the production of high-mobility graphene devices, reaching mobility up to $\approx 2 \times 10^5$ cm^2 V^{-1} s^{-1} (refs 32,33) and the reduction of graphene device noise[34] promise to provide an improvement in device sensitivity of three orders of magnitude. This would be sufficient to detect single guanine, cytosine and thymine on

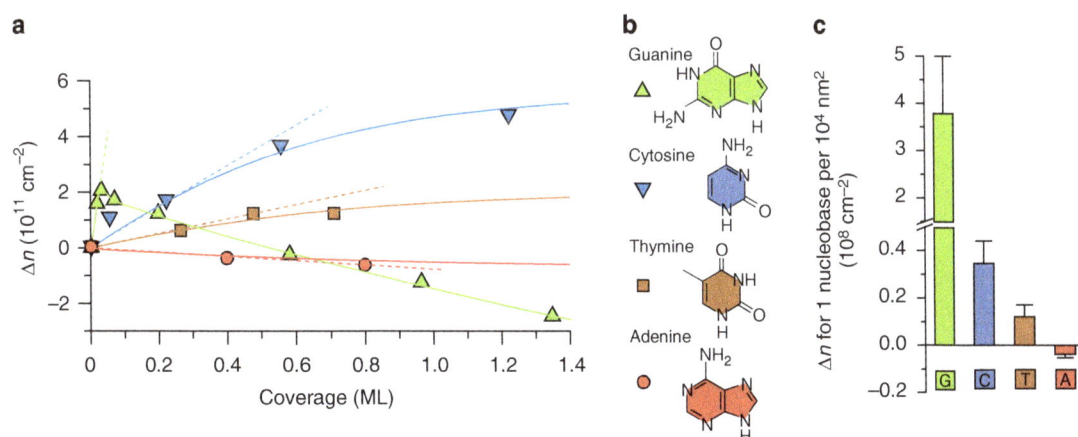

Figure 2 | Adsorbed nucleobase dependence of induced GFET CNP shifts. (a) Charge carrier density change (Δn) of GFETs induced by adsorption of guanine (G), cytosine (C), adenine (A) and thymine (T). The dashed lines represent linear fits to the low coverage regime used to calculate the expected single-molecule-induced shifts in (**b**). The solid lines are fits to C, A and T data using the electrostatic depolarization model (equation 2). The solid line for G is a guide to the eye only. (**c**) Calculated charge carrier density shifts of our GFETs induced by 1 nucleobase molecule per 10^4 nm^2 of graphene.

$100 \times 100 \, \text{nm}^2$ GFET devices. To sense adenine, however, an additional enhancement of the sensitivity would be required. This suggests that realistic DNA sequencing with GFET devices will require use of graphene nanoribbon devices, which might exhibit completely different response to the adsorption of DNA nucleobases than our bulk GFETs because of the existence of edge states[30,35]. The precise control of graphene edges will be extremely important in graphene nanoribbon FETs[35] but, as recent theoretical work predicts[4], there is a hope that 1-nm wide graphene nanoribbons with perfect arm chair edges would be able to discriminate the individual nucleobases and their sequence when a DNA strand is translocated across the ribbon.

Comparison between electric transport and XPS. To get a more detailed insight into the sensing mechanism of GFETs and the related molecule–graphene interactions, we have conducted, simultaneously with the electrical transport measurements, synchrotron-based XPS and NEXAFS measurements. Both of these highly surface-sensitive techniques have the capability of resolving chemical state information and conformation of adsorbed molecules on the graphene surface, information that is necessary for studying molecular adsorption processes on the atomically thin graphene samples. In addition, XPS analysis allows us to directly corroborate the measured changes in electrical data with respect to the changes in binding energy of the core level C 1s XPS peak of graphene, and thus changes in its Fermi level[36]. This was possible because the C 1s peak of graphene is well separated from the molecular C 1s peaks (Supplementary Fig. 2). To compare the magnitude of the shift between the XPS C 1s position and the charge carrier density changes (Δn) measured via CNP, it is convenient to convert induced carrier densities to Fermi level shifts via the tight binding model approximation of graphene's dispersion relation near the Dirac point[37]:

$$\Delta E = v_{\text{F}} \hbar \sqrt{\pi} \Delta \sqrt{n}, \tag{1}$$

where the Fermi velocity $v_{\text{F}} = 1.0936 \times 10^6 \, \text{ms}^{-1}$, as measured by cyclotron resonance[38]. To test the effect of multilayered graphene, XPS measurements were obtained on both SLG and bilayer graphene (BLG) samples, with A and T molecules on the former and C and G molecules on the latter. The comparison is shown in Fig. 3. Although the magnitude of the measured data differs, the C 1s shifts for both BLG and SLG show a qualitatively similar trend to electrical measurements. This provides strong evidence that these observations are not unique to our electrical devices and are a result of molecule-specific interactions with the bulk graphene on SiO_2. A comparison of the differences in the magnitude of the two measurement techniques shows the measured binding energy shifts to be consistently larger than measured CNP shifts. Unlike electrical CNP measurements of GFETs, we expect the measured C 1s shifts to be independent of any graphene contact work function alignment. This would indicate that the current choice of Ti/Au contacts is limiting the GFET sensitivity to adsorbed nucleobases.

The nature of the molecular adsorption. Carbon and nitrogen angle-resolved NEXAFS measurements were performed on SLG and BLG samples to obtain information about the geometry of the adsorbed molecules on graphene. NEXAFS spectra taken at a range of incident photon angles allows determination of the angle between the π^* and σ^* C and N orbitals and the electric field vector of the incoming light[39]. Using this technique, we observed a tilt angle of roughly 40° between each of the four nucleobases purine/pyridine rings and the graphene surface plane (Supplementary Fig. 3), independent of coverage. It is important to note that this analysis could not be reliably

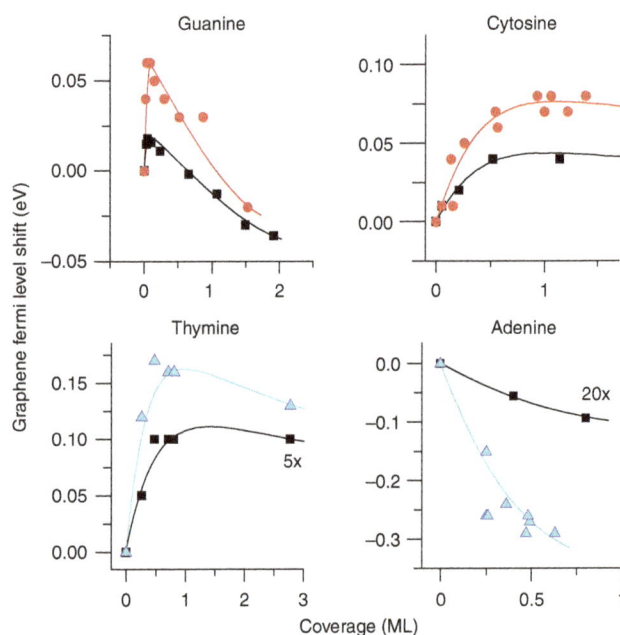

Figure 3 | Comparison of nucleobase-induced Fermi level shifts. Graphene Fermi level shifts of SLG GFETs (black squares) are calculated from CNP measurements using the tight binding model dispersion equation (1). XPS measurements of Fermi level shifts a determined by changes in the C 1s position of SLG (blue triangles) and BLG (red circles). The two techniques confirm the structure observed molecular coverage dependence of shifts in the graphene's Fermi level. The error for electrical measurements was 0.005 eV, the fitting error of XPS measurements was 0.02 eV and the energy resolution of the XPS was 0.05 eV (ref. 56).

applied to molecular coverages less than half a monolayer due to difficulties in resolving the molecular NEXAFS signal from the large graphene background. NEXAFS and XPS also allow for the direct detection of graphene–adsorbate (or substrate) hybridization[40]. No such hybridization of the graphene π^* orbital was observed after nucleobase deposition on our samples, confirming that the nucleobase–graphene interaction is moderated only by van der Waals forces. This strict physisorption is important for sensors that rely on a capture sense and release mechanism, such as those previously suggested for DNA sensing[4,7].

Discussion

The observed changes in graphene's electronic structure on the adsorption of four individual DNA nucleobases indicate molecule-specific interactions with a strong coverage dependency. To understand whether the sensing mechanism is purely related to the molecule-induced electronic structure modification of graphene, we need to look at the molecule–graphene interface electronic structure in more detail. When a molecule is brought into contact with graphene, there are several physical effects that can influence the energy alignment of the molecular and graphene levels at the interface[41]. Previous DFT calculations of DNA nucleobases adsorbed on graphene have predicted that nucleobases interact with graphene only weakly, predominantly by van der Waals interactions[4,13,16]. The frontier molecular orbitals (highest occupied and lowest unoccupied molecular orbitals) lie far from the Fermi level of graphene[13]. As a result there is only a very weak hybridization between the molecular levels of the nucleobases and the low-lying π-states of graphene, leading to a negligible charge transfer between the molecules and graphene[13]. Even though there is almost no direct charge transfer, the electronic structure of graphene can still be significantly influenced by the molecules.

It is commonly accepted that the adsorption of molecules onto a metal surface gives rise to the formation of molecular dipoles in the adsorbate layer. The effect of the electric field generated by such a dipole layer is a modulation of the surface potential of the metal, which can be measured as a change in the surface work function of the metal[26,27]. As graphene is a two-dimensional material the dipole electric field also modulates the number of charge carriers causing shifts in the Fermi level of the graphene, with the metal contacts acting as charge reservoirs[42]. This molecular gating is similar to the doping induced by a back gate contact when a voltage is applied. A similar mechanism has been previously reported for graphene on polymer substrates, where dipole fields of the polymer substrates caused both a work function shift and Fermi level shift in the graphene. In the case of polymer substrates, the Fermi level shift disappeared on samples without metal contacts, ruling out polymer to graphene charge transfer as the doping mechanism[43]. DFT modelling predicts that nucleobase adsorption onto graphene will also induce interfacial dipole formation, giving rise to a shift in graphene's work function by 0.22, 0.15, 0.13 and 0.01 eV for G, A, C and T, respectively[13]. This has been calculated for one nucleobase molecule adsorbed in a flat geometry in a 5×5 supercell of graphene. Our experimental results, however, differ from the DFT predictions. The magnitude of measured Fermi level shifts shows a different order with respect to deposited nucleobases, and that these changes vary significantly with coverage. The consistency of our experimental shifts has been confirmed by two independent techniques and using different graphene samples from two producers. In the lower coverage regime (below 0.1 ML) the observed order of magnitude of Fermi level shift was $G > C > T > A$, which gradually changed to a molecular sequence of $C > T > A > G$ at coverages greater than a monolayer. The magnitude of the shifts was in the range of 0–0.2 and 0–0.06 eV for C 1s and CNP shifts, respectively. There are several potential explanations for this strong coverage dependence and inconsistency with the previous DFT calculations. The most pointed difference being the observed angle at which the nucleobases adsorbed onto the graphene changing the strength of the molecular dipole[44].

The strong coverage dependence of DNA nucleobases induced CNP shifts in graphene can be partly explained by the aforementioned molecular gating mechanism generating shifts in CNP. The electrostatic depolarization model is commonly used to explain molecular modifications of the work function and electron affinity of semiconductor and metal surfaces due to the electric potential drop across a layer of dipolar molecules[26,27]. At high coverages, the mutual interaction between the formed dipole moments can lead to a strong coverage dependency[26]. This can happen with and without a direct charge transfer between molecules and substrate. The electric potential drop, which is proportional to a change in the Fermi level of graphene, can be described as a function of molecular coverage (N):

$$\Delta V = \frac{e\mu_z N}{\epsilon_0(1 + c\alpha N^{3/2})}, \qquad (2)$$

where μ_z is the z-component of the dipole moment of the molecule, α is polarizability of the molecule and c is a constant representing the geometry of the dipole lattice[26]. The fits obtained using this model to our electric transport data are shown by solid lines in Fig. 2. The model correctly predicts an initial increase of the carrier density with increasing molecular coverage and saturation as the intermolecular interactions depolarize the absorbed molecules. The model fits well the data for C, T and

A; however, the guanine data could not be properly fitted by this model due to a change of sign at high coverage. In spite of this, the depolarization model indicates quite correctly that guanine possess the highest dipole moment and polarizability, followed by C, T and A (ref. 44). It is important to note that the dipole moment sequence is exactly the same as the sequence observed in CNP and C 1s shifts in our experiments.

The adsorption geometry of molecules is an important factor that affects the interface electronic structure. Unlike the in-plane alignment considered in most of the DFT calculations[13-15,45], our NEXAFS measurements have shown a tilt of all four DNA nucleobases to the graphene surface of ~ 30–$45°$. The method used to calculate this tilt angle[39], however, is sensitive to disorder in the layer, and the measured values may represent partly ordered molecular layers with patches of disordered molecules. This structure is similar to the previously observed nucleobase films on graphite by scanning tunnelling microscopy[46,47]. Such disorder in the molecular layers could also be caused by defects and ripples ubiquitously present in the CVD graphene samples. A different nucleobase adsorption angle on graphene is expected to significantly alter the distance between the molecules and the graphene, and the hybridization of the molecular orbitals with graphene. Ahmed et al.[16] have recently demonstrated using DFT that the local density of states of DNA nucleobases on graphene exhibits a strong angle dependence for the molecular states within an energy range of a few electronvolts of the Fermi level. These states could be at the origin of the different sequence of nucleobase-induced shifts in our experiments compared with DFT calculations.

Another possibility for the difference is the interaction of DNA nucleobases with defects in graphene or even the SiO_2 substrate. Structural defects and surface contamination are known to produce localized electronic states close to the Fermi level and act as internal or external scattering centres for electron waves, greatly degrading transport properties of graphene[48,49]. CVD graphene contains a range of defects, such as grain boundaries and point defects[49]. Unfortunately, the exact type of defects, the electronic structure and the nature of interactions between these graphene defects and adsorbed DNA molecules is still unknown. Although defects may play a dominant role in the observed shifts, we can exclude that a unique defect or a contamination in our samples is responsible for the sensing mechanism by observing the same results on two different graphene samples from different sources. The SiO_2 substrate can also play an important role in the sensing mechanism of GFETs on SiO_2 as has been proposed recently[50]. Using DFT modelling it has been suggested that the sensitivity of graphene to gas adsorbates can be attributed to external defects in the insulating substrate[50]. However, at this point it is difficult to clearly identify the role of these defects in our measurements and further studies are needed to elucidate the dominant sensing mechanism in our GFET devices.

Although both electric transport and XPS measurements have shown qualitatively similar behaviour, there is a difference in the information obtained by these two methods. The magnitude of the observed C 1s binding energy shifts relative to Fermi level shifts calculated from CNP shifts has shown to be consistently higher for the same coverage of an adsorbed nucleobase. For instance, the C 1s shifts for thymine are five times larger than the effective shifts calculated from the electric transport using equation (1) (Fig. 3). The simplest explanation for this difference is the presence of metal contacts near the measured graphene in the GFETs. Metal–graphene interfaces have been identified to have a significant influence on the transport measurements in graphene electronic devices[51,52]. The alignment between the metal work function (4.3 and 4.7 eV for Ti and Au) and graphene (4.5 eV) can lead to the presence of potential steps around graphene/metal contacts that have been reported to extend up to a few micrometres from the metal electrodes[52]. The contacts can thus act as effective transport barriers, causing high resistance in GFET devices. As the measured GFET transport properties are dominated by the areas of highest resistance, molecule-induced changes in the contact resistance could dominate our observed results, despite the large device areas. This characteristic indicates that if proper metal contacts are made with an optimal alignment with graphene, the electrical transport can become substantially more sensitive than seen with our graphene devices. If the measured signals from XPS are taken as the upper bound, we can expect an improvement in the sensitivity of GFETs to DNA nucleobases by up to a factor of 10 with proper metal contact engineering.

While these results confirm the fundamental sensitivity of graphene for DNA sensing, the task of implementing this design of graphene sensor into any realizable DNA sequencing technology presents even further challenges not yet discussed. Water-induced doping of graphene on SiO_2 reduces device sensitivity and increases device noise[53]. As DNA sequencing is necessarily carried out in buffered ionic solutions, reduction of this effect becomes vital. Recent research has shown that using a different substrate will modify the graphene–water interactions[24,54], suggesting that optimized substrate management may provide a pathway to water-based DNA sensing. Another problem faced by the use of large area graphene sensors for DNA detection is the requirement to distinguish between the effects of adsorbed nucleobases and the charged phosphate backbone. Experiments that test DNA strand sensing will be vital in understanding any phosphate backbone graphene interactions. Moreover, the device sensitivity will be influenced by the nucleobase adsorption angle that modulates the strength of the adsorbed molecular dipoles, potentially resulting in a significant difference between free nucleobases compared with attached nucleobase in a DNA strand. The charge state of graphene may provide an effective tool for electrical control of the nucleobase–graphene angle and the conformation of the DNA strand[55].

In conclusion, we have shown that the electric transport measurements using GFETs fabricated on SiO_2 can probe distinct conductance signatures upon adsorption of four different DNA nucleobases. By decreasing the size of the graphene channel in these devices and taking advantage of recent improvements in graphene mobility and noise reduction techniques, we predict that single-molecule sensing of guanine, cytosine and thymine by bulk graphene devices can be achieved. Successful device designs will require the identification and deeper characterization of dominant sensing mechanisms, combined with optimized metal contact engineering, to amplify the sensitivity of graphene devices to individual molecules. We have also established that the different nucleobases have different magnitudes of interaction with graphene that strongly depend on molecular coverage. Further research into device design may enable this molecule discriminability to be exploited for selective nucleobase sensing.

Methods

Fabrication of GFETs. SLG and BLG grown by CVD on Cu and transferred to SiO_2 (90 nm)/Si substrates were obtained commercially from Graphene Laboratories Inc and ACS Materials LLC. Two suppliers were used to test the effect of different intrinsic defects and impurities in the CVD graphene samples. Repeated acetone rinsing followed by a 48-h anneal at 250 °C in pure Ar (99.999%) at atmospheric pressure was used to remove residual polymethyl methacrylate from the samples. The fabrication of GFET devices was done using a polymethyl methacrylate and oxygen plasma patterning method followed by shadow mask evaporation of gold contacts (100 nm). Each sample was divided into two regions: one for the GFET devices and the second larger region for XPS analysis. SLG samples used for XPS measurements underwent the same fabrication steps to ensure any unintentional contamination was not unique to the FETs. Once loaded

into the UHV system the samples were further annealed at temperatures of 250 °C in pressures $< 10^{-9}$ Torr. Annealing was continued until the measured CNP did not change more than 0.1 V between subsequent annealing steps, typically between 2 and 4 h.

Deposition of DNA nucleobases. Powders of pure DNA nucleobases, purchased from Sigma Aldrich, were evaporated from commercial low-temperature effusion cells (MBE Komponenten) under UHV conditions. Prior to deposition the sources of DNA nucleobases were properly degassed at 110 °C for 24 h to remove water. Evaporation of DNA nucleobases was done at 138 °C for adenine, 175 °C for cytosine, 235 °C for guanine and 124 °C for thymine at a pressure of 10^{-9} mbar, corresponding to a deposition rate of ~ 1 monolayer per 20 min. It was found that for adenine and thymine the graphene samples needed to be kept below -50 °C to ensure deposited molecules did not rapidly desorb from the surface during sample transfer in UHV. To further ensure no water contaminated the measurements, the residual gas analysis was used to verify that the water partial pressure stayed below 1×10^{-10} mbar during evaporation.

Determination of CNP from conductivity curves. The CNP was taken as the gate voltage directly corresponding to the minimum measured I_{SD} in transport measurements. The resolution limit of this method is of the order of 0.1 V due to a broad structure of the minimum conductivity region (Fig. 1a). A way around this problem is to estimate the CNP by the intersection of two linear fits made in the high carrier concentration regions above and below the minimum conductivity region[31]. However, as we observed a significant asymmetry between electron and hole conductivity that increased with molecular coverage, this technique would result in calculated CNPs being a function of both Fermi level shifts and the magnitude of this asymmetry and so it was not used. The changes in the device mobility with molecular adsorption were observed to be quite complex for the different nucleobases studied. Further analysis of this asymmetry and total mobility is presented in Supplementary Fig. 4 and Supplementary Discussion.

Determination of molecular coverage from XPS. The fractional monolayer coverages of DNA molecules on graphene were estimated using the following ratio between the XPS peak areas of the N 1s peak corresponding to a monolayer coverage and the C 1s peak area of SLG

$$\frac{A_{N1s}}{A_{C1s}} = \frac{n_N}{n_C} \frac{\sigma_{N450eV}}{\sigma_{C330eV}} \frac{I_{450eV}}{I_{330eV}}, \tag{3}$$

where $\frac{n_N}{n_C}$ is a geometric factor representing the ratio of the density of molecular nitrogen atoms to graphene carbon atoms, σ is the photoionization cross section of the element at a specific photon energy taken from the NIST database (http://srdata.nist.gov/xps/) and I is the intensity of photons incident on the sample at a specific photon energy. Tabled values can be found in Supplementary Table 1. Both spectra were taken at similar photoelectron kinetic energy to ensure the analyser transmission intensity was comparable for each peak. To estimate the ratio of molecular nitrogen to graphene carbon $\frac{n_N}{n_C}$, we used the molecular unit cell sizes of monolayer coverage of each DNA nucleobase on graphite measured by scanning tunnelling microscopy[46,47].

References

1. Soon, W. W., Hariharan, M. & Snyder, M. P. High-throughput sequencing for biology and medicine. *Mol. Syst. Biol.* **9,** 640 (2013).
2. McGinn, S. & Gut, I. G. DNA sequencing spanning the generations. *N. Biotechnol.* **30,** 366–372 (2013).
3. Traversi, F. *et al.* Detecting the translocation of DNA through a nanopore using graphene nanoribbons. *Nat. Nanotechnol.* **8,** 939–945 (2013).
4. Min, S. K., Kim, W. Y., Cho, Y. & Kim, K. S. Fast DNA sequencing with a graphene-based nanochannel device. *Nat. Nanotechnol.* **6,** 162–165 (2011).
5. Avdoshenko, S. M. *et al.* Dynamic and electronic transport properties of DNA translocation through graphene nanopores. *Nano Lett.* **13,** 1969–1976 (2013).
6. Freedman, K. J., Ahn, C. W. & Kim, M. J. Detection of long and short DNA using nanopores with graphitic polyhedral edges. *ACS Nano* **7,** 5008–5016 (2013).
7. Wells, D. B., Belkin, M., Comer, J. & Aksimentiev, A. Assessing graphene nanopores for sequencing DNA. *Nano Lett.* **12,** 4117–4123 (2012).
8. Guo, Y.-D., Yan, X.-H. & Xiao, Y. Computational investigation of DNA detection using single-electron transistor-based nanopore. *J. Phys. Chem. C* **116,** 21609–21614 (2012).
9. Connelly, L. S. *et al.* Graphene nanopore support system for simultaneous high-resolution afm imaging and conductance measurements. *ACS Appl. Mater. Interfaces* **6,** 5290–5296 (2014).
10. Haque, F., Li, J., Wu, H.-C., Liang, X.-J. & Guo, P. Solid-state and biological nanopore for real-time sensing of single chemical and sequencing of DNA. *Nano Today* **8,** 56–74 (2013).
11. Puster, M., Rodríguez-Manzo, J. A., Balan, A. & Drndi, M. Toward sensitive graphene nanoribbon-nanopore devices by preventing electron beam-induced damage. *ACS Nano* **7,** 11283–11289 (2013).

12. Rajan, A. C. *et al.* Two dimensional molecular electronics spectroscopy for molecular fingerprinting, DNA sequencing, and cancerous DNA recognition. *ACS Nano* **8**, 1827–1833 (2014).

13. Lee, J.-H., Choi, Y.-K., Kim, H.-J., Scheicher, R. H. & Cho, J.-H. Physisorption of DNA nucleobases on H-Bn and graphene: VDW-corrected DFT calculations. *J. Phys. Chem. C* **117**, 13435–13441 (2013).

14. Gowtham, S., Scheicher, R. H., Ahuja, R., Pandey, R. & Karna, S. P. Physisorption of nucleobases on graphene: density-functional calculations. *Phys. Rev. B* **76**, 033401 (2007).

15. Le, D., Kara, A., Schrder, E., Hyldgaard, P. & Rahman, T. S. Physisorption of nucleobases on graphene: a comparative van der Waals study. *J. Phys. Condens. Matter* **24**, 424210 (2012).

16. Ahmed, T. *et al.* Electronic fingerprints of DNA bases on graphene. *Nano Lett.* **12**, 927–931 (2012).

17. Shedin, F. *et al.* Detection of individual gas molecules adsorbed on graphene. *Nat. Mater.* **6**, 652–655 (2007).

18. Wehling, T. O. *et al.* Molecular doping of graphene. *Nano Lett.* **8**, 173–177 (2008).

19. Liu, G., Rumyantsev, S., Shur, M. S. & Balandin, A. A. Origin of 1/f noise in graphene multilayers: surface vs. volume. *Appl. Phys. Lett.* **102**, 093111 (2013).

20. Hess, L. H. *et al.* Graphene transistor arrays for recording action potentials from electrogenic cells. *Adv. Mater.* **23**, 5045–5049 (2011).

21. Jiang, S. *et al.* Real-time electrical detection of nitric oxide in biological systems with sub-nanomolar sensitivity. *Nat. Commun.* **5**, 2225 (2013).

22. Some, S. *et al.* Highly sensitive and selective gas sensor using hydrophilic and hydrophobic graphenes. *Sci. Rep.* **3**, 1868 (2013).

23. Kuila, T. *et al.* Chemical functionalization of graphene and its applications. *Prog. Mater. Sci.* **57**, 1061–1105 (2012).

24. Wehling, T. O., Lichtenstein, A. I. & Katsnelson, M. I. First-principles studies of water adsorption on graphene: the role of the substrate. *Appl. Phys. Lett.* **93**, 202110 (2008).

25. Li, Z. *et al.* Effect of airborne contaminants on the wettability of supported graphene and graphite. *Nat. Mater.* **12**, 925–931 (2013).

26. Monti, O. L. A. Understanding interfacial electronic structure and charge transfer: an electrostatic perspective. *J. Phys. Chem. Lett.* **3**, 2342–2351 (2012).

27. Piana, S. & Bilic, A. The nature of the adsorption of nucleobases on the gold [111] surface. *J. Phys. Chem. B* **110**, 23467–23471 (2006).

28. Xu, G. *et al.* Low-noise submicron channel graphene nanoribbons. *Appl. Phys. Lett.* **97**, 073107 (2010).

29. Xu, G. *et al.* Enhanced conductance fluctuation by quantum confinement effect in graphene nanoribbons. *Nano Lett.* **10**, 4590–4594 (2010).

30. Yang, Y. & Murali, R. Impact of size effect on graphene nanoribbon transport. *IEEE Elec. Dev. Lett.* **31**, 237–239 (2010).

31. Chen, J.-H. *et al.* Charged-impurity scattering in graphene. *Nat. Phys.* **4**, 377 (2008).

32. Bolotin, K. *et al.* Ultrahigh electron mobility in suspended graphene. *Solid State Commun.* **146**, 351–355 (2008).

33. Dean, C. R. *et al.* Boron nitride substrates for high-quality graphene electronics. *Nat. Nanotechnol.* **5**, 722–726 (2010).

34. Balandin, A. A. Low-frequency 1/f noise in graphene devices. *Nat. Nanotechnol.* **8**, 549555 (2013).

35. Basu, D., Gilbert, M. J., Register, L. F., Banerjee, S. K. & MacDonald, A. H. Effect of edge roughness on electronic transport in graphene nanoribbon channel metal-oxide-semiconductor field-effect transistors. *Appl. Phys. Lett.* **92**, 042114 (2008).

36. Hüfner, S. *Photoelectron Spectroscopy* (Springer, 2003).

37. Castro Neto, A. H., Guinea, F., Peres, N. M. R., Novoselov, K. S. & Geim, A. K. The electronic properties of graphene. *Rev. Mod. Phys.* **81**, 109–162 (2009).

38. Deacon, R. S., Chuang, K.-C., Nicholas, R. J., Novoselov, K. S. & Geim, A. K. Cyclotron resonance study of the electron and hole velocity in graphene monolayers. *Phys. Rev. B* **76**, 081406 (2007).

39. Stöhr, J. & Outka, D. A. Determination of molecular orientations on surfaces from the angular dependence of near-edge X-ray-absorption fine-structure spectra. *Phys. Rev. B* **36**, 7891–7905 (1987).

40. Lee, V. *et al.* Substrate hybridization and rippling of graphene evidenced by near-edge X-ray absorption fine structure spectroscopy. *J. Phys. Chem. Lett.* **1**, 1247–1253 (2010).

41. Braun, S., Salaneck, W. R. & Fahlman, M. Energy-level alignment at organic/metal and organic/organic interfaces. *Adv. Mater.* **21**, 1450–1472 (2009).

42. Cervenka, J. *et al.* Graphene field effect transistor as a probe of electronic structure and charge transfer at organic molecule-graphene interfaces. *Nanoscale* **7**, 1471–1478 (2014).

43. Lee, S. K. *et al.* Inverse transfer method using polymers with various functional groups for controllable graphene doping. *ACS Nano* **8**, 7968–7975 (2014).

44. Jasien, P. G. & Fitzgerald, G. Molecular dipole moments and polarizabilities from local density functional calculations: application to DNA base pairs. *J. Chem. Phys.* **93**, 2554–2560 (1990).

45. Berland, K., Chakarova-Kck, S. D., Cooper, V. R., Langreth, D. C. & Schrder, E. A van der Waals density functional study of adenine on graphene: single-molecular adsorption and overlayer binding. *J. Phys. Condens. Matter* **23**, 135001 (2011).

46. Tao, N. J. & Shi, Z. Monolayer guanine and adenine on graphite in NaCl solution: a comparative STM and AFM study. *J. Phys. Chem.* **98**, 1464–1471 (1994).

47. Sowerby, S. J. & Petersen, G. B. Scanning tunneling microscopy of uracil monolayers self-assembled at the solid|liquid interface. *J. Electroanal. Chem.* **433**, 85–90 (1997).

48. Song, H. S. *et al.* Origin of the relatively low transport mobility of graphene grown through chemical vapor deposition. *Sci. Rep.* **2**, 337 (2012).

49. Batzill, M. The surface science of graphene: metal interfaces, CVD synthesis, nanoribbons, chemical modifications, and defects. *Surf. Sci. Rep.* **67**, 83–115 (2012).

50. Kumar, B. *et al.* The role of external defects in chemical sensing of graphene field-effect transistors. *Nano Lett.* **13**, 1962–1968 (2013).

51. Xia, F., Perebeinos, V., Lin, Y., Wu, Y. & Avouris, P. The origins and limits of metalgraphene junction resistance. *Nat. Nanotechnol.* **6**, 179–184 (2011).

52. Lee, E. J. H., Balasubramanian, K., Weitz, R. T., Burghard, M. & Kern, K. Contact and edge effects in graphene devices. *Nat. Nanotechnol.* **3**, 486–490 (2008).

53. Kaverzin, A. A., Mayorov, A. S., Shytov, A. & Horsell, D. W. Impurities as a source of 1/f noise in graphene. *Phys. Rev. B* **85**, 075435 (2012).

54. Levesque, P. L. *et al.* Probing charge transfer at surfaces using graphene transistors. *Nano Lett.* **11**, 132–137 (2011).

55. Shankla, M. & Aksimentiev, A. Conformational transitions and stop-and-go nanopore transport of single-stranded DNA on charged graphene. *Nat. Commun.* **5**, 5171 (2014).

56. Cowie, B. C. C., Tadich, A. & Thomsen, L. The current performance of the wide range (902500 ev) soft X-ray beamline at the Australian synchrotron. *AIP Conf. Proc.* **1234**, 307–310 (2010).

Acknowledgements

This research was supported under Australian Research Council's Discovery Projects funding scheme (project number DE120101100). The authors acknowledge the facilities and the scientific assistance of the Soft X-ray beam line at the Australian Synchrotron, Victoria, Australia.

Author contributions

N.D. and J.C. prepared the manuscript, designed and coordinated the experiments. N.D., J.C. and A.S. discussed the results and analysed the data. All authors aided in carrying out the experiments and commented on the manuscript.

Additional information

Hydrogen-atom-mediated electrochemistry

Jin-Young Lee[1], Jae Gyeong Lee[1], Seok-Ha Lee[2], Minjee Seo[1], Lilin Piao[1], Je Hyun Bae[1], Sung Yul Lim[1], Young June Park[2] & Taek Dong Chung[1]

Silicon dioxide thin films are widely used as dielectric layers in microelectronics and can also be engineered on silicon wafers. It seems counterintuitive that electrochemical reactions could occur on such an insulator without relying on tunnelling current. Here we report electrochemistry based on electron transfer through a thin insulating layer of thermally grown silicon dioxide on highly n-doped silicon. Under a negative electrical bias, protons in the silicon dioxide layer were reduced to hydrogen atoms, which served as electron mediators for electrochemical reduction. Palladium nanoparticles were preferentially formed on the dielectric layer and enabled another hydrogen-atom-mediated electrochemistry, as their surfaces retained many electrogenerated hydrogen atoms to act as a 'hydrogen-atom reservoir' for subsequent electrochemical reduction. By harnessing the precisely controlled electrochemical generation of hydrogen atoms, palladium–copper nanocrystals were synthesized without any surfactant or stabilizer on the silicon dioxide layer.

[1] Department of Chemistry, Seoul National University, Seoul 151-747, Korea. [2] Department of Electrical and Computer Engineering, Seoul National University, Seoul 151-744, Korea. Correspondence and requests for materials should be addressed to T.D.C. (email: tdchung@snu.ac.kr).

Precise control of electrochemical reactions on dielectric layers is a key issue that has been increasingly studied in recent years, because it would enable patterned circuits, stacks of thin films and various nanoparticles (NPs) to be fabricated on dielectric layers by direct electroplating. Although electrochemistry on dielectric layers has the potential to offer new fundamental methodologies in various fields such as the semiconductor industry, energy conversion, sensors, catalysts and supercapacitors, there have been relatively few studies on this topic, mainly because it is counterintuitive that electrochemical reactions could occur on dielectric layers, that is, insulators. In fact, electrochemistry through insulators such as glass[1], metal oxide[2,3], organic monolayers[4] and rubbed Teflon[5] has been investigated. Previous studies have reported on electrochemistry through amorphous glass films[1], which have many pores that permit the penetration of redox species and ions. Because of the permeability of many chemical species, no electron mediator was required to be in the glass membrane for a faradaic reaction. A silicon (Si) electrode modified with an organic monolayer is another example that has been intensively investigated over the past decades. The reproducible fabrication of stable and sufficiently thick dielectric layers has been found to be crucial for informative and practical systems. Silver (Ag) NPs were electrodeposited on aluminum (Al) oxide at $-9\,V$ (ref. 3) and the direct electrochemical reduction[2] of TiO_2 to Ti was performed at $950\,°C$. Rubbed Teflon was also adopted as an insulator on which metal was deposited[5]. Although such studies have suggested new approaches to introducing electrochemistry to insulating materials (dielectrics), there are still challenges to adapting these approaches to practical applications, including the permeability, reproducibility, temperature and controllability of the electrochemical reaction. More importantly, only a limited number of studies have investigated electrochemistry processes that occur at the dielectric layer. In this work, we introduced thermally grown silicon dioxide (SiO_2), a dense and stable dielectric layer, and explored the novel phenomena that could be observed in this electrochemical system.

Thermal SiO_2 is a dielectric material used for conventional metal/insulator/metal capacitors[6], and SiO_2 films are widely used as gate insulators of metal/oxide/semiconductor (MOS) devices[7], in which the field-induced drift of protons to the Si/SiO_2 interface and their interactions with electrons can create hydrogen (H) atoms at the interface[8,9]. Shkrob et al.[10] demonstrated that protons were reduced to generate mobile H atoms in B_2O_3 glasses at room temperature. Energy calculations revealed that H atoms in SiO_2 are isolated by themselves without a strong interaction with oxygen (O_2) or Si atoms. H atoms reside in interstitial sites ~2.0 Å away from the nearest O_2 atoms and do not disturb the surrounding lattice[11]. Schrauben et al.[12] provided evidence for H atom transfer through reduced TiO_2 and ZnO NPs in solution using two H abstractors, namely, 2,4,6-tri-tert-butylphenoxyl radical ($^tBu_3ArO\bullet$) and 2,2,6,6-tetramethyl-piperidin-1-yl-oxyl. Revesz[13] studied on an electrolyte/oxide/Si (EOS) system to determine the concentration of certain active H-bearing species in the oxide.

The highly n-doped Si/thermal SiO_2 thin layers in an aqueous electrolyte considered in this work represent an EOS system suitable for producing H atoms electrochemically by finely tuning the applied potential bias and pH of the solution. The SiO_2 thin film, which is thermally grown on highly n-doped Si (the fabrication method and properties are described in detail in Methods and the Supplementary Information), has a highly dielectric structure with minimal defects. Therefore, it permits both the selective migration of protons and the diffusion of electrogenerated H atoms[9] in the middle of the electrical potential gradient. This result suggests that H atoms in the EOS system

could be used as electron mediators on dielectric oxide layers acting as 'chemical electrodes' for electrochemical reactions. Free H atoms are theoretically very strong reducing agents ($-2.106\,V$ versus normal hydrogen electrode)[14]. In the proposed system of highly n-doped Si/thermal SiO_2 thin layers, the reducing power of H atoms in SiO_2 should not be as strong as that of freestanding H atoms. However, the well-ordered SiO_2 structure provides a remarkably stable and inert environment. H atoms in titanium dioxide (TiO_2) and zinc oxide (ZnO) are strongly bound and act as a dopant but weakly interact with the neighbouring oxygen atoms in SiO_2 (refs 10,15). Thermally grown SiO_2 is not expected to modify the chemical properties of the interstitial H atoms to the same degree as TiO_2 (ref. 11) and ZnO (ref. 16). Therefore, in the proposed system, electrogenerated H atoms would serve as sufficiently strong reducing agents to reduce most reducible species in solution. Next, metal precursors could be electroplated by H atoms on the thermal SiO_2 surface to potentially form the corresponding metal NPs. This process could offer important inspiration for many interesting applications in that various metal NPs adsorbed on oxide surfaces are widely used for energy technology, pollution prevention and environmental clean-up efforts[17,18]. In particular, SiO_2 is commonly used as an inert support for heterogeneous catalysts.

Compared with chemical reduction, electrochemical methods enable fine and easy control by electronic devices and primarily require no reducing agents because of the direct injection of electrons[19]. Therefore, electroplating has been considered as an alternative method for the cost-effective and stabilizer-free NP preparation. However, long-term electrochemical operation requires appropriate electrodes that must be safely free from passivation to continue providing electrons for sustainable faradaic reactions. Many electrodes are subject to deactivation, which is mostly caused by the irreversible adsorption of adsorptive intermediates and/or the deterioration of electrode materials by reactive species.

Herein, we report electron transfer through a thin insulating layer of the 6-nm-thick thermal SiO_2 on highly n-doped silicon (n^+-Si), leading to unprecedented electrochemistry on the dielectric surface. Under a negative electrical bias, protons from the solution media migrate into the silicon dioxide layer and produce significant cathodic current. The asymmetric voltammograms from the proposed system suggest electrogeneration of H atoms as electron mediators for electrochemical reduction at the interface between the dielectric surface and solution. By precisely tuning the voltammetric conditions, palladium (Pd)–copper (Cu) nanocrystals (NCs) are synthesized without any surfactant or stabilizer, and both O_2 and carbon dioxide (CO_2) are electrochemically reduced on the SiO_2 layer.

Results

SiO_2 layers in the EOS system. The thermally grown SiO_2 considered in this work has an exact stoichiometric composition[20]. Supplementary Fig. S1 presents high-resolution transmission electron microscopy (HRTEM) cross-sectional images of the high n^+-Si/SiO_2 (6-nm thick). A compact SiO_2 layer was uniformly formed on Si; this layer was very stable as demonstrated by the tests of breakdown voltage and metal electroplating (Supplementary Figs S2 and S3; Supplementary Note 1).

The n^+-Si/SiO_2 immersed in an aqueous solution was used as the working electrode to investigate the electrochemistry in this system, as illustrated in Fig. 1. The linear sweep voltammogram (i–V curve) obtained in the proposed EOS system using the n^+-Si/SiO_2 electrode demonstrates that the 6-nm-thick thermal SiO_2 clearly differed from the native SiO_x on n^+-Si (Supplementary

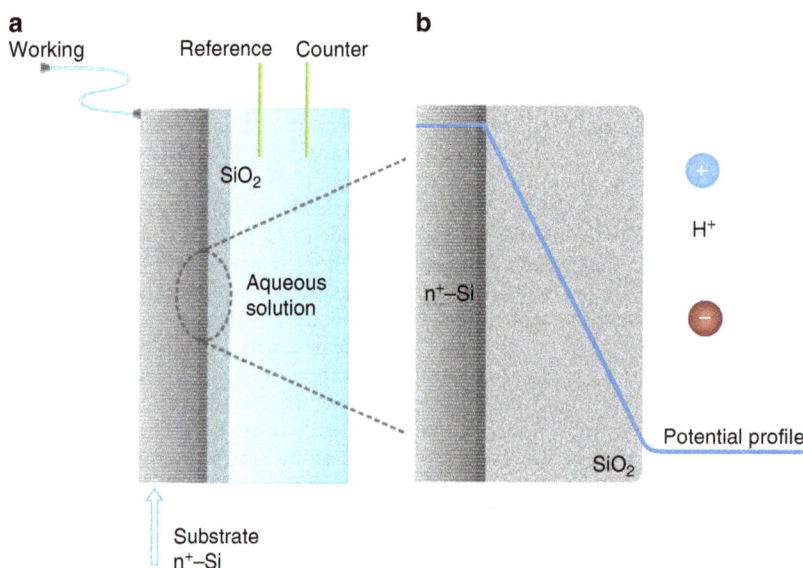

Figure 1 | Schematic view of the electrochemical experiment. (**a**) Experimental system and (**b**) schematic representation of the electrical potential profile across highly n-doped Si (n^+-Si)/thermal SiO_2/aqueous electrolyte.

Fig. S4). Compared with the native SiO_x layer, the structure of the thermal SiO_2 is sufficiently dense and uniform for the background current to be negligible, providing a reliable system for electrochemical study.

In previous studies, the electrical potential distribution across the Si/SiO_2 interface in an EOS system was not significantly different from that in a MOS system[21]. In contrast to the solution in an MOS configuration, the solution in an EOS system acts as an infinite source of ions that can be polarized near the oxide surface upon the application of an electric field as long as the dielectric layer is sufficiently thin to incompletely isolate the conducting Si phase from the solution, in electrical terms[7]. However, the applied electrical potential still drops mainly across the oxide layer because of its high resistance of $10^7\,\Omega\,cm^2$, for the thermally grown 6-nm-thick SiO_2 layer on n^+-Si, as measured by impedance analysis (Supplementary Note 2). The electrical potential drop at the 6-nm-thick SiO_2 layer was 2×10^5 times greater than that at the SiO_2/solution interface in this EOS system (Supplementary Fig. S5). Therefore, the potential difference at the interface between SiO_2 and the solution is negligible compared with the entire applied bias.

Electron transfer through the SiO_2 dielectric layer. Figure 2a presents the cyclic voltammograms recorded using an n^+-Si/SiO_2 electrode and the comparison of these voltammograms with that obtained using a glassy carbon (GC) electrode, in 0.1 M potassium phosphate buffer solution at pH 3. The electrochemical oxidation of $Ru(NH_3)_6^{2+}$ at the n^+-Si/SiO_2 required a considerably higher overpotential than the GC electrode. The onset potential was observed at approximately $+3.2\,V$ (versus Ag/AgCl), and the current increased gradually as the potential increased in the positive direction. In contrast, the overpotential for $Ru(NH_3)_6^{3+}$ reduction was far lower than expected from the oxidation results in the same system. The faradaic current for reduction sensitively responded to the electrical potential bias, despite the compact dielectric layer of thermally grown SiO_2. In addition, Fig. 2b,c demonstrates that the onset potential of an n^+-Si/SiO_2 electrode systematically depended on the pH of the aqueous electrolyte in the cathodic regime, whereas no pH dependence was observed in the anodic region. The

asymmetric voltammogram in Fig. 2a strongly implies that there are electrochemically reducible species inside the SiO_2 layer, preferentially taking up electrons from the highly doped Si phase.

These electrochemically reducible species can be identified from the following process. First, the cathodic onset potential of the n^+-Si/SiO_2 electrode in an aprotic organic electrolyte, such as acetonitrile, was observed at approximately $-2.2\,V$ (versus Ag/Ag^+; Supplementary Fig. S6). The cathodic current disappeared without a proton. Second, the behaviour in acetonitrile follows the Fowler–Nordheim tunnelling (Supplementary Fig. S7), which is similar to MOS systems[22]. When a small amount of proton sources, $HClO_4$, was added to the solution, the cathodic current began to flow at a more positive value of approximately $-1.2\,V$, whereas the background current was negligible in the same potential range (Supplementary Fig. S8). The proton sources clearly caused the system to deviate from the Fowler–Nordheim tunnelling model (Supplementary Note 3), revealing that the proposed system cannot be explained by a simple direct tunnelling model. Third, the thermal SiO_2 layer in this system completely blocks permeation of all chemical species, including $Ru(NH_3)_6^{2+/3+}$ ions and electrolytes, except protons, which can diffuse through the SiO_2 interfacial plane[23] and reversibly migrate back and forth in response to an external electric field[24]. This evidence unequivocally supports the conclusion that the cathodic currents in the linear sweep voltammograms in Fig. 2 were attributed to protons, which can migrate into the SiO_2 layer and be electrochemically reduced (Supplementary Note 4).

Figure 2a reveals that the current in the forward scan in the negative potential region was smaller than that in the reverse scan. This result again demonstrates that the thermal SiO_2 layer in the proposed system is not a simple insulator through which a tunnelling current predominantly flows. During the period in which a negative potential was applied, protons accumulated in the SiO_2 layer and were electrochemically reduced to H atoms. This result is ensured by a comparison of the consecutive scans in linear sweep voltammetry. The onset potential of the second scan, which was performed immediately after the first scan, shifted in the positive direction and the cathodic current increased (Supplementary Fig. S9). The H atoms remaining inside the SiO_2 layer, which were generated during the previous scan, could relay electrons to the protons originating from the solution via

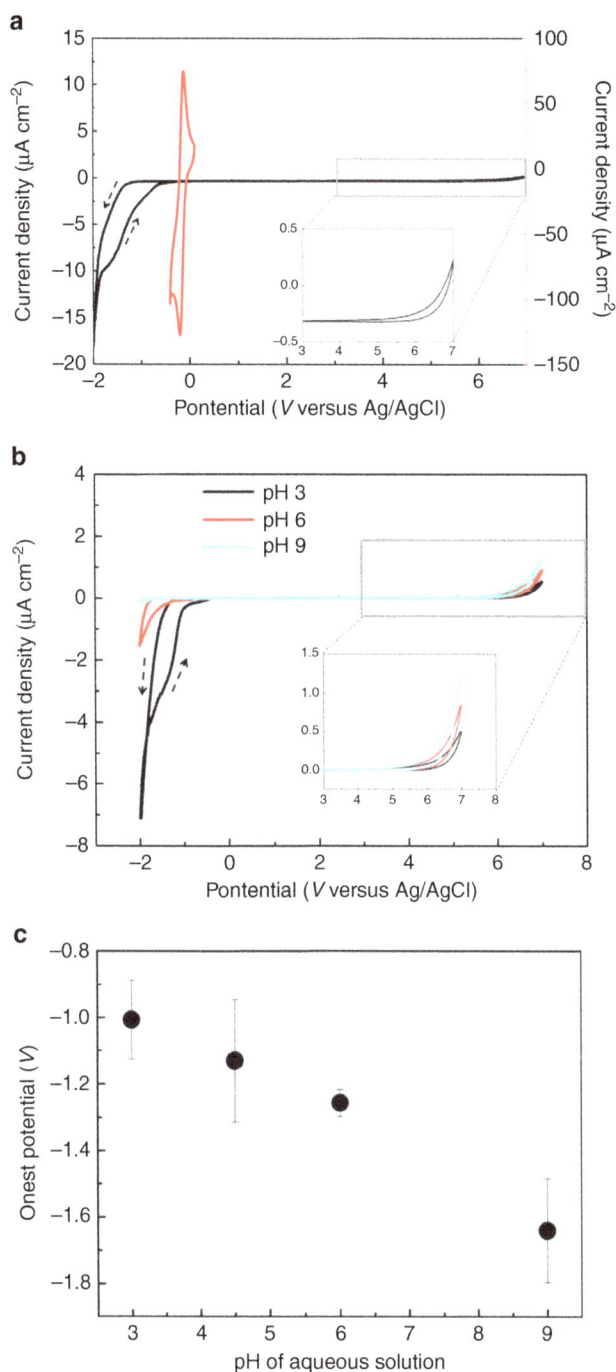

Figure 2 | Electrochemistry of the proposed EOS system. (**a**) Cyclic voltammograms (CVs) of the n^+-Si/SiO$_2$ electrode (black) and GC electrode (red) were recorded at room temperature in an N$_2$-purged solution of 1 mM Ru(NH$_3$)$_6$Cl$_3$ + Ru(NH$_3$)$_6$Cl$_2$ + 0.1 M potassium phosphate solution at pH 3 with a scan rate of 10 mV s^{-1}. (**b**) CV curves of the n^+-Si/SiO$_2$ (6-nm thick) electrode were obtained in 0.1 M phosphate buffer at pH 3 (black), pH 6 (red) and pH 9 (bright blue). The insets in **a** and **b** present the enlarged portion of the CV for the anodic potential at the n^+-Si/SiO$_2$ electrode. (**c**) The onset potential depending on the pH in **b**. The onset potential was determined at the potential at which the current density is above 30% compared with the baselines. The voltammograms were obtained at 10 mV s^{-1} at room temperature under N$_2$.

homogeneous electron exchange. The onset potential was restored after leaving the electrode in a solution for ~ 1 h without any electrical bias (Supplementary Fig. S9).

Figure 3a illustrates how the migration and reduction of protons in the SiO$_2$ layer can enable the reduction of redox species without a direct tunnelling current despite the presence of the dense insulating SiO$_2$ layer. The electrogenerated H atoms either directly serve as electron mediators or form H$_2$ molecules at the SiO$_2$/solution interface.

Electrochemical reduction of CO$_2$ and O$_2$ in the EOS system. The H-atom-mediated electrochemistry of this simple n^+-Si/SiO$_2$/aqueous electrolyte system could be used for the electrochemical reduction of CO$_2$ and O$_2$, which involves strongly reactive or adsorptive intermediates. In particular, the reductive transformation of CO$_2$ to fuels and commodity chemicals is one of the most important contemporary energy and environmental challenges. The initial step from CO$_2$ to CO$_2^{\bullet-}$ has been reported to be rate limiting in many cases. Either the overpotentials are exceedingly high or the metal surface becomes deactivated by the intermediates[25,26]. The long-term stability of the electrodes is a crucial challenge in making electrochemical CO$_2$ reduction economically feasible[26]. Figure 3b demonstrates the CO$_2$ reduction in this system. The product of CO$_2$ reduction in this system was mainly formic acid at -1.5 V (versus Ag/AgCl) in CO$_2$-saturated aqueous solution at pH 3 for longer than 4 h. H atoms that were not used for CO$_2$ reduction are expected to form molecular hydrogen. Both O$_2$ and CO$_2$ can be reduced in this system (Fig. 3c) without significant deterioration of the oxide structure by electrolysis over 1 h.

Bimetallic NCs synthesized electrochemically on the SiO$_2$ layer. The finely controllable H-atom-mediated electrochemistry of the n^+-Si/SiO$_2$ (6-nm thick)/aqueous electrolyte system was used for the electrodeposition of several types of metals. Direct electron tunnelling can reduce metal precursors on native silicon oxide, producing 1- to 2-nm-thick non-adherent powders that are readily washed off by rinsing[27]. In contrast, the Pd NPs (diameter of 10–40 nm) formed in this system were uniformly distributed on the oxide layer due to the reduction of Pd^{2+} ions to metallic Pd (Supplementary Fig. S10). The Pd NPs were anchored so strongly that very few were washed off by rinsing. H$_2$ molecules are likely to prevent the initial Pd seeds from aggregating, resulting in immobilization[28], and thus, H atoms, which are much stronger reducing agents, are assumed to have a pivotal role in the prompt reduction of Pd^{2+} ions. Thus, this system enables the single-step electrochemical fabrication of Pd NPs firmly adhered to the SiO$_2$ layer in aqueous solution at room temperature without surfactants or stabilizers. Pd NPs prepared in this manner are expected to have application in hydrogen storage as catalysts on solid supports and as sensing devices[29].

Metallic Pd is well known to retain H atoms on its surface. Once Pd NPs are formed on the SiO$_2$ layer, the excess H atoms are predicted to be adsorbed on the surfaces of the Pd NPs spontaneously at room temperature. Even if H$_2$ molecules are formed, they would be readily dissociated on the Pd surface[29]. Thus, the hydrogen-rich surfaces of the Pd NPs are expected to facilitate deposition of metals, such as Cu, which otherwise do not adhere to the SiO$_2$ surface[27]. The Pd NPs on the SiO$_2$ surface in this system act as 'H-atom lagoons' that collect and store as many H atoms as they can on their surfaces, as demonstrated in Fig. 3d. Such a novel H-atom lagoon leads to the subsequent electroplating of various metals, including bimetallic NCs.

Bimetallic NCs are of considerable interest because of their unique properties and potential applications[30,31]. H-atom-mediated electrochemistry employing H-atom lagoons, such as those formed by Pd NPs on SiO$_2$, can achieve facile bimetallic NC synthesis with a polyhedral alloy on an inert support. Figure 4a

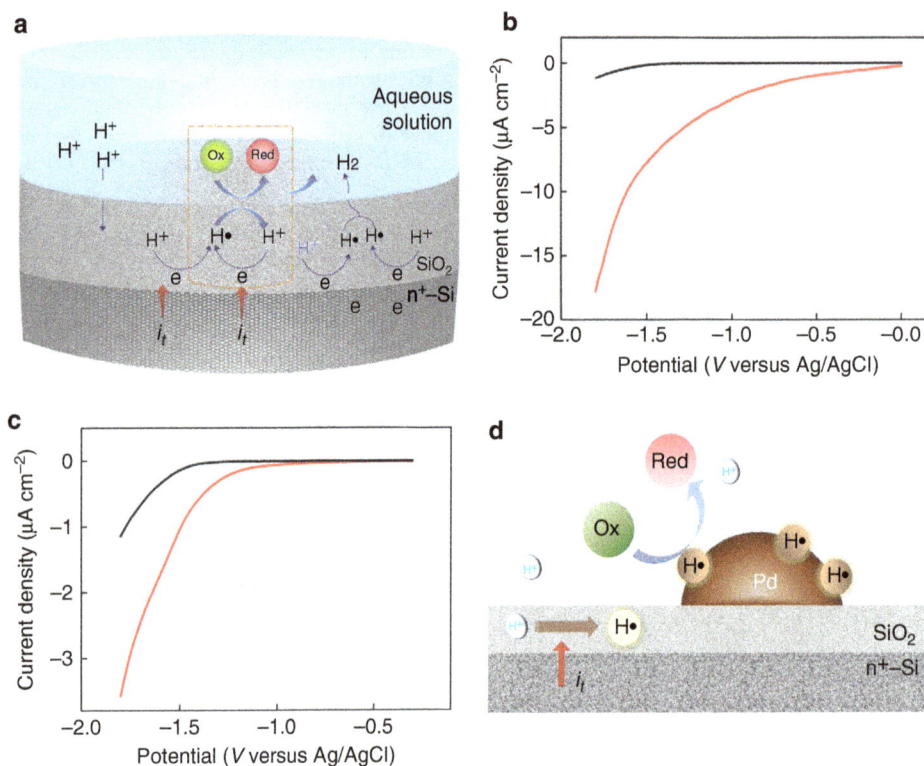

Figure 3 | CO_2 and O_2 reduction and Pd NPs acting as H-atom lagoons. (**a**) Schematic view of the reactions in the proposed system: H-atom-mediated electron transfer in n^+-Si/thermally oxidized SiO_2/aqueous solution. Protons in the aqueous solution drift into the SiO_2 layer and become H atoms by capturing electrons tunnelling through the Si/SiO_2 interface under a negative potential. (**b**) The linear sweep voltammogram of n^+-Si/thermally oxidized SiO_2 in black was obtained at room temperature in an N_2-purged solution of 0.1 M potassium phosphate at pH 3. The linear sweep voltammogram in red was obtained at room temperature in a CO_2-saturated solution of 0.1 M potassium phosphate at pH 3. (**c**) The linear sweep voltammogram of n^+-Si/thermally oxidized SiO_2 in black was obtained at room temperature in an N_2-purged solution of 0.1 M potassium phosphate at pH 3. The linear sweep voltammogram in red was obtained at room temperature in an O_2-saturated solution of 0.1 M potassium phosphate at pH 3. (**d**) Schematic illustration of the H-atom lagoon effect of Pd NPs on n^+-Si/SiO_2 (thermally oxidized) for the reduction of reducible species.

presents electron microscopic images of polyhedral Pd–Cu bimetallic NCs consisting of octahedron (32%), triangular prisms (38%) and others (30%). This study represents the first electrochemical shape-controlled synthesis of bimetallic NCs without any stabilizer or surfactant. Energy-dispersive spectroscopy (EDS) revealed that the electroplated Pd–Cu bimetallic NCs were a Pd–Cu mixed alloy (Supplementary Fig. S11). The TEM image and corresponding electron diffraction pattern of a single Pd–Cu NC in Fig. 4f indicate that the NC was a piece of a single crystal enclosed by {111}, {200} and {220} facets. In addition to Pd–Cu NCs, Pd–Au NCs and Pd–Co NCs were also synthesized on a thin SiO_2 layer at room temperature without the aid of a stabilizer or additive (Supplementary Figs S12 and S13).

Discussion

The remarkably low onset potential (Fig. 2a) for the cathodic reaction indicates that protons were reduced inside the SiO_2 layer. Thus, the observed voltammetric results can be understood by a scenario in which an external electric field causes protons to migrate in the oxide layer[24] and the protons are electrochemically reduced to produce H atoms[9,32] in a negative potential regime. Shkrob et al.[10] suggested that electrons react with network-bound protons at 180–530 K to produce H atoms.

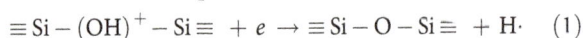

$$\equiv Si - (OH)^+ - Si \equiv\ +\ e \rightarrow\ \equiv Si - O - Si \equiv\ +\ H\cdot \quad (1)$$

We applied a negative potential to an n^+-Si/SiO_2 electrode in 0.1 M potassium phosphate buffer solution at pH 3 and acquired

surface-enhanced Raman scattering (SERS) spectra from the SiO_2 layer. The peak of the $Si-(OH^+)-Si$ complex[33] at $\sim 2,905\ cm^{-1}$ was observed (Supplementary Fig. S14) in accordance with the suggested mechanism[10]. This observation confirms the permeation of protons into the SiO_2 layer and supports the idea that H atoms can be generated because of proton reduction in the SiO_2 layer under an external electric field.

The presence and role of H atoms as electron mediators in the SiO_2 layer can be verified by independent voltammetric experiments using several redox-active species with different reduction potentials. Although redox-active species are characteristically reduced at different potentials on conventional conducting electrodes, such as GC and Au, their reduction behaviour in this system exhibited no correlation with the reduction potentials of their own (not shown). This result cannot be explained by a direct tunnelling current.

The electrochemical reduction of iridium (Ir) precursors (Supplementary Fig. S15) provides additional evidence of the electrogeneration of H atoms in the SiO_2 layer. Although the Ir precursors used in this study require a very negative potential or commensurately strong reducing agent to be reduced, metallic Ir NPs were electrodeposited on the SiO_2 surface at a moderate potential bias of $-1.3\ V$. Considering the monotonously linear and predominant drop of electrical potential across the SiO_2 film (Fig. 1), it is unlikely that a direct tunnelling current is responsible for this result. The experiment indicated that the Ir precursors are not reduced by H_2 molecules but may be reduced by H atoms[34,35].

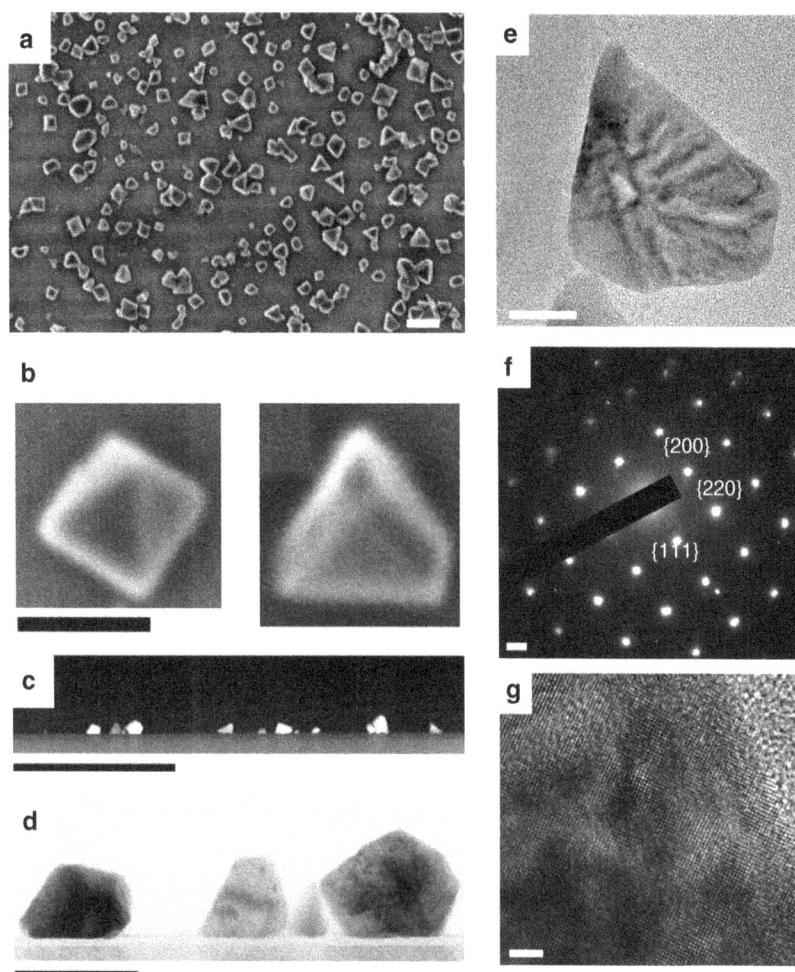

Figure 4 | Pd-Cu NCs. (**a**) FESEM image of Pd-Cu NCs prepared as follows. First, a constant potential of $-1.2\,V$ was applied to n^{+}-Si/SiO$_2$ for 30 min in 1 mM PdCl$_2$ + 0.1 M potassium phosphate solution at pH 3. Next, SiO$_2$-supported Pd NPs were immersed in 1 mM CuSO$_4$ + 0.1 M H$_2$SO$_4$ solution and a constant potential of $-1.3\,V$ was applied for 30 min. Scale bar, 200 nm. (**b**) High-magnification FESEM image of Pd-Cu NCs prepared as in **a**. Octahedron-shaped Pd-Cu NCs (left), triangular-prism-shaped Pd-Cu NCs (right). Scale bar, 100 nm. (**c**) High-angle annular dark field-scanning TEM image of SiO$_2$-supported Pd-Cu NCs prepared as in **a**. Scale bar, 1 µm. (**d**) TEM image of Pd-Cu NCs. Scale bar, 100 nm. (**e**) TEM image of Pd-Cu NCs. Scale bar, 20 nm. (**f**) Electron diffraction pattern of NCs. Scale bar, 2 nm^{-1}. (**g**) HRTEM image of a single Pd-Cu NC. Scale bar, 2 nm.

In this study, we propose and investigate H-atom-mediated electrochemistry in the system of n^{+}-Si/thermal SiO$_2$/aqueous solution in which a thin and dense SiO$_2$ layer acted as a proton-selective membrane and reservoir of atomic H. In voltammetry, the redox current was not linearly dependent on the external potential bias, and reduction occurred at a lower overpotential in proton-rich solutions compared with oxidation, resulting in asymmetric voltammetric behaviour. The experimental evidences unequivocally demonstrates that protons from the solution permeated into the SiO$_2$ layer and were reduced to form H atoms by the uptake of electrons from the highly doped Si phase under a negative potential. The electrogenerated H atoms diffuse into the SiO$_2$ layer and reduce the precursors, having the role of electron mediators. The suggested mechanism is in good agreement with the observed nonlinear and asymmetric voltammetric behaviour. The proposed H-atom-mediated electrochemistry can be used to synthesize Pd NPs with an average size of 10–30 nm without using stabilizers or additives (Supplementary Fig. S10). More importantly, the Pd NPs on the SiO$_2$ layer can act as a H-atom lagoon for the facile nanofabrication of bimetallic NCs such as Pd–Cu, Pd–Au and Pd–Co. In particular, the system

in this study successfully demonstrated its potential for producing various metallic nanostructures, including Pd–Cu bimetallic NCs, with such shapes as triangular prisms and octahedron on the SiO$_2$ substrate. Both bimetallic NCs and monometallic NPs were synthesized on the SiO$_2$ in the absence of any additive or stabilizer. H-atom-mediated electrochemistry will provide an impetus to the future fabrication of electronic circuits, sensory units and thin-film-based energy conversion devices on insulating layers.

Methods

Preparation of thermal oxidized SiO$_2$. N-type (n^{+}-Si), phosphorous-doped, $<100>$ oriented Si wafers with a resistivity as low as 0.005 Ω cm were obtained from STC (Japan). Highly n-doped Si with thermally oxidized SiO$_2$ (n^{+}-Si/SiO$_2$) was prepared via conventional thermal oxidation of highly doped Si wafers ($<100>$, n-type, 0.005 Ω cm). After cleaning with a mixture of H$_2$SO$_4$ and H$_2$O$_2$, the native oxide was stripped by HF dipping and the 200-Å-thick thermal oxide was produced at 850 °C in a furnace with dry O$_2$ blowing. Next, 200-Å-thick thermal oxides were stripped again by HF dipping. Cleaning was repeated and the target 50- or 60-Å-thick thermal SiO$_2$ layer was formed at 850 °C in a furnace with dry O$_2$ blowing. The thermal SiO$_2$ layers produced were uniformly flat. Ellipsometry and HRTEM (Supplementary Fig. S1) confirmed that the thicknesses of the thermal SiO$_2$ layer on Si was 60 Å.

Chemicals and electrochemical experiments. Before every experiment, the samples were degreased by sonication in acetone and isopropanol, and rinsed with methanol or deionized water. For back contact, the back side of the wafer sample was scratched with a diamond knife to remove the air-formed SiO_2 film followed by smearing with a Ga–In eutectic. This surface was then pressed to a metal support. The ohmic nature of the contact was verified by electrical measurements.

After this procedure, the samples were pressed against the O-ring of an electrochemical cell, leaving $0.066\,cm^2$ exposed to an aqueous solution of 0.1 M potassium phosphate (Daejung Chemicals & Metals Co., Ltd) in deionized water. A pH electrode (ROSS 8102, Orion) was used to adjust the pH of the aqueous solution for electrochemical measurement. The electrochemical experiments were performed at room temperature in a three-electrode electrochemical cell connected to a potentiostat (750, CHI Instruments, Inc.). An Ag/AgCl (in 3 M NaCl, BAS Inc.) and Pt wire (diameter 0.5 mm) were used as reference and counter electrodes, respectively. Cyclic voltammograms and linear sweep voltammograms were obtained with or without redox species after purging with N_2 for 20 min. Hexaammineruthenium (II) chloride ($Ru(NH_3)_6Cl_2$, $\geq$ 99.9% trace metal basis), hexaammineruthenium (III) chloride ($Ru(NH_3)_6Cl_3$, 98%) and Ir (III) chloride hydrate ($Ir_3Cl \cdot xH_2O$, $\geq$ 99.9% trace metal basis) were purchased from Sigma-Aldrich for use as redox species.

For CO_2 and O_2 reduction, Si (n^+-Si)/SiO_2 (6-nm thick) was used as a working electrode; this electrode was exposed to 0.1 M potassium phosphate buffer at pH 3 after rinsing with deionized water. The aqueous solution was purged with CO_2 or O_2 gas for at least 20 min. An Ag/AgCl (in 3 M NaCl, BAS, Inc.) and Pt mesh were used as reference and counter electrodes, respectively. For CO_2 reduction, after applying a constant potential of -1.5 V (versus Ag/AgCl) to the working electrode for 4–8 h, the solution was analysed using HPLC (Dionex, USA) at the National Instrumentation Center for Environmental Management of Seoul National University.

SERS on n^+-Si/SiO_2 in the EOS system. The 11-mercaptoundecanol-modified gold microshell was used for SERS monitoring[36]. The Raman spectra were obtained using a homemade Ramboss Micro-Raman system spectrometer with a TE cooled ($-60\,°C$) charged-coupled device camera ($1,024 \times 128$ pixels). The 632.8 nm line from a He–Ne laser (LASOS Lasertechnik GmbH, USA) was employed as an excitation source and focused through a $\times 50$ objective for *in situ* electrochemical SERS experiments. The grating (1,200 and 1,800 grooves mm^{-1}) and slit provided a spectral resolution of $4\,cm^{-1}$. The spectrometer was calibrated with the Raman bands of a silicon wafer at $520\,cm^{-1}$ and indene at 730.4, 1,018.3, 1,205.6, 1,552.7 and $1,610.2\,cm^{-1}$.

Ir deposition catalysed by H atoms present on the SiO_2 surface. The n^+-Si/SiO_2 (6-nm thick) electrode was immersed in an N_2-purged solution of iridium (III) chloride hydrate and a constant potential of -1.3 V was applied to the electrode. The presence of Ir particles on n^+-Si/SiO_2 was confirmed by field-emission scanning electron microscopy (FESEM) and X-ray photoelectron spectroscopy. Supplementary Fig. S11 presents the deposition of Ir, and X-ray photoelectron spectroscopy provided data characteristic of metallic Ir, namely Ir $4f_{7/2}$ and Ir $4f_{5/2}$ peaks at 61.0 and 63.9 eV, respectively.

Formation of NCs or NPs on SiO_2. Palladium (II) chloride ($PdCl_2$, $\geq$ 99.9%), potassium gold (III) chloride ($KAuCl_4$, 98%) and cobalt sulfate hydrate ($CoSO_4 \cdot \times H_2O$, $\geq$ 99.9% trace metal basis) were purchased from Sigma-Aldrich. Copper sulphate ($CuSO_4$) was purchased from Junsei Chemicals Co., Ltd. Deposition of Pd NPs on n^+-Si/SiO_2 (6-nm thick) was performed in a conventional three-electrode cell with an Ag/AgCl electrode (3 M NaCl, BAS Inc.) and Pt wire (diameter of 0.5 mm) as the reference and counter electrodes, respectively. All potentials refer to the Ag/AgCl reference electrode. The NPs and NCs were electrochemically synthesized in 0.1 M potassium phosphate buffer at pH 3 in an N_2 environment immediately after purging the solution with N_2 gas for 20 min at room temperature.

To deposit Pd NPs on the SiO_2 surface, an aqueous solution composed of 0.2–1 mM $PdCl_2$ in 0.1 M potassium phosphate buffer at pH 3 was prepared and a constant potential of -1.2 V or -1.5 V (versus Ag/AgCl) was applied to the n^+-Si/SiO_2. After Pd deposition on the SiO_2 surface, other metals, including Cu, Au and Co, were deposited on n^+-Si/SiO_2-supported Pd NP electrodes to determine whether Pd-based bimetallic fabrication occurred. For Pd–Cu bimetallic NCs, a 1 mM $CuSO_4$ aqueous solution in 0.1 M potassium phosphate buffer at pH 3 was prepared and the constant potential of -1.3 V (versus Ag/AgCl) was applied to the n^+-Si/SiO_2-supported Pd NPs electrode. For Pd–Au bimetallic NPs, a constant potential of -1.3 V (versus Ag/AgCl) was applied to the n^+-Si/SiO_2-supported Pd NPs electrode in an aqueous solution of 1 mM $KAuCl_4$ in 0.1 M potassium phosphate buffer at pH 3. Pd–Co bimetallic NPs were synthesized under -1.8 V (versus Ag/AgCl) applied to the n^+-Si/SiO_2-supported Pd NPs electrode for 30 min.

Characterization of NCs and NPs on SiO_2. The NCs and NPs on SiO_2 prepared in the manner described above were characterized by FESEM (Carl Zeiss, Germany) and EDS (Carl Zeiss, Germany) at the National Instrumentation Center

for Environmental Management of Seoul National University. For high-resolution images, HRTEM (JEOL, Japan) and EDS (JEOL, Japan) were used at the UNIST (Ulsan National Institute of Science and Technology) Central Research Facilities Center (UCRF) of the UNIST. Focused ion beam (FEI, USA) milling was used to prepare the TEM specimens. Mo grids were used for the HRTEM–EDS analysis of Pd–Cu NCs. The average atomic percentage of Pd:Cu was 93.1:6.9 (Supplementary Fig. S11). The average atomic percentage of Pd:Au was 5.2:1 according to a quantitative analysis of Pd–Au NPs by FESEM–EDS (Supplementary Fig. S12). The average atomic percentage of Pd:Co in the Pd–Co NPs was 94.7:5.3 according to FESEM–EDS (Supplementary Fig. S13).

References

1. Velmurugan, J., Zhan, D. & Mirkin, M. V. Electrochemistry through glass. *Nat. Chem.* **2**, 498–502 (2010).
2. Chen, G. Z., Fray, D. J. & Farthing, T. W. Direct electrochemical reduction of titanium dioxide to titanium in molten calcium chloride. *Nature* **407**, 361–364 (2000).
3. Wang, H.-H. *et al.* Highly Raman-enhancing substrates based on silver nanoparticle arrays with tunable sub-10 nm gaps. *Adv. Mater.* **18**, 491–495 (2006).
4. Gittins, D. I., Bethell, D., Schiffrin, D. J. & Nichols, R. J. A nanometer-scale electronic switch consisting of a metal cluster and redox-addressable groups. *Nature* **408**, 67–69 (2000).
5. Liu, C. & Bard, A. J. Electrostatic electrochemistry at insulators. *Nat. Mater.* **7**, 505–509 (2008).
6. Babcock, J. A. *et al.* Analog characteristics of metal-insulator-metal capacitors using PECVD nitride dielectrics. *IEEE Electron Device Lett.* **22**, 230–232 (2001).
7. Morrison, S. R. *Electrochemistry at Semiconductor and Oxidized Metal Electrodes.* pp 304 (Plenum Press, New York, 1980).
8. Griscom, D. L., Brown, D. B. & Sakes, N. S. in *The Physics and Chemistry of SiO_2 and the Si-SiO_2 Interface.* (eds Helms, C. R. & Deal, B. E.) 287–297 (Plenum, New York, 1988).
9. Nicollian, E. H., Berglund, C. N., Schmidt, P. F. & Andrews, J. M. Electrochemical charging of thermal SiO_2 films by injected electron currents. *J. Appl. Phys.* **42**, 5654–5663 (1971).
10. Shkrob, I. A., Tadjikov, B. M., Chemerisov, S. D. & Trifunac, A. D. Electron trapping and hydrogen atoms in oxide glasses. *J. Chem. Phys.* **111**, 5124–5140 (1999).
11. Xiong, K., Robertson, J. & Clark, S. J. Behavior of hydrogen in wide band gap oxides. *J. Appl. Phys.* **102**, 083710–083713 (2007).
12. Schrauben, J. N. *et al.* Titanium and zinc oxide nanoparticles are proton-coupled electron transfer agents. *Science* **336**, 1298–1301 (2012).
13. Revesz, A. G. On the mechanism of the ion sensitive field effect transistor. *Thin Solid Films* **41**, L43–L47 (1977).
14. Bard, A. J., Parsons, R. & Jordan, J. (eds) *Standard Electrochemical Potentials in Aqueous Solution* 39–48 (Marcel Dekker, New York, 1985).
15. Pineda, A. C. *et al.* The effect of network topology on proton trapping in amorphous SiO_2. *IEEE T. Nucl. Sci* **48**, 2081–2085 (2001).
16. van de Walle, C. G. Hydrogen as cause of doping in zinc oxide. *Phy. Rev. Lett.* **85**, 1012–1015 (2000).
17. Gao, F. & Goodman, D. W. Model catalysts: simulating the complexities of heterogeneous catalysts. *Annu. Rev. Phys. Chem.* **63**, 265–286 (2012).
18. Farmer, J. A. & Campbell, C. T. Ceria maintains smaller metal catalyst particles by strong metal-support bonding. *Science* **329**, 933–936 (2010).
19. Tian, N., Zhou, Z.-Y., Sun, S. -G., Ding, Y. & Wang, Z. L. Synthesis of tetrahexahedral platinum nanocrystals with high-index facets and high electro-oxidation activity. *Science* **316**, 732–735 (2007).
20. Zhang, X. G. *Electrochemistry of Silicon and Its Oxide* 3–93 (Kluwer Academic, New York, 2001).
21. Fisher, P. R., Daschbach, J. L., Gragson, D. E. & Richmond, G. L. Sensitivity of 2nd-harmonic generation to space-charge effects at Si(111) electrolyte and Si(111)/SiO_2 electrolyte interfaces. *J. Vac. Sci. Technol. A* **12**, 2617–2624 (1994).
22. Depas, M., Vermeire, B., Mertens, P. W., Meirhaeghe, R. L. V. & Heyns, M. M. Determination of tunneling parameters in ultra-thin oxide layer poly-Si/SiO_2/Si structures. *Electronics* **38**, 1465–1471 (1995).
23. Rashkeev, S. N., Fleetwood, D. M., Schrimpf, R. D. & Pantelides, S. T. Dual behavior of H$^+$ at Si-SiO_2 interfaces: mobility versus trapping. *Appl. Phys. Lett.* **81**, 1839–1841 (2002).
24. Vanheusden, K. *et al.* Non-volatile memory device based on mobile protons in SiO_2 thin films. *Nature* **386**, 587–589 (1997).
25. Gattrell, M., Gupta, N. & Co, A. A review of the aqueous electrochemical reduction of CO_2 to hydrocarbons at copper. *J. Electroanal. Chem.* **594**, 1–19 (2006).
26. Kumar, B. *et al.* Photochemical and photoelectrochemical reduction of CO_2. *Annu. Rev. Phys. Chem.* **63**, 541–569 (2012).

27. Arrington, D., Curry, M., Street, S., Pattanaik, G. & Zangari, G. Copper electrodeposition onto the dendrimer-modified native oxide of silicon substrates. *Electrochim. Acta* **53,** 2644–2649 (2008).

28. Salzemann, C. & Petit, C. Influence of hydrogen on the morphology of platinum and palladium nanocrystals. *Langmuir* **28,** 4835–4841 (2012).

29. Lee, J., Shim, W., Noh, J.-S. & Lee, W. Design rules for nanogap-based hydrogen gas sensors. *ChemPhysChem* **13,** 1395–1403 (2012).

30. Wang, D. & Li, Y. Bimetallic nanocrystals: liquid-phase synthesis and catalytic applications. *Adv. Mater.* **23,** 1044–1060 (2011).

31. Rodriguez, J. A. & Goodman, D. W. The nature of the metal-metal bond in bimetallic surfaces. *Science* **257,** 897–903 (1992).

32. Afanas'ev, V. V., Ciobanu, F., Pensl, G. & Stesmans, A. Proton trapping in SiO_2 layers thermally grown on Si and SiC. *Solid State Electron.* **46,** 1815–1823 (2002).

33. Yokozawa, A. & Miyamoto, Y. First-principles calculations for charged states of hydrogen atoms in SiO_2. *Phys. Rev. B* **55,** 13783–13788 (1997).

34. Sawy, E. N. E. & Birss, V. I. Nano-porous iridium and iridium oxide thin films formed by high efficiency electrodeposition. *J. Mater. Chem.* **19,** 8244–8252 (2009).

35. Vot, S. L., Roue, L. & Belanger, D. Electrodeposition of iridium onto glassy carbon and platinum electrodes. *Electrochim. Acta* **59,** 49–56 (2012).

36. Kim, S., Piao, L., Han, D., Kim, B. J. & Chung, T. D. Surface enhanced raman scattering on non-SERS active substrates and in situ electrochemical study based on single gold microshell. *Adv. Mater.* **25,** 2056–2061 (2013).

Acknowledgements

This work was supported by the Global Frontier R&D Program on Center for Multiscale Energy System funded by the National Research Foundation under the Ministry of Education, Science and Technology, Korea (number 2012M3A6A7055873), by the Nano Material Technology Development Program through the National Research Foundation of Korea (NRF) funded by the Ministry of Science, ICT and Future (number 2011-0030268), by the National Research Foundation of Korea (NRF) grant funded by the Korean government (MEST) (number 2012R1A2A1A03011289), and by the Center for Integrated Smart Sensors funded by the Ministry of Science, ICT & Future Planning as Global Frontier Project (number CISS-2011-0031845). Mr Young-Ki Kim (Ulsan National Institute of Science and Technology (UNIST) Central Research Facilities Center, Ulsan, Korea) is gratefully acknowledged for conducting the HRTEM characterization.

Author contributions

T.D.C. and J.-Y.L. designed the study and wrote the paper, J.-Y.L. performed the electrochemical experiments and collected data, J.G.L. conducted a portion of the electrochemical experiments and CO_2 reduction; M.S. prepared the bimetallic NCs, S.-H.L. and Y.J.P. fabricated and evaluated the Si/SiO_2 electrodes, L.P. recorded the SERS spectra, and J.H.B. and S.Y.L. assisted with the impedance analysis. All of the authors discussed the results and commented on the manuscript.

Additional information

4

Quantification of plasma HIV RNA using chemically engineered peptide nucleic acids

Chao Zhao[1], Travis Hoppe[2], Mohan Kumar Haleyur Giri Setty[3], Danielle Murray[4], Tae-Wook Chun[4], Indira Hewlett[3] & Daniel H. Appella[1]

The remarkable stability of peptide nucleic acids (PNAs) towards enzymatic degradation makes this class of molecules ideal to develop as part of a diagnostic device. Here we report the development of chemically engineered PNAs for the quantitative detection of HIV RNA at clinically relevant levels that are competitive with current PCR-based assays. Using a sandwich hybridization approach, chemical groups were systematically introduced into a surface PNA probe and a reporter PNA probe to achieve quantitative detection for HIV RNA as low as 20 copies per millilitre of plasma. For the surface PNA probe, four cyclopentane groups were incorporated to promote stronger binding to the target HIV RNA compared with PNA without the cyclopentanes. For the reporter PNA probe, 25 biotin groups were attached to promote strong signal amplification after binding to the target HIV RNA. These general approaches to engineer PNA probes may be used to detect other RNA target sequences.

[1] Laboratory of Bioorganic Chemistry, NIDDK, NIH, DHHS, 9000 Rockville Pike, Bethesda, Maryland 20892, USA. [2] Laboratory of Biochemistry and Genetics, NIDDK, NIH, DHHS, 9000 Rockville Pike, Bethesda, Maryland 20892, USA. [3] Laboratory of Molecular Virology, Center for Biologics Evaluation and Research, FDA, 9000 Rockville Pike, Bethesda, Maryland 20892, USA. [4] Laboratory of Immunoregulation, NIAID, NIH, DHHS, 9000 Rockville Pike, Bethesda, Maryland 20892, USA. Correspondence and requests for materials should be addressed to D.H.A. (email: appellad@niddk.nih.gov).

Diagnostic testing of HIV in both infected and non-infected patients is crucial to control the disease within the global population[1]. Moving such tests closer to the point-of-care (POC) for a patient helps doctors to quickly discern infection status and select the proper antiretroviral medication[2]. Obtaining this information more quickly and accurately promotes a test-and-treat strategy that has known benefits to limit spread of the virus[3]. HIV tests performed at the POC are typically antibody based, qualitative and require some type of follow-up testing to confirm infection[4–9]. Despite the popularity of antibody-based tests, nucleic acid testing (NAT) for HIV RNA continues to be the ultimate standard to confirm an infection and it is also the only method to quantify the viral load within infected patients[10,11]. Plasma HIV RNA levels correlate very well with the acute phase of infection, when the virus is most likely to be transmitted, and increases in viral load signal when the virus has developed resistance to a particular antiretroviral therapy[12]. Despite the benefits of tracking plasma HIV RNA in patients, there is no standard nucleic acid-based test for HIV that can be used at a patient's POC (ref. 13) and only about one-third of all public health laboratories in the US are equipped to perform NAT for HIV RNA[14]. There are even fewer public health facilities in developing countries that test for HIV RNA[15]. The lack of testing for HIV RNA reflects the complications of using reverse transcription-PCR (RT-PCR), the most common method to detect and quantify RNA[16–18]. The typical cost, instrumentation and expertise needed to perform RT-PCR prohibit its implementation at most public health settings in the US and abroad[10].

One approach to design NAT diagnostics without PCR amplification is to use a sandwich hybridization approach where target sequences are bound between two separate probes[19]. Normally one probe provides target segregation from the bulk solution (the surface probe), while the second probe imparts a measurable signal to the hybridization event (the reporter probe). The sandwich hybridization approach has been successfully implemented in a variety of nucleic acid sensing techniques, including: fluorescence imaging[20–23], electrochemical detection[24–26], template-mediated fluorescence activation[27,28] and surface-enhanced Raman scattering[29]. In most cases the requirement of two orthogonal binding events reduces background noise; however, the sensitivity of most sandwich assays is not as good as that achieved using a PCR-based method due to weak binding to the target or insufficient signal amplification.

The aminoethylglycyl (aeg) peptide nucleic acid molecule (aegPNA) has nucleobases attached to a simple, non-natural polyamide backbone that consists of alternating ethylene diamine and glycine units (Fig. 1a)[30]. In general, aegPNA binds to complementary DNA and RNA sequences following Watson–Crick hydrogen bond pairing rules and forms duplex structures that are often more stable than duplexes of two nucleic acids (Fig. 1b)[31,32]. Since it is synthetic and does not occur in nature, aegPNA sequences are remarkably stable to degradation by proteases and nucleases[33]. Previously, we have shown that sequential introduction of cyclopentane groups into the PNA backbone (to create cyclopentane PNA) systematically enhances binding to target nucleic acid sequences (Fig. 1c)[34–37].

In this manuscript we report the engineering principles to make a new type of RNA detection system that combines the accuracy of NAT with the practical convenience of enzyme-linked immunosorbent assay (ELISA) methodology. Key to the success of this merger is properly designed PNAs[30–32] that target HIV RNA sequences. By incorporating cyclopentanes into the surface PNA probe, we demonstrate that binding to the target RNA is significantly improved over aegPNA and subsequently improves

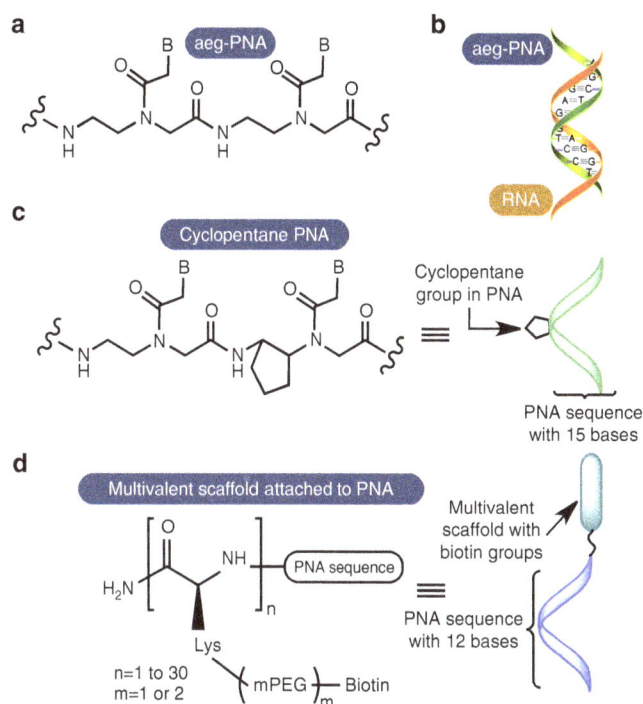

Figure 1 | Chemical structures and cartoon representations. (**a**) aegPNA showing two nucleobases (B = A, T, G or C). (**b**) aegPNA–RNA duplex. (**c**) Cyclopentane-modified PNA. The partial structure may be extended in either direction with additional aegPNA or cyclopentane PNA residues. (**d**) General strategy by which biotins are attached to the C-terminal of a PNA to create a multivalent scaffold with biotins.

nucleic acid detection. We have also designed a signal read-out system that can be coupled to the terminal end of a reporter PNA molecule to signal binding of a target RNA sequence. In this case, a unique chemical scaffold was developed to support the multivalent display of up to 30 biotin groups. The spacing of the biotins on the scaffold has been optimized to support assembly of multiple streptavidin-conjugated horseradish peroxidase enzymes around the terminal end of a PNA reporter probe (PNA-RP; Fig. 1d). The assembly of this nanostructure upon PNA binding to target RNA sequences allows for dramatic signal amplification so that low levels of RNA are detectable. Proper combination of cyclopentanes in the surface PNA and multivalent biotins in the reporter PNA results in a synergistic improvement in detection for HIV RNA. Furthermore, we show that the optimized system can detect HIV RNA across a wide range of groups and clades of the virus, is compatible with detection of virus in human plasma, and that the results correlate with a current PCR standard of detection (COBAS Ampliprep v2.0; ref. 38). We also demonstrate that the enzymatic amplification system is robust (at least a 30 day shelf life at room temperature (RT)), that it is cost effective ($0.67 per well in a 96-well plate), and that the signal reporting on the amount of HIV can be determined either spectrophotometrically or with the naked eye. To abbreviate our detection method we have coined the term NAT-PELA, which stands for NAT–PNA Enzyme-Linked Assay.

Results

Sandwich setup for NAT-PELA. In the current study, a sandwich hybridization assay using two PNAs is used to detect HIV RNA. The basic detection scheme is outlined in Fig. 2. In this protocol, the first step is to attach a PNA surface probe (PNA-SP, 15 base

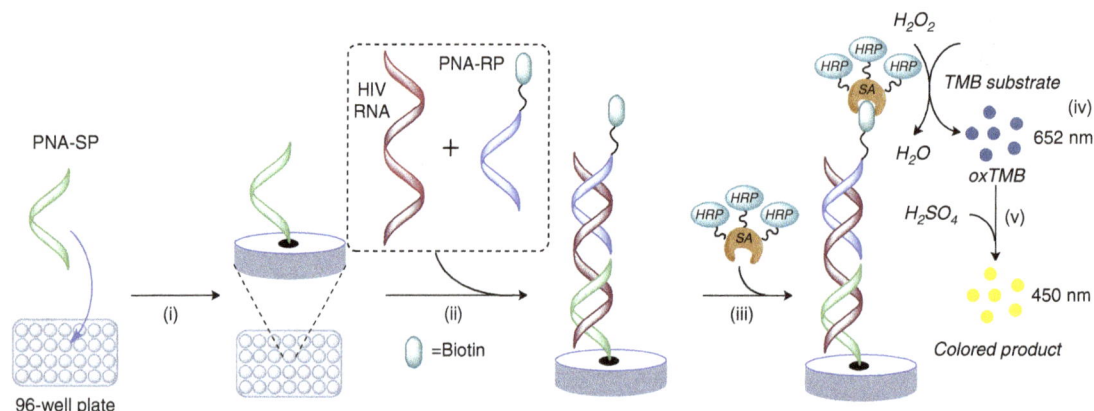

Figure 2 | General scheme for sandwich detection using a surface PNA (PNA-SP) and a reporter PNA (PNA-RP). (i) PNA-SP is covalently attached to the surface. (ii) If target HIV RNA is present, PNA-SP and PNA-RP both bind to their complementary sequences and a noncovalent complex is formed. In the absence of target RNA, PNA-RP is removed by subsequent washing steps. (iii) pHRP–SA is added. (iv) TMB is added and signal is monitored at 652 nm. (v) Enzymatic oxidation is stopped by the addition of sulphuric acid and the signal at 450 nm is measured.

sequence, see Supplementary Fig. 1 for all chemical structures and Supplementary Tables 1 and 2 for characterization) covalently via an amide bond to the plastic surface of a NUNC 96-well plate (i). PNA-RP (12 base sequence) is free in solution and contains biotin groups covalently attached to the C-terminal. The sequences of PNA-SP and PNA-RP are complementary to a 27-base sequence (5'-TTCTGCAGCTTCCTCATTGATGGTCTC-3') that is part of HIV RNA in the gag region. If the target RNA is present, PNA-SP and PNA-RP both bind to their complementary sequences and a noncovalent complex of PNA-SP + PNA-RP + RNA is formed on the plastic surface (ii). The presentation of biotins from PNA-RP allows basic ELISA methodology to be used to amplify a signal for detection[39]. In this study, polyhorseradishperoxidase–streptavidin (pHRP–SA) is used in step iii to signal that the RNA target is present[40]. In this step, the streptavidin portion of pHRP–SA forms a strong complex with biotins attached to PNA-RP. Next, introduction of a substrate (tetramethylbenzidine (TMB) and peroxide) for the horseradishperoxidase enzyme part of pHRP–SA (step iv) results in oxidation of the tetramethylbenzidine to give products with blue colour[41]. The enzymatic oxidation reaction can be stopped by the addition of acid to afford yellow products (step v). Both the blue and yellow products can be quantified (at 652 nm and 450 nm, respectively) to show the level of RNA originally present. If the RNA target is not present in step ii, then PNA-RP is removed in subsequent wash steps, pHRP–SA cannot attach to the plastic surface, and no signal results when the tetramethylbenzidine solution is added. Tests for the stability of pHRP–SA showed that there is no loss in enzymatic activity for 30 days at RT (Supplementary Fig. 2).

Engineering, optimization and modelling of PNA-SP and PNA-RP. A synthetic strand of HIV RNA with the 27-base target sequence was used to optimize the detection system. In all tests, the HIV RNA concentration in solution was varied and the data obtained at 652 nm and 450 nm were plotted as a function of absorbance versus time and absorbance versus concentration, respectively (Fig. 3). The absorbance data at 450 nm, obtained after quenching of enzymatic oxidation with H_2SO_4, were plotted and subsequently fit to a four-parameter logistic curve to show the quantitative limits of detection (Fig. 3d and Supplementary Fig. 3). To start, PNA-SP had no cyclopentane groups (PNA-SP0) (Supplementary Fig. 1a) and PNA-RP had one biotin (PNA-RP1) (Supplementary Fig. 1f). In this initial detection system, the lower limit of quantitative detection for the target RNA is in hundreds

of millions (10^8) of molecules (Table 1 and Supplementary Fig. 3a).

To improve this method so that it would be useful for clinical determination of HIV viral load, the limit of quantitative detection needed to be lowered to levels competitive with RT-PCR (which is ∼ 20 molecules of RNA)[16]. Therefore, we explored the sequential incorporation of cyclopentane groups into PNA-SP (Supplementary Fig. 1b–e) and the incremental addition of biotin groups to the C-terminal of PNA-RP (Supplementary Fig. 1g–m). Additional cyclopentane groups in PNA-SP incrementally increases the binding affinity to the complementary RNA sequence (Table 1, SP1 to SP4, and Supplementary Table 2), and lowers the detection limit. Increasing the number of biotin groups on the end of PNA-RP enhances the signal intensity of the sandwich complex on the surface (Fig. 4). Interestingly, there is a modest decrease in the binding of PNA-RP to its complementary RNA sequence as the number of biotins attached to the end increases (Table 1, RP1 to RP30, and Supplementary Table 2). This loss in binding, however, does not diminish the stability of the sandwich complex as the cyclopentanes in PNA-SP compensate for any loss in stability (Supplementary Table 2). Attempts to incorporate cyclopentane groups into PNA-RP were not successful when large numbers of biotin groups were required. Quantitative detection limits were determined for every combination of PNA-SP and PNA-RP, and the results are presented in Table 1. Both types of modifications work together to lower the limit of RNA detection. In the optimal system, PNA-SP with four cyclopentane groups (Fig. 3a) and PNA-RP with 25 biotin groups (Fig. 3b) was sufficient to detect HIV RNA in a region that would be competitive with PCR diagnostics (Fig. 3c,d), and the detection of 60 copies could also be observed visually (Fig. 3e,f). Additional biotins did not further enhance detection (PNA-RP30, Supplementary Fig. 1m and Table 1). Notably, these engineering principles were able to improve detection of the target RNA by about 8 orders of magnitude from the original system (PNA-SP0 and PNA-RP1).

A model was developed to explain the relative contributions of adding cyclopentanes and biotins into PNA-SP and PNA-RP. As input, the logs of the detection limits from Table 1 were represented on a three-dimensional plot (coloured regions in Fig. 4). Attempting to fit a plane to the detection limits resulted in a poor fit, with a root-mean-square deviation (RMSD) of 1.9. This implies that a simple linear relationship between detection limits, cyclopentanes and biotins does not exist (Supplementary Fig. 4). Considering that PNA-SP should retain RNA on the surface and

a

PNA-SP with four cyclopentane groups

PNA-SP4

b

PNA-RP with 25 biotin groups

PNA-RP25

c

d

e

f

Table 1 | Modifications to PNA-SP and PNA-RP and effects on T_m to complementary RNA and quantitative detection limit of detecting HIV target 27-base RNA (detection limits are presented in number of molecules of RNA).

T_m(°C) RNA:	66.1	66.0	65.5	63.5	61.9	60.1	58.0	55.5
T_m(°C) RNA:	RP1	RP2	RP6	RP 10	RP 16	RP 20	RP 25	RP 30
70.5 SP0	10^8	10^8	10^8	10^8	10^4	10^3	10^3	10^3
74.2 SP1	10^8	10^8	10^7	10^7	104	10^3	10^3	10^3
78.5 SP2	10^7	10^6	10^6	10^6	10^3	10^3	10^2	10^3
82.1 SP3	10^7	10^6	10^5	10^5	10^2	10^2	78	10^2
84.5 SP4	10^7	10^6	10^5	10^5	10^2	12	4	12

PNA-RP, peptide nucleic acid reporter probe; PNA-SP, peptide nucleic acid surface probe. Detection limits were calculated using the absorbance data at 450 nm as described in 'Analysis of NAT-PELA results' under the Methods section.

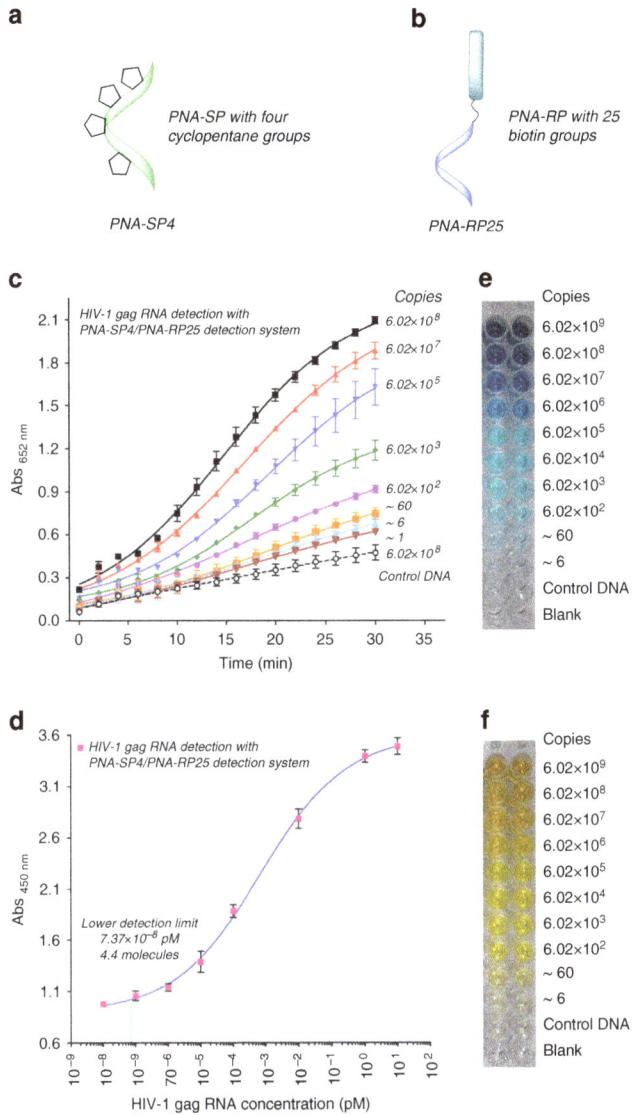

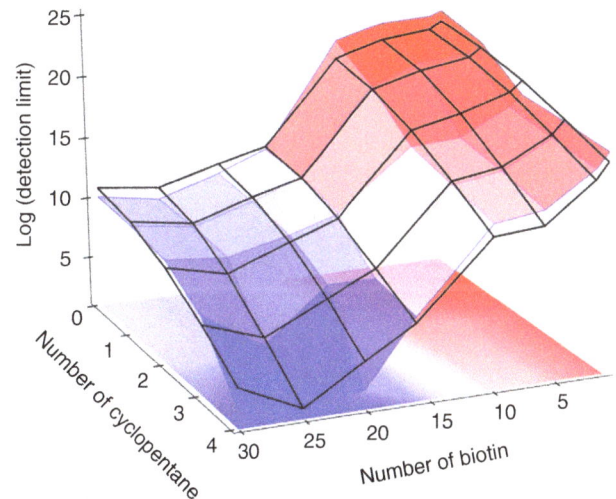

Figure 3 | Detection of the target HIV-1 gag RNA using the PNA-SP4/PNA-RP25 detection system. (**a**) Structure of PNA-SP4. (**b**) Structure of PNA-RP25. (**c**) Time-dependent absorbance (652 nm) changes upon analyzing different concentrations of HIV-1 gag RNA. (**d**) Signal response plots (absorbance at 450 nm versus concentration of HIV-1 gag RNA) obtained after quenching of enzymatic oxidation with H_2SO_4. The green line marks the minimal distinguishable differential concentration and represents the lowest limit of quantification. For **c** and **d**, values represent the mean ± s.d. of duplicate measurement ($n = 2$) for each sample. (**e,f**) Photographs of the 96-well plate show the variation in colour intensity for HIV-1 gag RNA detection before (**e**) and after (**f**) quenching of enzymatic oxidation with H_2SO_4.

directly influence the final concentration for detection, a simple model was developed where the detection limit is a function of the cyclopentanes plus a separable product of cyclopentane and biotin functions. This is represented by $f(x,y) = h_1(x) + h_2(x)h_3(y)$, where $f(x,y)$ is the log of the detection limit, x is the number of cyclopentanes in PNA-SP, and y is the number of biotins in PNA-RP. The form of our model is especially amenable to Principle Component Analysis (PCA)[42]. In this approach, a series of orthogonal vectors are fit to the data set with the goal of identifying a set of principle vectors that best represent the data with minimal variance. Prior to PCA, the data is mean centred with respect to cyclopentanes, $h_1(x)$. By retaining only the largest

Figure 4 | Detection limits as a function of biotin and cyclopentane. Coloured regions: three-dimensional representation of Table 1. Detection limits are represented on the z-axis as the $\log_{10}$ of the number of molecules of RNA, versus the number of cyclopentanes in PNA-SP and number of biotins in PNA-RP. The red colour indicates higher detection limits, while blue represents favourable, lower detection limits. The same data is projected as a contour map underneath. Black wireframe: the model equation derived from PCA (RMSD 0.67 relative to the experimental data) as described in Principle Component Analysis model in the Methods section.

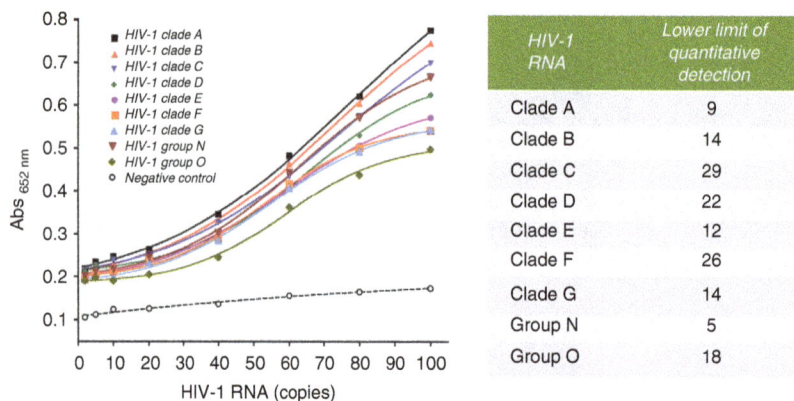

HIV-1 RNA	Lower limit of quantitative detection
Clade A	9
Clade B	14
Clade C	29
Clade D	22
Clade E	12
Clade F	26
Clade G	14
Group N	5
Group O	18

Figure 5 | Detection of seven different subtypes and two different groups of HIV-1 RNA using the PNA-SP4/PNA-RP25 detection system. The plot shows the signal response (absorbance at 652 nm versus concentration of HIV-1 RNA) obtained after 30 min. The table shows the calculated minimal distinguishable differential concentration, which represents the lowest limit of quantification. Each value in the table represents the smallest number of HIV-1 RNA molecules that could be detected based on the curve fit to the absorbance data versus RNA concentration at 90% confidence intervals. The negative control represents the signal obtained when virus-free plasma is subjected to the same NAT-PELA detection protocol.

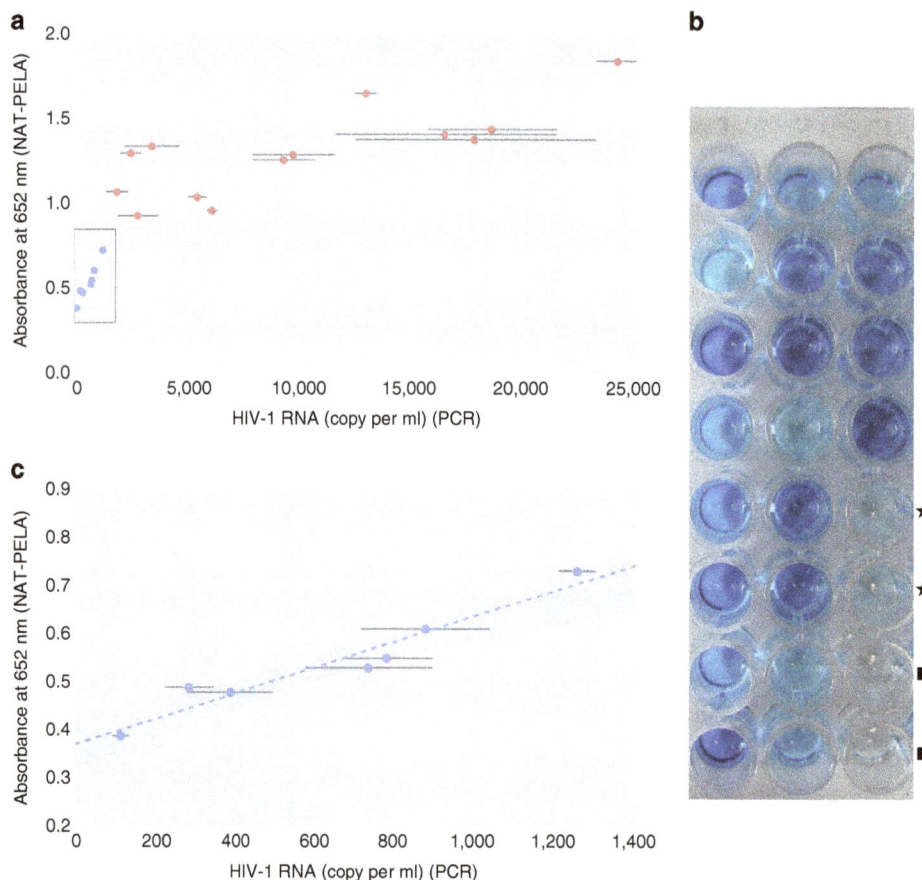

Figure 6 | HIV-1 quantification of patient samples with NAT-PELA (using PNA-SP4 and PNA-RP25) and PCR. (**a**) NAT-PELA data for all patient samples. Boxed region shows the samples with HIV viral loads <1,500 copies per ml of plasma. (**b**) Photograph of the 96-well plate, showing the variation in color intensity for HIV-1 RNA detection in patient samples by NAT-PELA. The wells designated by the star (★) are the results of using HIV-negative plasma, the wells designated by the square (■) are controls without RNA target. (**c**) Linear correlation between NAT-PELA and PCR viral load over 100 to 1,500 copies of HIV per ml of plasma. Linear fit to the data ($R^2 = 0.93$) with the shaded region representing a bootstrapped 95% confidence interval. There is a strong linear correlation of NAT-PELA signal with changes in HIV viral load (Pearson coefficient $P < 0.01$).

component, the principle component vector and its associated loading vector are $h_2(x)$ and $h_3(y)$, respectively. This model accounts for 90% of the variance of the original data (RMSD 0.67). The dark lines in Fig. 4 represent the model overlaid with the original data in colour. This decomposition was examined to gauge the contributions of cyclopentanes and biotins to the detection limit. As shown in Supplementary Fig. 5b, the monotonic increase in cyclopentane groups lowers detection (h_2).

For biotin there is a general trend that adding more biotins improves detection, but the functional form of improvement is more complex (h_3). As shown in Supplementary Fig. 5c, additional biotins can lower detection limits at specific intervals of biotins (such as those around 6 and 25 biotins) but in other ranges there is no improvement or even slight worsening of the detection limit (such as ~16 biotins).

Detection of HIV in plasma. The optimized PNAs (PNA-SP4 and PNA-RP25) were used to detect full length HIV RNA derived from samples where the virus was added into human plasma. For these sets of experiments, RNA extraction from the HIV-positive plasma samples was performed using the commercial reagent RNAzolRT following the manufacturer's procedures[43]. To determine whether the PNA probes could detect different genetic variants of HIV, seven different clades and two groups of the virus were tested under the same procedures. For all examples, the NAT-PELA system was able to quantify virus at levels (around 20 molecules of RNA) consistent with clinically useful levels (Fig. 5; Supplementary Fig. 6).

Titration curve for NAT-PELA with HIV-spiked plasma samples. To prepare NAT-PELA for detection of patient samples, a titration curve was determined using different concentrations of the virus. For these experiments, clade B HIV virus was spiked into human plasma and then subsequently diluted with additional human plasma to simulate a range of clinical concentrations for the virus. Each sample was then tested by both PCR (using COBAS Ampliprep v2.0; ref. 38) and by the NAT-PELA procedure. For the PCR tests, plasma samples were directly processed by the instrumentation. For the NAT-PELA tests, RNA extraction using RNAzolRT was performed for each plasma sample with a different HIV concentration[43]. The titration curve of NAT-PELA is shown in Supplementary Fig. 7. The quantitative values of virus in this experiment are determined by PCR. This experiment demonstrated that NAT-PELA provides clear signals over a range of HIV concentrations from 20 to 30,000 copies of HIV per ml of plasma.

Detection of HIV in patient samples by NAT-PELA and PCR. Finally, NAT-PELA was used to detect HIV in 20 plasma samples, each one from a different HIV-positive patient, representing a concentration range from 100 to 25,000 copies per ml of plasma. For each patient sample, there was sufficient volume for two PCR tests using the COBAS Ampliprep instrumentation and for one test by NAT-PELA. Processing of all samples was performed using the same protocol previously mentioned for the titration curve. For every HIV-positive patient sample, NAT-PELA showed higher absorbance signals over the control plasma samples that were HIV negative. The variation in signal intensity for NAT-PELA changes with concentration of the virus (Fig. 6), with the greatest change in signal occurring at the concentrations that are below ~1,500 copies of virus per ml of plasma (Fig. 6c, full data in Supplementary Fig. 8 and Supplementary Table 3). The NAT-PELA data for the samples with low viral loads showed a strong linear correlation with change in PCR-determined viral load (coefficient of determination $R^2 = 0.93$, Pearson product-moment correlation $P < 0.01$). At viral loads greater than 2,000 copies per ml of plasma, NAT-PELA shows very strong absorbance readings that are >1, which results in a loss of linear correlation with HIV concentration due to the high absorbance of these samples. Therefore, NAT-PELA is able to quantify HIV viral loads at 1,500 copies per ml of plasma or lower, and can be used qualitatively to signal higher viral loads.

Discussion

Antibody-based diagnostics using ELISA can provide a rapid readout of infection status as these tests can be coupled to automated engineering platforms that simplify routine testing[44]. Developing highly specific antibodies for biomarkers of diseases is crucial for any ELISA-based technology to be successful. We propose that NAT-PELA should similarly benefit NAT because automated ELISA technology could be readily adapted to use the PNA probes presented in this manuscript. In the same way that antibodies must be properly engineered for any ELISA-based assay, we find that PNAs also must be carefully designed for NAT-PELA to be successful and competitive with PCR. In the area of HIV diagnostics, the viral load of an infected individual must be routinely determined throughout therapeutic treatment to make certain that any medication maintains the virus at low levels[45]. The NAT-PELA protocol we have developed with PNA-SP4 and PNA-RP25 has sufficient sensitivity to detect and quantify HIV in patient plasma within a concentration range that would be useful for assisting with the selection of appropriate therapy. The same system may also be used to qualitatively signal a high level of virus in excess of 2,000 copies per ml of plasma. Samples that are too concentrated to be quantified could be diluted to a lower concentration that is within the range for quantification by the NAT-PELA system. While PCR-based methods to determine HIV viral loads are routine, new methods based on NAT-PELA may offer a more cost-effective alternative. In our laboratory, the reagent cost to perform one assay in a single well of a 96-well plate is about $0.67 (Supplementary Table 4). This cost includes the synthetic costs to make PNA-SP4 and PNA-RP25 as well as all additional reagents and buffers, but it does not include costs to purchase any instrumentation or any labour costs to perform the assay. The methods we present in this manuscript provide the basic concepts to design PNA probes for an automated NAT-PELA system for monitoring HIV or any other RNA-based markers of disease.

Methods

Reagents and materials. All *t*-butyloxycarbonyl (Boc)- and 9-fluorenylmethoxycarbonyl (Fmoc)-PNA monomers were purchased from PolyOrg, Inc. (Leominster, MA, USA); cyclopentane T and A PNA monomer was made following previously published procedures[37]. Acetonitrile, acetic anhydride (Ac₂O), pyridine, thioanisole, bovine serum albumin, dichloromethane (DCM), N,N-di-isopropylethylamine (DIEA), N,N-dimethylformamide (DMF), EDTA, diethyl ether (Et₂O), Kaiser test reagents, m-cresol, Na₂CO₃, N-methyl-2-pyrrolidinone (NMP), piperidine, sealing film for multi-well plates, trifluoroacetic acid (TFA), trifluoromethanesulfonic acid, polysorbate 20 (Tween 20) and single-stranded salmon sperm DNA were purchased from Sigma-Aldrich (St Louis, MO, USA). 1-Step Ultra-TMB, 96-well Nunc Immobilizer Amino plates, pHRP-SA was obtained from Thermo Scientific (Fairlawn, NJ, USA). RNA and DNA oligomers were ordered from IDT (Coralville, IA, USA). PBS was purchased from Quality Biological (Gaithersburg, MD, USA). High purity water (18 MΩ) was generated from a Millipore (Billerica, MA, USA) Milli-Q water system. Boc-8-amino-3,6-dioxaoctanoic acid (Boc-mPEG-OH), Fmoc-8-amino-3,6-dioxaoctanoic acid (Fmoc-mPEG-OH) and Boc-11-amino-3,6,9-trioxaundecanoic acid (Boc-mPEG3-OH) were purchased from Peptides International (Louisville, KY, USA). O-benzatriazole-N,N,N',N'-tetramethylammoniumhexafluorophosphate (HBTU), Fmoc-Lys(Boc)-OH, H-Lys-OH•HCl and methyl-benzhydrylamine (MBHA) resin were purchased from Advanced Chemtech (Louisville, KY, USA). H₂SO₄ was purchased from EMD Chemicals (Gibbstown, NJ, USA). RNAzolRT was obtained from Molecular Research Center, Inc. (Cincinnati, OH, USA).

Resin downloading protocol. MBHA resin was downloaded from 0.3 to 0.1 mmol g^{-1} using Boc-Lys(2-Cl-Z)-OH or Boc-G(Z)-OH. The MBHA resin (1.0 g) was swelled with DCM for 1 h in a peptide synthesis vessel. The following solutions were prepared: (a) 0.4 M Boc-Lys(2-Cl-Z)-OH in NMP, or 0.4 M Boc-G(Z)-OH in NMP, (b) 0.2 M HBTU in DMF and (c) 0.5 M DIEA in NMP. Next, 450 μl of solution a, 460 μl of solution c and 1.59 ml NMP were combined and mixed to make solution 1. Then, 550 μl of solution b were diluted with NMP (1.95 ml) to make solution 2. Solutions 1 and 2 were combined and pre-mixed for ~30 s before adding to the drained, swelled resin. The resin/coupling mixture was agitated for 1 h before draining and washing with DMF (4 ×), DCM (4 ×), 5% DIEA in DCM (1 × ~30 s) and finally DCM (4 ×). Any remaining active sites were capped using capping cocktail (1:25:25 Ac$_2$O:NMP:pyridine) for 20 min. The reaction was drained and rinsed with DMF (3 ×) and DCM (3 ×). The progress of the capping was followed by a qualitative Kaiser test. If the test was positive, the resin was resubmitted to capping. After a negative test for primary amines, the resin was washed with DCM (3 ×) and dried under vacuum for 30–60 min and then stored in a dessicator.

General method for PNA synthesis. PNAs were prepared on 5 µmol scale using either Boc- or Fmoc-solid phase peptide synthesis protocols on an Applied Biosystems 433a automated peptide synthesizer with HBTU as the amide-forming reagent. All PNA-SP were synthesized on Boc-G(Z)-OH downloaded MBHA resin, and all PNA-RP were synthesized on Boc-Lys(2-Cl-Z)-OH downloaded MBHA resin.

Cleavage and recovery of crude PNA from resin. The resin, in a peptide synthesis vessel, was first washed with TFA (2 × for 4 min). To the drained resin, cleavage cocktail (1.5 ml, 900 µl:300 µl:150 µl:150 µl TFA/trifluoromethanesulfonic acid/thioanisole/m-cresol), cooled over ice, was added and reacted for 1 h. The cleavage mixture was collected in a glass vial using N$_2$ pressure to drain the vessel. The resin was resubmitted to fresh cleavage cocktail and cleaved for 1 h, and was drained into the first cleavage fraction. The volatiles were removed by flowing dry N$_2$ over the solution to produce a yellow-brown oil. Approximately 10 ml of Et$_2$O was added to the cleavage oil to create a suspended white precipitate. The suspension was partitioned into five 2 ml microcentrifuge tubes and chilled over dry ice for 10 min. The tubes were centrifuged at 12,000 r.p.m. for 40 s to produce a white pellet. Et$_2$O was carefully decanted, leaving the white crude PNA solid.

Further washing was performed by adding ∼1.6 ml of Et$_2$O to each tube, mixing to resuspend the precipitate, then chilling on dry ice for 5 min. Following centrifugation and decanting, the washes were repeated twice without dry ice. After the final wash, the white precipitate was dried by carefully passing a stream of dry N$_2$ over the crude PNA.

Purification of crude PNA and characterization. Purification was performed on an Agilent (Santa Clara, CA) 1200 Series RP-HPLC with automatic fraction collection using ultraviolet detection at 260 nm. Waters (Milford, MA, USA) XBridge C18 (10 × 250 mm, 5 µm) columns were used in conjunction with Solvents A and B. Solvent A was 0.05% TFA in water and Solvent B consisted of 90% acetonitrile in water. PNA HPLC isolates were characterized using electrospray ionization-mass spectrometry on a Waters/Micromass LCT Premier time-of-flight mass spectrometer. The instrument was operated in W-mode at a nominal resolution of 10,000. The electrospray capillary voltage was 2 kV and the sample cone voltage was 60 V. The desolvation temperature was 275 °C and the desolvation gas was N$_2$ with a flow of 300 l h^{-1}. Accurate masses were obtained using the internal reference standard method. The sample was introduced into the mass spectrometer via the direct loop injection method. Deconvolution of multiply charged ions was performed with MaxEnt I. All PNA oligomers gave molecular ions consistent with the calculated theoretical product values (Supplementary Table 1).

Thermal melting analysis. Ultraviolet concentration was determined by adding 4 µl of RNA solution to 196 µl milli-Q water. If the signal was too intense the concentration was diluted by adding 198 µl of water to 2 µl of the original RNA solution. Water was blanked against the background at 80 °C on an Agilent 8453 UV/Vis spectrometer equipped with an Agilent 89090A Peltier temperature controller and a computer interface. Then the unknown solution was added to the quartz cell (Helma) and vigorously shaken, replaced in the spectrophotometer and the absorbance was read at 260 nm. The mixing and reading was repeated three times. Values were converted to concentration, based on average absorbance. After initial measurement by ultraviolet, the concentration was determined based on appropriate $\varepsilon_{260\,nm}$ (calculated on nearest neighbour approximation for PNA or provided by IDT or Thermo Scientific for oligonucleotides) and then used from that point forward for additional experiments. Thermal melting experiments were performed by preparing 2 µM oligonucleotide solution in 1 × PBS. Experiments traversed from 90 to 20 °C and back to 90 °C at 1 °C intervals while monitoring at 260 nm. An equilibration of 60 s at each temperature measurement step before readings was employed. Cooling and heating profiles were generated for each run with duplicates for each. The T_m (melting temperature) for duplexes was determined using the maximum first derivative of the cooling and heating curves, then taken as an average of both runs.

Buffers for NAT-PELA. Blocking buffer (BLB): 2% bovine serum albumin, 25 mM tris(hydroxymethyl) aminomethane (Tris), 150 mM NaCl, 0.05% TWEEN 20, 0.1 mM EDTA, pH 7.4; BLBs: BLB with 0.1 µg ml^{-1} single-stranded salmon sperm DNA; Capping solution (CAP): 25 mM Lys, 10 mM NaH$_2$PO$_4$, 100 mM NaCl, 0.1 mM EDTA, pH 8.0; immobilization buffer: 100 mM Na$_2$CO$_3$, pH 9.6; 1 × PBS: PBS, 9.0 g l^{-1} NaCl, 144 mg l^{-1} KH$_2$PO$_4$, 795 mg l^{-1} Na$_2$HPO$_4$, pH 7.4; 1 × PBST: 1 × PBS with 0.05% Tween 20.

Plate preparation. PNA-SP was dissolved in immobilization buffer (1.0 µM) and a volume of 100 µl was added to each well of the plate. Blank wells were left untouched throughout the process to blank against the plate. The plate was sealed with sealing tape for multi-well plates and agitated on a plate shaker (600–700 r.p.m.) for 2 h at RT. Then 50 µl of CAP was added to modified wells to give a final volume of 150 µl. This was again agitated at RT on a plate shaker for 30 min. Modified wells were then aspirated and washed four times with 300 µl of 1 × PBST and four times with 1 × PBS. A final addition of 250 µl of 1 × PBS was used to store the wells until used in experiments. When the assay was ready to be performed, the 1 × PBS was removed by aspiration and 200 µl BLBs was added to the wells. The plate was sealed and incubated with shaking for 30 min at 37 °C, the wells were immediately aspirated and ready for sample addition.

Preparation of synthetic HIV-1 gag RNA samples. All samples were prepared in glass vials. Synthetic RNA was ordered from IDT. The final volume for all samples used for RNA detection was 600 µl. Each of these samples had the PNA-RP concentration fixed at 15 nM. The specific concentration of the target RNA for detection was prepared by a serial dilution of a stock 50 nM solution of the RNA target in 1 × PBS buffer prepared from the commercial RNA. The initial (and highest concentration) RNA sample for detection was the first sample made, using the desired amount of RNA from the 50 nM stock solution and diluting with a solution of the reporter probe (PNA-RP, 15 nM) in 1 × PBS buffer. To dilute further from the initial RNA detection sample, a desired volume of RNA solution was removed and further diluted with the 15 nM reporter probe solution in 1 × PBS buffer. For example, to dilute an initial 450 pM RNA sample solution by three times, 200 µl of the initial RNA sample solution was removed and added to 400 µl of 15 nM reporter probe solution in 1 × PBS buffer to give a final RNA concentration of 150 pM. Once the final RNA concentration was obtained after dilution, the samples were heated to 95 °C in a sand bath for 5 min and then snap-cooled by placing the vials directly into an ice bath for 5 min. Then, the samples were directly used for RNA detection by removing 100 µl of each sample and individually adding the RNA solution to one well on the prepared 96-well plate (using a 100 µl pipettor). To construct a concentration-dependent response curve, 10 different RNA concentrations were prepared and each concentration was prepared in duplicate. Two wells of each plate were dedicated to the detection of one RNA concentration. After the samples were added to the wells, the plate was sealed and incubated with shaking at 37 °C for 3 h. Each well was aspirated following incubation and then washed with 300 µl of 1 × PBST four times followed by 300 µl of 1 × PBS four times.

HIV-1 RNA extraction by RNAzolRT. Analysis of patient samples was approved by the Office of Human Subjects Research at NIH (National Institutes of Health). Virus was isolated from peripheral blood mononuclear cell samples of HIV-1-positive individuals infected with nine different HIV-1 subtypes (Clades A, B, C, D, E, F, G, and Groups N and O). The culture supernatants were harvested and spiked in Base matrix. Viral loads were measured by the Abbott RealTime HIV-1 system (Abbott Molecular Inc. Des Plaines, IL, USA). One plasma sample from an HIV-1-negative individual was also included as a negative control. For extraction of HIV-1 RNA using RNAzolRT, 400 µl of plasma was mixed with 1 ml of RNAzolRT reagent followed by the addition of 200 µl RNAase free water and mixed well. Samples were incubated at RT for 15 min and centrifuged at 12,000 g, 4 °C for another 15 min. The upper layer was removed (∼1500 µl) and an equal volume of isopropanol was added and mixed well by tilting tubes. The tubes were left at RT for 15 min after which they were centrifuged at 12,000 g for 15 min, aspirated and washed with 70% ethanol twice. The pellets were dried and dissolved in 50 µl of RNAase free water.

Correlation of NAT-PELA with RT-PCR. HIV-1-positive plasma (Clade B) with viral loads of 1.5×10^9 copies per ml, stored at −80 °C, was diluted to 10,000, 5,000, 1,000, 500, 100, 50 and 10 copies per ml in HIV-1-negative plasma (base matrix). A total of 6 ml volume was prepared for each concentration, and divided into six tubes with 1 ml each. For each concentration, RNA extraction using RNAzolRT was independently performed on three 1 ml samples and the isolated RNA from each sample was subsequently used in the NAT-PELA assay. The remaining three 1 ml samples at the same concentration were subjected to the COBAS AmpliPrep/COBAS TaqMan HIV-1 Test (Roche Molecular Systems, Inc., Branchburg, NJ, USA).

HIV-1 RNA detection in patient plasma samples. Twenty plasma samples collected from 20 different HIV-positive patients were each tested by NAT-PELA and RT-PCR. Each plasma sample consisted of 3 ml total volume that was divided into three tubes with 1 ml of plasma in each tube. For each patient sample, one of the 1 ml tubes was tested using NAT-PELA (where the RNA was extracted using RNAzolRT), and the other two 1 ml tubes for each patient sample were tested using the COBAS AmpliPrep/COBAS TaqMan HIV-1 Test. To exclude the possibility of biased results due to non-specific interactions, two HIV-negative plasma samples were also included.

NAT-PELA protocol. The wells were incubated with 200 µl of BLB. The plate was sealed and shaken at 37 °C for 30 min. Wells were then aspirated, following by addition of 100 µl of 0.1 µg ml^{-1} pHRP–SA in BLB. The plate was sealed and incubated for 20 min at RT before aspirating. The wells were aspirated and followed a washing procedure that consisted of 300 µl of four times of 1 × PBST and four times of 1 × PBS.

NAT-PELA readout. Following washes, 100 µl 1-Step Ultra TMB solution was added via a multichannel pipette to facilitate nearly simultaneous initial starting points for all wells. The plate was immediately placed in a Molecular Devices (Sunnyvale, CA, USA) SpectraMax M5 multi-mode microplate reader and monitored at 652 nm over the course of 30 min at RT. The plate was then removed, followed by addition of 50 µl 2 M H$_2$SO$_4$ and mixed by hand for 10 s to quench the enzyme reaction. The plate was then returned to the plate reader and a final reading at 450 nm was performed. Plates were sealed with a film following reading and placed in the refrigerator.

Analysis of NAT-PELA results. *Analysis of absorbance (652 nm) kinetics.* The absorbance was monitored at 2 min intervals over 30 min (16 pts) at each concentration of target nucleic acid. The absorbance at 652 nm was plotted versus time to give the kinetic curve for each concentration using Prism 4.0 software (Fig. 3c), from which we can clearly see the qualitative detection limit.

Nonlinear regression analysis of quenched assay results. The protocol employing quenched TMB product that used end point absorbance values at 450 nm was used as a response value and plotted with respect to target nucleic acid concentration using Prism 4.0 software. The limits of quantitation for the HIV-1 RNA were determined using the 4-PL models with Prism 4.0 software by calculating the upper limit of the zero analyte concentration parameter error using a 90% confidence interval. These values provide a limit that is distinguishable from background noise with 96% certainty (Minimal distinguishable differential concentration). The resulting fits are graphically represented in Supplementary Fig. 3, and the quantitative detection limits are listed in Table 1.

Four-PL model equation: $y = A + ((D - A)/(1 + 10^{(\log C - x) B}))$, where A is the response at a concentration of zero (baseline); B is the slope factor; C is the inflection point (IC50); D is the response at infinite concentration; Y is the response; X is the analyte concentration.

Enzyme stability study. The pHRP–SA stability was determined by incubating the enzyme at RT for different times. Ten millilitres of 0.1 µg ml^{-1} pHRP–SA was prepared in BLB and was incubated at RT for 0, 1, 2, 4, 6, 8, 10, 15, 20, 25, 30, 35, 40, 45, 50, 55 and 60 days. Aliquots were withdrawn at each time point to test the remaining activity by SP0/ RP1/HIV-1 gag synthetic 27-nucleobase single-stranded DNA at standard conditions as described above, the target HIV-1 gag DNA concentration is 100 pM. The fresh prepared enzyme was considered to be a control and was assumed to have 100% activity.

PCA model. The data in Table 1 represent the detection limits as a function of the number of cyclopentanes in PNA-SP and biotins in PNA-RP. Attempts were made to develop a qualitative model that explains the dependence of the detection limits on these two variables. A simple linear response, where the dependence of the detection limit is fit to a simple plane, does not adequately explain the data (Supplementary Fig. 5). We next recast the data in terms of a function, f(x,y), where f is the detection limit, x is the number of cyclopentanes in PNA-SP and y is the number of biotins in PNA-RP. We hypothesize that the interaction of the two variables is largely separable, in which case the following equation: $f(x,y) = h_1(x) + h_2(x)h_3(y)$ should be true where $h_1(x) = <f(x,y)>_x$ is the signal at a fixed value of cyclopentane, averaged over all the observations of biotin. This formulation naturally leads to a decomposition with PCA. This is a classic algorithmic tool to reduce multidimensional data sets into a series of orthogonal vectors designed to maximize the explained variance in each direction. Ideally, after the analysis, the data can be reproduced with a subset of these principle component vectors. In addition, if the largest principal component is relatively dominant to the others, its components can give physical insight to the contributions of individual components (in this case the addition of cyclopentane groups and biotin groups). We find that a significant fraction of the variance in Table 1 can be explained using only the largest principal component from the PCA.

To perform the PCA, the first consideration is that the measured data points of Table 1 are not on a regular grid. To remove undue weight on closely centred observations, a linearly interpolated matrix at 500 regularly spaced intervals was constructed between the data range. We perform PCA on this matrix by first subtracting the mean for each row and taking the singular value decomposition of the remaining matrix, F, so that:

$$F = UDW^T$$

$$T = UD$$

where D is a square diagonal matrix of singular values and the columns of U and W are the left and right singular vectors. A subset of the L largest principal components can be constructed by noting $T_L = U_L D_L = FW_L$. The columns of U,

multiplied by the magnitude of the associated element of the diagonal matrix, $U_i D_{ii}$ are known as the principal components, while the columns of W are known as the loading vectors. The principal components are a linear combination of the input data while the loading vectors associate a weight to each variable in this linear combination. We find that the largest principal component can explain 90% of the variance of the data. An illustration of the signal projected using only the largest component is shown in Fig. 4. The directions of the principal components and the mean values, $h_1(x)$, $h_2(x)$ and $h_3(y)$ are shown in Supplementary Fig. 5.

It is evident that the effect of cyclopentane gives a favourable monotonic decrease in the signal detection limit. The effect of adding more biotins, while generally favourable, shows a more complicated contribution to the detection limit.

References

1. Wilson, E., Tanzosh, T. & Maldarelli, F. HIV diagnosis and testing: what every healthcare professional can do (and why they should). *Oral. Dis.* **19**, 431–439 (2012).
2. Klein, D. J., Bershteyn, A. & Eckhoff, P. A. Dropout and re-enrollment: implications for epidemiological projections of treatment programs. *AIDS* **28**(Suppl 1): S47–S59 (2014).
3. Castel, A. D. *et al.* Temporal association between expanded HIV testing and improvements in population-based HIV/AIDS clinical outcomes. *AIDS Care* **26**, 785–789 (2014).
4. Gaydos, C. & Hardick, J. Point of care diagnostics for sexually transmitted infections: perspectives and advances. *Expert Rev. Anti. Infect. Ther.* **12**, 657–672 (2014).
5. Pilcher, C. D. *et al.* Performance of rapid point-of-care and laboratory tests for acute and established HIV infection in San Francisco. *PLoS ONE* **8**, e80629 (2013).
6. Galli, R. A. *et al.* Evaluation of the accuracy and ease of use of a rapid HIV-1 Antibody Test performed by untrained operators at the point of care. *J. Clin. Virol.* **58**(Suppl 1): e65–e69 (2013).
7. Nasrullah, M. *et al.* Performance of a fourth-generation HIV screening assay and an alternative HIV diagnostic testing algorithm. *AIDS* **27**, 731–737 (2012).
8. Stekler, J. D. *et al.* Relative accuracy of serum, whole blood, and oral fluid HIV tests among Seattle men who have sex with men. *J. Clin. Virol.* **58**(Suppl 1): e119–e122 (2013).
9. Faraoni, S. *et al.* Evaluation of a rapid antigen and antibody combination test in acute HIV infection. *J. Clin. Virol.* **57**, 84–87 (2013).
10. Wang, S., Xu, F. & Demirci, U. Advances in developing HIV-1 viral load assays for resource-limited settings. *Biotechnol. Adv.* **28**, 770–781 (2010).
11. van Deursen, P. *et al.* Measuring human immunodeficiency virus type 1 RNA loads in dried blood spot specimens using NucliSENS EasyQ HIV-1 v2.0. *J. Clin. Virol.* **47**, 120–125 (2010).
12. Branson, B. M. The future of HIV testing. *J. Acquir. Immune Defic. Syndr.* **55**(Suppl 2): S102–S105 (2010).
13. Dineva, M. A., MahiLum-Tapay, L. & Lee, H. Sample preparation: a challenge in the development of point-of-care nucleic acid-based assays for resource-limited settings. *Analyst* **132**, 1193–1199 (2007).
14. HIV Diagnostics Survey. *Public Health Laboratory Issues in Brief Association of Public Health Laboratories* (Association of Public Health Laboratories, Silver Spring, 2012).
15. Alemnji, G., Nkengasong, J. N. & Parekh, B. S. HIV testing in developing countries: what is required? *Indian J. Med. Res.* **134**, 779–786 (2011).
16. Ouma, K. N. *et al.* Evaluation of quantification of HIV-1 RNA viral load in plasma and dried blood spots by use of the semiautomated cobas amplicor assay and the fully automated Cobas Ampliprep/TaqMan Assay, version 2.0, in Kisumu, Kenya. *J. Clin. Microbiol.* **51**, 1208–1218 (2013).
17. Shan, L. *et al.* A novel PCR assay for quantification of HIV-1 RNA. *J. Virol.* **87**, 6521–6525 (2013).
18. Jangam, S. R., Agarwal, A. K., Sur, K. & Kelso, D. M. A point-of-care PCR test for HIV-1 detection in resource-limited settings. *Biosens. Bioelectron.* **42**, 69–75 (2013).
19. Nicholls, P. J. & Malcolm, A. D. Nucleic acid analysis by sandwich hybridization. *J. Clin. Lab. Anal.* **3**, 122–135 (1989).
20. Zhao, X., Tapec-Dytioco, R. & Tan, W. Ultrasensitive DNA detection using highly fluorescent bioconjugated nanoparticles. *J. Am. Chem. Soc.* **125**, 11474–11475 (2003).
21. Chen, L. *et al.* DNA hybridization detection in a microfluidic channel using two fluorescently labelled nucleic acid probes. *Biosens. Bioelectron.* **23**, 1878–1882 (2008).
22. Li, S., Liu, H., Liu, L., Tian, L. & He, N. A novel automated assay with dual-color hybridization for single-nucleotide polymorphisms genotyping on gold magnetic nanoparticle array. *Anal. Biochem.* **405**, 141–143 (2010).
23. Zhang, C. Y. & Hu, J. Single quantum dot-based nanosensor for multiple DNA detection. *Anal. Chem.* **82**, 1921–1927 (2010).
24. Dequaire, M. & Heller, A. Screen printing of nucleic acid detecting carbon electrodes. *Anal. Chem.* **74**, 4370–4377 (2002).

25. Wang, J., Liu, G. & Merkoci, A. Electrochemical coding technology for simultaneous detection of multiple DNA targets. *J. Am. Chem. Soc.* **125,** 3214–3215 (2003).

26. Ferapontova, E. E. *et al.* Electrochemical DNA sandwich assay with a lipase label for attomole detection of DNA. *Chem. Commun. (Camb.)* **46,** 1836–1838 (2010).

27. Kleinbaum, D. J. & Kool, E. T. Sandwich probes: two simultaneous reactions for templated nucleic acid detection. *Chem. Commun. (Camb.)* **46,** 8154–8156 (2010).

28. Franzini, R. M. & Kool, E. T. Two successive reactions on a DNA template: a strategy for improving background fluorescence and specificity in nucleic acid detection. *Chemistry* **17,** 2168–2175 (2011).

29. Cao, Y. C., Jin, R. & Mirkin, C. A. Nanoparticles with Raman spectroscopic fingerprints for DNA and RNA detection. *Science* **297,** 1536–1540 (2002).

30. Nielsen, P. E., Egholm, M., Berg, R. H. & Buchardt, O. Sequence-selective recognition of DNA by strand displacement with a thymine-substituted polyamide. *Science* **254,** 1497–1500 (1991).

31. Egholm, M. *et al.* PNA hybridizes to complementary oligonucleotides obeying the Watson-Crick hydrogen-bonding rules. *Nature* **365,** 566–568 (1993).

32. Nielsen, P. E. & Appella, D. H. *Peptide Nucleic Acids Methods and Protocols* (Humana Press, 2014).

33. Demidov, V. V. *et al.* Stability of peptide nucleic acids in human serum and cellular extracts. *Biochem. Pharmacol.* **48,** 1310–1313 (1994).

34. Pokorski, J. K., Witschi, M. A., Purnell, B. L. & Appella, D. H. S,S)-trans-cyclopentane-constrained peptide nucleic acids. A general backbone modification that improves binding affinity and sequence specificity. *J. Am. Chem. Soc.* **126,** 15067–15073 (2004).

35. Zhang, N. & Appella, D. H. Colorimetric detection of anthrax DNA with a peptide nucleic acid sandwich-hybridization assay. *J. Am. Chem. Soc.* **129,** 8424–8425 (2007).

36. Micklitsch, C. M., Oquare, B. Y., Zhao, C. & Appella, D. H. Cyclopentane-peptide nucleic acids for qualitative, quantitative, and repetitive detection of nucleic acids. *Anal. Chem.* **85,** 251–257 (2013).

37. Englund, E. A., Zhang, N. & Appella, D. H. Cyclopentane peptide nucleic acids. *Methods Mol. Biol.* **1050,** 13–18 (2014).

38. Buraruk, T. A. & Stager, C. E. in *Modern Clinical Molecular Techniques* (eds Hu, P., Hegde, M. & Lennon, P. A.) 123–140 (Springer, 2012).

39. Wild, D. (ed.) *The Immunoassay Handbook* (Elsvier, 2005).

40. Marquette, C. A., Hezard, P., Degiuli, A. & Blum, L. J. Macro-molecular chemiluminescent complex for enhanced immuno-detection onto microtiter plate and protein biochip. *Sens. Actuator B-Chem.* **113,** 664–670 (2006).

41. Josephy, P. D., Eling, T. & Mason, R. P. The horseradish peroxidase-catalyzed oxidation of 3,5,3′,5′-tetramethylbenzidine. Free radical and charge-transfer complex intermediates. *J. Biol. Chem.* **257,** 3669–3675 (1982).

42. Jolliffe, I. T. *Principle Component Analysis*, 2nd edn (Springer, 2002).

43. Chomczynski, P., Wilfinger, W., Kennedy, A., Rymaszewski, M. & Mackey, K. RNAzol RT: a new single-step method for isolation of RNA. *Nat. Methods* an4–an5 (2010).

44. Chin, C. D., Linder, V. & Sia, S. K. Commercialization of microfluidic point-of-care diagnostic devices. *Lab Chip* **12,** 2118–2134 (2012).

45. McNairy, M. L. & El-Sadr, W. M. Antiretroviral therapy for the prevention of HIV transmission: What will it take? *Clin. Infect. Dis.* **58,** 1003–1011 (2014).

Acknowledgements

This research was supported by the Intramural Research Programme of NIDDK, NIH and the Intramural AIDS Targeted Antiviral Programme of NIH. We would like to thank Dr Frank Maldarelli for assistance in processing the patient samples.

Author contributions

C.Z. synthesized and optimized all the molecules and assays, T.H. developed the model and performed the PCA; M.K.H.G.S. and I.H. provided the different strains of HIV for testing; D.M. and T.-W.C. performed the PCR analysis, D.H.A. conceived the NAT-PELA strategy and wrote the manuscript.

Additional information

5

Concentration and chemical-state profiles at heterogeneous interfaces with sub-nm accuracy from standing-wave ambient-pressure photoemission

Slavomír Nemšák[1,2,†], Andrey Shavorskiy[3], Osman Karslioglu[3], Ioannis Zegkinoglou[3], Arunothai Rattanachata[1,2], Catherine S. Conlon[1,2], Armela Keqi[1,2], Peter K. Greene[1], Edward C. Burks[1], Farhad Salmassi[2], Eric M. Gullikson[2], See-Hun Yang[4], Kai Liu[1], Hendrik Bluhm[3] & Charles S. Fadley[1,2]

Heterogeneous processes at solid/gas, liquid/gas and solid/liquid interfaces are ubiquitous in modern devices and technologies but often difficult to study quantitatively. Full characterization requires measuring the depth profiles of chemical composition and state with enhanced sensitivity to narrow interfacial regions of a few to several nm in extent over those originating from the bulk phases on either side of the interface. We show for a model system of NaOH and CsOH in an ~ 1-nm thick hydrated layer on α-Fe_2O_3 (haematite) that combining ambient-pressure X-ray photoelectron spectroscopy and standing-wave photoemission spectroscopy provides the spatial arrangement of the bulk and interface chemical species, as well as local potential energy variations, along the direction perpendicular to the interface with sub-nm accuracy. Standing-wave ambient-pressure photoemission spectroscopy is thus a very promising technique for measuring such important interfaces, with relevance to energy research, heterogeneous catalysis, electrochemistry, and atmospheric and environmental science.

[1] Department of Physics, University of California Davis, Davis, California 95616, USA. [2] Materials Sciences Division, Lawrence Berkeley National Laboratory, Berkeley, California 94720, USA. [3] Chemical Sciences Division, Lawrence Berkeley National Laboratory, Berkeley, California 94720, USA. [4] IBM Almaden Research Center, San Jose, California 95120, USA. † Present address: Peter-Grünberg-Institut 6, Forschungszentrum Jülich GmbH - 52425 Jülich, Germany. Correspondence and requests for materials should be addressed to S.N. (email: s.nemsak@fz-juelich.de) or to H.B. (email: hbluhm@lbl.gov) or to C.S.F. (email: fadley@physics.ucdavis.edu).

Solid/gas, liquid/gas and solid/liquid interfaces play a major role in many important areas of science and technology, for example, in energy generation, electrochemistry, corrosion, environmental science and information technology, yet pose experimental challenges for the investigation of their chemical and structural properties on the molecular scale. For example, the electrical double layer has been studied for over 100 years, and yet is not fully understood[1,2]. Characterizing such interfaces requires enhancing the signal from the narrow interfacial regions over those originating from the layers on either side of them.

Ambient-pressure X-ray photoelectron spectroscopy (APXPS, APPS), in which the pressure near the sample is maintained in the multi-Torr regime, is a rapidly growing technique that permits studying surface chemical interactions in more realistic conditions compared with previous studies at or near ultrahigh vacuum[3–8]. APXPS has been applied to a wide range of interfaces, including recently solid–liquid interactions in an electrochemical cell in operando[9,10].

However, APXPS still cannot quantitatively investigate interfaces in an element- and chemical-state-specific way with sub-nanometre (nm) resolution, especially for more deeply buried interfaces, such as those in a solid–liquid system. Probing depths in APXPS are controlled by photoelectron inelastic mean free paths (IMFPs), which are, for typical kinetic energies of a few hundred to a thousand eV, a maximum of a few tens of Angstroms[11]. Varying the photon energy so as to change the IMFP is one way of providing depth resolution in APXPS[7], but the data analysis involves a convolution integral that may need additional compositional constraints to be solved uniquely[12]. Tailoring the exciting X-ray wave field into a standing wave is a method for achieving greater depth sensitivity in photoemission, and it has been applied to a number of solid/solid interfaces[13–17]. In fact this method has also been applied previously in X-ray reflectivity and emission measurements on multilayer systems and at the solid–liquid interface[2,18], but photoemission has the significant advantages of being more interface sensitive due to the shorter escape depths of electrons and of providing more direct chemical-state information from chemical shifts in core-level binding energies.

We here combine standing-wave and ambient-pressure photoemission (SWAPPS) to study the regions near a solid/liquid and a liquid/gas interface, represented by a hydrated, mixed NaOH/CsOH layer on a polycrystalline haematite surface. We show that SWAPPS can determine the chemical-state resolved depth profiles of every element in the sample with sub-nm resolution. For example, we show that the Na and Cs ion have distinctly different depth distributions. This work thus demonstrates the high future potential of SWAPPS for studying a broad range of interface phenomena.

Results

The sample. The sample configuration is shown in Fig. 1a. As discussed further in Methods, a strong standing wave (SW) was created in the sample by growing a haematite layer on a synthetic multilayer mirror for X-rays, which consists of a number of bilayers of low-Z/high-Z material (here Si/Mo), and then scanning the incidence angle over the first-order Bragg's reflection of the mirror. This scan sweeps the SW vertically through the sample over roughly one-half of its period, and generates what is termed a rocking curve (RC) of various individual core-level photoelectron peak intensities. We present below experimental results together with theoretical photoemission calculations that quantitatively incorporate the standing-wave field[19,20]. Comparison of the experimental RCs immediately provides some qualitative information as to relative depths of all

components of the interface and adjacent bulk regions, with the theoretical calculations finally providing a quantitative model of depth distributions for all of the chemical components present, with sub-nm accuracy.

Rocking curves. In Fig. 1b a typical O 1s spectrum is shown, with four distinct peaks resolved, which are due to water in the gas

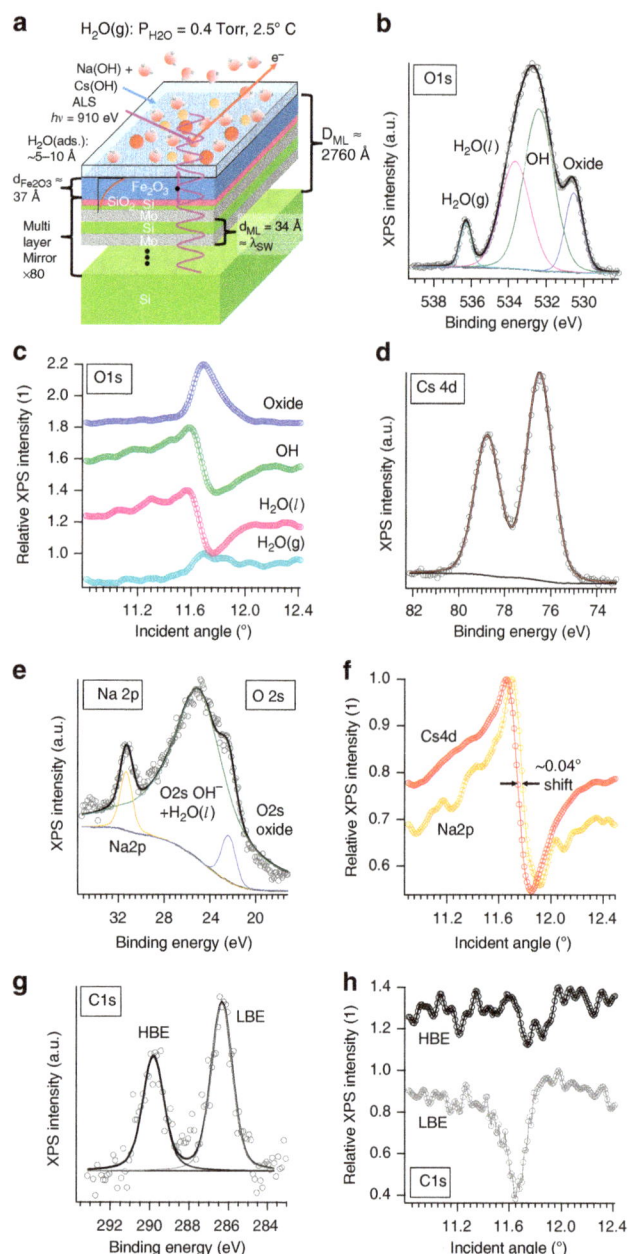

Figure 1 | The sample configuration, together with some key rocking curves for the various elements in the sample. (**a**) The sample configuration, with some relevant dimensions noted. (**b**) An O 1s spectrum, resolved into four components by peak fitting. The component labelled OH⁻ may also contain contributions from carboxyl and bicarbonate species. (**c**) The rocking curves derived from the four components of (**b**). (**d**) A Cs 4d spectrum, including peak fitting. (**e**) Analogous overlapping Na 2p and O 2s spectra, with peak fitting. (**f**) The rocking curves for Cs 4d and Na 2p derived from spectra such as those in (**d**) and (**e**). (**g**) A typical C 1s spectrum, showing the two components, one at low binding energy (LBE) and one at higher binding energy (HBE). (**h**) The rocking curves for the two C 1s components.

phase above the sample, liquid-like water adsorbed on and above the Fe_2O_3 surface, oxygen in the alkali hydroxides and possibly also in carboxyl- or carbonate species that we cannot resolve in this peak and that we designate simply as OH^- and oxide oxygen from the Fe_2O_3. The assignment of these peaks is based on previous APXPS studies[21]. Gas-phase peaks are seen in all APXPS spectra, as the X-ray beam is wide enough to pass through the gas above the sample in a region also seen by the spectrometer[6,7]. The distinction of the peaks from $H_2O(\ell)$ and OH^- is further justified by examining their different relative intensities through the rocking curve scan (results not shown here). From an analysis of experimental relative core peak intensities using the SESSA XPS simulation program[22,23] the adsorbed water layer is found to be about 5–10 Å thick, although we will later see that the analysis of the SWAPPS results provides a much more accurate value for this thickness of ~ 10 Å. This thickness is significantly larger than that of H_2O on pure metal oxide surfaces at our relative humidity of $\sim 8\%$ (refs 21,24). This is due to facilitation of water adsorption by Cs^+, Na^+ and OH^- ions and formation of their solvation shells. Figure 1c shows O 1s RC data, based on spectral deconvolution by peak fitting as shown in 1(b). All four peaks exhibit the effects of the SW scanning through the adsorbed layer, with relative magnitudes as high as 50%, including even the gas-phase water, which is modulated due to the changes in reflectivity that influence the total E-field strength above the sample. All RCs are normalized to have a maximum value of 1.0, with vertical shifts to permit overlapping them, so the fractional

changes can easily be discerned from the abscissa scale. The distinct shapes of the different O 1s RCs directly indicate that the chemical species associated with the respective photoemission peak have different average vertical positions in the sample. The O 1s(OH^-) and O 1s(H_2O) RCs do not show a significant shift, but nonetheless exhibit slightly different behaviour in the wings of the curves, with O 1s(H_2O) being more intense, suggesting not quite identical depth distributions, and we explore this further below. Figure 1d shows the Cs 4d spectrum whose peak-fit intensity was used to derive the RC in Fig. 1f. Figure 1e shows the overlapping Na 2p and O 2s valence spectra from which the Na 2p RC in Fig. 1f was derived. From Fig. 1f, we find a small but reliable shift of 0.04° between the Cs 4d and Na 2p RCs in the steeply sloping regions of the two curves near the Bragg's angle of the multilayer, as well as different shapes in the wings away from these regions, with Cs having higher intensity both below and above the Bragg's angle. Thus, the data alone suggest different depth distributions of Cs^+ and Na^+. The C 1s photoelectron spectrum contains two components (Fig. 1g), which follow distinctively different RC behaviour (Fig. 1h). The low binding energy (LBE) component, which we tentatively assign to hydrocarbon-like C, exhibits the largest fractional SW modulation, which is typically a sign of a thinner layer, with thicker layers averaging over the phase of the SW and thus reducing the intensity modulation in a RC. The modulation of the high binding energy (HBE) component, probably reflecting the presence of carboxylic and/or bicarbonate groups, shows a much smaller amplitude, and this in turn suggests a much broader depth profile for this species that thus averages over the phase of the SW. The presence of carbon in different oxidation states in the adsorbed layer on our sample under *in situ* measurement conditions at this high a relative humidity is due to both the sample preparation protocol as well as the inherent background contamination in experiments at elevated pressures, and has been observed in a number of experiments under similar conditions[21].

Depth sensitivity estimate. To provide some first-order semi-quantitative insight into the depth resolution of these RC measurements, Fig. 2 shows model calculations of the shapes of rocking curves for emission from a water 'delta layer' of 1 Å thickness that was successively moved through a 10 Å water layer, calculated at the average O 1s kinetic energy, but which could also represent Cs^+, Na^+ or other species, as a function of distance Δz above the Fe_2O_3 surface. Figure 2a illustrates the model geometry on top of the nominal structure of the mirror. These calculations have been carried out using a versatile X-ray optics program described elsewhere that has been used in a number of previous SW photoemission studies of multilayer solid systems[20]. The results of these calculations in Fig. 2b indicate that a 0.04° shift at the maximum slope region can be associated with a 4 Å difference in position, but more so that the behaviour in the wings of the rocking curve is also sensitive to the vertical position, showing even larger changes in angular width on the low-angle side of the rocking curve. Thus, we can further suggest that Cs^+ is on average a few Å farther from the Fe_2O_3 surface, and by the absence of subsidiary Kiessig interference fringes that are somewhat more clearly visible for Na^+, that its distribution is probably broader. A more detailed analysis incorporating some assumed distributions in depth for both Cs^+ and Na^+ is necessary to be fully quantitative. This quantitative analysis of our data is described in the following.

Quantitative analysis. The optimization of the structure was done by first carrying out calculations of the Fe 3p (oxide), O 1s(oxide), summed [O 1s(OH^-) + O 1s($H_2O(l)$)] RCs, and the

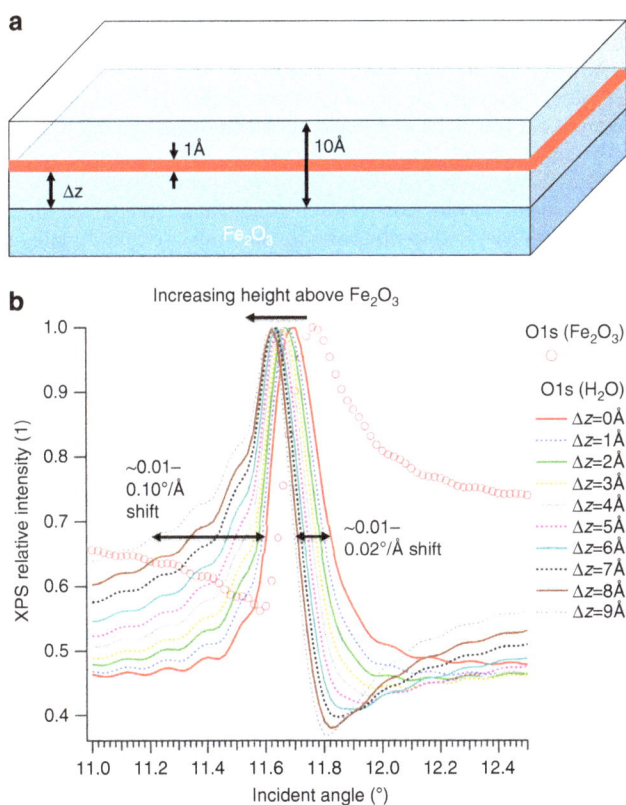

Figure 2 | A set of model calculations for assessing the sensitivity of SWAPPS rocking curves to depth. (a) The model sample geometry used for scanning a water 'delta layer' of 1 Å thickness through a water film of 10 Å thickness on top of Fe_2O_3 to assess the change in rocking curves. (b) A summary of calculated rocking curves for O1s from the delta layer, with steps in height Δz of 1 Å. A reference rocking curve for O 1s emission from Fe_2O_3 is also shown. The typical changes seen to the right and to the left of the curves are also indicated.

C 1s LBE RC, and treating as fitting parameters the vertical positions and thicknesses of the Fe_2O_3 layer, the liquid-like $H_2O + OH^-$ layer and the layer containing LBE C contaminants, together with the thicknesses of the mutual interdiffusion between these three layers, which were assumed to be linear in form. Self-consistency requires that Fe 3p(oxide) and O 1s(oxide) be associated with the same depth distribution for Fe_2O_3, thus increasing the final accuracy of the structure determination for these layers. At each step, the goodness-of-fit of theoretical predictions generated by the code was judged by a χ^2 sum, divided by the number of data points. A modified fast-converging Newton method with successive over-relaxation was used to find optimum parameters[25]. Figure 3a,b,f and g compares experiment and theory from this step, and Fig. 4a presents the resulting depth distribution. In the following step, the same technique was used to fit the Cs 4d and Na 2p RCs, and derive their depth distributions. Self-consistency again could be used in requiring that Na 2s (not shown) and Na 2p be associated with the same distribution for Na^+. The results from this second step are further shown in Fig. 3c–e, and summarized separately in Fig. 4b, c. As a final step, the HBE C rocking curve was fit, with results shown in Fig. 3h, and its much broader distribution given in Fig. 4d, which summarizes all of the depth distributions in the form of curves. Excellent agreement is found between experiment and X-ray optical theory for all of the core-level rocking curves, and our estimated accuracy in deriving these structural parameters is ± 1-2 Å.

Figure 4 further permits concluding that the Fe_2O_3 surface has an effective roughness and/or interface interdiffusion distance of about 6 Å. In fact, atomic force microscopy measurements on the Fe_2O_3 surface after measurement and after cleaning with distilled water showed a root mean squared roughness very close to this value. Figure 4 also shows that the Na^+ ions are located at an average distance of 5.5 Å above the mean of the Fe_2O_3 surface and extend over about 11 Å in depth, while the Cs^+ ions are located at a larger average distance of 9.5 Å above the mean Fe_2O_3 surface

and extend over about 12 Å in depth. Cs^+ is on average 4 Å further from the surface than Na^+, in qualitative agreement with our discussion of the model rocking curve simulations presented in Fig. 2b. The LBE C-containing contaminants are localized within a thin layer of about 5 Å thickness at the liquid/gas interface and thus appear to be hydrophobic, while the HBE C-containing contaminants, possibly carboxyl or bicarbonate species, are spread over the whole-depth range of the 'wet' layer without a significant concentration gradient. Through separate fitting of the O 1s OH^- and O 1s H_2O RCs (not shown here), the OH- ions and the water molecules are found to be virtually identical in depth distribution, with OH^- being deeper in its distribution by ~ 1 Å, although this is at the limit of our resolution and would require data of higher statistical accuracy to resolve more quantitatively. At this point, we thus conclude that OH^- and H_2O have very nearly the same depth distribution.

Quantitative concentrations. In the usual sense of XPS quantitative analysis, we can also make use of the relative core peak intensities to estimate the true concentrations of several important species in cm^{-3}. This was done by comparing the relative intensities of all of the peaks shown in Fig. 3a–e, as measured off the Bragg reflection to yield a uniform X-ray flux throughout the sample, with calculated relative intensities from the SESSA XPS simulation program[22,23]. The relevant concentrations in the layer structure of Fig. 4 were varied in the calculations to give the best agreement between experiment and theory. The Fe 3p and O 1s(oxide) intensities provide internal references for the concentration of the Fe and O atoms in Fe_2O_3 (3.95×10^{22} cm^{-3} for Fe and 5.93×10^{22} cm^{-3} for O). The IMFPs were assumed to be 18 Å for Fe 3p and 10 Å for O 1s in Fe_2O_3, and to be 37 Å for the complex overlayers, which are predominantly OH^- (including as before what might be carboxylic or bicarbonate species) + H_2O, with these IMFP numbers derived from a combination of the Tanuma-Powell-Penn formula[11] and theoretical and experimental results for H_2O[26]. The final results of these calculations yield concentrations in the 'bulk' central regions of each layer of Cs^+ and Na^+ are 1.36×10^{22} cm^{-3} and 0.64×10^{22} cm^{-3}, respectively. As a final result, the effective concentration of the 'solution' in this thin aqueous layer is thus very high at $\sim [Cs^+ + Na^+] \approx 22\,M + 10\,M = 32\,M$. Carrying this calculation one step further, we estimate about two monolayers total of Cs and one monolayer total of Na in the adsorbed layer on Fe_2O_3. The fact that the Cs^+ and Na^+ concentrations are not equal is easily explained by differing degrees of hydration of the solid CsOH and NaOH that were weighed in making up the initial solutions. The concentrations of other species based on O 1s and C 1s spectra could not be determined accurately due to a lack of knowledge of the spectrometer transmission as a function of kinetic energy, but could be in future studies using the SWAPPS method with proper calibration.

Depth-dependent binding energies. As a final aspect of our data, we show in Fig. 5 binding energy changes with depth that can be observed as the SW is scanned through the sample. In Fig. 5a is plotted the variation of the electric field intensity with depth and incident angle, with the profiles of the different chemical species indicated on top of it. The positions of the SW maxima are obvious from the bright roughly elliptical spots, and the scanning in vertical position is obvious from the slopes of the main axis of these spots. The blue arrows indicate extrema in the SW depth from nearer to the Fe_2O_3 surface to nearer to the top surface of the sample. Figure 5b now shows the binding energy of several core levels as an RC is scanned, and it is clear that, as the SW

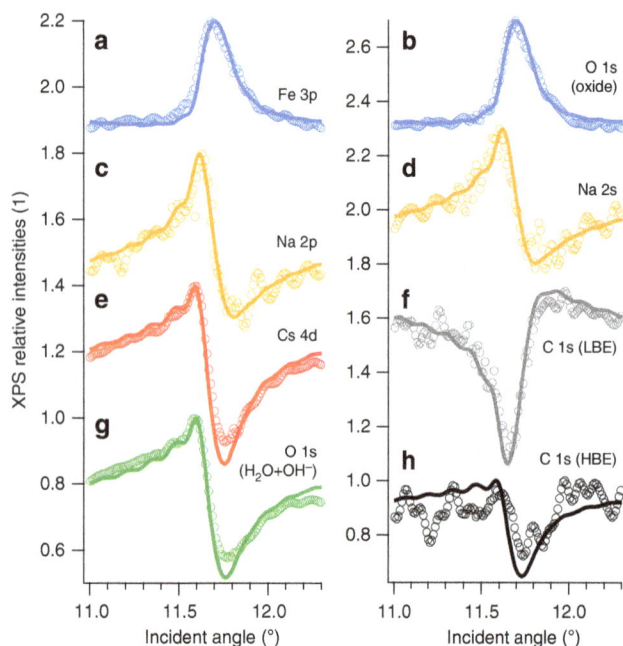

Figure 3 | A comparison of the experimental rocking curves for various core-level intensities with theoretical calculations based on the optimized sample configuration defined in Fig. 4. (**a**) Fe 3p, (**b**) O 1s (oxide), (**c**) Na 2p, (**d**) Na 2s, (**e**) Cs 4d, (**f**) low binding energy C 1s, (**g**) O 1s ($OH^- + H_2O$) (**h**) high binding energy C 1s.

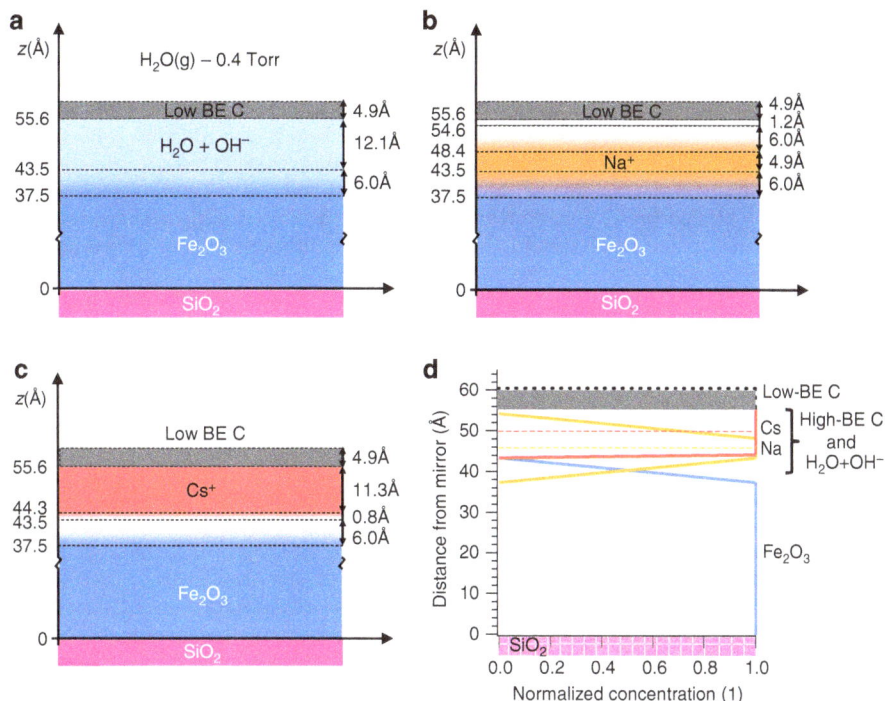

Figure 4 | The final depth distributions in the sample as derived by fitting experimental rocking curves to theory. Depth distributions for (**a**) Fe_2O_3, the $OH^- + H_2O$ electrolyte layer treated as a unit, and the low binding energy C, (**b**) the Na^+ layer and (**c**) the Cs^+ layer. In (**d**), the concentration distributions of all species are shown overplotted as curves, with the horizontal scale in normalized atomic concentrations that are unity at the maximum for each species.

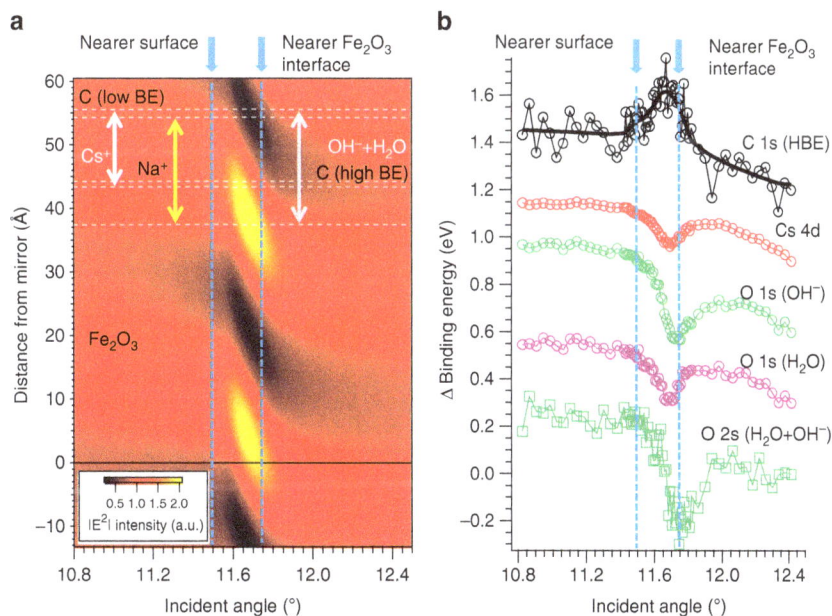

Figure 5 | Comparison of calculated standing-wave electric field depth profiles as a function of X-ray incidence angle to core-level binding energies also as a function of incidence angle. (**a**) The calculated standing-wave profile of electric field intensity ($|E^2|$) as a function of depth and incident angle. The two arrows and dashed lines indicate locations at which the field will focus more on the Fe_2O_3 interface or more towards to the top surface of the sample. (**b**) The variation in binding energy for some selected core-level peaks as incidence angle is scanned through a rocking curve. The curves are all on the same relative scale, but have been offset for visibility. These changes are different in position and sign, thus indicating that they cannot be due to sample charging or purely instrumental effects.

moves through the sample, there are significant shifts of 0.2–0.4 eV. The general downward slope in these curves is a purely instrumental effect that is found in all binding energies. The fact that all of these binding energy curves show different positions and shapes, although with OH^- and H_2O being very

similar to one another, and that they are not all of the same sign, argues strongly that these are due to bonding or local potential changes with depth in the sample. A detailed analysis of these shifts would require theoretical calculations beyond the scope of this paper, but these results thus point out an additional strength of the SWAPPS method, being able to study chemical states or electrostatic potentials in a depth-resolved way near an interface.

Discussion

We now comment on the relationship of our results to what is known about the adsorption of alkali ions on haematite and other oxides. Previous studies have demonstrated preferential adsorption of Na^+ over Cs^+ ions on haematite and other metal oxide surfaces[27–31]. The ionic adsorption sequence: $Li^+ > Na^+ > K^+ > Cs^+$, known as the inverse lyotropic adsorption sequence, was first observed for rutile TiO_2 by Bérubé et al. based on indirect capacity measurements[27] and later confirmed for various other metal oxides, including Al_2O_3 (ref. 29) and Fe_2O_3 (refs 30,31). The sequence has been explained on the basis of the ionic double-layer model, often referred to as the 'structure-making–structure-breaking' model. According to this picture, the adsorption of inorganic ions on the surface of a metal oxide in contact with an aqueous liquid phase is governed by the extent to which the surface and the ions disrupt or promote the structural order of water at the surface. Ions that cause a similar effect on the water structure as the surface tend to be easily adsorbed, while ions that affect the water structure differently from the surface tend to be excluded. Thus, when the surface is characterized by a strong affinity for water, then small, heavily hydrated ions, such as Li^+ and Na^+, which do not significantly affect the water structure in the surface region, are preferentially adsorbed over larger, less hydrated ions, such as K^+ and $Cs^{27,31}$. Haematite, along with Al_2O_3, TiO_2 and ZnO, belongs to this class of 'structure promoting' materials, exhibiting preferential adsorption of Na^+ over Cs^+. The opposite is the case for 'structure breaking' materials, such as SiO_2, V_2O_5 and WO_3, which exhibit preferential adsorption for larger ions such as Cs^+ rather than small ones such as Li^+ and Na^+ (ref. 28). Dumont et al.[28] have in fact shown that the 'structure promoting' or 'structure breaking' properties of a surface can be predicted based on the heat immersion value, or equivalently the point of zero charge (pzc) of the material, with pzc = 4 roughly being the value separating the two classes of compounds. Our study thus confirms the preferential adsorption of Na^+ over Cs^+ on Fe_2O_3, providing for the first time a direct probe of the distribution of different chemically distinct species through a solid/liquid interface, including the detection of their chemical states and potential evolution near the interface. We conclude that the Cs^+ ions are practically excluded from the Fe_2O_3 surface, populating predominantly the outer atomic layers closer to the liquid/gas interface. The intrinsic roughness of the Fe_2O_3 surface affects mostly the Na^+-containing layer and less so the Cs^+ and LBE C-containing layers, which have sharper interfaces. This might indicate that the outer layers are predominantly influenced by the properties of the liquid/gas interface, with the roughness of the solid surface partly 'screened' by the inner Na^+ layers. SWAPPS thus provides quantitative information on the thickness of the various layers and the roughness of each interface.

We have thus demonstrated that combining ambient-pressure photoemission and standing-wave photoemission can yield depth profiles of chemical species and chemical states in a liquid-like environment above a solid surface with unprecedented detail. For the specific study case of Fe_2O_3 haematite in contact with an aqueous NaOH/CsOH electrolyte, the spatial distributions of the electrolyte ions and the carbon contaminants across the solid/liquid and liquid/gas interfaces were directly probed. Absolute concentrations of species were determined. Binding energy shifts with depth were observed that provide additional information on the detailed bonding and/or depth-dependent potentials in the

system. We have thus obtained quantitative insight into the roughness and microscopic nature of the interfaces and the atomic layers forming between them, demonstrating significant spatial separation of the Na^+ and Cs^+ ions, as well as of the different carbon species. The results for Na^+ and Cs^+ are furthermore consistent with previous studies of the relative adsorption strengths of these ions on metal oxide surfaces.

Although the technique requires growing the sample on some kind of multilayer mirror, there are many possible material choices for this[32], and the sample layers involved are often polycrystalline or amorphous and so often simple to grow on such mirrors, so we do not view sample synthesis as a major limitation for future applications. A very well-defined beam of at most ~ 0.01 degrees in angular divergence is also required, which probably requires using synchrotron radiation or free-electron lasers sources, but many APXPS facilities at such sources are available by now. Depending on the photon energy, interfaces below covering layers of a few nm (soft X-rays up to $\sim 1\,keV$) or of a few 10 s of nm (~ 5–$10\,keV$) should be capable of study.

Carrying out such SWAPPS measurements with harder X-ray excitation in the multi-keV regime should also permit looking through relatively thick liquid electrolyte films with in operando condition[9,10], thus providing a truly quantitative picture of the solid/liquid interface and the electrochemical double layer for the first time. Ambient pressures that may go up to 1 atm should also be possible, in particular with hard X-ray excitation. Future time-resolved SWAPPS studies using free-electron laser[33] or high-harmonic generation light sources[34] would also permit, via pump-probe methods, looking at the timescales of processes at interfaces on the femtosecond timescale. Finally, doing photoelectron microscopy (PEEM) has already been demonstrated in APXPS[35] and separately with SW excitation for enhanced depth resolution[36], so future SWAPPS-PEEM studies should provide a true three-dimensional picture of complex interfaces. The range of future applications and measurement scenarios for SWAPPS is thus enormous.

Methods

Sample preparation. We have obtained our SWAPPS data for the different layers and interfaces of hydrated NaOH and CsOH on the surface of a thin film of α-Fe_2O_3, which was grown on a Si/Mo multilayer mirror. Figure 1a shows the sample configuration. The Si/Mo mirror was grown at the Lawrence Berkeley National Laboratory (LBNL) Center for X-Ray Optics. The multilayer mirror consisted of 80 repeats of a Si-15 Å/Mo-19 Å bilayer, yielding a repeat distance of $d_{ML} = 34$ Å that can also be shown to be very close to the SW period λ_{SW} at the Bragg reflection angle[13]. A layer of Fe_2O_3 of nominal 37 Å thickness was grown on top of this mirror in the Liu Laboratory at UC Davis by reactive magnetron sputtering in an ultrahigh vacuum chamber with a base pressure of 9×10^{-9} torr. X-ray diffraction showed the layer to be either amorphous or nanocrystalline, and Fe 2p core X-ray photoemission, Fe L-edge X-ray absorption and magnetometry confirmed it to be α-Fe_2O_3 haematite.

CsOH and NaOH were chosen as the electrolytes for two reasons: they are much different in size, and so might be expected to show different depth distributions. OH^- anions are also expected to be less prone to beam damage than, for example, halides[37], due to the presence of water vapour. The CsOH/NaOH layer was prepared by depositing a droplet of 0.01 M CsOH/0.01 M NaOH solution onto the sample. The bulk of the solution was then blown away by dry N_2, leaving a few monolayers of CsOH + NaOH on the Fe_2O_3 surface. The surface was then rehydrated inside the measurement chamber by exposing it to 0.4 Torr water vapour at a temperature of 2.5 °C, resulting in a relative humidity of $\sim 8\%$ with respect to the sample. Na^+ and Cs^+ were chosen as cations for their identical charge and different ionic radii. The influence of beam damage was checked in our experiments and found to be negligible.

Photoemission experiments. The photoemission measurements were carried out at the LBNL Advanced Light Source, using the ambient-pressure photoemission system at the Molecular Environmental Science beamline 11.0.2 (ref. 38). The SWAPPS measurements were carried out using a photon energy of 910 eV, corresponding to an X-ray wavelength of $\lambda_x = 13.6$ Å. With d_{ML} equal to the mirror bilayer spacing, this yields a first-order Bragg angle of $arcsin[\lambda_x/2d_{ML}] = 11.6°$, about which the incidence angle was scanned to yield RCs for core levels of every element in the sample. For reference, the IMFP of Fe 3p electrons emitted from Fe_2O_3 at this photon energy is estimated to be 18 Å; thus, we see through the Fe_2O_3 layer to some degree and signal is also detected from the Si and Mo on which the sample was grown. We also expect to see weaker Kiessig fringe features due to reflection from the top and bottom of the sample adjacent to the Bragg effects. These have oscillations given by $arcsin[m\lambda_x/2D_{ML}]$, where m is the order and D_{ML} is the total thickness of the sample, in this case $\sim 80 \times 34 + 37 \approx 2,760$ Å. A simple calculation[13] indicates these fringes are separated by about 0.14°, with these fringes in fact being observed in theoretical calculations, and to some degree also in experiment.

References

1. Ohno, H. *Electrochemical Aspects of Ionic Liquids* (John Wiley & Sons, 2011).
2. Brown, G. E. & Calas, G. Mineral-aqueous solution interfaces and their impact on the environment. *Geochem. Perspect.* **1**, 483–742 (2012) with special discussion of the mineral-electrolyte double layer and prior standing-wave X-ray reflectivity and emission studies over, 552–557.
3. Starr, D. E., Liu, Z., Hävecker, M., Knop-Gericke, A. & Bluhm, H. Investigation of solid/vapor interfaces using ambient pressure X-ray photoelectron spectroscopy. *Chem. Soc. Rev.* **42**, 5833–5857 (2013).
4. Salmeron, M. & Schlögl, R. Ambient Pressure Photoelectron Spectroscopy: A new tool for surface science and nanotechnology. *Surf. Sci. Rep.* **63**, 169–199 (2008).
5. Siegbahn, H. Electron spectroscopy for chemical analysis of liquids and solutions. *J. Phys. Chem.* **89**, 897–909 (1985).
6. Knop-Gericke, A. *et al.* High-pressure X-ray photoelectron spectroscopy: a tool to investigate heterogeneous catalytic processes. *Adv. Catal.* **52**, 213–272 (2009).
7. Ghosal, S. *et al.* Direct Observation of Halide Enhancement at the Vapor/Alkali Halide Solution Interface. *Science* **307**, 563–566 (2005).
8. Zhang, C. *et al.* Measuring fundamental properties in operating solid oxide electrochemical cells by using in situ X-ray photoelectron spectroscopy. *Nat. Mater.* **9**, 944–949 (2010).
9. Masuda, T. *et al.* In situ X-ray photoelectron spectroscopy for electrochemical reactions in ordinary solvents. *Appl. Phys. Lett.* **103**, 111605 (2013).
10. Axnanda, S. *et al. Probing the solid-liquid interface with ambient pressure X-ray photoelectron spectroscopy using hard X-rays, private communication* (2014).
11. Tanuma, S., Powell, C. J. & Penn, D. R. Calculations of electron inelastic mean free paths. IX. Data for 41 elemental solids over the 50eV to 30keV range. *Surf. Interface Anal.* **43**, 689–713 (2011).
12. Macak, K. Encoding of stoichiometric constraints in the composition depth profile reconstruction from angle resolved X-ray photoelectron spectroscopy data. *Surf. Interface Anal.* **43**, 1581–1604 (2011).
13. Gray, A. X. *et al.* Interface properties of magnetic tunnel junction La0.7Sr0.3MnO3/SrTiO3 superlattices studied by standing-wave excited photoemission spectroscopy. *Phys. Rev. B* **82**, 205116 (2010).
14. Greer, A. A. *et al.* Observation of boron diffusion in an annealed Ta/CoFeB/ MgO magnetic tunnel junction with standing-wave hard X-ray photoemission. *Appl. Phys. Lett.* **101**, 202402 (2012).
15. Papp, C. *et al.* Nondestructive characterization of a TiN metal gate: chemical and structural properties by means of standing-wave hard X-ray photoemission spectroscopy. *J. Appl. Phys.* **112**, 114501 (2012).
16. Fadley, C. S. Hard X-ray photoemission with angular resolution and standing-wave excitation. *J. Electron Spectrosc. Relat. Phenom.* **190**, 165–179 (2013).
17. Gray, A. X. Future directions in standing-wave photoemission. *J. Electron Spectrosc. Relat. Phenom.* **195**, 399–408 (2014).
18. Bedzyk, M. J. X-ray standing wave studies of the liquid/solid interface and ultrathin organicfilms. In: *Surface X-ray and Neutron Scattering.* (Vol. 61 of Springer Proceedings in Physics). (eds Zabel H. and Robinson I.K.) Vol. 51, 113–117 (Springer-Verlag, 1992) and references therein.
19. Yang, S.-H. *et al.* Making use of X-ray optical effects in photoelectron-, Auger electron-, and X-ray emission spectroscopies: total reflection, standing-waves, and resonant excitation. *J. Appl. Phys.* **113**, 073513 (2013).
20. Yang, S.-H. X-Ray Optics (YXRO) program for simulating standing-wave photoemission above multilayers, https://sites.google.com/a/lbl.gov/yxro/home (2011).
21. Yamamoto, S. *et al.* Water adsorption on α-Fe2O3(0001) at near ambient conditions. *J. Phys. Chem. C* **114**, 2256–2266 (2010).
22. Smekal, W., Werner, W. S. M. & Powell, C. J. Simulation of electron spectra for surface analysis (SESSA): a novel software tool for quantitative Auger-electron spectroscopy and X-ray photoelectron Spectroscopy. *Surf. Interface Anal.* **3**, 1059–1067 (2005).
23. Werner, W. S.M., Smekal, W., Hisch, T., Himmelsbach, J. & Powell, C. J. Simulation of Electron Spectra for Surface Analysis (SESSA) for quantitative interpretation of (hard) X-ray photoelectron spectra (HAXPES). *J. Electron Spectrosc.* **190**, 137–143 (2013).
24. Newberg, J. T. *et al.* "Formation of hydroxyl and water layers on MgO films studied with ambient pressure XPS,". *Surf. Sci.* **605**, 89–94 (2011).
25. Press, W. H., Teukolsky, S. A., Vetterling, W. T. & Flannery, B. P. *Numerical Recipes in C-The Art of Scientific Computing* (Cambridge University Press, 1992).
26. Emfietzoglou, D. & Nikjoo, H. Accurate electron inelastic cross sections and stopping powers for liquid water over the 0.1-10keV range based on an improved dielectric description of the bethe surface. *Radiat. Res.* **167**, 110–120 (2007).
27. Bérubé, Y. G. & de Bruyn, P. L. Adsorption at the rutile-solution interface: II. model of the electrochemical double layer. *J. Colloid Interface Sci.* **28**, 92–105 (1968).
28. Dumont, F., Van Tan, D. & Watillon, A. Study of ferric oxide hydrosols from electrophoresis, coagulation, and peptization measurements. *J. Colloid Interface Sci.* **55**, 678–687 (1976).
29. Johnson, S., Scales, P. & Healy, T. The binding of monovalent electrolyte ions on α-Alumina. I. Electroacoustic studies at high electrolyte concentrations. *Langmuir.* **15**, 2836–2843 (1999).
30. Amhamdi, H., Dumont, F. & Buess-Herman, C. Effect of urea on the stability of ferric oxide hydrosols. *Colloids Surfaces A: Physicochem. Eng. Asp.* **125**, 1–3 (1997).
31. Kirwan, L. J., Fawell, P. D. & van Bronswijk, W. An in situ FTIR-ATR study of polyacrylate adsorbed onto hematite at high ph and high ionic strength. *Langmuir* **20**, 4093–4100 (2004).
32. Center for X-Ray Optics, Lawrence Berkeley National Laboratory, X-Ray Interactions With Matter website, table of multilayer mirror materials and reflectivities, http://henke.lbl.gov/cgi-bin/mldata.pl.
33. Hellmann, S. *et al.* Ultrafast melting of a charge-density wave in the Mott insulator 1T-TaS2. *Phys. Rev. Lett.* **105**, 187401 (2010).
34. Eich, S. *et al.* Time- and angle-resolved photoemission spectroscopy with optimized high-harmonic pulses using frequency-doubled Ti:Sapphire lasers. *J. Electron Spectrosc. Relat. Phenom.* **195**, 231–236 (2014).
35. Zhang, C. *et al.* Multi-element activity mapping and potential mapping in solid oxide electrochemical cells through the use of operando XPS. *ACS Catalysis* **2**, 2297–2304 (2012).
36. Gray, A. X. *et al.* Standing-wave excited soft X-ray photoemission microscopy: Application to Co microdot magnetic arrays. *Appl. Phys. Lett.* **97**, 062503 (2010).
37. Arima, K., Jiang, P., Deng, X., Bluhm, H. & Salmeron, M. *Water adsorption, solvation and deliquescence of alkali halide thin films on SiO2 studied using ambient pressure photoelectron spectroscopy. J. Phys. Chem. C* **114**, 14900–14906 (2010).
38. Ogletree, D. F., Bluhm, H., Hebenstreit, E. L.D. & Salmeron, M. Photoelectron spectroscopy under ambient pressure and temperature conditions. *Nucl. Instrum. Methods A* **601**, 151–160 (2009).

Acknowledgements

The Advanced Light Source and the Molecular Environmental Science beamline 11.0.2 are supported by the Director, Office of Science, Office of Basic Energy Sciences, Division of Chemical Sciences, Geosciences, and Biosciences and Materials Sciences Division of the US Department of Energy at the Lawrence Berkeley National Laboratory under Contract No. DE-AC02-05CH11231. C.S.F. and S.N. also acknowledge support from the Jülich Research Center, Peter Grünberg Institute, PGI-6; and from the APTCOM Project of a grant from the 'Laboratoire d'Excellence Physics Atom Light Matter' (LabEx PALM) overseen by the French National Research Agency (ANR) as part of the 'Investissements d'Avenir' program (reference: ANR-10-LABX-0039). P.K.G., E.C.B. and K.L. acknowledges support from the National Science Foundation (DMR-1008791).

Author contributions

S.N., A.S., O.K., I.Z., A.R., C.S.C. and A.K. carried out the experiments under the supervision of H.B. and C.S.F. Data analysis and X-ray optical simulations were performed by S.N., using a software package written by S.H.Y. The multilayer mirrors were grown by F.S. and E.M.G., and the haematite layers by P.K.G., E.C.B. and K.L., including atomic force microscopy characterization of the surface profile.

Additional information

6

Nanoporous gold supported cobalt oxide microelectrodes as high-performance electrochemical biosensors

Xing-You Lang[1,*], Hong-Ying Fu[1,*], Chao Hou[1], Gao-Feng Han[1], Ping Yang[2], Yong-Bing Liu[1] & Qing Jiang[1]

Tremendous demands for electrochemical biosensors with high sensitivity and reliability, fast response and excellent selectivity have stimulated intensive research on developing versatile materials with ultrahigh electrocatalytic activity. Here we report flexible and self-supported microelectrodes with a seamless solid/nanoporous gold/cobalt oxide hybrid structure for electrochemical nonenzymatic glucose biosensors. As a result of synergistic electrocatalytic activity of the gold skeleton and cobalt oxide nanoparticles towards glucose oxidation, amperometric glucose biosensors based on the hybrid microelectrodes exhibit multi-linear detection ranges with ultrahigh sensitivities at a low potential of 0.26 V (versus Ag/AgCl). The sensitivity up to 12.5 mA mM^{-1} cm^{-2} with a short response time of less than 1 s gives rise to ultralow detection limit of 5 nM. The outstanding performance originates from a novel nanoarchitecture in which the cobalt oxide nanoparticles are incorporated into pore channels of the seamless solid/nanoporous Au microwires, providing excellent electronic/ionic conductivity and mass transport for the enhanced electrocatalysis.

[1] Key Laboratory of Automobile Materials, Ministry of Education, and School of Materials Science and Engineering, Jilin University, Changchun 130022, China. [2] Cardiovascular medicine, Sino-Japan Friendship Hospital, Jilin University, Changchun 130033, China. * These authors contributed equally to this work. Correspondence and requests for materials should be addressed to Q.J. (email: jiangq@jlu.edu.cn).

Electrochemical biosensors, which have been widely employed in clinical, environmental, industrial and agricultural applications, recognize biological analytes through a catalytic or binding event occurring at the interface of electrodes[1-5]. Intimate correlation of biosensing performance and the electrocatalytic and structural properties of electrodes has stimulated considerable efforts devoted to innovational materials to coordinate mass- and charge-transport and electron-transfer kinetics for realizing simultaneous minimization of primary resistances in biosensing: electrochemical reaction occurring at electrolyte/electrode interface, mass transport of analyte in electrolyte and electrode, and the electron conduction in electrode and current collector[6-9]. Nanostructured metal oxides have promising applications as nonenzymatic catalysts in new generation of miniaturized glucose biosensors owing to their low cost, high biocompatibility and electrocatalytic activity, as well as enhanced electron-transfer and adsorption capacities[10,11]. Judicious use of nanostructured metal oxides is expected to circumvent the key limitations of expensive enzymes, typically glucose oxidase (GOx), with compromise of the sensitivity, reproducibility and stability of glucose detection as the catalytic activity of GOx is intrinsically susceptible to environmental conditions such as temperature, humidity, pH and toxic chemicals[2,11-13]. However, the poor electronic conductivity of metal oxides (for example, Co_3O_4 in $\sim 10^{-5}\,S\,m^{-1}$ at room temperature) significantly impedes them from wide use in electrochemical biosensing devices with high sensitivity and reliability, fast response and excellent selectivity[10,14,15]. One of major strategies to enhance charge transport in electrochemical biosensors is to design composite materials by combining highly electrocatalytic materials with a conductive substance[16-20], whereas single- and multi-walled carbon nanotubes (CNTs)[21-25], graphene[26,27] and noble metal nanoparticles[28] have been explored to serve as conductive pathways of metal oxides (TiO_2, MnO_2, RuO_2, Co_3O_4 and NiO). Although these low-dimensional composite nanostructures could provide extremely large specific surface area of the electrode/electrolyte interface[16-28], the assembled electrodes exhibit an undesirably low electronic conductance as a consequence of exceptionally low electron transport in the nanomaterials as well as the high contact resistances within nanomaterials and between the current collector and electrodes[29,30], greatly hindering their potential applications in electrochemical biosensors at ultralow concentrations in unconventional body fluids.

Here we report multifunctional seamless solid/nanoporous Au/Co_3O_4 (S/NPG/Co_3O_4) hybrid microelectrodes with three-dimensional (3D) bicontinuous nanoporosity, facilely fabricated by incorporating Co_3O_4 nanoparticles onto seamless solid/nanoporous Au (S/NPG) microwires, for nonenzymatic electrochemical glucose biosensors. Amperometric measurements at a low potential demonstrate that the seamless S/NPG/Co_3O_4 microelectrodes can detect ultralow-concentration glucose with an ultrahigh sensitivity of $\sim 12.5\,mA\,mM^{-1}\,cm^{-2}$ and a fast response time of $< 1\,s$. The outstanding sensing performance results from the unique nanoarchitecture, which simultaneously minimizes the contact resistances among current collector of solid Au wire, nanoporous Au skeleton and Co_3O_4 nanoparticles. Furthermore, the nanoarchitecture provides interconnected nanoporous channels and Au ligaments, concurrently enhancing the transports of glucose and electron, and offers a large specific surface area of electrode/electrolyte interface, facilitating the full use of the enhanced electrocatalysis of Au/Co_3O_4 hybrid.

Results

Synthesis and characterization. Our fabrication strategy briefly illustrated in Fig. 1 is to combine the alloying/dealloying and

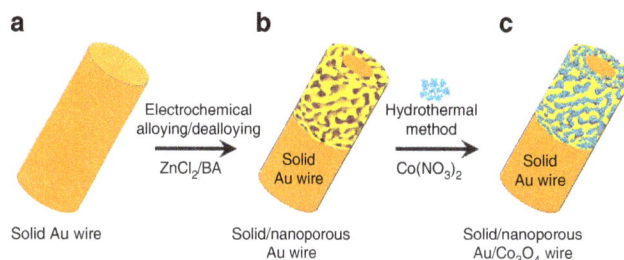

Figure 1 | Scheme for fabrication of hybrid microelectrodes. (**a**) Solid gold microwire. (**b**) 3D bicontinuous nanoporous gold layer is formed directly on the solid gold microwire by electrochemical alloying/dealloying in a mixed electrolyte of BA and $ZnCl_2$. (**c**) Co_3O_4 nanoparticles are decorated onto nanopore channels of seamless S/NPG microwire by a hydrothermal method in a mixed solution containing $Co(NO_3)_2$.

hydrothermal methods, through which the preparation of seamless S/NPG microwires and the incorporation of Co_3O_4 nanoparticles into the nanopore channels are consecutively carried out. The Au microwires of 200-μm-diameter with seamless solid/nanoporous architecture (Fig. 2a) are produced by *in situ* electrochemical alloying/dealloying protocol in an electrolyte composed of $ZnCl_2$ and benzyl alcohol (BA)[31], during which Au–Zn alloys are firstly formed via the electrodeposition of zinc, and the less noble Zn is then selectively dissolved in the dealloying process for the formation of nanoporous layer with high surface area (Supplementary Fig. S1)[32]. The alloying/dealloying conditions (Supplementary Table S1) are adjusted to manipulate the resulting microstructure of nanoporous layer (Fig. 2 and Supplementary Figs S2–S5). Figure 2b,c shows representative cross-sectional and top-view scanning electron microscope images of the seamless S/NPG microwire, respectively, that is fabricated at the optimized alloying/dealloying conditions in a mixture of BA and 1.5 M $ZnCl_2$ at a scan rate of $10\,mV\,s^{-1}$ and 120 °C (the first conditions in Supplementary Table S1). The seamless solid/nanoporous structure possesses $\sim 1\,\mu m$ thick nanoporous layer (Fig. 2b), where a uniform and 3D bicontinuous nanoporous structure consists of quasi-periodic nanopore channels and Au ligaments with characteristic length of $\sim 300\,nm$ (Fig. 2c and Supplementary Fig. S6)[32,33]. The decreases of the cycle number, temperature, $ZnCl_2$ concentration and scan rate in the alloying/dealloying process (Supplementary Table S1) will reduce the thickness and characteristic length of nanoporous layer and change the real surface area of S/NPG skeleton, as demonstrated by scanning electron microscope micrographs and *in situ* electrochemical measurements (Supplementary Figs S2–S5,S7a and Supplementary Notes 1,2)[34].

Seamless S/NPG/Co_3O_4 hybrid microwires are synthesized by the hydrothermal method (Supplementary Table S1), wherein the loading amount and morphology of Co_3O_4 can be tailored by the $Co(NO_3)_2$ concentration (Supplementary Fig. S8) and the heating temperature (Supplementary Fig. S9), respectively. At a constant hydrothermal temperature of 180 °C, the loading amount of Co_3O_4 increases with the increasing concentration of $Co(NO_3)_2$ (Supplementary Fig. S8). In the mixed solution containing 6 mM $Co(NO_3)_2$, the well-crystallized Co_3O_4 nanoparticles uniformly and directly grow into nanopores along the Au ligaments of the seamless S/NPG microwires (Fig. 2d and Supplementary Fig. S10). The rough Co_3O_4 nanoparticles are composed of nanocrystals with diameter of $\sim 5–20\,nm$, giving rise to a larger real surface area of S/NPG/Co_3O_4 than that of bare S/NPG (Fig. 2e and Supplementary Fig. S7). When the hydrothermal temperature increases to 260 °C, nanosheet-shaped Co_3O_4 grows vertically on Au ligaments (Supplementary Fig. S9a), whereas at

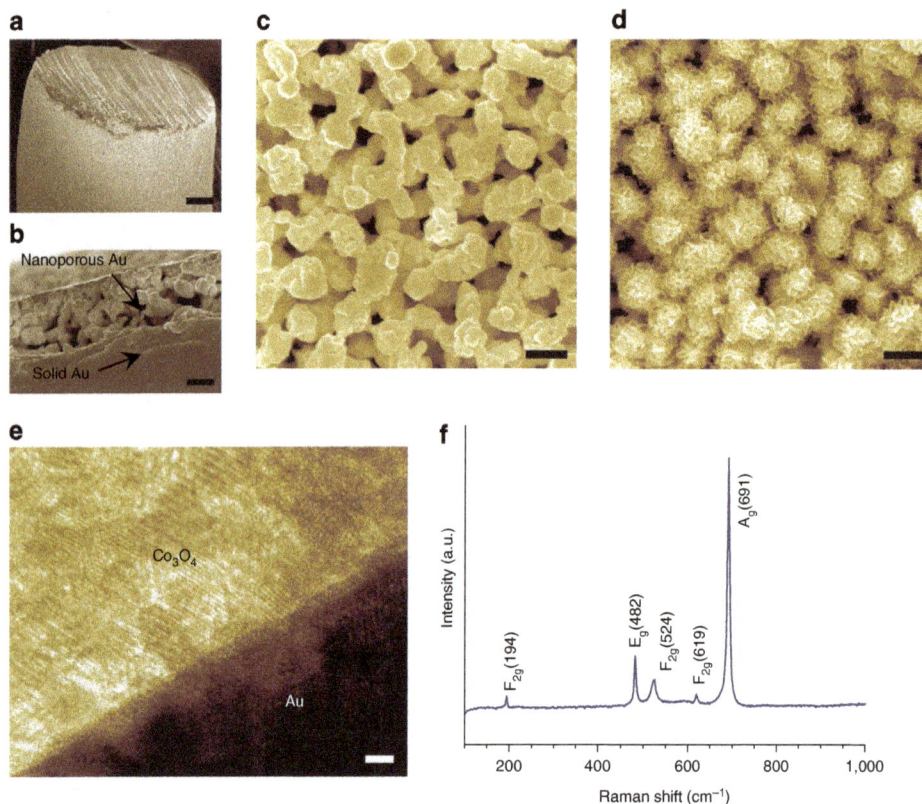

Figure 2 | Microstructure characterization. (**a**) Low-magnification and (**b**) cross-section SEM images of seamless S/NPG microwire. (**a**) Scale bar, 25 μm; (**b**) scale bar, 300 nm. Top-view SEM images of (**c**) bare S/NPG and (**d**) S/NPG/Co$_3$O$_4$ hybrid with the characteristic length of ~300 nm. (**c,d**) Scale bar, 300 nm. (**e**) High-resolution transmission electron microscope (HRTEM) image of Au/Co$_3$O$_4$ interfacial structure. (**e**) Scale bar, 2 nm. (**f**) Surface-enhanced Raman spectrum of seamless S/NPG/Co$_3$O$_4$ hybrid microwire.

100 °C, the formation of Co$_3$O$_4$ film seals the nanoporous structure of hybrid microelectrode (Supplementary Fig. S9c), significantly reducing the available electrolyte/electrode interface (Supplementary Fig. S7b and Supplementary Note 1). The high-resolution transmission electron microscope image of Au/Co$_3$O$_4$ interface (Fig. 2e) indicates that the Co$_3$O$_4$ nano-particles have end-bonded contacts with the Au ligaments[29], offering excellent electrical conductivity between Au and Co$_3$O$_4$. The characteristic peaks at 194, 482, 524, 619 and 691 cm^{-1} in the surface-enhanced Raman spectrum of Co$_3$O$_4$, which are assigned to the F$_{2g}$, E$_g$, F$_{2g}$, F$_{2g}$ and A$_{1g}$ vibrational modes, respectively, unveils a spinel-type crystalline structure (Fig. 2f)[35,36]. This is further confirmed by the obvious diffraction peaks in X-ray diffraction pattern of S/NPG/Co$_3$O$_4$ hybrid corresponding to the (220), (311), (222), (400), (422), (511) and (440) planes of spinel-type Co$_3$O$_4$ (JCPDS 42–1467)[27], apart from two significant diffraction peaks at $2\theta = 44.4°$ and 77.6° attributed to the (200) and (311) reflections of Au (JCPDS 65–2870) (Supplementary Fig. S11). The direct decoration of 3D bicontinuous nanoporous Au layer on the Au microwires facilitates the integration of nanoporous metal/oxide composites with current collectors without any additional contact resistance[37]. By this technology, the entire assembled electrode not only retains excellent electrical conductivity up to the Au bulk value (~4 × 10^5 S cm^{-1}), much higher than those of conventional electrodes based on conducting polymers and various carbon materials, but exhibits exceptional mechanical flexibility (Supplementary Fig. S12).

Electrocatalytic activity for glucose oxidization. To assess their electrocatalytic activity, self-supported seamless S/NPG and

S/NPG/Co$_3$O$_4$ microwires as well as Co$_3$O$_4$ nanoparticles (size ~5–20 nm, similar to those grown on seamless S/NPG micro-wire) on an ITO glass substrate (Supplementary Fig. S13a) are directly used as working electrodes for cyclic voltammetry measurements. In a N$_2$-bubbled mixture of 0.5 M KOH and 10 mM glucose at a scan rate of 20 mV s^{-1}, the cyclic voltammograms (CVs) for the S/NPG/Co$_3$O$_4$ microelectrodes synthesized under various alloying/dealloying and hydrothermal conditions (Supplementary Table S1) are acquired, as shown in Fig. 3 and Supplementary Figs S2d,S3–S5c,S8–S9d. Figure 3a illustrates the typical CV for S/NPG/Co$_3$O$_4$ hybrid microelectrode fabricated under the optimized alloying/dealloying and hydro-thermal conditions (the first ones in Supplementary Table S1), in distinct contrast with its blank voltammetry (without glucose) (Fig. 3b) due to the glucose oxidation. In the glucose-free alkaline solution, S/NPG/Co$_3$O$_4$ concurrently presents characteristic features of both Au and Co$_3$O$_4$ nanoparticles. These include the current-density peaks of the oxidation of Au and the more negative subsequent reduction of Au oxides than that of the bare S/NPG microwire (Fig. 3b)[38]. In addition, two redox pairs of Co$_3$O$_4$ involve the reversible transition between Co$_3$O$_4$ and CoOOH via Co$_3$O$_4$ + OH$^-$ + H$_2$O $\leftrightarrow$ 3CoOOH + e$^-$ and the further conversion between CoOOH and CoO$_2$ via CoOOH + OH$^-$ $\leftrightarrow$ CoO$_2$ + H$_2$O + e$^-$ (Supplementary Fig. S13b)[15,27]. For comparison, the CVs of the corresponding S/NPG microwire and Co$_3$O$_4$ nanoparticles in 0.5 M KOH with and without 10 mM glucose are also included in Fig. 3a,b, and Supplementary Fig. S13b, respectively, presenting low electrocatalytic activities[15,27,38,39]. Remarkably, the CV of the seamless S/NPG/Co$_3$O$_4$ hybrid microelectrode shows a much more negative onset potential and higher anodic current density

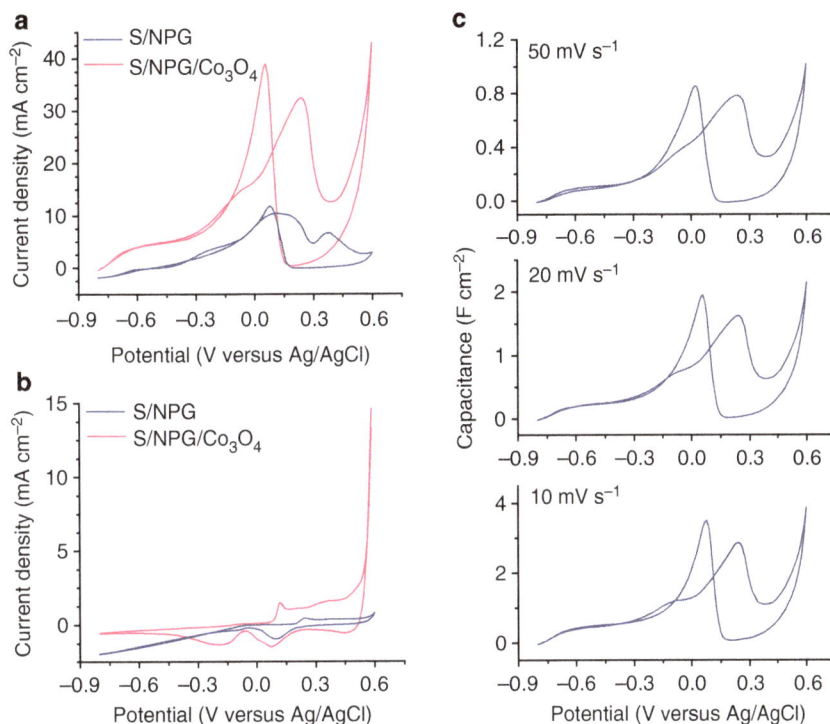

Figure 3 | Electrochemical characterization. Cyclic voltammetry (CV) curves of seamless S/NPG and S/NPG/Co$_3$O$_4$ hybrid microelectrodes in 0.5 M KOH (**a**) with and (**b**) without 10 mM glucose at a scan rate of 20 mV s^{-1}. (**c**) CV profiles of the seamless S/NPG/Co$_3$O$_4$ wire electrode at different scan rates. Current axis in **c** is scaled by the scan rate.

than those of the bare seamless S/NPG microwire and the Co$_3$O$_4$ nanoparticles (Fig. 3a and Supplementary Fig. S13b), suggesting synergistic electrocatalytic activity of S/NPG and Co$_3$O$_4$ in the hybrid towards glucose electrooxidation. As shown in Fig. 3a, the voltammetric behaviour of the seamless S/NPG/Co$_3$O$_4$ hybrid microelectrode in the low potential range ($< \sim -0.06$ V) produces a substantially enhanced current response as a result of glucose electrosorption[28,38,40]. This takes place on the significantly increased electroactive surface sites to form the adsorbed intermediates by releasing one proton per glucose molecule at an onset potential of ~ -0.78 V (much more negative than ~ -0.49 V for bare S/NPG microwire). Accompanying by electrosorption of glucose, the accumulation of the concomitant intermediates on the electrode surface inhibits direct oxidation of glucose on the active sites of Au hence gives rise to the small anodic current peak at ~ -0.06 V (refs 28,38,40). Meanwhile, Co$_3$O$_4$ nanoparticles hybridized with seamless S/NPG microwire intervene in a catalytical promotion of glucose oxidation (Supplementary Fig. S13b), leading to the continuous increase of the current. At higher positive potentials, both AuOH and CoOOH form on the internal surface of seamless S/NPG/Co$_3$O$_4$ electrode by the partial discharge of OH$^-$, and consequently catalyse the oxidation of intermediates and the direct oxidation of glucose[27,28,38,40] by the conversions of AuOH→Au and CoOOH→Co$_3$O$_4$, respectively. The corresponding broad anodic current peak is at ~ 0.25 V. When the potential sweeps more positively, the current density decreases in the potential range of 0.25–0.39 V as formation of Au oxides suppresses the direct electrooxidation of glucose[28,38,40]. Upon the potential higher than ~ 0.39 V, the current density dramatically increases as a result of the formation of CoO$_2$ on which direct glucose oxidation occurs via CoO$_2$ + C$_6$H$_{12}$O$_6$→CoOOH + C$_6$H$_{10}$O$_6$ (refs 15,27). In the negative scan, Au oxides are reduced and a large number of active sites on the internal surface of seamless S/NPG/Co$_3$O$_4$ electrode are

regenerated for the direct glucose oxidation, leading to an anodic current peak at ~ 0.05 V (refs 28,38). When Co$_3$O$_4$ nanoparticles are incorporated into the S/NPG skeleton with a larger characteristic-length and thinner NPG layer (Supplementary Note 2) at the same hydrothermal conditions (Supplementary Figs S2–S5 and Supplementary Table S1), the CV of S/NPG/ Co$_3$O$_4$ hybrid microelectrode shows a lower anodic current density and a less negative onset potential (Supplementary Figs S2d,S3c–S5c) because of less surface area for the formation of highly active Au/Co$_3$O$_4$ hybrid (Supplementary Fig. S7a). This demonstrates the important role of high nanoporosity of S/NPG skeleton in producing a large amount of Au/Co$_3$O$_4$ hybrid which significantly enhances the electrocatalytic activity of the whole hybrid microelectrodes.

Furthermore, the enhanced electrocatalytic behaviour of seamless S/NPG/Co$_3$O$_4$ hybrid microelectrodes is found to rely significantly on the scan rate (Fig. 3c). For lower scan rates, comparatively more glucose is electrooxidized at the surface, indicating a surface-controlled electrochemical process. This is further verified by the observation that the smaller the surface area of S/NPG/Co$_3$O$_4$ microelectrodes, the lower the current densities (Supplementary Figs S7b,S9d). It is noteworthy that the change of the surface area results from the evolution of morphology and loading amount of Co$_3$O$_4$ on the same S/NPG microwires at different hydrothermal conditions (Supplementary Table S1). The current density of S/NPG/Co$_3$O$_4$ hydrothermally synthesized at 100 °C falls dramatically as the nanoporous structure is sealed by the Co$_3$O$_4$ film (Supplementary Fig. S9c,d), which restrains ion and glucose transports as well as the availability of electrolyte/electrode interface for glucose electrooxidation.

Amperometric glucose detection. For all S/NPG and S/NPG/ Co$_3$O$_4$ microelectrodes (Supplementary Table S1), the glucose sensing performance is evaluated by amperometric measurements, and their current-time curves for successive addition of glucose with concentrations ranging from $1\,\mu M$ to $10\,mM$ are recorded at an applied potential of $0.26\,V$ (Fig. 4 and Supplementary Fig. S14). As a result of the enhanced electrocatalytic activities of hybrid nanoarchitecture towards glucose oxidation, the current responses and sensitivities of S/NPG/Co_3O_4 microelectrodes are enhanced by increasing the surface areas of both the S/NPG skeletons and their Co_3O_4 hybrids (Supplementary Figs S14,S15). The former affords more Au/Co_3O_4 interface for enhancing the electrochemical activity of Co_3O_4 and improving the electron transport of Co_3O_4 to Au network, whereas the latter enables to trap more molecules for electrooxidation and to provide the large surface area of electrolyte/electrode interface for the full use of highly active Au/Co_3O_4 hybrid.

Figure 4 gives the typical comparisons of amperometric responses between bare S/NPG and S/NPG/Co_3O_4 hybrid microelectrodes synthesized at the optimized alloying/dealloying and hydrothermal conditions (the first ones in Supplementary Table S1). Owing to the synergistic electrocatalytic activity of Au/Co_3O_4 hybrid nanostructure and the excellent conductivity of seamless solid/nanoporous architecture, the amperometric response of the seamless S/NPG/Co_3O_4 hybrid microelectrode exhibits perfect and stable step curve with dramatically enhanced current density in comparison with that of bare S/NPG (Fig. 4a). It should be noted that a response time no more than 1 s to research 95% of steady-state current (Fig. 4c and Supplementary Fig. S16a) is much faster than 2 s for nanometal-decorated graphite nanoplatelet[19], 6 s for GOx-CNT networks[41] or 7 s for porous Co_3O_4 nanofibers[15]. In consequence, incorporating Co_3O_4 nanoparticles into the 3D bicontinuous nanoporous gold of

seamless solid/nanoporous structure remarkably improves the glucose sensing ability compared with the bare seamless S/NPG microwire as well as free Co_3O_4 nanoparticles (Fig. 4a and Supplementary Fig. S13c), which is further verified by their calibration curves (Fig. 4b and Supplementary Fig. S13d). As illustrated in the current density versus concentration plots (Supplementary Figs S16–S18), the glucose biosensors based on Co_3O_4 nanoparticles, seamless S/NPG and S/NPG/Co_3O_4 hybrid microelectrodes show multi-linear detection ranges with different sensitivities up to 7, 30 and $70\,mM$, respectively, displaying a strong dependence on the concentration of added glucose (Fig. 4d). As the concentration decreases to $1\,\mu M$, the sensitivity of seamless S/NPG/Co_3O_4 hybrid increases up to $\sim 12.5\,mA\,mM^{-1}\,cm^{-2}$, much higher than the values of bare seamless S/NPG ($0.72\,mA\,mM^{-1}\,cm^{-2}$) (Fig. 4d) and Co_3O_4 nanoparticles (Supplementary Fig. S13). Such an ultrahigh sensitivity offers the seamless S/NPG/Co_3O_4 microelectrode a remarkably low detection limit of $5\,nM$ (Fig. 5a), much lower than some of the lowest values reported previously: $970\,nM$ for electrospun Co_3O_4 nanofibers[15], $25\,nM$ for 3D graphene/Co_3O_4 composite[27] or $1.29\,\mu M$ for the GOx-immobilized one-dimensional TiO_2 nanostructure[21]. Moreover, the seamless S/NPG/Co_3O_4 microelectrode provides superb selectivity for glucose detection at the low applied potential of $0.26\,V$, which minimizes the responses of common interference species such as uric acid, acetamidophenol and ascorbic acid, in physiological levels[38,42], as well as other sugars including fructose, mannose, maltose, sucrose and lactose[42]. Addition of $0.02\,mM$ uric acid, $0.1\,mM$ acetamidophenol, ascorbic acid or other sugars to a 1-mM glucose solution results in only ~ 0.5–5% increase in the current density, and does not interfere with the glucose detection even without the permselective coatings (Fig. 5b and

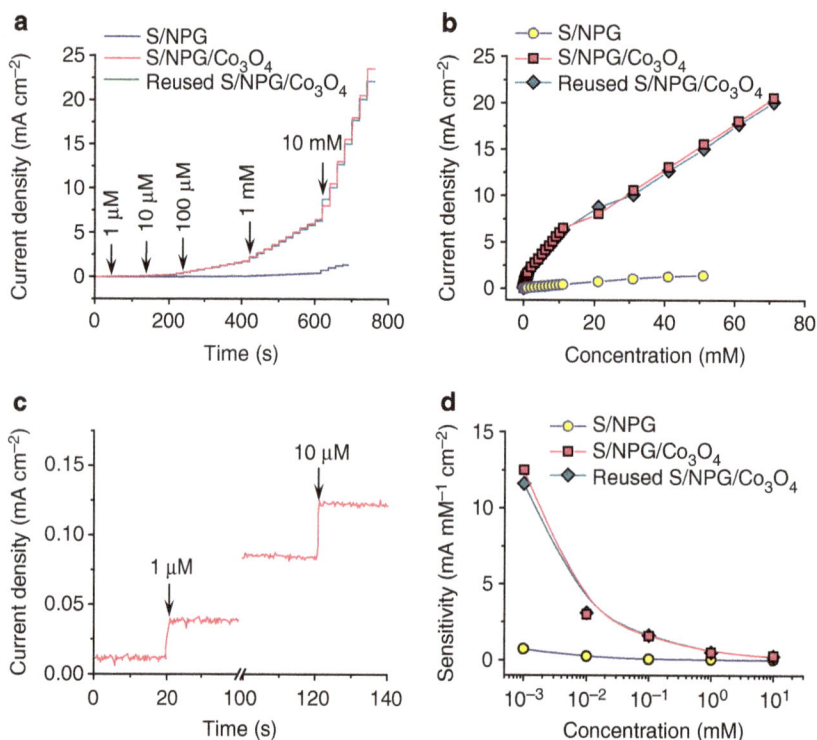

Figure 4 | Amperometric glucose detection. (a) Chronoamperometry curves and (b) the calibration curves (current density versus glucose concentration) for the amperometric responses of seamless S/NPG and fresh and reused S/NPG/Co_3O_4 microelectrodes to the successive addition of glucose with the concentrations from $1\,\mu M$ to $10\,mM$ in $0.5\,M$ KOH solution at a constant potential of $0.26\,V$ (versus Ag/AgCl). (c) Fast response of seamless S/NPG/Co_3O_4 hybrid microelectrode to 1 and $10\,\mu M$ glucose. (d) Glucose sensitivities of seamless S/NPG and fresh and reused S/NPG/Co_3O_4 microelectrodes as a function of the concentration of glucose.

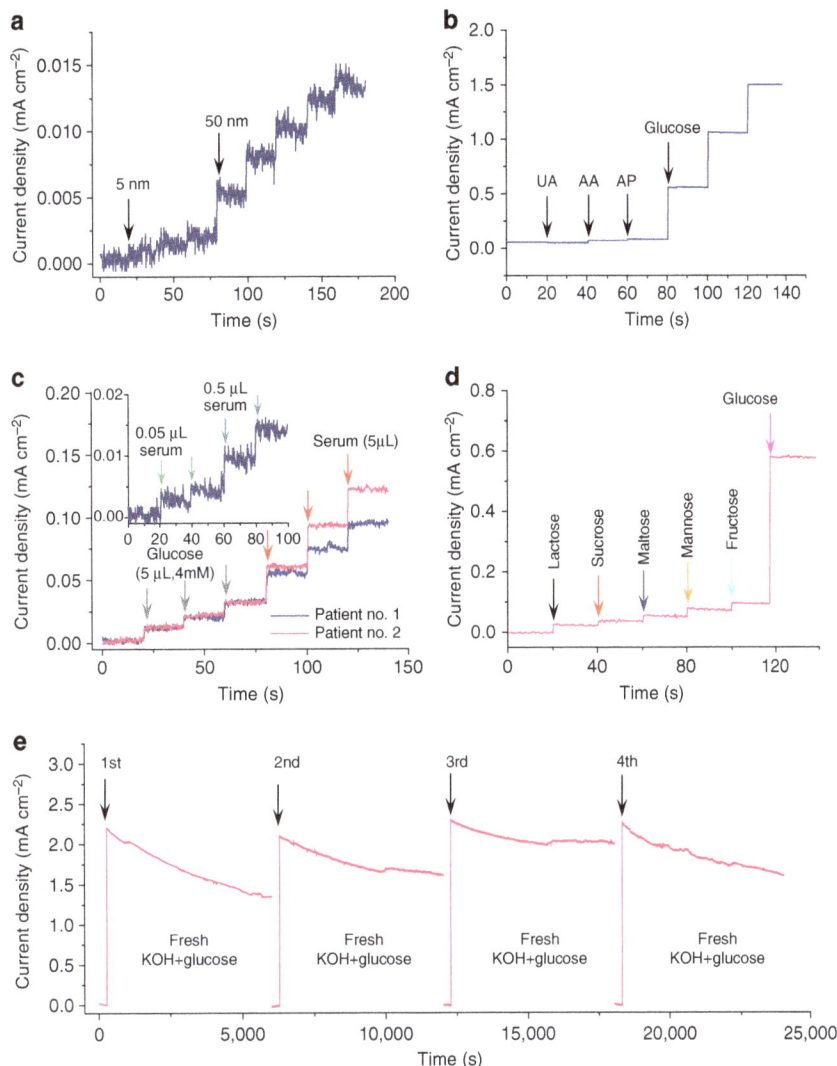

Figure 5 | Selectivity and stability. (**a**) Amperometric response to glucose with the low limit concentrations of 5 and 50 nM. Amperometric response to the successive addition of interfering compounds such as (**b**) 0.02 mM uric acid (UA), 0.1 mM acetamidophenol (AP), 0.1 mM ascorbic acid (AA) and (**d**) 0.1 mM fructose, mannose, maltose, sucrose, lactose, as well as 1 mM glucose at a potential of 0.26 V. (**c**) Current response of the S/NPG/Co$_3$O$_4$ microelectrode to successive three-time additions of 5 µl glucose (4 mM) and blood serum of two diabetic patients into 20 ml 0.5 M electrolyte. Inset: Current response of the S/NPG/Co$_3$O$_4$ microelectrode stored at room temperature for 4 months to ultralow concentration of serum. (**e**) Current-time curve for the S/NPG/Co$_3$O$_4$ at 0.26 V in four fresh electrolytes containing 0.5 M KOH and 10 mM glucose.

Supplementary Table S2). These predominant properties enlist this hybrid microelectrode to exhibit high reliability for analysing blood serums. Figure 5c demonstrates the current-time curves for the successive additions of 5 µl 4 mM glucose standard solution and 5 µl blood serums of diabetic patients into 20 ml 0.5 M KOH solution, according to which the measured blood-glucose concentrations of diabetic patients are consistent with the values reported by the clinical analysis in the hospital (Supplementary Table S3).

The electrochemical stability of the S/NPG/Co$_3$O$_4$ hybrid microelectrode is investigated by the chronoamperometric method in four fresh solutions containing 0.5 M KOH and 10 mM glucose (Fig. 5d). In spite of the slow decrease of the current with time, the starting current response maintains stable in each fresh electrolyte, revealing the good stability of the hybrid microelectrodes during the electrochemical measurements. This is also confirmed by UV–visible spectrum of the measurement solution, where no Co or Au ion is detected

(Supplementary Fig. S19). Moreover, the aging tolerance ensures the S/NPG/Co$_3$O$_4$ hybrid microelectrode to retain ~99.5% of its original current response over a storage period of 15 days at room temperature (Supplementary Fig. S20), indicating the excellent reproducibility with a relative s.d. of ~0.6%. Even the S/NPG/Co$_3$O$_4$ hybrid microelectrode stored at room temperature for 4 months still maintains the high capability to analyse serum sample at ultralow concentrations (inset of Fig. 5c).

To demonstrate the reusability, an S/NPG/Co$_3$O$_4$ hybrid microelectrode is reconstructed via a procedure: etching the used S/NPG/Co$_3$O$_4$ microelectrode in concentrated HNO$_3$ to remove Co$_3$O$_4$ (Supplementary Fig. S21a), and then reincorporating Co$_3$O$_4$ nanoparticles onto the used S/NPG microwire by hydrothermal method (Supplementary Fig. S21b). The amperometric response of the reused S/NPG/Co$_3$O$_4$ microelectrode to the successive addition of glucose shows outstanding glucose sensing performance comparable to the as-prepared one (Fig. 4a,b,d, and Supplementary Fig. S22), indicating its superior reusability and reproducibility. Thus, the seamless S/NPG/Co$_3$O$_4$ hybrid microelectrode is an excellent material for the enhanced electrochemical sensing.

Discussion

In the perspective of the Langmuir isothermal theory for surface electrocatalytic reaction[15], the sensitivity (m) of seamless S/NPG/Co$_3$O$_4$ microelectrode is governed by the glucose concentration (C_g) (as shown in Fig. 4d), namely, $m = di_T/dC_g = K_B K_A C_a/(1 + K_A C_g)^2$, where the total current density $i_T = K_B C_{gS}$ with $C_{gS} = K_A C_g C_a/(1 + K_A C_g)$ being the concentration of glucose adsorbed on the catalyst surface, C_a denotes the molar concentration of active sites on the catalyst, K_A and K_B are the adsorption equilibrium constant and the reaction rate constant, respectively[15]. As C_g is low enough (1 μM), that is, $1 \ll K_A C_g$, $m_L \approx K_B K_A C_a = 12.5$ mA mM^{-1} cm^{-2}, indicating the intrinsically ultrahigh electrocatalytic activity of seamless S/NPG/Co$_3$O$_4$ hybrid microelectrode towards glucose oxidation due to the unique nanoarchitecture of seamless S/NPG skeleton and synergistic Au/Co$_3$O$_4$ hybrid. The seamless S/NPG framework, ingeniously integrating NPG layer with current collector by directly alloying/dealloying process on Au wire, minimizes the contact resistance between NPG layer and current collector, and thus optimizes the sensing performance in comparison with NPG species binded to current collector by polymer binders or adhesives (Supplementary Fig. S23)[38]. Meanwhile, the 3D open and bicontinuous nanoporous structure provides a large room for loading Co$_3$O$_4$ nanoparticles, and the produced nanoporous Au/Co$_3$O$_4$ hybrid with extremely large specific surface area enhances the mass transport kinetics and facilitates sufficient use of the highly active Au/Co$_3$O$_4$ hybrid[6,7,9]. Therefore, the thicker the NPG layer with the same characteristic length (Supplementary Note 2), the higher the glucose sensitivity of S/NPG/Co$_3$O$_4$ hybrid microelectrode (Supplementary Fig. S24). Although free nanostructured Co$_3$O$_4$ has intrinsically ultralow conductivity that limits its application in biosensing (Supplementary Figs S13,S18)[15], the nanoporous Au/Co$_3$O$_4$ hybrid substantially enhances the electron transfer and improves the electrocatalytic activity of Co$_3$O$_4$ (m_{Co3O4}), which is approximately calculated by subtracting the value of NPG (m_{NPG}) according to the equation $m_{Co3O4} = m_L - m_{NPG} = 11.72$ mA mM^{-1} cm^{-2}, two orders of magnitude higher than that of porous Co$_3$O$_4$ nanofibers[15]. This great enhancement allows the seamless S/NPG/Co$_3$O$_4$ microelectrode to detect glucose at ultralow concentrations with exquisite selectivity for potential applications in monitoring glucose in conventional and unconventional body fluids, as well as complementing or even replacing the GOx-based biosensors outside their rigorous operational environments (high temperature, high pH and so on). Additionally, the facile collection and reusability of S/NPG and S/NPG/Co$_3$O$_4$ microwires can dramatically reduce the cost of Au/Co$_3$O$_4$-based biosensing devices, meeting low-cost requirements for the large-scale biosensing use.

In summary, by the facile two-step procedure involving electrochemical alloying/dealloying and hydrothermal synthesis we have developed a seamless microelectrode with hybrid nanoarchitecture of S/NPG/Co$_3$O$_4$ as a promising electrode material for high-performance electrochemical biosensors with ultrahigh sensitivity and ultrafast response time. The impressive glucose sensing behaviour of seamless S/NPG/Co$_3$O$_4$ hybrid microelectrodes originates from the substantially enhanced electrocatalytic activity of the Au/Co$_3$O$_4$ hybrid and the unique seamless solid/porous nanoarchitecture, which simultaneously minimizes the primary resistances in biosensing, viz. the electrochemical reaction occurring at electrode/electrolyte interfaces, the mass transport of analyte in electrolyte and electrode, and the electron conduction in electrode and current collector.

Methods

Fabrication of seamless microwires. The seamless S/NPG wires were prepared by a multicyclic electrochemical alloying/dealloying method based on a three-electrode electrochemical cell (Iviumstat electrochemical analyzer, Ivium Technology), in which Zn plate and wire were used as the auxiliary electrode and the reference electrode in a BA electrolyte containing 1.5 M ZnCl$_2$, respectively[31]. 3D bicontinuous nanoporous layer with tunable microstructural features was decorated on polished Au wires with diameters of ∼200 μm after the electrochemical cycles in the potential range from − 0.72 to 1.88 V (versus Zn) at different conditions (Supplementary Table S1)[31]. After acetone, ethanol and deionized water (18.2 MΩ cm) rinsing in sequence, the seamless S/NPG wire fixed on a plastic clamp was moved into the Teflon-lined stainless steel autoclave filled with the mixture of 3, 6, 9 mM Co(NO$_3$)$_2$, 2 mM cetyltrimethylammonium bromide, 5 ml pure water and 30 ml ethanol. The sealed autoclave was heated in an electric oven at 100, 180 and 260 °C for 90 min for synthesis of the seamless Au/NPG/Co$_3$O$_4$ (Supplementary Table 1S). After cooled to room temperature, the hybrid wire was washed with deionized water. The S/NPG microwire was collected from the tested S/NPG/Co$_3$O$_4$ microelectrode via chemical etching in 60% HNO$_3$ for the further reconstruction of S/NPG/Co$_3$O$_4$ microelectrode via hydrothermal method at the first conditions in Supplementary Table S1.

Characterization. The microstructure and chemical composition of the specimens were investigated using a field-emission transmission electron microscope (JEOL JEM-2100F, 200 keV), and a field-emission scanning electron microscope (JEOL, JSM-6700F, 15 keV) equipped with an X-ray energy-dispersive microscopy. Raman spectrum was collected using a micro-Raman spectrometer (Renishaw) with a laser of 532 nm wavelength. X-ray diffraction measurement was carried out on a D/max2500pc diffractometer using Cu $K\alpha$ radiation.

Electrochemical Measurement. All cyclic voltammetry and amperometric experiments were performed in a three-electrode setup, which employed gold wire, seamless Au/NPG, Au/NPG/Co$_3$O$_4$ microwires and Co$_3$O$_4$ nanoparticles supported on ITO glass as the working electrodes, a Pt foil electrode as the counter electrode and a Ag/AgCl electrode as the reference electrode, in the 0.5 M KOH electrolyte bubbled by nitrogen gas for 30 min. The CVs were collected in the aqueous electrolyte with and without 10 mM glucose. The amperometric responses of microelectrodes to serum and glucose with different concentrations ranging from 1 μM to 10 mM were recorded under steady-state conditions at a potential of 0.26 V (versus Ag/AgCl). The concentration of blood glucose of diabetic patients was measured by automatic biochemical analyzer (Bakeman Coulter DXC800).

References

1. Ronkainen, N. J., Halsall, H. B. & Heineman, W. R. Electrochemical biosensors. *Chem. Soc. Rev.* **39**, 1747–1763 (2010).
2. Wang, J. Electrochemical glucose biosensors. *Chem. Rev.* **108**, 814–825 (2008).
3. Chen, R. J. *et al.* An investigation of the mechanisms of electronic sensing of protein adsorption on carbon nanotube devices. *J. Am. Chem. Soc.* **126**, 1563–1568 (2004).
4. Im, H., Huang, X. J., Gu, B. & Choi, Y. K. A dielectric-modulated field-effect transistor for biosensing. *Nat. Nanotech.* **2**, 430–434 (2007).
5. Xiang, Y. & Lu, Y. Using personal glucose meters and functional DNA sensors to quantify a variety of analytical targets. *Nat. Chem.* **3**, 697–703 (2011).
6. Kissinger, P. T. & Heineman, W. R. *Laboratory Techniques in Electroanalytical Chemistry* (Marcel Dekker Inc, New York, 1996).
7. Yang, W. *et al.* Carbon nanomaterials in biosensors: should you use nanotubes or graphene? *Angew. Chem. Int. Ed.* **49**, 2114–2138 (2010).
8. Šljukić, B., Banks, C. E. & Compton, R. G. Iron oxide particles are the active sites for hydrogen peroxide sensing at multiwalled carbon nanotube modified electrodes. *Nano. Lett.* **6**, 1556–1558 (2006).
9. Minteer, S. D., Atanassov, P., Luckarift, H. R. & Johnson, G. R. New materials for biological fuel cells. *Mater. Today* **15**, 166–173 (2012).
10. Solanki, P. R., Kaushik, A., Agrawal, V. V. & Malhotra, B. D. Nanostructured metal oxide-based biosensors. *NPG Asia Mater.* **3**, 17–24 (2011).
11. Kimmel, D. W., LeBlanc, G., Meschievitz, M. E. & Cliffel, D. E. Electrochemical sensors and biosensors. *Anal. Chem.* **84**, 685–707 (2012).
12. Wilson, R. & Turner, A. P. F. Glucose oxidase: an ideal enzyme. *Biosens. Bioelectron.* **7**, 165–185 (1992).
13. Wang, J. P., Thomas, D. F. & Chen, A. C. Nonenzymatic electrochemical glucose sensor based on nanoporous PtPb networks. *Anal. Chem.* **80**, 997–1004 (2008).
14. Duan, H. M., Xu, D. Y., Li, W. Z. & Xu, H. Y. Study of the redox properties of noble metal/Co$_3$O$_4$ by electrical conductivity measurements. *Catal. Lett.* **124**, 318–323 (2008).

15. Ding, Y. *et al.* Electrospun Co_3O_4 nanofibers for sensitive and selective glucose detection. *Biosens. Bioelectron.* **26**, 542–548 (2008).

16. Kros, A., Nolte, R. J. M. & Sommerdijk, N. A. J. M. Conducting polymers with confined dimensions: track-etch membranes for amperometric biosensor applications. *Adv. Mater.* **14**, 1779–1782 (2002).

17. Besteman, K., Lee, J., Wiertz, F. G. M., Heering, H. A. & Dekker, C. Enzyme-coated carbon nanotubes as single-molecule biosensors. *Nano Lett.* **3**, 727–730 (2003).

18. Willner, I. & Willner, B. Biomolecule-based nanomaterials and nanostructures. *Nano Lett.* **10**, 3805–3815 (2010).

19. Lu, J., Do, I., Drzal, L. T., Worden, R. M. & Lee, I. Nanometal-decorated exfoliated graphite nanoplatelet based glucose biosensors with high sensitivity and fast response. *ACS Nano* **2**, 1825–1832 (2008).

20. Guo, S. J., Wen, D., Zhai, Y. M., Dong, S. J. & Wang, E. K. Platinum nanoparticle ensemble-on-graphene hybrid nanosheet: one-pot, rapid synthesis, and used as new electrode materials for electrochemical sensing. *ACS Nano* **4**, 3959–3968 (2010).

21. Si, P., Ding, S. J., Yuan, J., Lou, X. W. & Kim, D. H. Hierarchically structured one-dimensional TiO_2 for protein immobilization, direct electrochemistry, and mediator-free glucose sensing. *ACS Nano* **5**, 7617–7626 (2011).

22. Chen, J., Zhang, W. D. & Ye, J. S. Nonenzymatic electrochemical glucose sensor based on MnO_2/MWNTs nanocomposite. *Electrochem. Commun.* **10**, 1268–1271 (2008).

23. Tehrani, R. M. A. & Ghani, S. A. MWCNT-ruthenium oxide composite paste electrode as non-enzymatic glucose sensor. *Biosens. Bioelectron.* **38**, 278–283 (2012).

24. Campuzano, S. & Wang, J. Nanobioelectroanalysis based on carbon/inorganic hybrid nanoarchitectures. *Electroanalysis* **23**, 1289–1300 (2011).

25. Reddy, A. L. M., Shaijumon, M. M., Gowda, S. R. & Ajayan, P. M. Coaxial MnO_2/carbon nanotube array electrodes for high-performance lithium batteries. *Nano Lett.* **9**, 1002–1006 (2009).

26. Ye, D. X., Luo, L. Q., Ding, Y. P., Liu, B. D. & Liu, X. Fabrication of Co_3O_4 nanoparticles-decorated graphene composite for determination of L-tryptophan. *Analyst* **137**, 2840–2845 (2012).

27. Dong, X. C. *et al.* 3D graphene-cobalt oxide electrode for high-performance supercapacitor and enzymeless glucose detection. *ACS Nano* **6**, 3206–3213 (2012).

28. Ding, Y., Liu, Y. X., Parisi, J., Zhang, L. C. & Lei, Y. A novel NiO-Au hybrid nanobelts based sensor for sensitivie and selective glucose detection. *Biosens. Bioelectron.* **28**, 393–398 (2011).

29. Léonard, F. & Talin, A. A. Electrical contacts to one- and two-dimensional nanomaterials. *Nat. Nanotech.* **6**, 773–783 (2011).

30. Gowda, S. R., Reddy, A. L. M., Zhan, X. B., Jafry, H. R. & Ajayan, P. M. 3D nanoporous nanowire current collectors for thin films microbatteries. *Nano Lett.* **12**, 1198–1202 (2012).

31. Yu, C. F., Jia, F. L., Ai, Z. H. & Zhang, L. Z. Direct oxidation of methanol on self-supported nanoporous gold film electrodes with high catalytic activity and stability. *Chem. Mater.* **19**, 6065–6067 (2007).

32. Erlebacher, J., Aziz, M. J., Karma, A., Dimitrov, N. & Sieradzki, K. Evolution of nanoporosity in dealloying. *Nature* **410**, 450–453 (2001).

33. Fujita, T. *et al.* Unusually small electrical resistance of three-dimensional nanoporous gold in external magnetic fields. *Phys. Rev. Lett.* **101**, 166601 (2008).

34. Trasatti, S. & Petrii, O. A. Real surface area measurements in electrochemistry. *Pure Appl. Chem.* **63**, 711–734 (1991).

35. Yeo, B. S. & Bell, A. T. Enhanced activity of gold-supported cobalt oxide for the electrochemical evolution of oxygen. *J. Am. Chem. Soc.* **133**, 5587–5593 (2011).

36. Yang, J., Liu, H. W., Martens, W. N. & Frost, R. L. Synthesis and characterization of cobalt hydroxide, cobalt oxyhydroxide, and cobalt oxide nanodics. *J. Phys. Chem. C* **114**, 111–119 (2010).

37. Lang, X. Y., Hirata, A., Fujita, T. & Chen, M. W. Nanoporous metal/oxide hybrid electrodes for electrochemical supercapacitors. *Nat. Nanotech.* **6**, 232–236 (2011).

38. Chen, L. Y., Fujita, T., Ding, Y. & Chen, M. W. A three-dimensional gold-decorated nanoporous copper core-shell composite for electrocatalysis and nonenzymatic biosensing. *Adv. Funct. Mater.* **20**, 2279–2285 (2010).

39. Fujita, T. *et al.* Atomic origins of the high catalytic activity of nanoporous gold. *Nat. Mater.* **11**, 775–780 (2012).

40. Aoun, S. B. *et al.* Effect of metal ad-layers on Au(111) electrodes on electrocatalytic oxidation of glucose in an alkaline solution. *J. Electroanal. Chem.* **567**, 175–183 (2004).

41. Claussen, J. C., Franklin, A. D., Haque, A. U., Porterfield, D. M. & Fisher, T. S. Electrochemical biosensor of nanocube-augmented carbon nanotube networks. *ACS Nano* **3**, 37–44 (2009).

42. Toghill, E. E. & Compton, R. G. Electrochemical non-enzymatic glucose sensors: a perspective and evaluation. *Int. J. Electrochem. Soc.* **5**, 1246–1301 (2010).

Acknowledgements

This work was supported by the National Natural Science Foundation of China (no. 51201069, 81270315), the National Key Basic Research Development Program (no. 2010CB631001), the Keygrant Project of Chinese Ministry of Education (no. 313026), the Program for New Century Excellent Talents in University (no. NCET-10-0437) and the Research Fund for the Doctoral Program of Higher Education of China (no. 20120061120042).

Author contributions

X.-Y.L., H.-Y.F. and Q.J. conceived and designed the experiments. X.-Y.L., H.-Y.F., C.H., G.-F.H. carried out the fabrication of materials and performed electrochemical and microstructural characterizations. X.-Y.L., H.-Y.F. and Q.J. wrote the paper, and all authors discussed the results and commented on the manuscript.

Additional information

7

Rational design of a chalcogenopyrylium-based surface-enhanced resonance Raman scattering nanoprobe with attomolar sensitivity

Stefan Harmsen[1,*], Matthew A. Bedics[2,*], Matthew A. Wall[1,3], Ruimin Huang[1], Michael R. Detty[2] & Moritz F. Kircher[1,4,5]

High sensitivity and specificity are two desirable features in biomedical imaging. Raman imaging has surfaced as a promising optical modality that offers both. Here we report the design and synthesis of a group of near-infrared absorbing 2-thienyl-substituted chalcogenopyrylium dyes tailored to have high affinity for gold. When adsorbed onto gold nanoparticles, these dyes produce biocompatible SERRS nanoprobes with attomolar limits of detection amenable to ultrasensitive *in vivo* multiplexed tumour and disease marker detection.

[1] Department of Radiology, Memorial Sloan Kettering Cancer Center, 1275 York Avenue, New York, New York 10065, USA. [2] Department of Chemistry, University at Buffalo, The State University of New York, Buffalo, New York 14260-3000, USA. [3] Department of Chemistry, Hunter College of the City University of New York, 695 Park Avenue, New York, New York 10065, USA. [4] Center for Molecular Imaging and Nanotechnology (CMINT), Memorial Sloan Kettering Cancer Center, 1275 York Avenue, New York, New York 10065, USA. [5] Department of Radiology, Weill Cornell Medical College, 445 East 69th Street, New York, New York 10021, USA. * These authors contributed equally to this work. Correspondence and requests for materials should be addressed to M.R.D. (email: mdetty@buffalo.edu) or to M.F.K. (email: kircherm@mskcc.org).

Surface-enhanced Raman scattering (SERS) is rapidly gaining interest in the field of biomedical imaging[1-3]. By adsorbing a molecule on a noble metal surface, the weak Raman scattering of a molecule (only 1 in $\sim 10^7$ photons induces Raman scattering) is massively amplified (enhancement factor $10^7 - 10^{10}$)[4-6]. This phenomenon creates a spectroscopic technique that not only has high sensitivity ($10^{-9}-10^{-12}$ M limits of detectability), but also the potential for multiplexing capabilities due to the unique vibrational structure of adsorbed molecules[7-9]. These characteristics have prompted the use of SERS in a wide array of biomedical imaging applications[2,10-17].

Orders-of-magnitude higher sensitivities ($10^{-12}-10^{-14}$ M) can be achieved utilizing Raman reporters that are in resonance with the incident laser, thereby producing surface-enhanced resonance Raman scattering (SERRS) nanoprobes[18-20]. Absorption of light by biological tissue is minimal in the near-infrared (NIR) window and, as a consequence, most optical biomedical applications use NIR detection lasers. While a great deal of attention has been given to dye molecules that absorb light in the visible region, less work has been devoted to developing Raman reporters with absorption maxima that are resonant with NIR detection lasers. The most common Raman reporters are members of the cyanine class of dyes[21].

Herein we report thiophene-substituted chalcogenopyrylium (CP) dyes as a new class of ultra bright, NIR-absorbing Raman reporters. One notable feature of the pyrylium dyes is the ease in which a broad range of absorptivities can be accessed, and consequently be matched with the NIR light source by careful tuning of the dye's optical properties. Specifically, the large differences in absorption maxima introduced by switching the chalcogen atom is a useful property of this dye class[22]. Another important consideration is the affinity of the reporter for the surface of gold. Since the SERS effect decreases exponentially as a function of distance from the nanoparticle[23], it is important that the Raman reporter is near the gold surface. The 2-thienyl substituent provides a novel attachment point to gold for Raman reporters. The 2-thienyl group is not only part of the dye chromophore, but also can be rigorously coplanar with the rest of the chromophore[24]. This allows the dye molecules to be in close proximity to the nanoparticle surface, creating a brighter SERRS signal.

Results

Chalcogenopyrylium dye synthesis and characterization.
Cationic CP dyes 1–3, with absorption maxima near the 785-nm emission of the detection laser were synthesized as outlined in Fig. 1a. The addition of MeMgBr to the known chalcogenopyranones[25] (4, 6), followed by dehydration with the appropriate acid (HZ), yields 4-methyl pyrylium compounds (5, 7) with the desired counterion (PF_6^- or ClO_4^-)[22,26,27]. The condensation of 7 with N,N-dimethylthioformamide in Ac_2O, and subsequent hydrolysis of the intermediate iminium salt yields the (chalcogenopyranylidene)acetaldehyde 8, the penultimate compound leading to trimethine CP dyes[22]. Condensation of 4-methylpyrylium salt 5 and the (chalcogeno-pyranylidene)acetaldehyde 8 bearing the desired R groups and chalcogen atom in hot Ac_2O (ref. 26) forms the final dye compounds 1 − 3 that are substituted with 2-phenyl or 2-thienyl groups, and different combinations of chalcogen atoms (S or Se) (Table 1). The Cl^- and Br^- counterions of dye 1a were accessed by treating the PF_6^- salts with an Amberlite ion exchange resin.

SERRS nanoprobe synthesis and characterization.
CP dyes 1 − 3 were dissolved in dry N,N-dimethylformamide (DMF), at a concentration between 1.0 and 10 mM, and were subsequently used to generate the SERRS nanoprobes. The SERRS nanoprobes consist of a gold core onto which the SERRS reporter is adsorbed, which is then protected by an encapsulating silica layer (Fig. 1b, Table 1). The pyrylium-based SERRS nanoprobes were synthesized by encapsulating 60-nm spherical citrate-capped gold nanoparticles via a modified Stöber procedure[28-30] in the presence of the reporter. After 25 min, the reaction was quenched by the addition of ethanol and the SERRS nanoprobes were collected through centrifugation. Typically, the as-synthesized SERRS nanoprobes had a mean diameter of ~ 100 nm (non-aggregated; Supplementary Figs 1 and 2).

Effect of counterion on colloidal stability and SERRS signal.
In previous reports, the dye counterion was shown to affect the structural and electronic properties of polymethine dyes[31] and the solubility of CP dyes[32]. Since SERRS is highly dependent on these factors, we evaluated the effect of the counterion (Z^-) on the SERRS spectrum, intensity and colloidal stability of the pyrylium-based SERRS nanoprobes. We compared chloride (Cl^-), bromide (Br^-), perchlorate (ClO_4^-) and hexafluorophosphate (PF_6^-) as counterions for CP dye 1a. The SERRS nanoprobes were synthesized in the presence of equimolar amounts (10 μM) of CP dye $1a \cdot Z^-$ (where $Z^- = Cl^-$, Br^-, ClO_4^- or PF_6^-). The counterion introduces almost no difference in optical properties such as absorption maxima and extinction coefficient. Furthermore, with the exception of the chloride counterion, the Raman shifts and intensity of 1a were minimally affected by the different counterions (Fig. 2a). The colloidal stability, however, was shown to be highly counterion dependent (Fig. 2b, Supplementary Fig. 1 and Supplementary Table 1). The least chaotropic counterion, Cl^-, strongly destabilized the gold colloids and caused aggregation of SERRS nanoprobes utilizing 1a as a reporter as evidenced by the strong absorption between 700 and 900 nm. The strongest chaotropic anion, PF_6^-, did not affect colloidal stability during the synthesis of SERRS nanoprobes as evidenced by the strong absorption at 540 nm and low absorbance between 700 and 900 nm (monomeric 60-nm spherical gold nanoparticles have an absorption maximum around 540 nm). Since the PF_6^- anion induced the least nanoparticle aggregation, it was used for further SERRS experiments.

Effect of increased affinity on colloidal stability and SERRS signal.
We also examined the SERRS signal intensity as a function of the number of sulphur atoms in the dye. Sulphur-containing functionality has been used frequently to adhere molecules to gold[33], with several reports using thiol or lipoic acid functional groups to add sulphur-containing functionality[21,34]. In our structures, 2-thienyl groups attached to the 2- and 6- positions of the dye were used to bind the dyes to the gold surface. We also explored the impact of the chalcogen atoms in the CP core, switching a Se (1a and 2a) to S (1b and 2b). The chalcogen switch was used to increase semicovalent interactions with the gold surface, and also to create a chromophore that had a more resonant absorption with the 785-nm detection laser (Table 1). CP dyes 1 − 3 were used at a final concentration of 1.0 μM, which prevented nanoparticle aggregation for dye 3. Fig. 3a shows the molecular structures of the CP dyes. The SERRS intensity of the different as-synthesized pyrylium-based SERRS nanoprobes, which were synthesized at equimolar reporter concentrations, were measured at equimolar SERRS nanoprobe concentrations at low laser power to prevent CCD-saturation (50 μW, 1.0 s acquisition time, 5 × objective). We specifically focused on the 1,600 cm^{-1} peak, which corresponds to aromatic ring stretching modes; and is a mode shared by CP dyes 1–3. The SERRS signal

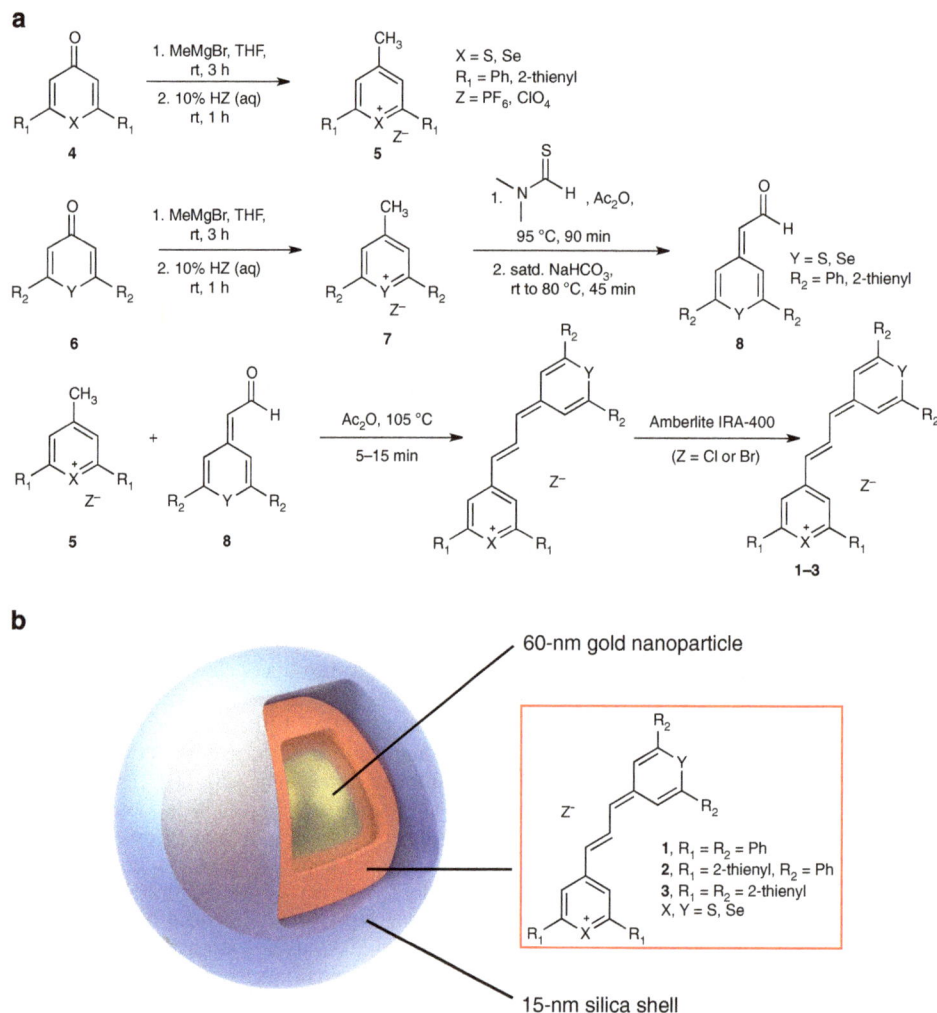

Figure 1 | Synthesis and structure of the SERRS reporters and SERRS nanoprobe. (**a**) Reaction scheme for the synthesis of pyrylium-based SERRS reporters (**1** – **3**). (**b**) A 60-nm gold core (yellow) encapsulated in a 15-nm thick chalcogenopyrylium dye (red; Ph, phenyl)-containing silica shell (blue). The structure, yields and optical properties of the different chalcogenopyrylium-based Raman reporters are shown in the table.

Table 1 | Chalcogenopyrylium dye structural and optical characteristics.

Dye	X	Y	λ_{max} (CH$_2$Cl$_2$)	Log (ε)	Yield (%)
1a	Se	Se	806 nm	5.40	86
1b	S	Se	784 nm	5.30	86
2a	S	Se	810 nm	5.40	87
2b	S	S	789 nm	5.34	88
3	S	S	813 nm	5.45	94

intensity of the 1,600 cm^{-1} peak increased significantly as the number of 2-thienyl substituents increased (Fig. 3b, Supplementary Table 1) without causing significant aggregation (Fig. 3c, Supplementary Fig. 2 and Supplementary Table 1). Thus, **3** produced the highest SERRS signal, which was significantly more intense than **2a/2b** or **1a/1b** ($P < 0.05$) and **2a/2b** were significantly more intense than **1a/1b** ($P < 0.05$). There was a less noticeable, but significant, increase from the chalcogen switch in the core (**1a/1b** and **2a/2b** being significantly different ($P < 0.05$)). This strongly supports the hypothesis that 2-thienyl groups are an effective means of adhering dyes to gold, resulting in brighter SERRS nanoprobes.

Comparison of CP-dye 3 with a cyanine-based SERRS reporter. To assess the quality of our optimized nanoprobe, thiopyrylium dye **3** and commercially available **IR792** (Fig. 4a), which has been previously used to generate surface-enhanced resonance Raman scattering nanoprobes[35], were studied. A direct comparison of the nanoprobes synthesized in the presence of equimolar (1.0 μM) amounts of **3** and **IR792** shows a five- to sixfold higher signal for nanoprobes generated with dye **3** (Fig. 4b). It is interesting to note that a fluorescence background is minimal in the SERRS spectra of the CP- and cyanine-based SERRS nanoprobes (Supplementary Fig. 3). Whereas fluorescence interference would not be expected from CP dyes containing heavy chalcogens that enhance intersystem crossing[36], fluorescence interference could be expected for the cyanine dye **IR792**. In fact, when equimolar amounts of the CP dyes **1** – **3** and **IR792** were incorporated in silica (without gold nanoparticle), **IR792** demonstrated strong fluorescence when excited at 785 nm (50 μW, 1.0 s acquisition time), while minimal fluorescence was observed for CP **1**–**3**. As shown in Fig. 4b and Supplementary Fig. 3, the fluorescence interference of the cyanine dye **IR792** is minimal in its SERRS spectrum. This is due to quenching effects near the surface of the nanoparticle[37].

A concentration series of the as-synthesized SERRS nanoprobes was generated in triplicate manner (Supplementary Fig. 4) to determine the limit of detection (LOD) of both nanoprobes.

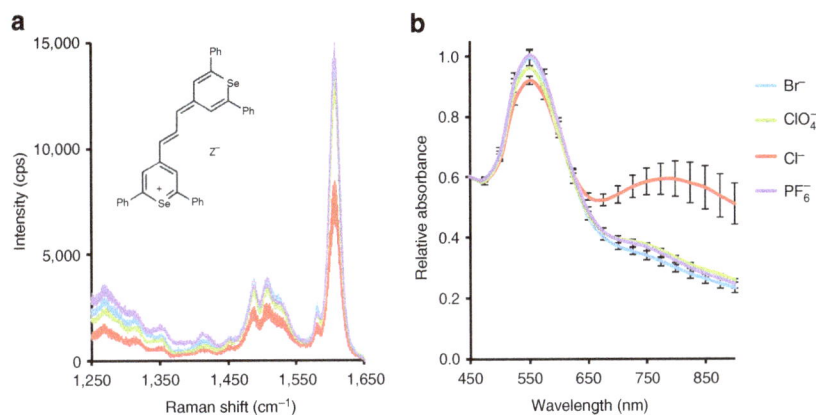

Figure 2 | The effect of the counterion on colloidal stability. (a) The effect of the counterion (Z^-) on SERRS intensity (785 nm, 50 μW, 1.0 s acquisition time, 5 × objective). Inset: Structure of CP dye **1a** (Ph, phenyl). **(b)** Effect of counterion on the colloidal stability of CP-dye **1a**-based SERRS nanoprobes ($n = 3$, error bars represent s.d.; See also Supplementary Fig. 1; Supplementary Table 1).

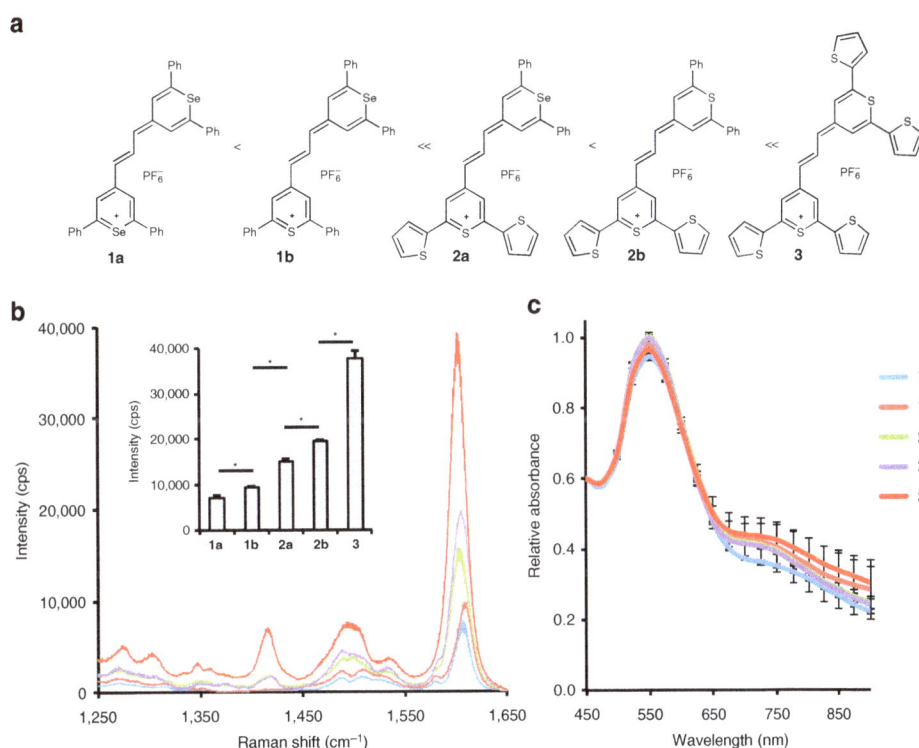

Figure 3 | The SERRS intensity as a function of dye affinity for the gold surface. (a) Molecular structures of the adsorbed CP dyes (**1-3**) arranged by increased number of 2-thienyl substituents (Ph, phenyl). **(b)** SERRS spectra of the CP-based SERRS nanoprobes. The SERRS spectra were baseline corrected to allow proper comparison. (For the non-baseline corrected spectra see Supplementary Fig. 3). Inset: intensity of the 1,600 cm^{-1} peak ($n = 3$; error bars represent s.d., *$P < 0.05$; an unpaired Student's t-test was performed). **(c)** Colloidal stability of the CP-based SERRS nanoprobes as determined by LSPR measurements ($n = 3$; error bars represent s.d.; See also Supplementary Fig. 2; Supplementary Table 1).

Fig. 4c shows the LOD for **IR792**-based nanoprobes to be 1.0 fM, while **3**-based nanoprobes had a 10-fold lower LOD, 100 aM. To our knowledge this is the lowest reported LOD utilizing a biologically relevant NIR excitation source. We also evaluated the serum stability of the **3**-based SERRS nanoprobe. The SERRS nanoprobe was shown to be serum stable (no significant difference between $t = 1$ h and $t = 48$ h) for at least 48 h (Supplementary Fig. 5; Supplementary Methods). This is supported by a study by Thakor *et al.*[38] who have shown that SERS nanoparticles of similar size and composition remain stable *in vivo* for >2 weeks.

***In vivo* comparison of EGFR-targeted 3- or IR792-SERRS nanoprobes.** The ability of our SERRS nanoprobe to delineate tumour tissue *in vivo* was assessed by utilizing CP dye **3** and **IR792**-based SERRS nanoprobes functionalized with an epidermal growth factor receptor (EGFR)-targeting antibody. Equimolar amounts (15 fmol g^{-1}) of these two EGFR-targeted nanoprobes were injected intravenously into athymic nude mice that had been inoculated 2 weeks prior with the EGFR-overexpressing cell line A431 (1×10^6 cells). After 18 h, the skin around the tumour was carefully peeled back and multiplexed Raman imaging of the tumour site and surrounding tissue was

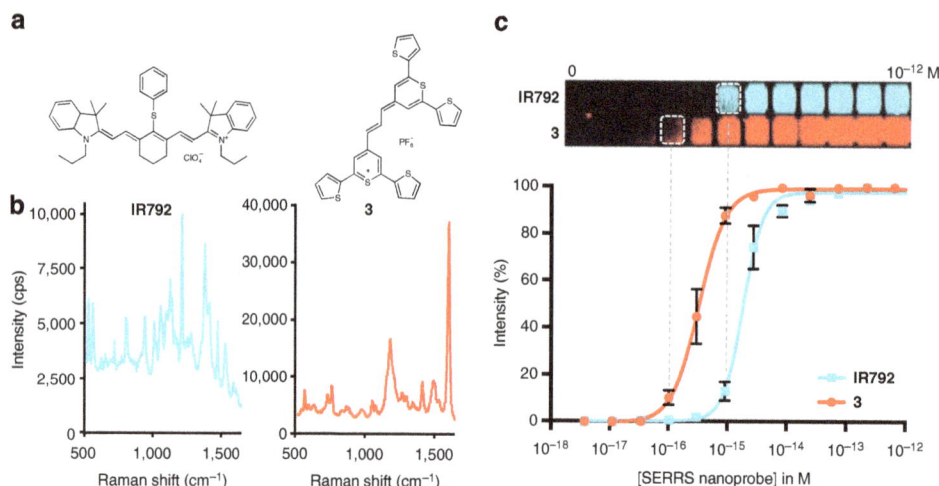

Figure 4 | Comparison of the SERRS signal intensity of the optimized CP-dye 3 versus a widely used resonant dye IR792. (a) Structure of the resonant dye **IR792** and chalcogenopyrylium dye **3**. (b) SERRS intensity of an equimolar amount of an **IR792**-based SERRS nanoprobe and a **3**-based SERRS nanoprobe that were synthesized with an equimolar amount of the dyes. (c) Limits of detection of the **IR792**- (cyan) and **3**- (red) based SERRS nanoprobes were performed in triplicate and determined to be 1.0 fM and 100 aM, respectively (error bars represent s.d.; see also Supplementary Fig. 4).

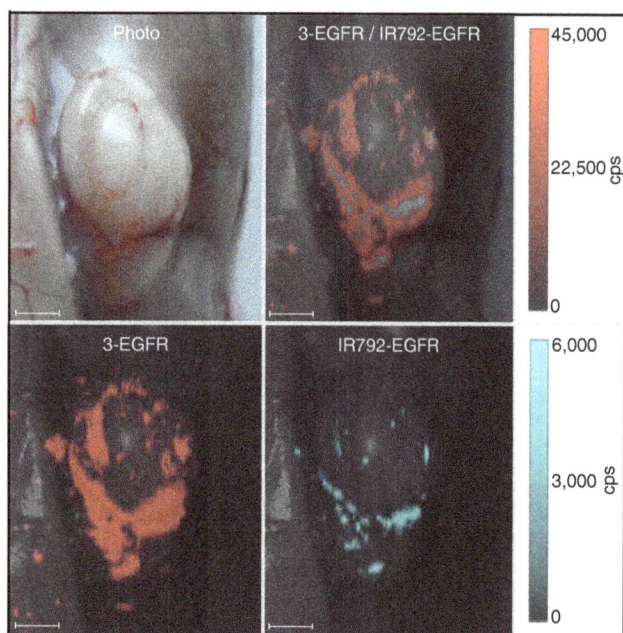

Figure 5 | Comparison between EGFR-targeted IR792- or 3-based SERRS nanoprobes in an A431 tumour xenograft. Female nude mice ($n = 5$) bearing A431 xenograft tumours were injected intravenously via tail vein with an equimolar amount of EGFR-antibody (cetuximab)-conjugated **IR792**- and CP **3**-based SERRS nanoprobes (15 fmol g^{-1} per probe; total injected dose: 30 fmol g^{-1}). After 18 h, the tumours were imaged in situ by Raman (10 mW, 1.5 s acquisition time, 5 × objective). The chalcogenopyrylium dye **3**-based SERRS nanoprobe (red) provided ~3 × higher signal than the **IR792**-based SERRS nanoprobe (cyan; 22.442 versus 7.313 cps cm^{-2}, respectively). All scale bars represent 2.0 mm.

performed (Figs 5 and 6). A Raman map was generated and the signals from the multiplexed SERRS nanoprobes were deconvoluted by applying a direct classical least square algorithm (DCLS)[7]. The SERRS signal from both nanoprobes was more intense for the tumour site than for the surrounding tissue, showing that the EGFR-targeted SERRS nanoprobes had selectively localized at the tumour site. The SERRS signal

intensity in the tumour was approximately 3 × higher for the **3**-based SERRS nanoprobe than for the otherwise identical **IR792**-based SERRS nanoprobe. Ex vivo multiplexed Raman imaging of the tumour showed Raman signal of the EGFR-targeted SERRS nanoprobes throughout the tumour with the exception of a hypointense Raman region in the centre of the tumour. Hematoxyolin and eosin and immunohistochemical staining for EGFR was performed (Supplementary Methods) and revealed that the hypointense Raman region corresponded with an area of necrosis, which explains the lack of SERRS nanoprobe accumulation and decreased Raman signal. In addition, to validate EGFR targeting, we injected A431-tumour-bearing mice with cetuximab (50 pmol g^{-1}) 3 h before injection with the EGFR-targeted SERRS nanoprobes. Preblocking of EGFR by cetuximab resulted in decreased accumulation of the EGFR-targeted SERRS nanoprobes within the tumours of animals that were injected with cetuximab before EGFR-targeted SERRS nanoprobe injection as compared with animals that were injected with EGFR-targeted SERRS nanoprobes and were not preinjected with cetuximab (Supplementary Fig. 6).

Discussion

Effective biomedical imaging requires low limits of detection and high specificity for biological targets. Raman imaging has surfaced as an optical imaging modality that has the promise to enable both. While the Raman effect is relatively weak (1 in 10^7 photons), the Raman scattering cross-section of a molecule can be massively amplified by noble metal surfaces[4–6]. Here we demonstrated that rational SERRS reporter design afforded SERRS nanoprobes with unprecedented limits of detection: 100 aM. This is, to the best of our knowledge, the lowest reported limit of detection at near-real-time detection (≤ 2.0 s acquisition times) for SERRS nanoprobes that are compatible with a NIR light source. As a comparison, non-resonant SERS nanoprobes are in the $0.1 - 1.0$ pM range (1,000 − 10,000-fold less sensitive)[2], while reported detection limits of SERRS nanoprobes are > 17 fM at near-real-time detection[39]. Others have reported a 0.4 fM detection limit; however, this was acquired through cumulative data acquisition with an acquisition time ≥ 60 s, which is not practical for biomedical imaging applications[35].

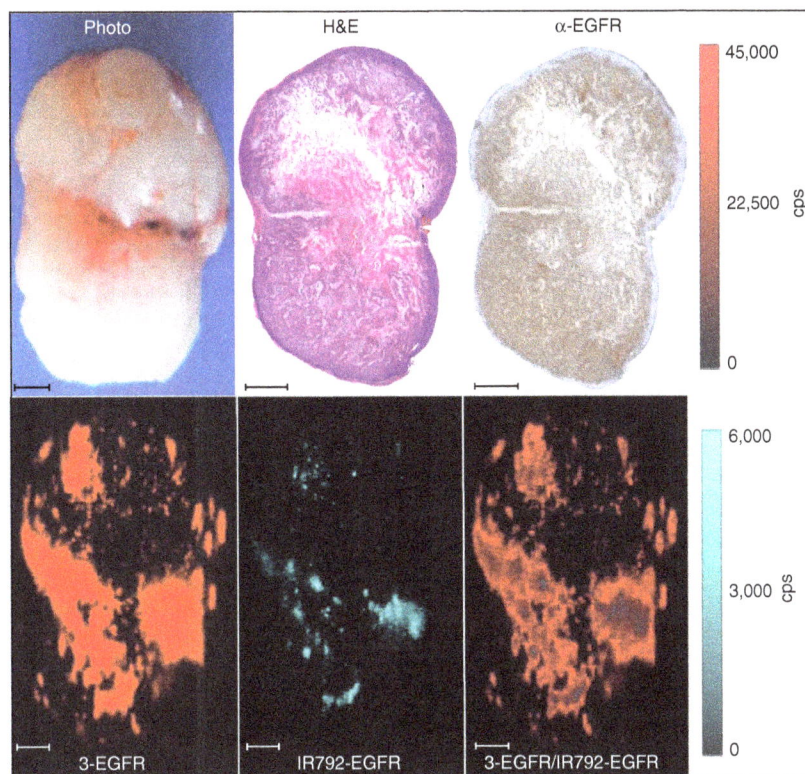

Figure 6 | Immunohistochemistry and *ex vivo* Raman imaging of the A431 tumour. The excised tumour was scanned by Raman imaging (10 mW, 1.5 s acquisition time, 5 × objective), fixed in 4% paraformaldehyde, and processed for hematoxyolin and eosin staining and anti-EGFR immunohistochemistry. With the exception of a hypointense Raman region in the centre of the tumour, the tumour homogenously expressed EGFR and the EGFR-targeted SERRS nanoprobes had accumulated throughout the tumour. The hypointense Raman area corresponds to a highly necrotic region within the tumour, which explains the lack of SERRS nanoprobe accumulation and decreased Raman signal. All scale bars represent 1.0 mm.

We believe the unprecedented limit of detection of our novel SERRS nanoprobe is due to several factors. First, we demonstrate that rational design and optimization of the SERRS reporter is important to achieve efficient 'loading' on the nanoparticle. Our results demonstrate that the counterion and gold surface affinity are important considerations. For instance, while the chaotropic PF_6^- anions stabilized the dye–nanoparticle system during silica shell formation in ethanol, the system becomes more destabilized with Cl^- (more kosmotropic) ions present. Chloride-induced aggregation of colloidal dispersions in relation to SERS has been studied. Natan and colleagues[40] demonstrated that the strongest enhancements were obtained from aggregates with effective diameters of <200 nm and aggregates with sizes >200 nm did not generate appreciable SERS intensities. The aggregates that were induced by the chloride counterion in our system were >200 nm (Supplementary Fig. 1), which might explain the reduced SERRS signal when chloride is used as a counterion. Others have shown that the kosmotropic chloride ion could induce reorientation of the dye on the surface, which could also contribute to the reduced SERRS intensities[41]. However, while we did observe a decrease in the SERRS signal intensity when chloride is present, we did not find any appreciable differences between the Raman spectra of the dyes when different counterions were used, which would have been expected if the molecule had reoriented on the surface. Since the most chaotropic counterion, PF_6^-, induced the least aggregation and generated robust SERRS signal intensities, we used PF_6^- as a counterion.

Next, we showed that an increase in affinity of the SERRS reporter for the gold nanoparticle surface via incorporation of 2-thienyl functional groups considerably increased the SERRS

signal without inducing aggregation. Others have reported the functionalization of NIR dyes with thiol or lipoic acid functional groups. In contrast to a 2-thienyl substituent, thiol and lipoic acid functional groups offer no benefit to the optical properties of the dye, and as a tether, do not allow the dye to be as close to the gold surface[34]. Moreover, based on the absorption spectra of reported lipoic acid-modified cyanine dye–gold nanoparticle conjugates, it is clear that lipoic acid-modified dyes promote aggregation[21].

Finally, the strategy chosen to stabilize the SERRS nanoprobe is a key factor. Others have reported using either surfactants or thiolated polymers to stabilize their SERRS nanoparticles[35,39,42]. However, such stabilizing agents compete with the SERRS reporter for the surface of the nanoparticle, which leads to lower SE(R)RS-signal. We achieved very low limits of detection by using a primerless silication procedure in which the silica not only served as a stabilizing agent, but also as a matrix to contain our optimized CP-based SERRS reporter. Since silica has much lower affinity for the gold than the applied SERRS reporters, attomolar limits of detection were achieved.

The CP dyes represent a new class of SERRS reporters. Selection of the right combination of chaotropic counterions and increased affinity of the SERRS reporter for the gold nanoparticle's surface produces stable SERRS nanoprobes with exceptionally low limits of detection (attomolar range). The low limit of detection (that is, close to single nanoparticle detection) in combination with the high resolution of Raman imaging enables highly sensitive and specific near real-time tumour delineation and, as a result of the fingerprint-like spectra of the different SERRS nanoprobes, can offer multiplexed disease marker detection *in vivo*.

Methods

Materials. *Chalcogenopyrylium dye synthesis and characterization.* All reactions were performed open to air. Concentration *in vacuo* was performed on a rotary evaporator. NMR spectra were recorded at 300 or 500 MHz for ^1H and at 75.5 MHz for ^{13}C with residual solvent signal as internal standard. If a mixture of CD_2Cl_2 and CD_3OD was used to acquire ^1H NMR, the peak for CH_2Cl_2 was used as the internal standard. UV/VIS-near-IR spectra were recorded in quartz cuvettes with a 1-cm path length. Melting points were determined with a capillary melting point apparatus and are uncorrected. Non-hygroscopic compounds have a purity of ≥95% as determined by elemental analyses for C, H and N. Experimental values of C, H and N are within 0.3% of theoretical values. ^{13}C NMR was not recorded for pyrylium dyes due to limited solubility in common NMR solvents. Chalcogenopyranones[25] **4** and **6** were made according to literature procedures as were 4-methylchalcogenopyrylium compounds **5** and **7** (refs 22,25–27).

*Preparation of 4-(2,6-di(thiophen-2-yl)-4H-thiopyran-4ylidene)acetaldehyde (**8**, Y = S, R₂ = 2-thienyl).* 4-Methyl-2,6-di(thiophen-2-yl)thiopyrylium hexafluorophosphate (0.350 g, 0.833 mmol), *N,N*-dimethylthioformamide (0.213 ml, 2.50 mmol) and Ac_2O (3.0 ml) were combined in a small round-bottom flask and heated at 95 °C for 1 h. After cooling to ambient temperature an additional portion of Ac_2O (2.0 ml) was added and the solution diluted with ether. The formed iminium salt was allowed to precipitate in the freezer overnight, and then isolated by filtration to yield a bright orange solid. This solid was dissolved in CH_3CN (3.0 ml) and saturated aqueous $NaHCO_3$ (3.0 ml) was added. This mixture was heated to 80 °C over 15 min, and kept at that temperature for 30 min. After diluting with H_2O (30 ml) the product was extracted with CH_2Cl_2 (3 × 50 ml), dried with Na_2SO_4 and purified on SiO_2 with a 10% $EtOAc/CH_2Cl_2$ eluent (R_f = 0.71) to yield a yellow oil that was recrystallized in CH_2Cl_2/hexanes to yield 0.219 g (87%) of a yellow crystalline solid, mp 143–144 °C: ^1H NMR (500 MHz, $CDCl_3$) δ 9.84 (d, 1 H, *J* = 6.0 Hz), 8.26 (s, 1 H), 7.45-7.39 (m, 4 H), 7.13-7.11 (m, 2 H), 6.88 (s, 1 H), 5.72 (d, 1 H, *J* = 6.5 Hz); ^{13}C NMR (75.5 MHz, $CDCl_3$) δ 188.05, 146.43, 139.36, 139.07, 137.33, 136.65, 128.16, 127.78, 127.58, 126.30, 126.01, 122.48, 117.63, 117.48; HRMS (ESI) *m/z* 302.9971 (calculated for $C_{15}H_{10}OS_3 + H^+$: 302.9967).

*Preparation of 4-(2,6-diphenyl-4H-selenopyran-4ylidene)acetaldehyde (**8**, Y = Se, R₂ = Ph).* 4-Methyl-2,6-di(phenyl)selenopyrylium hexafluorophosphate (0.200 g, 0.439 mmol), *N,N*-dimethylthioformamide (0.112 ml, 1.32 mmol) and Ac_2O (4.0 ml) were added to a round-bottom flask and heated at 95 °C for 90 min. After cooling to room temperature CH_3CN (4.0 ml) was added and the product precipitated by addition of ether and chilling overnight in the freezer. The iminium salt was isolated by filtration, and hydrolysed by dissolving in CH_3CN (4.0 ml), adding saturated aqueous $NaHCO_3$ (4.0 ml) and heating the mixture to 80 °C over a 15 min period. The reaction was maintained at this temperature for 30 min, after which the reaction was diluted with H_2O (50 ml), the product extracted with CH_2Cl_2 (3 × 30 ml), dried with $MgSO_4$ and after concentration purified on SiO_2 with first a CH_2Cl_2 and then a 10% $EtOAc/CH_2Cl_2$ (R_f = 0.70) eluent to yield 0.122 g (82%) of an orange oil: ^1H NMR (500 MHz, $CDCl_3$) δ 10.11 (d, 1 H, *J* = 10.5 Hz), 8.32 (s, 1 H), 7.62-7.46 (m, 9 H), 7.00 (s, 1 H), 5.88 (d, 1 H, *J* = 11.0 Hz); ^{13}C NMR (75.5 MHz, $CDCl_3$) δ 188.69, 148.44, 147.01, 146.03, 138.65, 138.40, 129.96, 129.90, 129.14, 126.62, 126.44, 126.37, 126.31, 125.67, 120.57, 120.48; HRMS (ESI) *m/z* 339.0292 (calculated for $C_{19}H_{14}O^{80}Se + H^+$: 339.0283).

*Preparation of 4-(3-(2,6-diphenyl-4H-selenopyran-4-ylidene)prop-1-enyl)-2,6-diphenylselenopyrylium (**1a**) (CAS Registry Number: 51848-65-8).* **PF_6^-** 4-Methyl-2,6-di(phenyl)selenopyrylium hexafluorophosphate (0.190 g, 0.417 mmol), 4-(2,6-diphenyl-4H-selenopyran-4-ylidene)acetaldehyde (0.155 g, 0.459 mmol) and Ac_2O (3.0 ml) were combined in a round-bottom flask and heated at 105 °C for 10 min. The reaction was cooled to ambient temperature, precipitated with ether and the collected solid recrystallized from CH_3CN/ether to yield 0.278 g (86%) of a golden-green solid: ^1H NMR (500 MHz, CD_2Cl_2) δ 8.59 (t, 1 H, *J* = 13.5 Hz), 8.40–7.80 (br s, 4 H), 7.71 (d, 8 H, *J* = 7.0 Hz), 7.63-7.59 (m, 12 H), 6.85 (d, 2 H, *J* = 13.0 Hz); analytically calculated for $C_{37}H_{27}Se_2 \cdot PF_6$: C, 57.38; H, 3.51; F, 14.72. Found: C, 57.34; H, 3.48; F, 14.76; LRMS (ESI) *m/z* 631.2 (calculated for $C_{37}H_{27}^{80}Se_2$: 631.0); λ_{max} (CH_2Cl_2) = 806 nm, ε = 2.5 × 10^5 M^{-1} cm^{-1}.

ClO_4^- 4-Methyl-2,6-di(phenyl)selenopyrylium perchlorate (50.0 mg, 0.122 mol), 4-(2,6-diphenyl-4H-selenopyran-4-ylidene)acetaldehyde (81.4 mg, 0.241 mmol) and Ac_2O (2.0 ml) were treated as described for the PF_6^- salt to yield 82.0 mg (90%) of a golden-green solid: ^1H NMR (500 MHz, 1:1 CD_2Cl_2:CD_3OD) δ 8.77 (t, 1 H, *J* = 13.5 Hz), 8.60–7.80 (m, 4 H), 7.71 (d, 8 H, *J* = 7.0 Hz), 7.66–7.54 (m, 12 H), 6.83 (d, 2 H, *J* = 14.0 Hz); analytically calculated for $C_{37}H_{27}Se_2 \cdot ClO_4$: C, 60.96; H, 3.73. Found: C, 60.69; H, 3.83; λ_{max} (CH_2Cl_2) = 806 nm, ε = 2.5 × 10^5 M^{-1} cm^{-1}.

Cl^- The hexafluorophosphate salt (50 mg) was converted to the chloride salt by treating with Amberlite IRA-400 chloride form (200 mg) in a 1:1 CH_2Cl_2:MeOH mixture (3.0 ml). This process was repeated two more times after which the product was dissolved in CH_2Cl_2, washed with water, the organic layer dried with Na_2SO_4, filtered over Celite and after concentration recrystallized from CH_3CN/ether to yield a bronze solid: ^1H NMR (500 MHz, 3:1 CD_3OD:CD_2Cl_2) δ 8.87 (t, 1 H, *J* = 13.0 Hz), 8.40–7.80 (m, 4 H), 7.70 (d, 8 H, *J* = 7.0 Hz), 7.56-7.50 (m, 12 H), 6.84 (d, 2 H, *J* = 13.0 Hz); analytically calculated for $C_{37}H_{27}Se_2 \cdot Cl \cdot 4/3H_2O$: C, 64.50; H, 4.34; Cl, 5.15. Found: C, 64.54; H, 4.42; Cl, 4.98; λ_{max} (CH_2Cl_2) = 806 nm, ε = 2.3 × 10^5 M^{-1} cm^{-1}.

Br^- The hexafluorophosphate salt (50 mg) was converted to the bromide salt by treating with Amberlite IRA-400 bromide form (200 mg) in a 1:1 CH_2Cl_2:MeOH mixture (3.0 ml). This process was repeated two more times after which the product was dissolved in CH_2Cl_2, washed with water, the organic layer dried with Na_2SO_4, filtered over Celite and after concentration recrystallized from CH_3CN/ether to yield a bronze solid: ^1H NMR (500 MHz, 3:2 CD_2Cl_2:CD_3OD) δ 8.79 (t, 1 H, *J* = 13.5 Hz), 8.40–7.80 (br s, 4 H), 7.71 (d, 8 H, *J* = 7.0 Hz), 7.60–7.54 (m, 12 H), 8.50 (d, 2 H, *J* = 13.0 Hz); analytically calculated for $C_{37}H_{27}Se_2 \cdot Br \cdot H_2O$: C, 61.09; H, 4.02; Br, 10.98. Found: C, 61.08; H, 3.89; Br, 10.77; λ_{max} (CH_2Cl_2) = 806 nm, ε = 2.3 × 10^5 M^{-1} cm^{-1}.

*Preparation of 4-(3-(2,6-diphenyl-4H-thiopyran-4-ylidene)prop-1-enyl)-2,6-diphenylselenopyrylium hexafluorophosphate (**1b**) (CAS Registry Number: 79054-92-5).* 4-Methyl-2,6-di(phenyl)thiopyrylium hexafluorophosphate (0.128 g, 0.312 mmol), 4-(2,6-diphenyl-4H-selenopyran-4-ylidene)acetaldehyde (0.157 g, 0.344 mmol) and Ac_2O (2.0 ml) were combined in a round-bottom flask and heated at 105 °C for 10 min. The reaction was cooled to ambient temperature, CH_3CN (2.0 ml) was added and ether was used to precipitate product from solution to yield 0.196 g (86%) of a copper–bronze solid: ^1H NMR (500 MHz, CD_2Cl_2) δ 8.54 (t, 1 H, *J* = 13.0 Hz), 8.20-7.80 (br s, 4 H), 7.78 (d, 4 H, *J* = 8.0 Hz), 7.70 (d, 4 H, *J* = 7.5 Hz), 7.66–7.58 (m, 12 H), 6.78 (d, 2 H, *J* = 13.5 Hz); analytically calculated for $C_{39}H_{34}O_3Se_2 \cdot PF_6$: C, 61.08; H, 3.74. Found: C, 61.10; H, 3.68; LRMS (ESI) *m/z* 583.3 (calculated for $C_{37}H_{27}S^{80}Se$: 583.1); λ_{max} (CH_2Cl_2) = 784 nm, ε = 2.0 × 10^5 M^{-1} cm^{-1}.

*Preparation of 4-(3-(2,6-dithiophen-2-yl-4H-thiopyran-4-ylidene)prop-1-enyl)-2,6-diphenylselenopyrylium (**2a**).* 4-Methyl-2,6-di(phenyl)selenopyrylium hexafluorophosphate (0.102 g, 0.225 mmol), 4-(2,6-(thiophen-2-yl)-4H-thiopyran-4-ylidene)acetaldehyde (75.0 mg, 0.248 mmol) and Ac_2O (3.0 ml) were combined in a round-bottom flask and heated at 105 °C for 5 min. The reaction was cooled to ambient temperature, precipitated with ether and the collected solid recrystallized from CH_3CN/ether to yield 0.145 g (87%) of a bronze solid, mp 229–231 °C: ^1H NMR [500 MHz, CD_2Cl_2] δ 8.46 (t, 1 H, *J* = 13.0 Hz), 7.71–7.58 (m, 18 H), 7.26 (t, 2 H, *J* = 4.0 Hz), 6.77 (d, 1 H, *J* = 13.0 Hz), 6.70 (d, 1 H, *J* = 14.0 Hz); analytically calculated for $C_{33}H_{23}S_3Se \cdot PF_6$: C, 53.59; H, 3.13; F, 15.41. Found: C, 53.79; H, 3.13; F, 15.19; HRMS (ESI) *m/z* 595.0125(calculated for $C_{33}H_{23}S_3^{80}Se$: 595.0122); λ_{max} (CH_2Cl_2) = 810 nm, ε = 2.5 × 10^5 M^{-1} cm^{-1}.

*Preparation of 4-(3-(2,6-dithiophen-2-yl-4H-thiopyran-4-ylidene)prop-1-enyl)-2,6-diphenylthiopyrylium hexafluorophosphate (**2b**).* 4-Methyl-2,6-diphenylthiopyrylium hexafluorophosphate (30.0 mg, 73.0 μmol), 4-(2,6-(thiophen-2-yl)-4H-thiopyran-4-ylidene)acetaldehyde (24.4 mg, 81.0 μmol) and Ac_2O (1.0 ml) were combined in a round-bottom flask and heated at 105 °C for 5 min. The reaction was cooled to ambient temperature, CH_3CN (4.0 ml) was added and ether was used to precipitate product from solution to yield 45.0 mg (88%) of a bronze solid, mp > 260 °C: ^1H NMR (500 MHz, CD_2Cl_2) δ 8.44 (t, 1 H, *J* = 13.0 Hz), 8.40–7.80 (br s, 4 H), 7.78 (d, 4 H, *J* = 7.0 Hz), 7.67–7.59 (m, 10 H), 7.24 (t, 2 H, *J* = 4.5 Hz), 6.71 (d, 1 H, *J* = 13.0 Hz), 6.63 (d, 1 H, *J* = 13.5 Hz); analytically calculated for $C_{33}H_{23}S_4 \cdot PF_6$: C, 57.21; H, 3.35. Found: C, 56.97; H, 3.36; HRMS (ESI) *m/z* 547.0674 (calculated for $C_{33}H_{23}S_4$: 547.0677); λ_{max} (CH_2Cl_2) = 789 nm, ε = 2.2 × 10^5 M^{-1} cm^{-1}.

*Preparation of 4-(3-(2,6-dithiophen-2-yl-4H-thiopyran-4-ylidene)prop-1-enyl)-(2,6- dithiophen-2-yl)thiopyrylium hexafluorophosphate (**3**) (CAS Registry Number: 95410-36-9).* 4-Methyl-2,6-di(thiophen-2-yl)thiopyrylium hexafluorophosphate (11.0 mg, 26.2 μmol), 4-(2,6-(thiophen-2-yl)-4H-thiopyran-4-ylidene)acetaldehyde (9.5 mg, 31.4 μmol) and Ac_2O (1.0 ml) were combined in a round-bottom flask and heated at 105 °C for 5 min. The reaction was cooled to ambient temperature, CH_2Cl_2 (2.0 ml) was added and ether was used to precipitate product from solution to yield 17.8 mg (94%) of a bronze solid, mp > 260 °C: ^1H NMR (500 MHz, CD_3CN) δ 8.32 (t, 1 H, *J* = 13.5 Hz), 7.68 (d, 2 H, *J* = 4 Hz), 7.56 (br. s, 4 H) 7.14 (t, 4 H, *J* = 4.5 Hz), 6.48 (d, 2 H, *J* = 13.0 Hz); analytically calculated for $C_{29}H_{19}S_6 \cdot PF_6$: C, 49.42; H, 2.72. Found: C, 49.19; H, 2.79; HRMS (ESI) *m/z* 558.9805 (calculated for $C_{29}H_{19}S_6$: 558.9806); λ_{max} (CH_2Cl_2) = 813 nm, ε = 2.8 × 10^5 M^{-1} cm^{-1}. Spectral data agree with published spectra[43].

SERRS nanoprobe synthesis. Gold nanoparticles were synthesized by adding 7.5 ml 1% (w/v) sodium citrate to 1.0 l boiling 0.25 mM $HAuCl_4$. The as-synthesized gold nanoparticles were concentrated by centrifugation (10 min, 7,500 × *g*, 4 °C) and dialysed overnight (3.5 kDa MWCO; 51 18.2 MΩ cm). The dialysed gold nanoparticles (100 μl; 2.0 nM) were added to 1,000 μl absolute ethanol in the presence of 30 μl 99.999% tetraethyl orthosilicate (Sigma Aldrich), 15 μl 28% (v/v) ammonium hydroxide (Sigma Aldrich) and 1 μl CP dye (1 – 10 mM) in DMF. After shaking (375 rpm) for 25 min at ambient conditions in a plastic container, the SERRS nanoprobes were collected by centrifugation, washed with ethanol and redispersed in water to yield 2.0 nM SERRS nanoprobes.

SERRS nanoprobe characterization. The as-synthesized SERRS nanoprobes were characterized by transmission electron microscopy (TEM; JEOL 1200ex-II, 80 kV, 150,000 × magnification) to study the SERRS nanoprobe structural morphology. The size and concentration of the SERRS nanoprobes were determined on a Nanoparticle Tracking Analyzer (Malvern Instruments, Malvern, UK). Absorption spectra to determine possible nanoparticle aggregation (typically detectable at wavelengths > 600 nm) were measured on an M1000Pro spectrophotometer

(Tecan Systems Inc., San Jose, CA). Finally Raman spectra were acquired on a Renishaw InVIA system equipped with a 785-nm laser (Renishaw Inc., Hoffman Estates, IL). All measurements were performed at a laser power of $50\,\mu W$ (1.0 s acquisition time, $5\times$ objective).

SERRS nanoprobe limit of detection. SERRS nanoprobes were synthesized as described above in the presence of an equimolar $(1.0\,\mu M)$ amount of **3** or **IR792**. SERRS imaging to determine the limit of detection was performed at 100 mW (2.0 s acquisition time (StreamLine), $5\times$ objective) on a phantom that consisted of a serially diluted **IR792**- or CP dye **3**-based SERRS nanoprobe redispersed in $10\,\mu l$ water (concentration range $3,000 - 0.003\,fM$; $n = 3$). The Raman maps were generated using WiRE 3.4 software (Renishaw) by applying a DCLS algorithm. The Raman images were analysed with ImageJ software and plotted in GraphPad Prism (GraphPad Software Inc., La Jolla, CA).

Animal studies. All animal experiments were approved by the Institutional Animal Care and Use Committees of Memorial Sloan Kettering Cancer Center.

In vivo comparison of EGFR-targeted 3- or IR792-SERRS nanoprobes. Eight-to-ten week-old female athymic nude mice ($n = 5$; Hsd:Athymic Nude-$Foxn1^{nu}$; Harlan Laboratories) were subcutaneously inoculated with the EGFR-over-expressing cell line A431 (1×10^6 cells; ATCC CRL-1555). After 2 weeks, the mice were injected with an equimolar amount ($15\,fmol\,g^{-1}$) of EGFR-targeted **IR792**- and **3**-based SERRS nanoprobes. The EGFR-targeted SERRS nanoprobes were synthesized as described above in the presence of an equimolar $(1.0\,\mu M)$ amount of **3** or **IR792**. The as-synthesized SERRS nanoprobes were subsequently functionalized with sulfhydryl groups by heating the SERRS nanoprobes in 5 ml 2% (v/v) mercaptotrimethoxysilane in ethanol at $70\,°C$ for 2 h. The sulfhydryl-functionalized SERRS nanoprobes were washed with water, redispersed in 10 mM MES buffer (pH 7.1), and conjugated to an EGFR-targeting antibody (cetuximab; Genentech, South San Francisco, CA) with a 4,000 Da heterobifunctional maleimide/N-hydroxysuccinimide polyethylene glycol linker[44]. Eighteen hours later, the mice were sacrificed by CO_2 asphyxiation. The tumour was exposed and scanned by Raman imaging (10 mW, 1.5 s acquisition time (StreamLine), $5\times$ objective). The Raman maps were generated using WiRE 3.4 software (Renishaw) by applying a DCLS algorithm.

References

1. Wang, Y., Yan, B. & Chen, L. SERS tags: novel optical nanoprobes for bioanalysis. *Chem. Rev.* **113,** 1391–1428 (2013).
2. Kircher, M. F. *et al.* A brain tumor molecular imaging strategy using a new triple-modality MRI-photoacoustic-Raman nanoparticle. *Nat. Med.* **18,** 829–834 (2012).
3. Qian, X. *et al.* In vivo tumor targeting and spectroscopic detection with surface-enhanced Raman nanoparticle tags. *Nat. Biotechnol.* **26,** 83–90 (2008).
4. Kneipp, K. *et al.* Single molecule detection using surface-enhanced Raman scattering (SERS). *Phys. Rev. Lett.* **78,** 1667–1670 (1997).
5. Nie, S. & Emory, S. R. Probing single molecules and single nanoparticles by surface-enhanced Raman scattering. *Science* **275,** 1102–1106 (1997).
6. Jeanmaire, D. L. & Van Duyne, R. P. Surface Raman spectroelectrochemistry. Part I. Heterocyclic, aromatic, and aliphatic amines adsorbed on the anodized silver electrode. *J. Electroanal. Chem. Interfacial Electrochem.* **84,** 1–20 (1977).
7. Zavaleta, C. L. *et al.* Multiplexed imaging of surface enhanced Raman scattering nanotags in living mice using noninvasive Raman spectroscopy. *Proc. Natl Acad. Sci. USA* **106,** 13511–13516 (2009).
8. Faulds, K., Jarvis, R., Smith, W. E., Graham, D. & Goodacre, R. Multiplexed detection of six labeled oligonucleotides using surface enhanced resonance Raman scattering (SERRS). *Analyst* **133,** 1505–1512 (2008).
9. Gellner, M., Koempe, K. & Schluecker, S. Multiplexing with SERS labels using mixed SAMs of Raman reporter molecules. *Anal. Bioanal. Chem.* **394,** 1839–1844 (2009).
10. Craig, D. *et al.* Confocal SERS mapping of glycan expression for the identification of cancerous cells. *Anal. Chem.* **86,** 4775–4782 (2014).
11. Gracie, K. *et al.* Simultaneous detection and quantification of three bacterial meningitis pathogens by SERS. *Chem. Sci.* **5,** 1030–1040 (2014).
12. McAughtrie, S., Lau, K., Faulds, K. & Graham, D. 3D optical imaging of multiple SERS nanotags in cells. *Chem. Sci.* **4,** 3566–3572 (2013).
13. Sha, M. Y., Xu, H., Natan, M. J. & Cromer, R. Surface-enhanced Raman scattering tags for rapid and homogeneous detection of circulating tumor cells in the presence of human whole blood. *J. Am. Chem. Soc.* **130,** 17214–17215 (2008).
14. Cao, Y. C., Jin, R. & Mirkin, C. A. Nanoparticles with Raman spectroscopic fingerprints for DNA and RNA detection. *Science* **297,** 1536–1540 (2002).
15. Yuen, J. M., Shah, N. C., Walsh, Jr. J. T., Glucksberg, M. R. & Van Duyne, R. P. Transcutaneous glucose sensing by surface-enhanced spatially offset Raman spectroscopy in a rat model. *Anal. Chem.* **82,** 8382–8385 (2010).
16. McQueenie, R. *et al.* Detection of inflammation in vivo by surface-enhanced Raman scattering provides higher sensitivity than conventional fluorescence imaging. *Anal. Chem.* **84,** 5968–5975 (2012).
17. Graham, D., Mallinder, B. J., Whitcombe, D. & Smith, W. E. Surface enhanced resonance Raman scattering (SERRS) - a first example of its use in multiplex genotyping. *Chemphyschem* **2,** 746–748 (2001).
18. Hildebrandt, P. & Stockburger, M. Surface-enhanced resonance Raman spectroscopy of Rhodamine 6G adsorbed on colloidal silver. *J. Phys. Chem.* **88,** 5935–5944 (1984).
19. Dieringer, J. A. *et al.* Surface-enhanced Raman excitation spectroscopy of a single rhodamine 6G molecule. *J. Am. Chem. Soc.* **131,** 849–854 (2009).
20. Harmsen, S. *et al.* Surface-enhanced resonance Raman scattering nanostars for high-precision cancer imaging. *Sci. Transl. Med.* **7,** 271ra277 (2015).
21. Samanta, A. *et al.* Ultrasensitive near-infrared Raman reporters for SERS-based in vivo cancer detection. *Angew. Chem. Int. Ed. Engl.* **50,** 6089–6092 (2011).
22. Detty, M. R. & Murray, B. J. Telluropyrylium dyes. 1. 2,6-Diphenyltelluropyrylium dyes. *J. Org. Chem.* **47,** 5235–5239 (1982).
23. Stiles, P. L., Dieringer, J. A., Shah, N. C. & Van Duyne, R. P. Surface-enhanced Raman spectroscopy. *Annu. Rev. Anal. Chem.* **1,** 601–626 (2008).
24. Detty, M. R. *et al.* Electron transport in 4H-1,1-dioxo-4-(dicyanomethylidene)thiopyrans. Investigation of x-ray structures of neutral molecules, electrochemical reduction to the anion radicals, and absorption properties and epr spectra of the anion radicals. *J. Org. Chem.* **60,** 1674–1685 (1995).
25. Leonard, K., Nelen, M., Raghu, M. & Detty, M. R. Chalcogenopyranones from disodium chalcogenide additions to 1,4-pentadiyn-3-ones. The role of enol ethers as intermediates. *J. Heterocycl. Chem.* **36,** 707–717 (1999).
26. Detty, M. R., McKelvey, J. M. & Luss, H. R. Tellurapyrylium dyes. 2. The electron-donating properties of the chalcogen atoms to the chalcogenapyrylium nuclei and their radical dications, neutral radicals, and anions. *Organometallics* **7,** 1131–1147 (1988).
27. Panda, J., Virkler, P. R. & Detty, M. R. A comparison of linear optical properties and redox properties in chalcogenopyrylium dyes bearing ortho-substituted aryl substituents and tert-butyl substituents. *J. Org. Chem.* **68,** 1804–1809 (2003).
28. Stoeber, W., Fink, A. & Bohn, E. Controlled growth of monodisperse silica spheres in the micron size range. *J. Colloid Interface Sci.* **26,** 62–69 (1968).
29. Liz-Marzán, L. M., Giersig, M. & Mulvaney, P. Synthesis of nanosized gold – silica core – shell particles. *Langmuir* **12,** 4329–4335 (1996).
30. Mulvaney, S. P., Musick, M. D., Keating, C. D. & Natan, M. J. Glass-coated, analyte-tagged nanoparticles: a new tagging system based on detection with surface-enhanced Raman scattering. *Langmuir* **19,** 4784–4790 (2003).
31. Bouit, P.-A. *et al.* A "cyanine-cyanine" salt exhibiting photovoltaic properties. *Org. Lett.* **11,** 4806–4809 (2009).
32. Detty, M. R., Merkel, P. B., Hilf, R., Gibson, S. L. & Powers, S. K. Chalcogenapyrylium dyes as photochemotherapeutic agents. 2. Tumor uptake, mitochondrial targeting, and singlet-oxygen-induced inhibition of cytochrome c oxidase. *J. Med. Chem.* **33,** 1108–1116 (1990).
33. Haekkinen, H. The gold-sulfur interface at the nanoscale. *Nat. Chem.* **4,** 443–455 (2012).
34. Mahajan, S., Baumberg, J. J., Russell, A. E. & Bartlett, P. N. Reproducible SERRS from structured gold surfaces. *Phys. Chem. Chem. Phys.* **9,** 6016–6020 (2007).
35. von Maltzahn, G. *et al.* SERS-coded gold nanorods as a multifunctional platform for densely multiplexed near-infrared imaging and photothermal heating. *Adv. Mater.* **21,** 3175–3180 (2009).
36. Detty, M. R. & Merkel, P. B. Chalcogenapyrylium dyes as potential photochemotherapeutic agents. Solution studies of heavy atom effects on triplet yields, quantum efficiencies of singlet oxygen generation, rates of reaction with singlet oxygen, and emission quantum yields. *J. Am. Chem. Soc.* **112,** 3845–3855 (1990).
37. Dulkeith, E. *et al.* Gold nanoparticles quench fluorescence by phase induced radiative rate suppression. *Nano Lett.* **5,** 585–589 (2005).
38. Thakor, A. S. *et al.* The fate and toxicity of Raman-active silica-gold nanoparticles in mice. *Sci. Transl. Med.* **3,** 79ra33 (2011).
39. Jokerst, J. V., Cole, A. J., Van de Sompel, D. & Gambhir, S. S. Gold nanorods for ovarian cancer detection with photoacoustic imaging and resection guidance via Raman imaging in living mice. *ACS Nano* **6,** 10366–10377 (2012).
40. Freeman, R. G., Bright, R. M., Hommer, M. B. & Natan, M. J. Size selection of colloidal gold aggregates by filtration: effect on surface-enhanced Raman scattering intensities. *J. Raman Spectrosc.* **30,** 733–738 (1999).
41. Grochala, W., Kudelski, A. & Bukowska, J. Anion-induced charge-transfer enhancement in SERS and SERRS spectra of Rhodamine 6G on a silver electrode: how important is it? *J. Raman Spectrosc.* **29,** 681 (1998).
42. Yuan, H. *et al.* Quantitative surface-enhanced resonant Raman scattering multiplexing of biocompatible gold nanostars for in vitro and ex vivo detection. *Anal. Chem.* **85,** 208–212 (2013).

43. Kudinova, M. A., Kurdyukov, V. V., Kachkovski, A. V. & Tolmachev, A. I. Pyrylocyanines. 36. alpha-Thienyl-substituted pyrylo- and thiopyrylocyanines. *Khim Geterotsikl* 494–500 (1998).

44. Chung, E. *et al.* Use of surface-enhanced Raman scattering to quantify EGFR markers uninhibited by cetuximab antibodies. *Biosens. Bioelectron.* **60,** 358–365 (2014).

Acknowledgements

We thank the Electron Microscopy and Molecular Cytology Core Facility at Memorial Sloan Kettering Cancer Center (MSKCC). This work was supported in part by the following grants: NIH R01 EB017748 (M.F.K.); NIH K08 CA163961 (M.F.K.); M.F.K. is a Damon Runyon–Rachleff Innovator supported (in part) by the Damon Runyon Cancer Research Foundation (DRR-29-14); MSKCC Center for Experimental Therapeutics Grant (M.F.K.). MSKCC Center for Molecular Imaging and Nano-technology Grant (M.F.K.); MSKCC Technology Development Fund Grant (M.F.K.); Geoffrey Beene Cancer Research Center at MSKCC Grant Award (M.F.K.) and Shared Resources Award (M.F.K.); The Dana Foundation Brain and Immuno-Imaging Grant (M.F.K.); Dana Neuroscience Scholar Award (M.F.K.); Bayer HealthCare Pharmaceuticals/RSNA Research Scholar Grant (M.F.K.); MSKCC Brain Tumor Center Grant (M.F.K.); Society of MSKCC Research Grant (M.F.K.); NIH GM-94367 (M.R.D.); and the National Science Foundation (CHE-1151379, M.R.D.). M.A.W. is supported by a National Science Foundation Integrative Graduate Education and Research Traineeship Grant (NSF, IGERT 0965983 at Hunter College). Acknowledgements are also extended to the grant-funding support provided by the MSKCC Core Grant (P30 CA008748).

Author contributions

S.H. and M.A.B. performed the experiments, analysed the data and wrote the paper. M.A.W. and R.H. participated in performing the experiments. M.F.K. and M.R.D. supervised the study, advised on experimental design, analysed the data and edited the paper.

Additional information

Competing financial interests: The authors declare no competing financial interests.

8

Single-molecule diffusion and conformational dynamics by spatial integration of temporal fluctuations

Maged F. Serag[1], Maram Abadi[1] & Satoshi Habuchi[1]

Single-molecule localization and tracking has been used to translate spatiotemporal information of individual molecules to map their diffusion behaviours. However, accurate analysis of diffusion behaviours and including other parameters, such as the conformation and size of molecules, remain as limitations to the method. Here, we report a method that addresses the limitations of existing single-molecular localization methods. The method is based on temporal tracking of the cumulative area occupied by molecules. These temporal fluctuations are tied to molecular size, rates of diffusion and conformational changes. By analysing fluorescent nanospheres and double-stranded DNA molecules of different lengths and topological forms, we demonstrate that our cumulative-area method surpasses the conventional single-molecule localization method in terms of the accuracy of determined diffusion coefficients. Furthermore, the cumulative-area method provides conformational relaxation times of structurally flexible chains along with diffusion coefficients, which together are relevant to work in a wide spectrum of scientific fields.

[1] Biological and Environmental Sciences and Engineering Division, King Abdullah University of Science and Technology (KAUST), Thuwal 23955-6900, Saudi Arabia. Correspondence and requests for materials should be addressed to S.H. (email: Satoshi.Habuchi@kaust.edu.sa).

Providing unprecedented insights into the diffusional dynamics of heterogeneous systems in materials and life sciences, single-molecule localization and tracking (SMLT) has been regarded as a powerful tool that offers spatiotemporal resolution not limited by diffraction of light and visualizes the dynamics of individual molecules not masked by ensemble averaging[1-4]. Notably, SMLT has become a standard technique in the study of subcellular dynamics in live cells, such as analyses of receptors on cell surfaces[5,6], viral infection of cells[7] and the entire pathway of gene transcription[8]. Furthermore, SMLT has been useful in the study of molecular mechanisms that affect the properties of polymers in materials processing and fabrication[9]. Interestingly, SMLT data can be interpreted to model the standard types of diffusional dynamics, including confined diffusion, random diffusion, directed diffusion and anomalous subdiffusion[10-20].

To analyse the diffusion of single molecules by SMLT, the emitted photon distribution of a fluorescent molecule, at least a few hundred nanometers in size, is fitted by a Gaussian function in both the x and y dimensions. This fitting aims at highly precise localization of the molecule, typically within a single nanometer to tens of nanometers, by determining the centre of its point-spread function (PSF)[1,21-23]. The statistical fit of the ideal Gaussian is repeated with time spacing over a longer period of time, typically a few seconds. The spatiotemporal positions of the molecule are then obtained and spatially connected to generate a temporal diffusion trajectory. The diffusion can be quantitatively characterized from both the spatial and temporal components of the trajectory by calculating the mean square displacement (MSD; Supplementary Fig. 1).

Although the well-established SMLT technology provides accurate spatiotemporal locations of single molecules[24,25], enabling their translational speed to be estimated, it fails to provide essential information about the shape and size of the molecule. Access to shape and size information and integrating it with translational speed measurements would provide important details about molecular diffusion as a crucial life process and reptation as an important physical phenomenon in polymer physics and analytical science. In addition, SMLT introduces several inherent limitations that in practice hinder proper data interpretation. Indeed, SMLT is inefficient and even fails to work in some cases. For example, shape fluctuation of flexible polymer molecules results in large localization error if the molecule is larger than the diffraction-limited size. In the extreme case, the shape of a long molecule temporally fluctuates in space, hampering the accurate localization of its centre-of-mass. Any out-of-focus motion results in large statistical error due to the limited length of the obtained diffusion trajectory. Although recent technologies such as three-dimensional-tracking and electrokinetic trapping offer remarkable performance enhancement of SMLT, these technologies require sophisticated analyses and have low throughput[26-29]. Statistical and localization errors result in broader distribution of diffusion coefficients of individual molecules. Such errors could affect the interpretation of diffusion data especially in complex analyses such as multimode diffusion studies[30,31]. Therefore, there remain many drawbacks to single-molecule localization-based methods that could be overcome by the development of new methods.

Here, we present a method to measure single-molecule diffusion that provides an efficient solution to the limitations described above. Typically, SMLT analyses express molecular motion in terms of accurate spatiotemporal positions of the molecule. Our new method, on the contrary, expresses molecular motion in terms of the increase of the cumulative area occupied by the molecule in space over time, which we call the cumulative-area (CA) method. Through careful adjustment of the number of pixels detected per each time-lapse image, information on translational diffusion, molecular size and frequency of conformational changes can be obtained.

We validate our approach by analysing the statistical distribution of diffusion coefficients of fluorescent nanospheres as calculated by the CA method and the distribution as calculated by SMLT. We further extend the potential usefulness of our approach by analysing the diffusion and conformational dynamics of double-stranded (ds)DNA of different lengths and topological forms, measurements that are critically sensitive to molecular size and conformational changes.

Results

Schematic principle of the CA method. The CA approach is based on calculating the increase in the cumulative fluorescent area of moving particles that occurs between time frames t_i and t_{i+1} ($i = 1,2,\ldots,n$) (Fig. 1). Given a stack of frames of a time-lapse image sequence of a moving particle, we first remove the background by setting an initial threshold. The threshold is set by fitting the frequency distribution of the intensity of all pixels in each frame with a Gaussian function (Fig. 2, Supplementary Fig. 2). We set the background threshold to $m + 4s$ for the diffusion measurements in which m and s denote the mean and the s.d. of the background intensity distribution determined by the Gaussian fitting. This makes independent background adjustment possible for each frame. The background subtraction is accompanied by removal of noise pixels in all frames. Then, we gradually increase the background threshold to define five pixels (see 'Validation of the CA analysis on simulated tracks' in the results section) as the area of space occupied by the molecule in each frame. We note that the number of pixels can be increased to define a complete molecular shape for conformational dynamics analysis (see 'Methods' section). The movement of these pixels results in an increase in the area occupied by the molecules across the stack over time (Fig. 1) and can be used to describe the lateral diffusion of the molecule. The quality of the stack is judged based on both the ratio of the number of noise pixels in all frames to the total number of pixels in the stack and the ratio of the number of the empty frames resulting from brief absences of the particle from the field of view to the total number of frames (we call these ratios the internal noise ratio and the dropped frames ratio, respectively; Supplementary Fig. 3). A stack is considered for analysis when the internal noise ratio and the dropped frames ratio do not exceed 20 and 1%, respectively.

To obtain the cumulative area (M_i) at each time frame, t_i, all frames are superimposed starting at time frame t_1 and continued in sequential order until frame t_i. Given the cumulative area throughout the time-lapse image sequence (Fig. 2), the increment of the cumulative area occupied by the molecule that occurs between time frames t_i and t_{i+1} (ΔA_i), because of random molecular motion (Fig. 1), is calculated by the following formula:

$$\Delta A_i = M_{i+1} - M_i. \tag{1}$$

Occasionally, the area difference (ΔA_i) is masked by the growing cumulative area. Consequently, the cumulative-area increment could drop to zero (Supplementary Fig. 4). We correct this by reversing the order of the superimposition (see 'Validation of the CA analysis on simulated tracks' in the Results section). In such a case, we obtain the cumulative area at time t_i and t_{i+1} by superimposing all the frames between t_n and t_i and t_n and t_{i+1}. ΔA_i is calculated by the equation,

$$\Delta A_i = M_i - M_{i+1}. \tag{2}$$

Any zero values in the forward superimposition are replaced with the correct values obtained from the backward

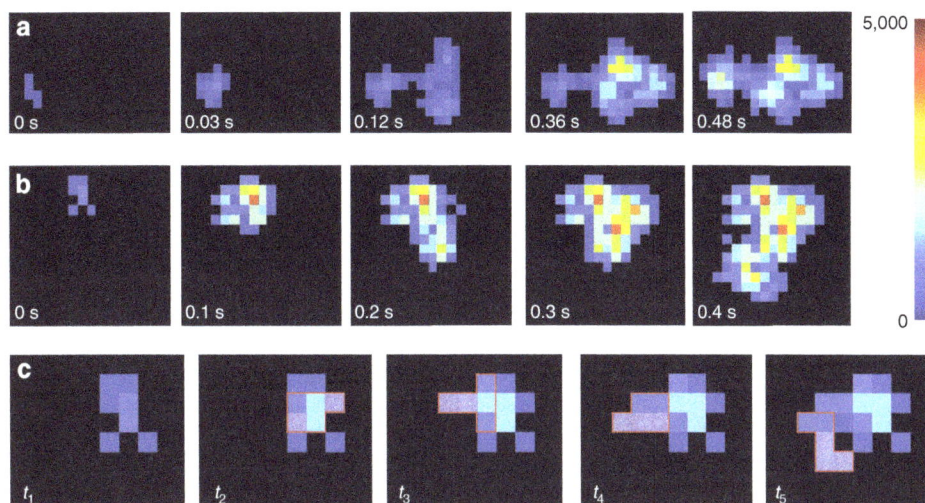

Figure 1 | The time-dependent increase in the cumulative area occupied by diffusing molecules. (**a**) The cumulative area occupied by diffusing fluorescent nanospheres and (**b**) YOYO-I labelled linear ColE$_1$ DNA gradually increases. Numbers at the bottom indicate time in seconds. The jet colour map represents a.u. values of the cumulative fluorescence intensity of individual pixels. (**c**) Increase in the cumulative area occupied by linear ColE$_1$ DNA at time frames t_1 (0 s), t_2 (0.0064 s), t_3 (0.0128 s), t_4 (0.0192 s) and t_5 (0.0256 s) over time. The red-bordered region defines the area occupied by the molecule at each time frame. The dashed pixels represent the area increment at each time frame as a result of random movement of the molecules. Pixel size = 0.16 μm.

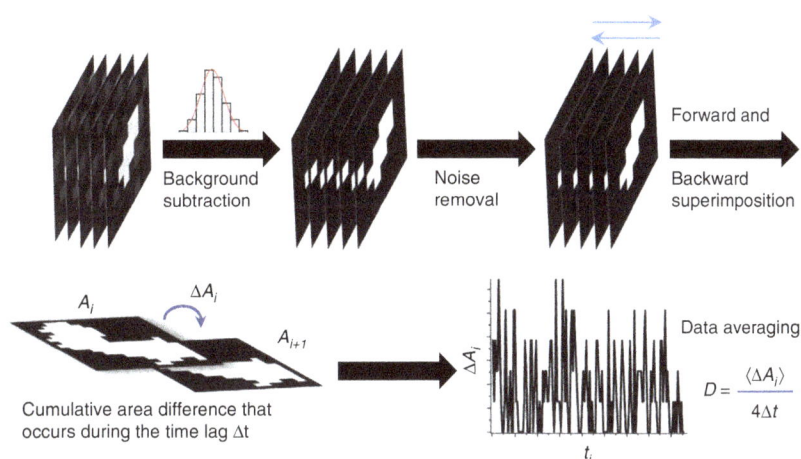

Figure 2 | A schematic illustration of the CA method. First, automatic thresholding is done by fitting the frequency distribution of all pixel values in each frame to a Gaussian function. This process is followed by pixel noise removal. Then, the frames are superimposed to generate the cumulative area (A_i) at each time frame t_i followed by subtraction to calculate the cumulative-area difference at the time lapse between the adjacent frames (Δt). Finally, the cumulative-area difference is used to calculate the diffusion coefficient using the formula $D = <\Delta A_i>/4\Delta t$.

superimposition. Finally, the lateral diffusion coefficient (D) is obtained from the following formula:

$$\langle \Delta A_i \rangle = 4D\Delta t, \qquad (3)$$

where $<\Delta A_i>$ is the average area increment at $1\Delta t$ and Δt is the time lapse between adjacent frames.

Diffusion of polymer nanospheres in a buffer. We validate our approach experimentally by comparing the distribution of diffusion coefficients of approximately 100 cross-linked fluorescent polystyrene nanospheres with a mean diameter of 0.19 μm suspended in 1 mM Tris buffer measured by means of single-molecule fluorescence imaging (see 'Methods' section) and analysed by both the CA and SMLT-MSD methods. The expected diffusion coefficient calculated from the Stokes–Einstein equation is 1.74 μm^2 s^{-1} at 20 °C. Results presented in Fig. 3

show that after fitting the data to the ideal Gaussian, the diffusion coefficient measured by the CA method (1.73 ± 0.24 μm^2 s^{-1}) is in close approximation with that measured by SMLT-MSD (1.63 ± 0.34 μm^2 s^{-1}). Although the distribution appears to be symmetrical around the mean in both analyses, the width of distribution varies to some extent. The s.d. of the diffusion coefficients (σ_D) measured by the CA method ($\sigma_D = 0.24$) was two-thirds of that measured by the SMLT-MSD method ($\sigma_D = 0.35$). The distribution of the diffusion coefficient obtained by the SMLT-MSD analysis can be reproduced reasonably by a theoretical statistical probability distribution of diffusion coefficients in a homogeneous environment (see 'Methods' section and Supplementary Fig. 5).

The good agreement of the two distributions confirms that the width of the frequency distribution is associated with statistical error in SMLT-MSD. The σ_D obtained by the CA method (14% of

a

Mean $D = 1.7\ \mu m^2 s^{-1}$
$\sigma_D = 0.24\ \mu m^2 s^{-1}$

b

Mean $D = 1.6\ \mu m^2 s^{-1}$
$\sigma_D = 0.35\ \mu m^2 s^{-1}$

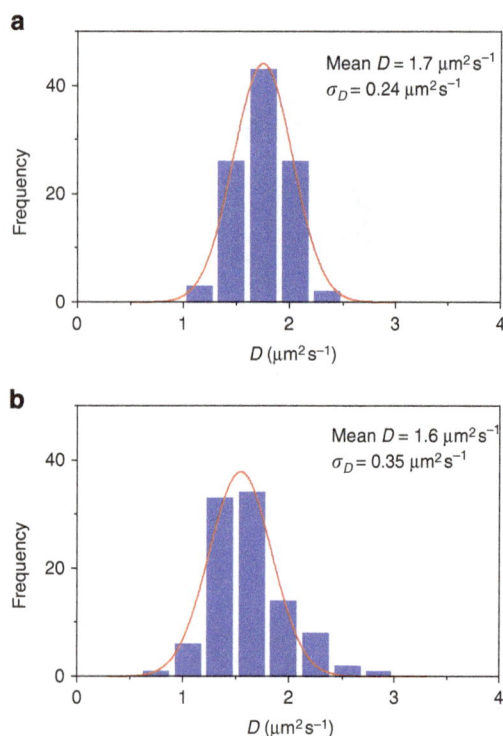

Figure 3 | Frequency histograms of the diffusion coefficient determined for nanospheres. The diffusion coefficient values of 0.19 μm fluorescent nanospheres were determined via (**a**) the CA and (**b**) the SMLT-MSD methods. The red lines in **a** and **b** are Gaussian fittings of the frequency distributions.

the mean of the D) is similar to that obtained by simulation (13–17% of the mean of the D, see 'Simulation of diffusion coefficient of single particles' section), suggesting that the D distribution obtained by the CA method also captures statistical error. Thus, the smaller σ_D obtained by the CA method as compared with that obtained by SMLT-MSD analysis corresponds to the smaller statistical error in the former method. The smaller statistical error originates from bypassing the localization of the molecule in the CA method

Simulation of diffusion coefficients of single particles. We evaluate the performance of the CA method using simulated random walk data of single particles. Random diffusion trajectories of 100 particles are generated in two-dimensional (2D) space (see 'Methods' section and Fig. 4a). Diffusion trajectories with three different diffusion coefficients ($D = 1.0$, 1.5 and $2.0\ \mu m^2 s^{-1}$) are simulated and then the D values are calculated using the CA method (Fig. 4b). Note that the D values and trajectory lengths in the simulation lie within the range of experimentally obtained diffusion coefficients and the trajectory lengths of the DNA molecules used in this study. The frame rate (156 frames s^{-1}) is also set at the value for the single-molecule imaging experiments on DNA molecules. Good agreement between the simulated diffusion values (1.0, 1.5 and $2.0\ \mu m^2 s^{-1}$) and the calculated values (1.10, 1.51 and $1.98\ \mu m^2 s^{-1}$) is obtained, demonstrating the accuracy of the CA method (Fig. 4c–e). Much smaller diffusion coefficients can also be reliably calculated provided that a suitable sampling rate is chosen (Supplementary Fig. 6). The results of the simulation shows that the sampling rate should be associated with a large-enough displacement to result in a detectable area increment over time. This ensures that the short displacements that could be generated at

fast sampling rates are not masked by the area occupied by the molecule in each frame. A similar prerequisite is crucial in SMLT-MSD analysis where the sampling rate should result in a displacement that is much larger than the localization accuracy obtained after the 2D Gaussian fitting.

Diffusional dynamics of dsDNA. Owing to ease, reliability and precise control of its synthesis, DNA is well established as a model biopolymer for understanding fundamental phenomena in polymer physics[32]. Therefore, studying the diffusional dynamics of single DNA molecules is a useful endeavour to determine the diffusion behaviour of isolated polymers in infinite dilutions.

To provide further evidence of the reliability and validity of our method, we measure the distribution of diffusion coefficients of YOYO-I labelled ColE$_1$ (6 kbp) and Charomid (42 kbp) dsDNA of different topologies (Table 1 and Fig. 5). The peaks of the D distribution of supercoiled and relaxed circular ColE$_1$ DNA measured by the CA method are in close approximation with values obtained by SMLT-MSD analyses (Fig. 5a,b). A slight deviation in the peak is observed in the D distributions of the linear form of ColE$_1$ DNA obtained by both methods (Fig. 5c). In contrast to the analogous peak values obtained by the two methods, a marked difference in the width of the distribution is observed. The widths of the D distributions obtained by the CA method (σ_D is approximately 15% of the mean of the D) are similar to those obtained using fluorescent nanospheres and a simulated diffusion trajectory (Table 1). On the other hand, much broader distributions (σ_D is approximately 40–65% of the mean of the D) are obtained by the SMLT-MSD method (Table 1). The distributions are indeed much broader than the theoretically calculated statistical probability distribution of diffusion coefficients in a homogeneous environment (see 'Methods' section and Supplementary Fig. 7). The calculated radii of gyration (R_G) of supercoiled, relaxed circular and linear ColE$_1$ DNA are 0.1, 0.13 and 0.21 μm, respectively[33,34]. Although the R_G of ColE$_1$ DNA in different topological forms are comparable to the width of the PSF of our experimental setup, the broadening of the D distribution obtained by the SMLT-MSD analyses indicates that the conformational changes even at this length scale affect the analysis of the diffusion behaviour in the SMLT-MSD method. The absence of the broadening of the D distribution in the CA analysis highlights the advantage of the CA method over the SMLT-MSD analysis.

The difference between the D distributions calculated by both methods is more remarkable for the much longer 42 kbp Charomid DNA ($R_G \approx 0.65\ \mu m$ (refs 35,36), Fig. 5d–f). This is associated with significant broadening of the distributions measured via SMLT-MSD and inconsistent peak values. Since the R_G of Charomid DNA is much larger than the PSF of the microscope, the fluctuations in conformational changes of the DNA are evident within each trajectory (Fig. 6a). This hampers accurate determination of the centre-of-mass of the molecules. Furthermore, the anisotropic shape of the fluorescently labelled Charomid molecules introduces considerable errors in the 2D Gaussian fitting. This, in turn, makes the tracking process fail quite quickly because of large fitting outliers. Although several algorithms have been successfully applied to optimize the tracking conditions of small and elongated molecules[37–39], these methods rely on the fitting of the image with a 2D function that limit their applicability to conformationally unstable molecules. The CA method circumvents the localization algorithm and shows narrower and much better symmetrical distribution around the mean diffusion coefficient (Fig. 5, Supplementary Fig. 8). Indeed, the widths of the D distribution obtained by the CA analysis (σ_D is approximately 20 % of the mean of the D) are

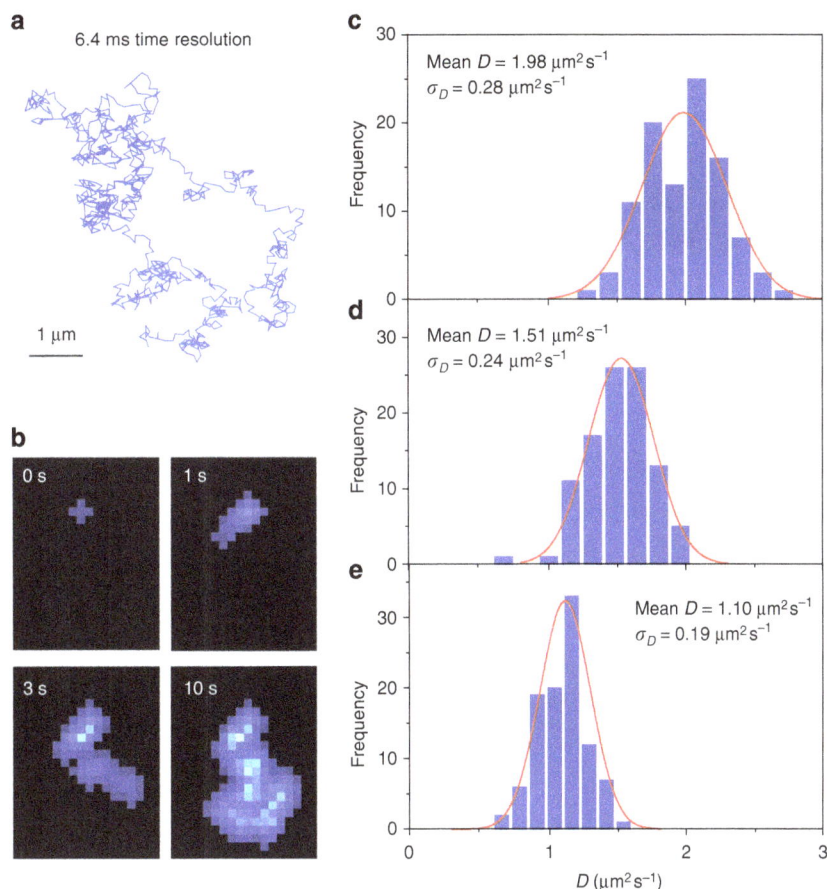

Figure 4 | Simulation of the diffusion coefficients of single particles in two dimensions. (**a**) Simulated 2D trajectories of a particle diffusing at $D = 1.0\ \mu m^2\ s^{-1}$. (**b**) Cumulative areas at time 0, 1, 3 and 10 s calculated using the diffusion trajectory shown in **a**. Frequency histograms of D calculated using the simulated trajectories diffusing at (**c**) $D = 2.0\ \mu m^2\ s^{-1}$, (**d**) $D = 1.5\ \mu m^2\ s^{-1}$ and (**e**) $D = 1.0\ \mu m^2\ s^{-1}$ by CA method. The lengths of trajectories in the simulation were set to 50, 100 and 300 frames for particles diffusing at $2.0\ \mu m^2\ s^{-1}$, $1.5\ \mu m^2\ s^{-1}$ and $1.0\ \mu m^2\ s^{-1}$, respectively. The red lines show Gaussian fittings of the histograms.

Table 1 | Measured diffusion coefficients for DNA molecules using the cumulative-area and SMLT-MSD methods.

DNA construct	Diffusion coefficients (mean ± s.d.), $\mu m^2\ s^{-1}$*					
	Supercoiled		Circular		Linear	
	CA	SMLT	CA	SMLT	CA	SMLT
ColE₁ (6 kbp)	1.80 ± 0.30	2.62 ± 1.72	1.74 ± 0.24	2.52 ± 1.32	1.61 ± 0.25	2.38 ± 0.97
Charomid (42 kbp)	1.14 ± 0.24	2.41 ± 1.20	1.10 ± 0.24	4.19 ± 1.81	0.99 ± 0.22	2.30 ± 1.25

CA, cumulative area; MSD, mean square displacement; SMLT, single-molecule localization and tracking.
*Diffusion coefficients were determined based on calculating the arithmetic mean of values in Fig. 5.

comparable to those obtained for the nanospheres and shorter 6 kbp ColE₁ DNA. Furthermore, the mean D obtained by the CA analysis ($0.99\ \mu m^2\ s^{-1}$ for the linear form of 42 kbp Charomid DNA) is comparable to the ensemble-averaged D obtained for a similar length of DNA ($0.80\ \mu m^2\ s^{-1}$ for the 48.5 kbp λ DNA)[40]. These results demonstrate the robustness of the CA method and its applicability to a wide range of the samples including large, conformationally unstable molecules.

Validation of the CA analysis on simulated tracks. We validate the input parameters of the CA method using simulated data to benchmark its performance to provide reliable estimates of the diffusion coefficients. First, we calculate the diffusion coefficients in the absence of the backward superimposition correction at

different tracking times (0.32 s ~32 s; frame rate = 6.4 ms; Supplementary Fig. 9). We found that the calculations underestimate the simulated diffusion coefficients (Supplementary Fig. 9c) because the growing area masks the cumulative-area increment ($\Delta A_i = 0$). As the backward superimposition correction is applied, reliable estimates of diffusion coefficients are obtained (Supplementary Fig. 9c). The observed gradual decrease of the calculated diffusion coefficients by increasing the tracking time (Supplementary Fig. 9c) is attributed to an increasing probability that the masked area ($\Delta A_i = 0$) could coincide in both the forward and the backward superimpositions as the tracking time is increased. We calculate the mean diffusion coefficients (D_μ) of 100 simulated molecules after varying three parameters to evaluate the validity of the CA method (Supplementary Tables 1 and 2): (1) the number of pixels used to represent the area

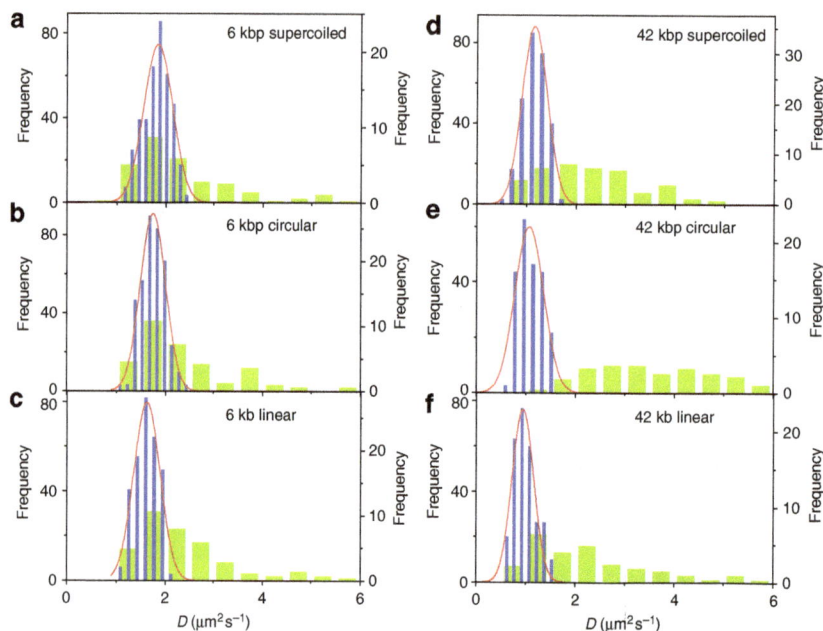

Figure 5 | Frequency histograms of the diffusion coefficients determined for DNA. Different topological forms of (**a**–**c**) ColE$_1$ (6 kbp) and (**d**–**f**) Charomid (42 kbp) DNA are analysed and the diffusion coefficient values are determined via the cumualtive area (blue bars) and the SMLT-MSD (green bars) methods for supercoiled (**a,d**), relaxed circular (**b,e**) and linear (**c,f**) forms. The red lines are Gaussian fittings of the respective frequency distributions.

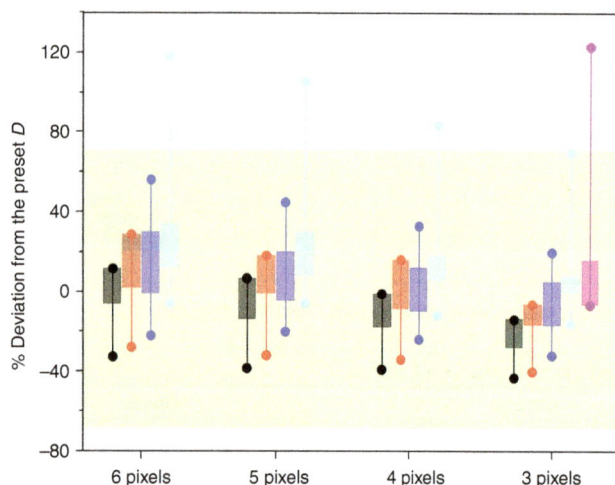

Figure 6 | Accuracy of the calculated diffusion coefficients. Random diffusion trajectories with the pixel size of 160 nm are generated and analysed by the CA method. The green zone represents the statistical error (±68%) of the diffusion coefficient calculated by the SMLT-MSD analysis of 100 particles. The yellow zone represents the threshold range (±20%) we set to validate the calculated diffusion coefficients. The values between the two circles on the same line represents the range of the calculated diffusion coefficients obtained for the simulated trajectories with preset diffusion coefficients of 2.0 (black), 1.5 (red), 1.0 (blue), 0.5 (cyan) and 0.2 μm^2 s^{-1} (magenta) by varying the tracking time from 0.32 to 32 s (Supplementary Table 1). The percentage deviations of the calculated means obtained within the range of the experimental tracking time (see Supplementary Table 3) are plotted as bars for each preset diffusion coefficient.

occupied by the molecule in space (three to six pixels), (2) the preset displacements (d = 50–240 nm corresponding to 0.3–1.5 × of the pixel size and corresponding to preset diffusion coefficients (D_p) = 0.1–2.25 μm^2 s^{-1} at frame rate = 6.4 ms) and (3) the

tracking time (0.32 to 32 s corresponding to the typical tracking times for the molecules diffusing at the preset displacements). To judge the reliability of the calculated $D_μ$ for each of the parameters, we set an accuracy threshold range, equal to ±20% of the D_p (Fig. 6 and Supplementary Tables 1 and 2). When the $D_μ$ values are outside of this range, we consider them as unreliably estimated values. It is important to note that statistical errors of ±68% (±3.29 s) of the calculated mean are obtained by the SMLT-MSD analysis for 100 particles (Fig. 3b). Therefore, the range of the accuracy threshold (±20%) represents only 30% of the statistical errors (Fig. 6). Our simulations predict that the calculated $D_μ$ values with the systematic uncertainty originating from the probabilistic coincidence of the masked area ($\Delta A_i = 0$) in both superimpositions lie within the ±20% threshold limit for the simulated particles. This explains the narrow distribution of the diffusion coefficients obtained by the CA method. The simulations further indicates that accurate calculations requires a trade-off between pixel size and the preset displacement. According to the simulations, the diffusion coefficient of a particle can be reliably estimated using the CA method by representing the particle with three to five pixels and within the range of displacements equals to 0.55–1.4 × of 160 nm pixel size. Shorter displacements (88–50 nm), can be accurately measured by the CA method using pixel size of 100 nm (Supplementary Table 2). It is important to underline that a wide range of diffusion coefficients can be analysed by choosing an appropriate frame rate, similar to SMLT-MSD.

We further investigate the effect of the shape fluctuation of the molecule on the CA analysis. In the CA analysis, the area occupied by the molecule is described by five brightest pixels in the image after the removal of noise pixels (Supplementary Fig. 3). The floppy nature of DNA causes the continuous fluctuation of the shape of the five pixels. Indeed, the shape of the five pixels obtained for the DNA molecule (Supplementary Fig. 10b) shows more fluctuations as compared with that obtained for the nanospheres (Supplementary Fig. 10a). Furthermore, the DNA molecule occasionally shows splitting of the five pixels into two parts due to the slight defocusing of the molecule (Supplementary Fig. 10b). To study the effects, we generated simulated diffusion trajectories with randomly varied shapes of the five pixels (Supplementary Fig. 11a). The splitting of the pixels at variable splitting distances is further introduced to the simulated trajectories in varying proportions (Supplementary Fig. 11b). The CA analysis on the simulated trajectories (Fig. 7 and Supplementary Fig. 12) showed that the calculated diffusion coefficients were located within the 20% accuracy threshold we set to judge the validity of the method (Fig. 7a). The result

demonstrates that the CA analysis tolerates the shape fluctuations of the five pixels.

While the diffusion coefficient is slightly overestimated when the splitting of the five pixels occurs (Fig. 7a), the CA analysis is valid as long as the splitting is smaller than three pixels and the proportion of the split image is less than 5% of the total frames (that is, the diffusion coefficient is within the 20% accuracy threshold, Fig. 7a). Since the splitting of the pixels is related to the defocusing of the image, this result demonstrates that the CA analysis tolerates the slight defocusing. In addition, the s.d. of the diffusion coefficients calculated from 1,000 simulated trajectories remain almost constant regardless the splitting distance and the proportion of the split image (Fig. 7b). This reinforces the special performance of the CA method in analysing the diffusion coefficient of conformationally unstable molecules as compared with SMLT-MSD.

Random, directed and confined diffusion can be probed with the CA method in similar way to MSD analyses (Fig. 8 and Supplementary Fig. 13). Since particles, moving in a pure Brownian fashion, occasionally revisit the pre-explored areas of space, the overall cumulative area resulting from the sequential superimposition between time frames t_1 and t_n tends to show

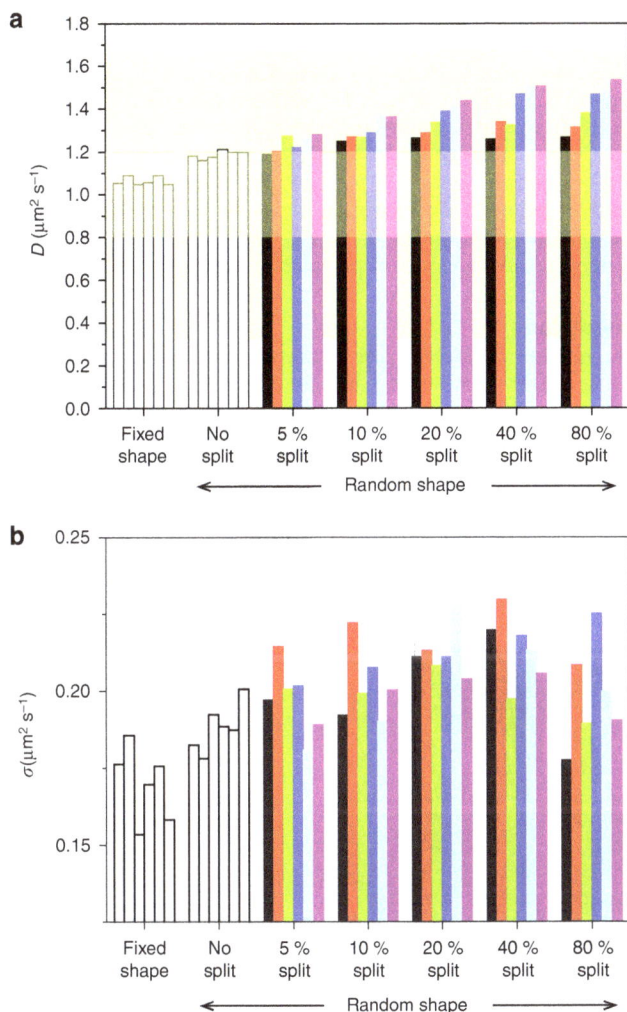

Figure 7 | Effect of the fluctuations of the molecular shape on the calculated diffusion coefficients. 2D trajectories of five pixels diffusing at $1 \mu m^2 s^{-1}$ are simulated (1,000 tracks, each consists of 500 steps with 0.0064 s time resolutions). The diffusion coefficients are calculated using the CA method. (**a**) Calculated diffusion coefficients of the simulated tracks using: (1) five pixels of fixed shape (the six columns show results of six runs of the simulation), (2) five pixels of random shape without splitting (the six columns show results of six runs of the simulation) and (3) five pixels of random shape with 5 (25 frames), 10 (50 frames), 20 (100 frames), 40 (200 frames), 80% (400 frames) of the total frames showing splitting. The colour code illustrates the split distance in pixels in both x and y directions: 2 (black), 3 (red), 4 (green), 5 (blue), 6 (cyan) and 7 pixels (magenta). The green zone represents the s.e. (±68%) of the diffusion coefficient calculated by the SMLT-MSD analysis of 100 particles. The yellow zone represents the threshold range (±20%) we set to validate the calculated diffusion coefficients. (**b**) s.d. calculated for the diffusion coefficients in **a**.

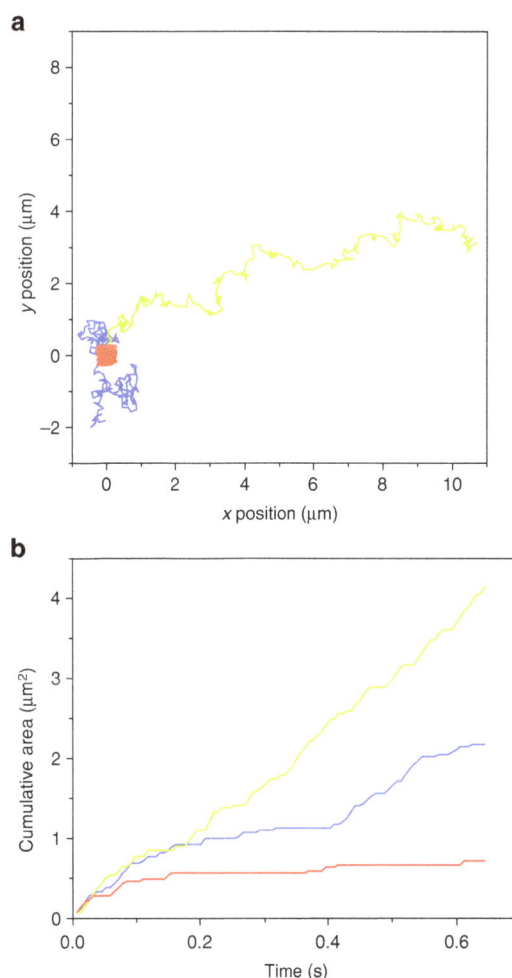

Figure 8 | Simulation of different diffusion modes of single particles. (**a**) Simulated 2D trajectories of a particle diffusing in a random (blue), directed (green) and confined fashion (red) at $D = 1.0 \mu m^2 s^{-1}$ (frame rate = 6.4 ms). (**b**) Cumulative areas at time 0–0.64 s calculated using the trajectories of random (blue), directed (green) and confined diffusion (red) displayed in **a**. Enlarged view of **a** is shown in Supplementary Fig. 13.

irregular, non-linearly increasing cumulative pattern (Fig. 8b, blue line). When the particles show a directed motion, a linear increase is observed because the particles tend to move away from its original location rather than re-exploring the space (Fig. 8b, red line). When the diffusion is restricted to a confined space, the cumulative area rapidly reaches an asymptote where further area increase is constrained (Fig. 8b, green line). Altogether, these results demonstrate the versatile applicability of the CA method in probing random as well as anomalous diffusion.

The conformational dynamics of Charomid DNA. Diffusion of a long and flexible chain (for example, Charomid DNA) is believed to be coupled with its conformational relaxation dynamics. However, simultaneous detection of these two parameters has been challenging due to the lack of suitable methods. One of the advantages of using the CA method over other methods is that it can provide direct information about conformational changes in parallel with diffusion dynamics.

The spontaneous conformational relaxation of spatially isolated DNA can yield dramatic change in the area occupied by the molecule in space (Fig. 9a). To measure the timescale of this change, the area occupied by the molecule (A_i) at a given time, t_i, is calculated (Fig. 9b, inset). The relaxation time of the molecule (τ_R) is estimated by the autocorrelation analysis of A_i using the following formula:

$$G(\tau) = \frac{\langle A_i \cdot A_{i+\tau} \rangle}{\langle A_i \rangle^2} = B \exp\left(-\frac{t}{\tau_R}\right), \tag{4}$$

where τ denotes the time lag. The autocorrelation plot obtained for the linear form of Charomid DNA molecules (42 kbp) with a 1-s sampling time can be fitted well with a single-exponential decay function (Fig. 9b). The average correlation time of 144 ms

was obtained (Fig. 9c), which is in close approximation with the reported relaxation time ($\tau_R \approx 200$ ms) of λ DNA (48.5 kbp) (ref. 40). The relaxation time of the cyclic form of the Charomid DNA calculated in the same way ($\tau_R \approx 80$ ms) is shorter than that of its linear counterpart, consistent with the topology-dependent relaxation time obtained in simulation studies[41,42]. These results imply a direct connection between the relaxation time obtained by the CA method and the conformational relaxation of the chains.

The anti-Brownian electrokinetic trap method and the CA method produce similar results[35]. In the anti-Brownian method, however, the diffusion coefficient is deduced from the applied force on the molecule. This makes the analysis cumbersome and complicated, especially when the diffusion is complex and inhomogeneous[31]. In contrast, the CA method simultaneously reports the diffusion coefficient and the relaxation time of the molecule in a straightforward way (Supplementary Figs 14 and 15) in the absence of external forces that might affect the conformational state of the trapped molecule. Combining reliable estimates of relaxation times with accurate diffusion measurements suggests a new way to investigate the heterogeneous dynamics of polymers in a crowded environment.

Discussion
The method we present here circumvents molecular localization, the cornerstone of single-molecule diffusion measurements and overcomes its inherent limitations. We validate the method using fluorescent nanospheres and dsDNA molecules of different lengths and topological forms. We found that the performance of the CA method surpasses the SMLT method in terms of reproducible measurement of relaxation times and diffusion dynamics with a narrow distribution of the values obtained for a particular DNA size and topological form.

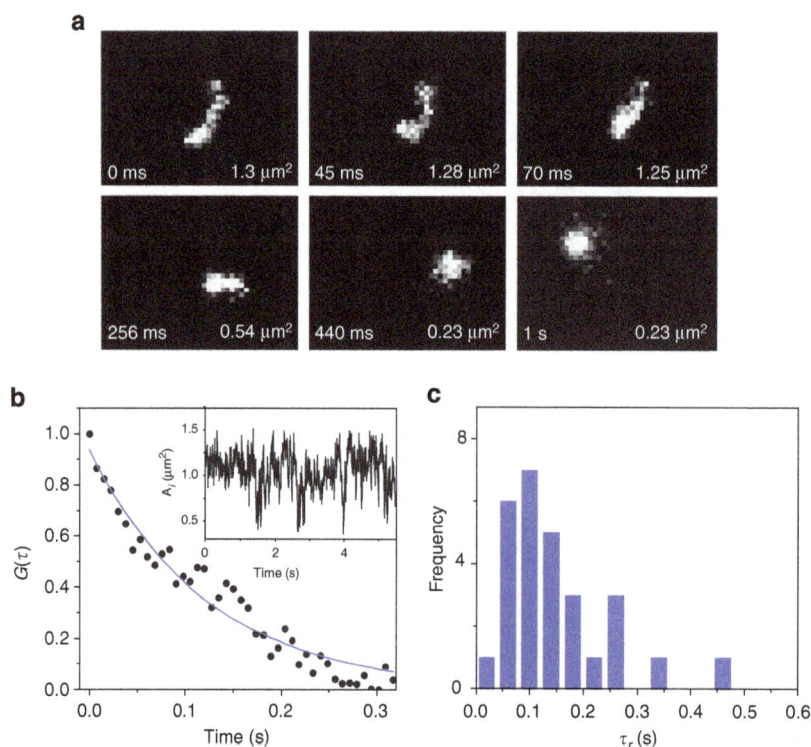

Figure 9 | Conformational dynamics of Charomid DNA. (**a**) Fluorescence images of single DNA molecules (42 kbp) at different time frames. The molecular area in μm^2 is illustrated for each image. (**b**) An autocorrelation plot of the area occupied by a single Charomid DNA. The blue line shows a fitting with a single-exponential decaying function. (Inset) Time lapse of the area occupied by a single Charomid DNA (42 kbp) molecule. (**c**) Frequency histogram of the calculated relaxation time.

Our new method holds promise to advance single-molecule fluorescence microscopy studies, which are relevant to a wide spectrum of scientific fields. From fundamental physics to materials science to chemistry, biology and analytical science, translational diffusion and conformational dynamics are, indeed, central to multidisciplinary studies of polymers. In closing, we offer suggestions on the immediate applicability of our method in these divergent fields.

In chemistry, physics and biological science, there is considerable fundamental interest in the effect of molecular topology on entangled biopolymer dynamics. In particular, our method can be reliably used to unveil nanoscale heterogeneity and molecular dynamics in complex systems. Our method, for example, can provide insights into anomalous diffusion including directional flow and constraint diffusion (see Fig. 8). In the field of soft condensed matter physics and gel electrophoresis, the flow behaviour of entangled polymers, viscous and elastic deformation and orientation in electrophoretic flow and the effect of molecular topology on electrophoretic mobility have rapidly become foci of research activity. We conceive of no better strategy to track such behaviours than by using the CA method, which promises significant analytical improvement over other methods. In polymer physics and materials science, understanding the diffusional dynamics of polymers is the key to characterizing the rheological properties of synthetic polymers and nanocomposites. Indeed, the viscoelastic properties of these materials can be discerned after diffusional and conformational change measurements using the CA method. Work in this area is motivated by the need to develop new materials that are relevant to industrial products, including plastics, fibres, emulsifiers and countless others. Additional methodological developments, including three-dimensional area tacking and multicolour imaging will broaden the horizon of applications of the CA method. Finally, we believe that our method will work in conjunction with super-resolution microscopy to unravel the dynamics of complex biosystems that continue to remain beyond experimental realization.

Methods

Fluorescent materials. Suncoast yellow fluorescent polymer nanospheres (excitation/emission maxima 540/600 nm) of nominal size (0.19 μm; 2.653×10^{12} nanspheres ml^{-1}) were purchased from Bangs Laboratories, Inc. (Fishers, IN, USA). The nanospheres were diluted with 10 mM TRIS buffer (pH 8) to a concentration of 1.5×10^6.

ColE$_1$ (6 kbp) and Charomid (42 kbp) DNA were purchased from Nippon Gene (Tokyo, Japan). The restriction enzyme SmaI (New England Biolabs (UK)) was used to prepare the linear form, and Topoisomerase I (New England Biolabs) was used to prepare the relaxed form. The DNA molecules were labelled with YOYO-1 (Molecular Probes, Carlsbad, CA) (dye molecules to base pair ratio of 1:5) as described in ref. 32.

Single-molecule fluorescence imaging setup. The single-molecule imaging experiments were carried out on a custom-built widefield epifluorescence microscope setup. Two solid-state lasers were used for illumination: a CW 60 mW 488 nm (MLD, Cobolt, Solna, Sweden) and a CW 100 mW 532 nm (Samba, Cobolt). In this setup, the laser beams pass through an acousto-optic tunable filter (AOTF; AA Optoelectronic, Orsay, France), which allows the intensity of the individual laser lines to be independently controlled. The AOTF is synchronized to an iXon3 897 EMCCD camera (Andor Technology, Belfast, Ireland) to illuminate the sample only during the light-integrating phase and thus to reduce photobleaching. The collimated laser beams are directed through a × 5 beam expander (Thorlab, Newton, NJ, USA), which in combination with an achromat lens in front of the microscope entrance provides widefield Kohler illumination of the sample. The fluorescence images were recorded using an inverted IX71 microscope (Olympus, Japan) with an Olympus Plan Apochromat × 100 NA 1.49 oil immersion objective lens. The fluorescence light coming from the sample was collected again by the objective lens and sent through the side port of the microscope towards the EMCCD camera. The fluorescence light is separated from the laser excitation light using a dichroic mirror (FF506-Di03) and an emission filter (FF01-609/181) obtained from Semrock (Lake Forest, IL, USA). The fluorescence images were recorded for 6.4 ms sampling time with a pixel size of 160 nm.

The image acquisition was done using the Andor iQ imaging software. Approximately, 100 single-molecule diffusion trajectories were analysed in each experiment.

Single-molecule fluorescence image analysis. Analyses of the diffusion trajectories by the area-based method and SMLT-MSD method were performed off-line using a routine written in Matlab. In the SMLT-MSD analysis, the positions of the molecules in each image were determined by using two-dimensional Gaussian fitting:

$$z = z_0 + A \exp\left[-\frac{(x - x_c)^2}{2w_x^2}\right] \exp\left[-\frac{(y - y_c)^2}{2w_y^2}\right]. \quad (5)$$

The MSD was calculated by using the following formula:

$$\text{MSD}(\Delta t) = \left\langle (x_{i+n} - x_i)^2 + (y_{i+n} - y_i)^2 \right\rangle = 4D\Delta t, \quad (6)$$

where x_{i+n} and y_{i+n} describe the molecular spatial position following a time interval, Δt, given by n frame rate after starting at positions x_i and y_i. D is the lateral diffusion coefficient and is calculated from the slope of the MSD plot (Supplementary Fig. 1).

Probability distribution of diffusion coefficients. The statistical probability distribution of the diffusion coefficient in a homogeneous environment is described by,

$$p(D)dD = \frac{1}{(n-1)!} \left(\frac{N}{D_0}\right)^N D^{N-1} \exp\left(\frac{-ND}{D_0}\right) dD \quad (7)$$

where N is the number of independent pairs (number of displacements), D_0 is the true mean diffusion coefficient and D is the diffusion coefficient for an individual trajectory.

Simulation of the random-walk of single molecules. The random-walk trajectory was constructed using a routine written in Matlab starting at $(x,y) = (0,0)$. The random displacements were generated using a distribution function (q) expected from the normal diffusion theory:

$$q = \frac{2r}{\langle r^2(\Delta t)\rangle} \exp\left[\frac{-r^2}{\langle r^2(\Delta t)\rangle}\right] \quad (8)$$

where r and Δt denote the displacements and the time lag. Thus, $\langle r^2(\Delta t)\rangle$ corresponds to the MSD at the time lag Δt. The angles between two successive displacements were generated based on random values between 0 to 360°. The obtained (x,y) positions were then used to define the centre of the same number of pixels (five pixels) used in our experiments. The frequency distribution calculated for 100 simulations is shown in Fig. 4.

Conformational dynamics of DNA. To measure the relaxation time of single DNA molecules, the molecular area was defined by specifying an upper threshold of 40–60 pixels. The upper threshold was determined based on the largest number of pixels obtained after background subtraction in each video recording. The largest number of pixels in each recording represents the fully extended DNA molecules. The area fluctuations of approximately 100 Charomid DNA molecules (42 kbp, 0.1 mg ml^{-1}) were autocorrelated and the autocorrelation plot was fitted to a single-exponential decay using Origin (originlab Version 9.0).

References

1. Yildiz, A. *et al.* Myosin V walks hand-over-hand: single fluorophore imaging with 1.5-nm localization. *Science* **300**, 2061–2065 (2003).
2. Douglass, A. D. & Vale, R. D. Single-molecule microscopy reveals plasma membrane microdomains created by protein-protein networks that exclude or trap signaling molecules in T cells. *Cell* **121**, 937–950 (2005).
3. Teramura, Y. *et al.* Single-molecule analysis of epidermal growth factor binding on the surface of living cells. *EMBO J.* **25**, 4215–4222 (2006).
4. Kirstein, J. *et al.* Exploration of nanostructured channel systems with single-molecule probes. *Nat. Mater.* **6**, 303–310 (2007).
5. Baker, A. *et al.* Functional membrane diffusion of G-protein coupled receptors. *Eur. Biophys. J.* **36**, 849–860 (2007).
6. Kusumi, A., Suzuki, K. G. N., Kasai, R. S., Ritchie, K. & Fujiwara, T. K. Hierarchical mesoscale domain organization of the plasma membrane. *Trends Biochem. Sci.* **36**, 604–615 (2011).
7. Brandenburg, B. & Zhuang, X. W. Virus trafficking - learning from single-virus tracking. *Nat. Rev. Microbiol.* **5**, 197–208 (2007).
8. Goldman, R. D., Swedlow, J. R. & Spector, D. L. *Live Cell Imaging: A Laboratory Manual*, 2nd edn (Cold Springer Harbor Laboratory Press, 2010).
9. Woll, D. *et al.* Polymers and single molecule fluorescence spectroscopy, what can we learn? *Chem. Soc. Rev.* **38**, 313–328 (2009).

10. Edidin, M., Kuo, S. C. & Sheetz, M. P. Lateral movements of membrane-glycoproteins restricted by dynamic cytoplasmic barriers. *Science* **254**, 1379–1382 (1991).

11. Kusumi, A., Sako, Y. & Yamamoto, M. Confined lateral diffusion of membrane-receptors as studied by single-particle tracking (nanovid microscopy)-effects of calcium-induced differentiation in cultured epithelial-cells. *Biophys. J.* **65**, 2021–2040 (1993).

12. Jacobson, K., Sheets, E. D. & Simson, R. Revisiting the fluid masaic model of membranes. *Science* **268**, 1441–1442 (1995).

13. Feder, T. J., BrustMascher, I., Slattery, J. P., Baird, B. & Webb, W. W. Constrained diffusion or immobile fraction on cell surfaces: a new interpretation. *Biophys. J.* **70**, 2767–2773 (1996).

14. Saxton, M. J. & Jacobson, K. Single-particle tracking: applications to membrane dynamics. *Annu. Rev. Biophys. Biomol. Struct.* **26**, 373–399 (1997).

15. Schutz, G. J., Schindler, H. & Schmidt, T. Single-molecule microscopy on model membranes reveals anomalous diffusion. *Biophys. J.* **73**, 1073–1080 (1997).

16. Daumas, F. *et al.* Confined diffusion without fences of a G-protein-coupled receptor as revealed by single particle tracking. *Biophys. J.* **84**, 356–366 (2003).

17. Hellriegel, C., Kirstein, J. & Brauchle, C. Tracking of single molecules as a powerful method to characterize diffusivity of organic species in mesoporous materials. *New J. Phys.* **7**, 23 (2005).

18. Robertson, R. M. & Smith, D. E. Self-diffusion of entangled linear and circular DNA molecules: dependence on length and concentration. *Macromolecules* **40**, 3373–3377 (2007).

19. Saxton, M. J. A biological interpretation of transient anomalous subdiffusion. I. Qualitative model. *Biophys. J.* **92**, 1178–1191 (2007).

20. Serag, M. F. *et al.* Spatiotemporal visualization of subcellular dynamics of carbon nanotubes. *Nano Lett.* **12**, 6145–6151 (2012).

21. Thompson, R. E., Larson, D. R. & Webb, W. W. Precise nanometer localization analysis for individual fluorescent probes. *Biophys. J.* **82**, 2775–2783 (2002).

22. Habuchi, S., Onda, S. & Vacha, M. Mapping the emitting sites within a single conjugated polymer molecule. *Chem. Commun.* 4868–4870 (2009).

23. Habuchi, S., Onda, S. & Vacha, M. Molecular weight dependence of emission intensity and emitting sites distribution within single conjugated polymer molecules. *Phys. Chem. Chem. Phys.* **13**, 1743–1753 (2011).

24. Michalet, X. Mean square displacement analysis of single-particle trajectories with localization error: Brownian motion in an isotropic medium. *Phys. Rev. E* **82**, 041914 (2010).

25. Michalet, X. & Berglund, A. J. Optimal diffusion coefficient estimation in single-particle tracking. *Phys. Rev. E* **85**, 061916 (2012).

26. Speidel, M., Jonas, A. & Florin, E. L. Three-dimensional tracking of fluorescent nanoparticles with subnanometer precision by use of off-focus imaging. *Opt. Lett.* **28**, 69–71 (2003).

27. Lessard, G. A., Goodwin, P. M. & Werner, J. H. Three-dimensional tracking of individual quantum dots. *Appl. Phys. Lett.* **91**, 224106 (2007).

28. Ram, S., Prabhat, P., Chao, J., Ward, E. S. & Ober, R. J. High accuracy 3D quantum dot tracking with multifocal plane microscopy for the study of fast intracellular dynamics in live cells. *Biophys. J.* **95**, 6025–6043 (2008).

29. Fields, A. P. & Cohen, A. E. Electrokinetic trapping at the one nanometer limit. *Proc. Natl Acad. Sci. USA* **108**, 8937–8942 (2011).

30. Habuchi, S., Fujiwara, S., Yamamoto, T., Vacha, M. & Tezuka, Y. Single-molecule study on polymer diffusion in a melt state: effect of chain topology. *Anal. Chem.* **85**, 7369–7376 (2013).

31. Habuchi, S., Satoh, N., Yamamoto, T., Tezuka, Y. & Vacha, M. Multimode diffusion of ring polymer molecules revealed by a single-molecule study. *Angew. Chem. Int. Ed.* **49**, 1418–1421 (2010).

32. Pecora, R. DNA-a model-compound for solutions studies of macromolecules. *Science* **251**, 893–898 (1991).

33. Voordouw, G., Kam, Z., Borochov, N. & Eisenberg, H. Isolation and physical studies of intact supercoiled, open circular and linear forms of ColE1 plasmid DNA. *Biophys. Chem.* **8**, 171–189 (1978).

34. Robertson, R. M., Laib, S. & Smith, D. E. Diffusion of isolated DNA molecules: dependence on length and topology. *Proc. Natl Acad. Sci. USA* **103**, 7310–7314 (2006).

35. Cohen, A. E. & Moerner, W. E. Internal mechanical response of a polymer in solution. *Phys. Rev. Lett.* **98**, 116001 (2007).

36. Rubinstein, M. & Colby, R. H. *Polymer Physics* (Oxford Univ. Press, 2003).

37. Jaqaman, K. *et al.* Robust single-particle tracking in live-cell time-lapse sequences. *Nat. Methods* **5**, 695–702 (2008).

38. Ruhnow, F., Zwicker, D. & Diez, S. Tracking single particles and elongated filaments with nanometer precision. *Biophys. J.* **100**, 2820–2828 (2011).

39. Serge, A., Bertaux, N., Rigneault, H. & Marguet, D. Dynamic multiple-target tracing to probe spatiotemporal cartography of cell membranes. *Nat. Methods* **5**, 687–694 (2008).

40. McHale, K. & Mabuchi, H. Precise characterization of the conformation fluctuations of freely diffusing DNA: beyond Rouse and Zimm. *J. Am. Chem. Soc.* **131**, 17901–17907 (2009).

41. Brown, S. & Szamel, G. Computer simulation study of the structure and dynamics of ring polymers. *J. Chem. Phys.* **109**, 6184–6192 (1998).

42. Fu, C. L., Ouyang, W., Sun, Z. Y. & An, L. Influence of molecular topology on the static and dynamic properties of single polymer chain in solution. *J. Chem. Phys.* **127**, 044903 (2007).

Acknowledgements

The research reported in this publication was supported by the King Abdullah University of Science and Technology, Thuwal, Saudi Arabia.

Author contributions

M.F.S. introduced the concept, designed the simulations and the analytical approach, conducted the experiments, wrote the Matlab scripts, analysed the data and wrote the manuscript. M.A. assisted on the experiments and data analysis. S.H. supervised the study and wrote the manuscript.

Additional information

9

Ultrasensitive visual read-out of nucleic acids using electrocatalytic fluid displacement

Justin D. Besant[1], Jagotamoy Das[2], Ian B. Burgess[2], Wenhan Liu[1], Edward H. Sargent[3] & Shana O. Kelley[1,2,4]

Diagnosis of disease outside of sophisticated laboratories urgently requires low-cost, user-friendly devices. Disposable, instrument-free testing devices are used for home and physician office testing, but are limited in applicability to a small class of highly abundant analytes. Direct, unambiguous visual read-out is an ideal way to deliver a result on a disposable device; however, existing strategies that deliver appropriate sensitivity produce only subtle colour changes. Here we report a new approach, which we term electrocatalytic fluid displacement, where a molecular binding event is transduced into an electrochemical current, which drives the electrodeposition of a metal catalyst. The catalyst promotes bubble formation that displaces a fluid to reveal a high contrast change. We couple the read-out system to a nanostructured microelectrode and demonstrate direct visual detection of 100 fM DNA in 10 min. This represents the lowest limit of detection of nucleic acids reported using high contrast visual read-out.

[1]Institute for Biomaterials and Biomedical Engineering, University of Toronto, Toronto, Canada M5S 3G9. [2]Department of Pharmaceutical Science, Leslie Dan Faculty of Pharmacy, University of Toronto, Toronto, Canada M5S 3M2. [3]Department of Electrical and Computer Engineering, Faculty of Engineering, University of Toronto, Toronto, Canada M5S 3G4. [4]Department of Biochemistry, Faculty of Medicine, University of Toronto, Toronto, Canada M5S 1A8. Correspondence and requests for materials should be addressed to E.H.S. (email: ted.sargent@utoronto.ca) or to S.O.K. (email: shana.kelley@utoronto.ca).

ow-cost, user-friendly diagnostics have the potential to expand the ubiquity of molecular testing in clinical medicine[1-11]. Disposable, instrument-free devices are used today, but have so far only achieved the detection of certain analytes that happen to be highly abundant. A key feature of these devices is the use of an easy-to-interpret visual read-out strategy. Existing read-out approaches require the accumulation of a high level of an analyte, and therefore only abundant analytes have been detected visually. Developing ways to link a visible, unambiguous colour change to rare biological molecules remains an unmet need. Recently, a variety of direct, colorimetric read-out strategies have been reported: these include approaches based on nanoparticles[12,13], plasmonic nanomaterials[14], 2D materials[15] and enzymatic reactions[7]. Unfortunately, these approaches require interpretation of subtle colour changes. This can make analyses operator-dependent, or, in other cases, diminishes the benefits of a test being instrument-free by requiring a scanner device.

Developing new, easy-to-interpret interfaces that convey diagnostic results obtained with low-abundance analytes would enable the development of low-cost diagnostics for a spectrum of new diseases. Motivating this work are rapid recent advances in biosensors that produce nanoampere electrical current changes as a function of specific biomarkers present in a sample[16-18]. New strategies to transduce extremely small electrochemical currents into easily perceived, high-contrast visual changes would allow the visual detection of low abundance analytes using electrochemical biosensors. In addition, low-cost current-to-colour conversion is of broad interest in displays and in sensors for non-medical applications.

Strategies for direct colorimetric read-out of electric currents include paper-based electrochromism[19], electrochromic polymers[20], metal oxides[21] and fluorescent dyes[22]. Electrochromic polymers and dyes allow for rapid and reversible colour switching in response to electrical currents, but the currents required to switch areas detectable to the naked eye are above the threshold necessary for sensitive electrochemical detection. Inducing visible colour changes using currents below 1 µA is a fundamental challenge, for such currents fail to supply enough electrons to electrochemically reduce a visibly perceptible quantity of electrochromic material. Directly translating such low currents into visible changes has yet to be achieved without the aid of costly, power-consumptive active electronics such as amplifiers.

We here develop an approach to amplify the changes to optical density triggered by the levels of electrochemical current generated at a nucleic acid sensor. We term our new approach electrocatalytic fluid displacement (EFD). An electrochemical current drives the deposition of a catalyst, which promotes the growth of a bubble that actuates a fluid. Specifically, the electrochemical current drives the electrodeposition of a metal catalyst for hydrogen peroxide decomposition. On the introduction of hydrogen peroxide liquid, a bubble catalytically forms, and this displaces a fluid. The bubble displaces a dye, or, in the alternative, modifies the index of refraction to reveal a structural colour change. We begin by providing a conceptual basis for our approach, and we benchmark it against other colorimetric read-out strategies. After optimizing the device parameters and geometry, we determine the minimum current necessary for successful colorimetric read-out. To showcase this approach, we demonstrate sensitive colorimetric detection of ssDNA by coupling the read-out to a nanostructured microelectrode (NME) and a novel electrocatalytic assay.

Results

Overview of electrocatalytic fluid displacement. The electrocatalytic fluid displacement (EFD) approach is based on the

electrodeposition of platinum, which catalyses the evolution of a bubble that actuates a fluid (Fig. 1). An electrochemical sensor is connected to a read-out chamber by a metallic bridge (Fig. 2a). On the introduction of the sample, the target analyte hybridizes to the complementary probe functionalized on the surface of the sensor (Fig. 2a). After hybridization, the electrocatalytic solution is introduced into the sensing chamber (Fig. 2b). On the application of a potential at 250 mV for 10 s, ruthenium is oxidized at the sensing electrode and platinum is simultaneously electrodeposited at a mesh electrode in the read-out chamber. The current is further amplified by two additional reducing agents in the electrocatalytic solution. After the application of the potential, hydrogen peroxide is introduced into the read-out chamber (Fig. 2c). The deposited platinum catalyses the decomposition of hydrogen peroxide into water and oxygen, which forms a bubble. The growing bubble displaces a fluid to reveal a colour change. In the case of read-out based on dye displacement, the opaque dye is displaced and the read-out window becomes transparent revealing a blue spot underneath the device (Fig. 2d). In the case of read-out based on a structural colour change, the growing bubble causes an index mismatch that unveils the diffraction grating patterned in the underside of the PDMS chamber lid, which causes light to diffract into its component colours (Fig. 2e).

To sense nucleic acids, the EFD read-out system is connected to a nanostructured microelectrode (NME), which acts as an ultrasensitive electrochemical biosensor (Fig. 2a)[17,23]. The NME sensors were fabricated on silicon substrates using a two-step electrodeposition process as previously described[23]. The gold microstructures protrude from the surface and reach into solution, which increases the probability of interaction with the target molecules[23]. The sensors are decorated with a second layer of finely nanostructured gold. The nanoscale roughness maximizes sensitivity by enhancing the hybridization efficiency of the probe and target[24,25]. These sensors have been used previously to detect a variety of chemical and biomolecular targets[26-28].

We use a multi-pronged strategy to minimize the current in the absence of target nucleic acid. We functionalize the sensors using a charge-neutral probe, and we read the current using a novel electrochemical assay. Specifically, the sensors are functionalized with thiolated peptide nucleic acid (PNA) probes complementary to the target sequence. PNA is a synthetic nucleic acid analogue that has a neutral charge. This neutral charge minimizes the background current and increases the signal-to-noise ratio.

After target hybridization and washing (Fig. 2a), we read-out the electrochemical signals using an electrochemical-chemical-chemical (ECC) redox cycle reporter system, which radically amplifies the current. It is worth noting that this is the first

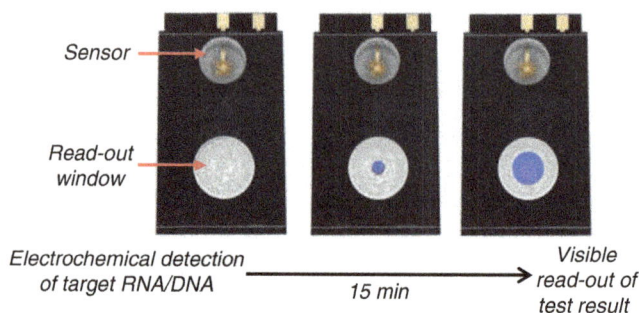

Electrochemical detection of target RNA/DNA → 15 min → Visible read-out of test result

Figure 1 | Schematic of an integrated device for electrocatalytic fluid displacement. An electrochemical current from a nanostructured microelectrode is converted into a visible change through the deposition of a catalyst that catalyses bubble formation. As the bubble grows, the white dye is displaced to reveal a blue colour.

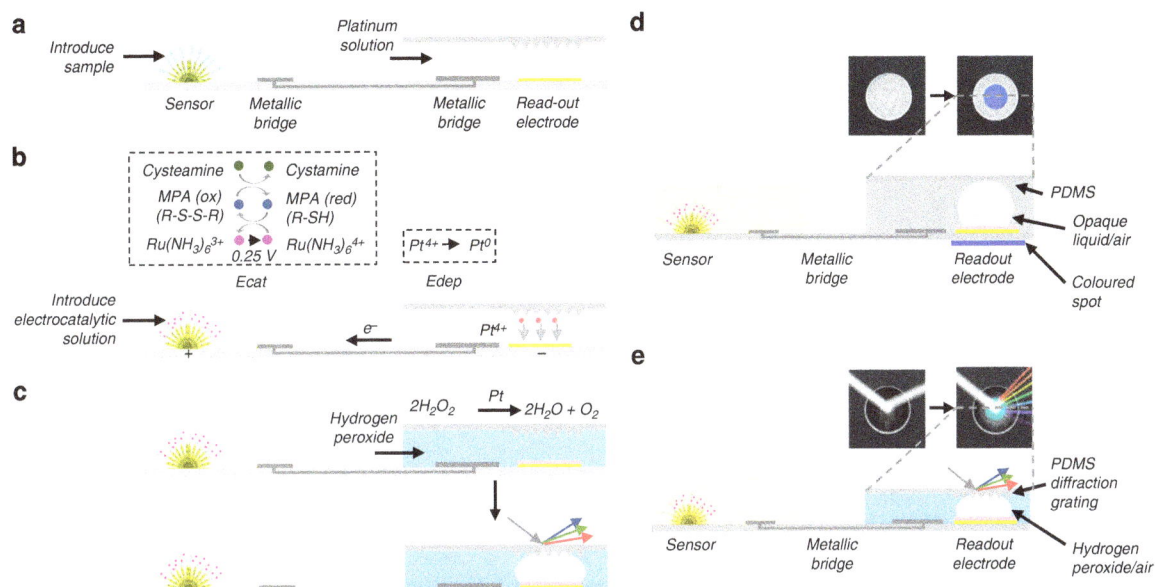

Figure 2 | Colorimetric detection of DNA using electrocatalytic fluid displacement. (**a**) Target hybridization. The analyte hybridizes to a complementary PNA probe. $Ru(NH_3)_6^{3+}$ is electrostatically attracted to the negatively charged backbone of the target nucleic acid. (**b**) Signal transduction. A potential is applied to the NME which oxidizes $Ru(NH_3)_6^{3+}$. The current is amplified using an electrochemical-chemical-chemical (ECC) reporter system. $Ru(NH_3)_6^{3+}$ is regenerated by MPA, which is in turn regenerated by cysteamine. The electrochemical current drives the deposition of platinum, a catalyst for hydrogen peroxide decomposition, on a mesh electrode immersed in platinum solution. (**c**) Colorimetric read-out. After the introduction of peroxide, a bubble forms as the platinum catalyses the decomposition of peroxide. The growing bubble is transduced into a colour change either through an optical density change or a structural colour change. In the optical density approach, the bubble displaces a white dye to reveal the blue spot. To induce a structural colour change, the bubble displaces peroxide that causes an index mismatch at a diffraction grating patterned in the underside of the chamber lid. Incident white light is diffracted into its component colours. (**d**) Optical density change. A cross-section of the electrocatalytic fluid displacement approach with read-out based on dye displacement. (**e**) Structural colour change. A cross-section of the electrocatalytic fluid displacement approach with read-out based on a structural colour change. In the case of electrocatalytic fluid displacement based on a structural colour change, the underside of the PDMS lid of the read-out chamber is patterned with a diffraction grating.

reported use of ECC for the detection nucleic acids[29]. To simplify the electronics in a disposable device, we use a DC potential for read-out as opposed to voltammetry, which requires a potential sweep and thus more complicated electronics. Using a DC potential, it is not possible to resolve the contribution from unwanted redox reactions occurring at nearby potentials to the overall signal. In the past we have used an electrocatalytic redox reporter system consisting of ruthenium hexamine and ferricyanide for nucleic acid detection using differential pulse voltammetry[30], but this gave high background currents using DC potential amperometry. At the reduction potential of ruthenium hexamine, ferricyanide is reduced as well, which contributes to the overall current even in the absence of bound-target nucleic acids. Therefore, we designed the ECC redox reporter system such that there are no interfering redox reactions near the potential of interest.

Our new assay employs $Ru(NH_3)_6^{3+}$, mercaptopropionic acid (MPA) and cysteamine. $Ru(NH_3)_6^{3+}$ is electrostatically attracted to the negatively charged phosphate backbone of the bound target nucleic acids. On the application of a potential at 250 mV, $Ru(NH_3)_6^{3+}$ is oxidized to $Ru(NH_3)_6^{4+}$ (ref. 31). The MPA present in solution chemically reduces $Ru(NH_3)_6^{4+}$ back to $Ru(NH_3)_6^{3+}$, allowing for multiple turnovers of $Ru(NH_3)_6^{3+}$, which generates a high electrocatalytic current. This signal is further amplified by cysteamine, another reducing agent, which is chemically oxidized to cystamine by reducing the oxidized form of MPA (R-S-S-R) back to its reduced form (R-SH).

This electrical current from the sensor is coupled to the EFD electrode immersed in a platinum solution to drive the deposition of the catalyst. When the NME is challenged with the target analyte, the current drives the electrodeposition of platinum on the EFD electrode, which catalytically forms a bubble that displaces the dye to reveal the blue spot. When the target sequence is not present, the current is too low to deposit a sufficient amount of platinum to catalyse bubble growth and no colour change occurs.

A comparative model of colour change. Catalytic electrochromic transduction methods offer significant signal amplification needed for transducing the ultra-low currents generated by the ECC assay compared with direct electrochromic reduction. To study the prospective performance of this approach, we calculated the predicted time required to induce a visible colour change using a variety of transduction strategies.

We illustrate the challenge of directly inducing a colour change by considering the example of electrodepositing an optically discernible quantity of metal. A 1 nA current applied for 10 s supplies 6×10^{10} electrons, which can turnover a maximum of 6×10^{10} molecules. Even under the generous assumption that a single molecular layer is visible, given an atomic radius of 1 Å, this yields a spot of only 50 μm × 50 μm. This is too small to be easily visible to the naked eye as the spatial resolution of human eyesight is ~100–200 μm (ref. 32).

We hypothesized that we could instead develop a means to amplify, by orders of magnitude, the colour change per charge. We would electrodeposit a catalyst, such as platinum, to turn on the colorimetric reaction[19]. By depositing a catalyst, each electron effectively converts multiple molecules, amplifying the colour transformation. However, as Fig. 3 shows, even the catalytic reduction of an electrochromic compound in bulk solution requires exceedingly long times to induce a visible change. Assuming a 50-μm tall chamber with a 200 μm diameter window filled with enough pigment, with the absorbance of malachite green, to give an OD of 1, it would take over 4 h to turnover the compound using the platinum deposited from a 1 nA current.

Thus, instead of catalytic reduction of a solution-based pigment, we considered catalytic evolution of gas as an equivalent molar amount of gas occupies a much larger volume than a liquid. At STP, the volume of one mole of gas is 22.4 l, which is three orders of magnitude larger than a mole of liquid H_2O (18 ml). Platinum is an excellent catalyst for the decomposition of hydrogen peroxide to form oxygen and water[33]. As shown in Fig. 3, the catalytic production of a visible bubble that fills the same window requires under 3 min, over 80 times faster than catalytic reduction of an electrochromic dye in solution.

We hypothesized that electrocatalytic bubble formation could be converted into a colorimetric change by actuating a fluid to modulate the optical density (OD) of the read-out window. This is a central step in EFD.

Optimization of device geometry. Motivated by our calculations, we sought to experimentally validate the electrocatalytic fluidic displacement approach. We patterned a rectangular gold electrode on a glass substrate, which sits at the bottom of a 50-μm tall by 1.5-mm wide circular chamber. After depositing platinum for 10 s at 1 nA, we introduced hydrogen peroxide and measured the rate of bubble growth using a microscope (Fig. 4a). Although we did not observe rapid bubble growth, we noticed that bubbles formed preferentially at the electrode edges.

To test the enhancement provided by edges, we designed mesh-shaped electrodes with increased ratios of edges to surface area. We applied 1 nA for 10 s to deposit platinum and recorded the rate of bubble growth (Fig. 4a). We found that the rate of bubble evolution increased with increasing numbers of edges. The highest density mesh, with 3.4 × the edge to surface area ratio of the rectangular electrode, provided the fastest bubble growth. No bubbles formed when no current was applied as no platinum was electrodeposited. Bubble growth was not observed after immersing the device in platinum solution for 25 min, indicating that platinum is not deposited via electroless deposition (Supplementary Fig. 1).

Using the high-density mesh electrodes, we measured the average growth of the bubble for various applied currents. Figure 4b shows the average bubble area measured after 20 min as a function of electrodeposition current while Fig. 4c shows the bubble growth over time. After 20 min, a 1 nA current applied for 10 s yields a bubble with an area of $0.25\,mm^2$, which is visibly detectable.

Electrocatalytic fluidic dye displacement. To induce a visible colour change that is easily interpretable by the end-user, we utilize the bubble to displace an opaque dye that obscures a blue spot beneath the read-out window. As the chamber fills with oxygen, the blue spot is revealed.

Increasing the dye concentration increases the opacity of the dye, but also increases its viscosity. We found that at higher viscosities, bubble formation was inhibited (Fig. 4d). We optimized the dye concentration and found that using a concentration of $25\,\mu g\,ml^{-1}$ allowed for sufficient optical density to conceal the blue spot while promoting bubble growth (Fig. 4d).

To determine the minimum visibly detectable current, we deposited platinum at various rates for 10 s and measured the exposed area of the blue spot (Fig. 4e). Using a 1 nA deposition current, the spot area grows to $0.09\,mm^2$ in 5 min. The exposed area expands to $0.24\,mm^2$ by 20 min. No bubble growth is observed when platinum is not electrodeposited. As the spatial resolution of human eyesight is about 200 μm (ref. 32) the smallest visible area ∼200 μm × 200 μm or $0.04\,mm^2$. Thus, the spot area of $0.09\,mm^2$ obtained from a 1 nA current after 5 min is visible to the naked eye.

To quantify the performance of our device we calculated the colouration efficiency, a metric that quantifies the efficiency of converting an electrical current into a colorimetric change. Colouration efficiency, CE, is given by:

$$CE = \frac{\Delta OD \cdot A}{Q} \tag{1}$$

Where ΔOD is the change in optical density, Q is the charge required for switching [C] and A is the spot area [cm^2]. We measured the optical density before and after switching and found a ΔOD of 0.27 (Fig. 4f). Given a switchable area of $0.24\,mm^2$ after 20 min using a 1 nA current applied for 10 s, this device has a colouration efficiency of $6.48 \times 10^4\,cm^2\,C^{-1}$. Figure 5a compares the switchable area as a function of charge for devices with the highest reported colouration efficiencies for a range of read-out strategies. Given the previous records of $2.6 \times 10^4\,cm^2\,C^{-1}$ for fluorescent polymers[22] and $9.3 \times 10^2\,cm^2\,C^{-1}$ for non-fluorescent electrochromic compounds[20], a colouration efficiency of $6.48 \times 10^4\,cm^2\,C^{-1}$ is, to our knowledge, the highest reported value in the literature for an electrochromic device.

Induction of a structural colour change. As optical absorbance increases with path length, the read-out window must be sufficiently tall for the dye to obscure the coloured spot beneath. This limits the response time of a colorimetric device based on dye displacement, as the bubble must grow large enough to reach the chamber ceiling.

By patterning substrates with feature sizes on the order of the wavelength of light, it is possible to produce vibrant structural colours[34]. Examples of this include diffraction gratings and iridescence. The colour of the substrate can be modified by matching the index refraction between a second medium and the substrate[35].

We hypothesized that we could exploit a structural colour change to decrease the read-out turnaround time. As structural colour changes rely on the index matching at an interface, the

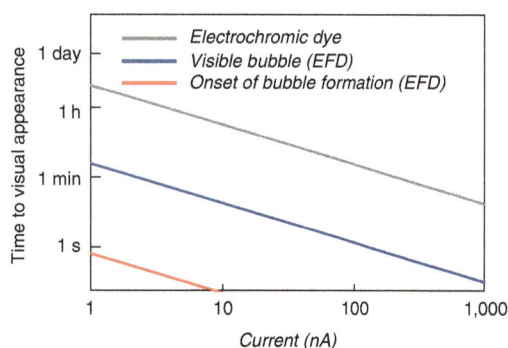

Figure 3 | Calculations of the time to visual appearance. We calculated the time to visual appearance using electrocatalytic fluid displacement and reduction of an electrochromic compound. We assume the read-out window is a 200 μm × 200 μm × 50 μm chamber and the current is applied for 10 s. The onset of bubble formation occurs as the solution is saturated with oxygen. A bubble is defined as visible once it reaches the volume of the chamber. We assumed the electrochromic dye had the absorbance of malachite green and a visible change corresponds to a ΔOD of 1.

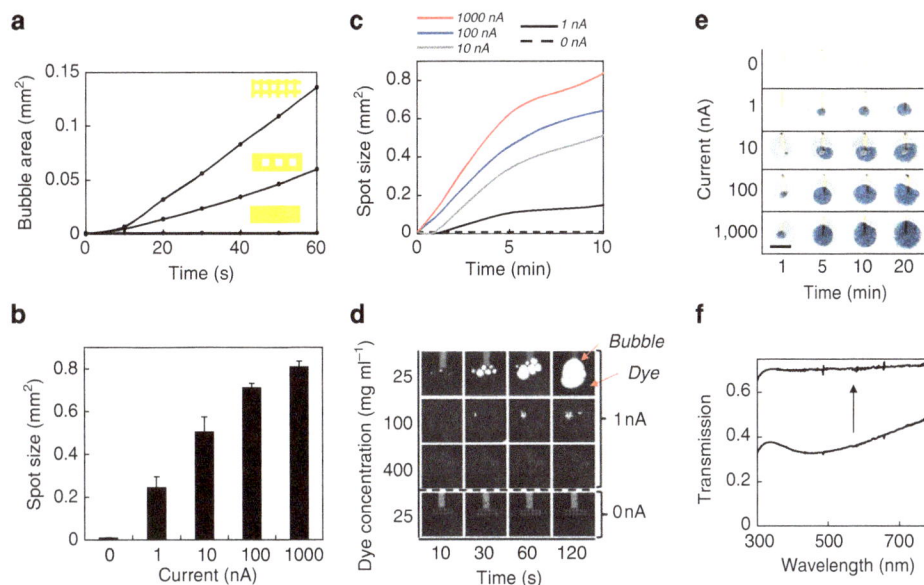

Figure 4 | Electrocatalytic fluid dye displacement. (a) Bubble evolution as a function of time for various electrode geometries. Platinum was deposited using a 1 nA current for 10 s. Bubble growth increases with the ratio of edges to surface area. **(b)** Average bubble area after 20 min as a function of applied current using the electrodes with the highest mesh density. The bubble is confined to a 50-μm tall channel. Error bars represent s.e.; all measurements represent n > 5 trials. **(c)** Bubble growth as a function of time for various deposition currents using electrodes with the highest mesh density. Bubbles do not form when no current is applied. **(d)** Images of bubble growth as a function of dye concentration acquired using an optical microscope. **(e)** Images of colorimetric read-out as a function of deposition current and time. One-nA currents are detectable in 5 min. Scale bar, 1 mm. **(f)** Transmission spectrum of the read-out window before and after bubble growth.

colour change is largely independent of the path length through the index-matching medium. Thus, we could expect a vibrant colour change using a device with a much smaller channel height than required when using dye displacement. As the substrate provides the colour, there is no need to increase the opacity of the peroxide by introducing additional compounds, which might interfere with the reaction.

To prove out this approach, we patterned a diffraction grating into the underside of the PDMS lid affixed to the top of the device with a 7-μm tall channel. As the index of refraction of peroxide ($n = 1.35$) is similar to that of PDMS ($n = 1.4$), the diffraction grating is invisible to incoming light when the device is initially loaded with peroxide. As the bubble forms, the peroxide is replaced with O_2, which has an index of refraction of 1. This index mismatch between the bubble and PDMS unveils the diffraction grating. The incident white light is diffracted into its component colours to reveal the circular spot.

Figure 5b shows the growth of the coloured spot using the diffraction grating approach while Fig. 5c shows the corresponding images of the spot over time. As the bubble grows, white light begins to diffract into its component colours. The window turns from optically transparent (which appears as black due to a black background) to cyan as light at that wavelength is diffracted towards the camera. Using a 1 nA deposition current, the spot size is 0.06 mm² after 1 min, which is visible by eye. This spot grows to 0.36 mm² and 1.1 mm² by 5 and 15 min, respectively. Given a spot size of only 0.1 mm² after 5 min using electrocatalytic fluidic displacement of a dye, the structural colour spot of 0.36 mm² is over three times larger in the same time frame. No spot forms when no current is applied Fig. 5c.

Colorimetric read-out of ssDNA. To test the capability of the EFD device to detect biomarkers, we connected the NME sensors in serial (Supplementary Fig. 2) to the EFD read-out chip and

challenged the NMEs with ssDNA. As an initial characterization of the ECC assay, we challenged the sensors with serial dilutions ssDNA. We then measured the corresponding currents after applying 250 mV (Fig. 6a). The average peak current decreases with decreasing target ssDNA concentration giving a detection limit of 1 fM (Fig. 6b). The current generated from 100 nM non-complementary ssDNA is less than 2 nA, which is similar to the background current, indicating this read-out method is specific.

To demonstrate colorimetric read-out of biomarkers, we coupled the assay to our read-out device and challenged the sensors with serial dilutions of ssDNA. To connect the sensors to our EFD device, we immersed the NME sensors in the ECC solution and the EFD read-out device in the platinum electrodeposition solution. To bridge electronically the sensor and read-out device, a platinum wire electrode immersed in the ECC solution is connected to a second platinum electrode in the electrodeposition bath. The EFD read-out device acts as the counter electrode for the entire system (Fig. 2). After applying 250 mV for 10 s to the NME, we introduced peroxide into the EFD chip and measured the rate of colour formation (Fig. 6c). We found a detection limit of 1 pM after 10 min with an average spot size of 0.068 mm². To our knowledge, a detection limit of 1 pM is the lowest reported limit of detection for colorimetric detection of ssDNA using an electrochemical sensor. No visible spot was observed when the sensors were challenged with 100 nM of non-complementary ssDNA indicating a specificity discrimination ratio of 1×10^5 (Fig. 6d).

We studied the performance of the diffraction grating approach for colorimetric ssDNA detection. First, we optimized the peroxide concentration to minimize bubble formation from currents at the background level. We found that bubble growth at low currents could be suppressed using 10% peroxide (Supplementary Fig. 3). We challenged the devices with ssDNA and measured the growth of the diffracting area (Fig. 6e). Figure 6f shows the corresponding images of the growth of

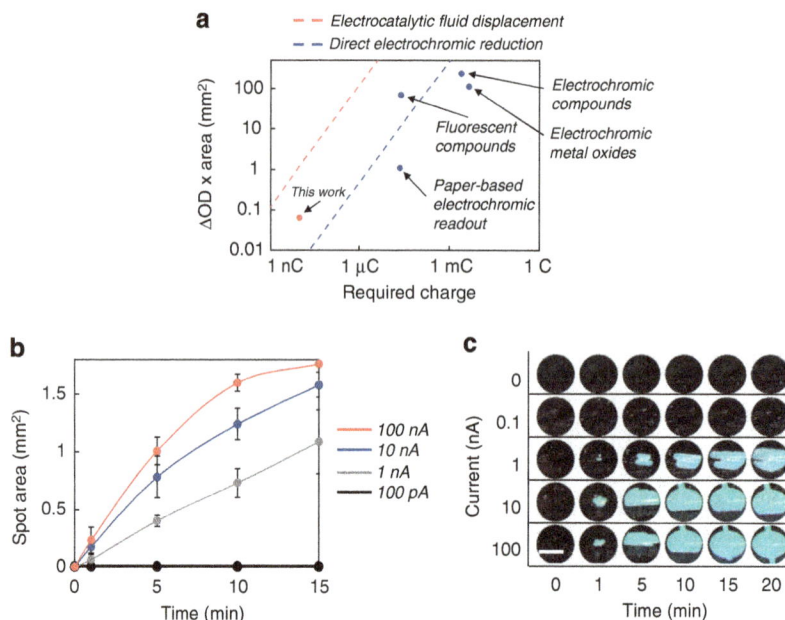

Figure 5 | Electrocatalytic fluid displacement reveals a structural colour change. (**a**) Comparison of the charge required to induce a visible colour of a certain area and optical density change for a variety of read-out strategies. The dashed red line represents the calculated exposed area of a bubble generated using electrocatalytic fluid displacement. We assume the bubble is confined to a 50-μm tall chamber, the reaction proceeds for 10 min, and the ΔOD is 1. The dashed blue line represents the area of a monoatomic layer of platinum directly reducible by the current. We assume the ΔOD is 1 and thus this represents an upper bound using this strategy. (**b**) Spot size as a function of time for various deposition currents using electrodes with the highest mesh density. Bubbles do not form when no current is applied. Error bars represent s.e. ($n = 3$). (**c**) Images of colorimetric read-out as a function of deposition current and time using a diffraction grating. The window turns from optically transparent (which appears as black due to a black background) to cyan as light at that wavelength is diffracted towards the camera. One-nA currents are detectable in 1 min. Scale bar, 1 mm.

the visible spot over time. Using 1 pM complementary ssDNA, the spot size was 0.15 mm^2 after 10 min. In that same time frame, the spot using dye displacement was 0.068 mm^2, which is about two times smaller. This is expected as the chamber height using the diffraction grating approach was five times smaller than the chamber in the dye displacement device, and thus the bubble can rapidly grow laterally. Using this method, 100 fM of ssDNA was also detectable by eye with an average spot size of 0.085 mm^2 (Fig. 6e). No spot was visible with 100 nM non-complementary ssDNA (Fig. 6f).

Discussion

This colorimetric read-out approach is an inexpensive, disposable and low-power alternative to using electronics for read-out of electrochemical currents. We use three stages of amplification to transform ultra-low currents into colorimetric changes. The ECC redox assay uses two chemical catalysis steps to amplify the electrodeposition current. Thus, each bound nucleic acid is converted into multiple deposited platinum atoms. Next, by electrodepositing a catalyst, the colorimetric reaction continues long after the initial application of the electrochemical current, obviating the need for high currents to induce colorimetric changes. Last, colorimetric read-out is accelerated by exploiting the fact that a gas occupies a much larger volume than an equivalent molar amount of liquid. A catalytic reaction can evolve a large volume of gas much faster than it can turnover a visible amount of dye of the same volume.

Even though spot sizes as small as 100–200 μm are visible under perfect conditions to those with 20/20 vision, spot sizes <1 mm^2 may be difficult to see for some. Small spot sizes could be easily magnified using inexpensive lenses fabricated from elastomers such as PDMS.

A fully integrated device would require the timed introduction of reagents with automated flow. This could be integrated onto an instrument-free device using passive fluidic systems such as paper microfluidics, capillary pumps or on-chip vacuum pumps[3,36,37]. As this strategy only requires the application of a DC potential, the potential could be applied using a DC power source such as a battery as opposed to a potentiostat.

In summary, we introduce a strategy for rapid and sensitive colorimetric read-out of electrochemical currents based on electrocatalytic fluid displacement. This approach relies on the electrochemical mediated deposition of platinum which catalyses the growth of a fluid displacing bubble. We present two strategies for converting this fluidic displacement into a visible colour change using a dye and a structural colour change. We demonstrate successful colorimetric detection of a 1 nA current in 1 min and calculate a colouration efficiency of 6.48×10^4 C cm^{-2}, which to our knowledge is the highest value reported in the literature. We showcase this approach by coupling our device to a novel electrocatalytic assay and nanostructured microelectrode sensor to demonstrate successful and specific colorimetric detection of 100 fM of ssDNA in 10 min with a discrimination ratio of over 1×10^5.

Methods

Calculation of time required for visible bubble formation. To calculate the rate of colour change using direct electrochromic colorimetric read-out, we assume a channel 50-μm tall by 200-μm wide filled with enough electrochromic dye to give an OD of 1. We assumed a high molar absorptivity of 1×10^7 M^{-1} m^{-1}, which is similar to that of malachite green. Using the catalysis rate of platinum, we calculate the time needed to turn over the dye in the channel. To calculate the rate of colour change using electrocatalytic fluidics, we assume a channel a chamber that is 50-μm tall with a 200-μm width. Using the catalysis rate of platinum, we calculated the rate of oxygen formation. The onset of bubble formation occurs as peroxide in the chamber is saturated with oxygen. We assume the bubble is visible once it grows to

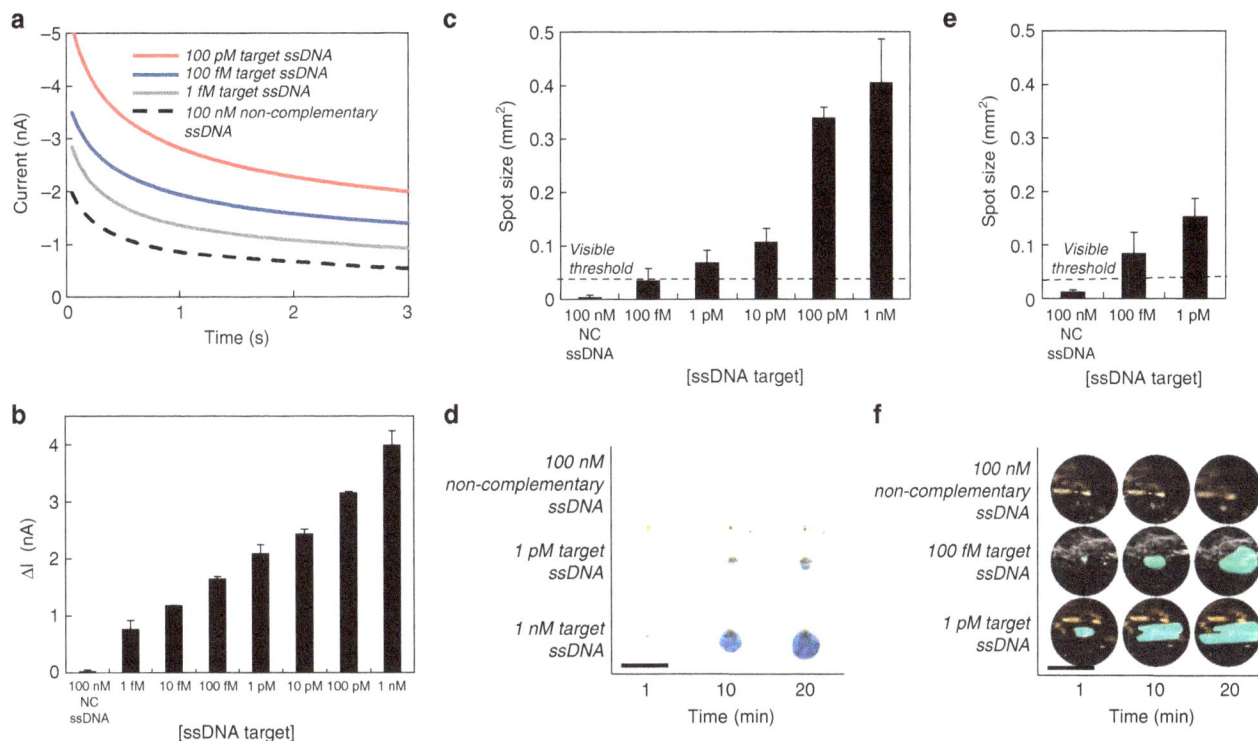

Figure 6 | Colorimetric detection of DNA. (a) Electrochemical current as a function of time for various analyte concentrations after applying 250 mV w.r.t. to a Ag/AgCl reference electrode for 3 s. **(b)** Average peak electrochemical current as a function of analyte concentration. Negative control is non-complementary (NC) DNA. **(c)** Spot size as a function of target DNA concentration after 10 min using dye displacement. One-pM ssDNA is detectable by eye. The visible threshold is defined as an area of 200 mm × 200 mm. **(d)** Images of the EFD device showing growth of the bubble over time as a function of ssDNA concentration using dye displacement. **(e)** Spot size as a function of target DNA concentration after 10 min using a structural colour change. **(f)** Images of the EFD device showing growth of the bubble over time as a function of ssDNA concentration using a structural colour change. Scale bar, 1 mm. Error bars represent s.e.; all measurements represent $n > 5$ trials.

the volume of the chamber. Parameters used in the calculations are listed in Supplementary Table 1.

Device fabrication. The device was fabricated using standard photolithographic methods. In brief, electrodes were patterned on a glass substrate. The device was passivated using SU-8 2002 (Microchem, Newton, MA) and apertures were patterned to expose the electrodes below. The channel was fabricated by patterning SU-8 3050 (Microchem, Newton, MA).

Platinum electrodeposition. The electrode was immersed in K_2PtCl_4 (Sigma-Aldrich, MO) and connected to an Epsilon potentiostat (BASi West Lafayette, IN) using a three-electrode set-up with a Ag/AgCl reference electrode and a Pt counter electrode. Using chronopotentiometry, various currents were applied for 10 s. After electrodeposition, the device was washed thoroughly with H_2O and covered with a PDMS (Dow Chemical, MI) lid.

Colorimetric read-out using a dye. Hundred microlitres of white dye (Liquitex Titanium White Ink) was centrifuged for 5 min at 15 000 g. The supernatant was removed and replaced with 400 µl of 30% H_2O_2 (Sigma-Aldrich, MO). The dye ($25 \mu g\,ml^{-1}$) was introduced into the channel and the amount of bubble generation was measured over time using a camera (Canon).

Colorimetric read-out using structural colour. A diffraction grating was patterned in PDMS by curing PDMS on a DVD-R. The PDMS diffraction grating lid was removed and attached to the device with a 7-µm tall channel patterned using SU-8 2010 (Microchem, Newton, MA). Twenty-seven percent H_2O_2 with 1% pluoronic (Sigma-Aldrich, MO) was introduced into the device, and colour changes were measured over time using a camera (Canon).

Sensor chip fabrication. Six-inch silicon wafers (University Wafer, MA) were passivated using a thick layer of thermally grown silicon dioxide and coated with a 25 nm Ti adhesion layer. A 350-nm gold layer was deposited on the chip using electron-beam-assisted gold evaporation, which was again coated with 5 nm of Ti. The electrodes were patterned in the metal layers using standard photolithography

and a lift-off process. A 500 nm layer of insulating Si_3N_4 was deposited using chemical vapour deposition. The 5-µm apertures were etched at the tips of the metal leads using standard photolithography. Contact pads (0.4 mm × 2 mm contact) were patterned using wet etching as well.

Fabrication of sensors. Chips were cleaned by sonication in acetone for 5 min, rinsed with isopropyl alcohol and DI water, and dried with nitrogen. Electro-deposition was performed at room temperature. The 5-µm apertures on the fabricated electrodes were used as the working electrodes and were contacted using the exposed contact pads. Nanostructured microelectrode sensors were electro-deposited in a solution of 50 mM $HAuCl_4$ (Sigma-Aldrich, MO) and 0.5 M HCl (Sigma-Aldrich, MO) using DC potential amperometry at 0 mV for 100 s. After washing with DI water and drying, the sensors were coated again with a thin layer of Au to form nanostructures by plating at $-450\,mV$ for 10 s.

Functionalization of sensors. An aqueous solution containing 1 µM of probe (5′-GGT CAG ATC GTT GGT GGA GT-3′) (PNA Bio, CA) was mixed with 10 µM of aqueous Tris(2-carboxyethyl)phosphine hydrochloride solution (Sigma-Aldrich, MO) and then the mixture was left for overnight to cleave disulphide bonds. After mixing 100 nM of 6-mercaptohexanol (MCH) (Sigma-Aldrich, MO) to this probe solution mixture, 20 µl was pipetted onto the chips and incubated for 3 h in a dark humidity chamber at room temperature for probe immobilization. The chips were then washed thrice for 5 min with 0.1 × PBS (Life Technologies, CA) at room temperature. The chips were then treated with 1 mM MCH for an hour at room temperature for back filling. After washing, the chips were challenged with different concentration of targets for 30 min at room temperature. After hybridization, the chips were washed thrice for 5 min with 0.1 × PBS at room temperature and the electrochemical scans were acquired.

Electrochemical detection of ssDNA. All electrochemical experiments were carried out using a Bioanalytical Systems Epsilon potentiostat with a three-electrode system featuring a Ag/AgCl reference electrode and a platinum wire auxiliary electrode. Electrochemical signals were measured in a Tris buffer solution (50 mM, pH 9) containing 10 µM $[Ru(NH_3)_6]Cl_3$ (Sigma-Aldrich, MO), 0.5 mM 3-mer-captopropionoic acid (MPA) (Sigma-Aldrich, MO) and 0.5 mM cysteamine (Cys)

(Sigma-Aldrich, MO). DC potential amperometry (DCPA) signals were obtained at $+250$ mV for 10 s. Signal changes, ΔI, were calculated with $\Delta I = I_c - I_0$ (where I_c is the current at a given concentration and I_0 is the current without analyte).

References

1. Kelley, S. O. *et al.* Advancing the speed, sensitivity and accuracy of biomolecular detection with multi-length scale engineering. *Nat. Nanotechnol.* **9,** 969–980 (2014).
2. Patterson, A. S., Hsieh, K., Soh, H. T. & Plaxco, K. W. Electrochemical real-time nucleic acid amplification: towards point-of-care quantification of pathogens. *Trends Biotechnol.* **31,** 704–712 (2013).
3. Begolo, S., Zhukov, D. V., Selck, D. A., Li, L. & Ismagilov, R. F. The pumping lid: investigating multi-material 3D printing for equipment-free, programmable generation of positive and negative pressures for microfluidic applications. *Lab. Chip.* **14,** 4616–4628 (2014).
4. Funes-Huacca, M. *et al.* Portable self-contained cultures for phage and bacteria made of paper and tape. *Lab. Chip.* **12,** 4269–4278 (2012).
5. Deiss, F., Funes-Huacca, M. E., Bal, J., Tjhung, K. F. & Derda, R. Antimicrobial susceptibility assays in paper-based portable culture devices. *Lab. Chip.* **14,** 167–171 (2014).
6. Nie, S. *et al.* An automated integrated platform for rapid and sensitive multiplexed protein profiling using human saliva samples. *Lab. Chip.* **14,** 1087–1098 (2014).
7. Song, Y. *et al.* Multiplexed volumetric bar-chart chip for point-of-care diagnostics. *Nat. Commun.* **3,** 1283 (2012).
8. Scida, K., Li, B., Ellington, A. D. & Crooks, R. M. DNA detection using origami paper analytical devices. *Anal. Chem.* **85,** 9713–9720 (2013).
9. Cunningham, J. C., Brenes, N. J. & Crooks, R. M. Paper electrochemical device for detection of DNA and thrombin by target-induced conformational switching. *Anal. Chem.* **86,** 6166–6170 (2014).
10. Scida, K., Cunningham, J. C., Renault, C., Richards, I. & Crooks, R. M. Simple, sensitive, and quantitative electrochemical detection method for paper analytical devices. *Anal. Chem.* **86,** 6501–6507 (2014).
11. Yang, F. *et al.* A bubble-mediated intelligent microscale electrochemical device for single-step quantitative bioassays. *Adv. Mater.* **26,** 4671–4676 (2014).
12. Xia, F. *et al.* Colorimetric detection of DNA, small molecules, proteins, and ions using unmodified gold nanoparticles and conjugated polyelectrolytes. *Proc. Natl Acad. Sci. USA* **107,** 10837–10841 (2010).
13. Taton, T. A., Mirkin, C. A. & Letsinger, R. L. Scanometric DNA array detection with nanoparticle probes. *Science* **289,** 1757–1760 (2000).
14. De la Rica, R. & Stevens, M. M. Plasmonic ELISA for the ultrasensitive detection of disease biomarkers with the naked eye. *Nat. Nanotechnol.* **7,** 821–824 (2012).
15. Guo, Y. *et al.* Hemin − graphene hybrid nanosheets with intrinsic peroxidase-like activity for label-free colorimetric detection of single-nucleotide polymorphism. *ACS Nano* **5,** 1282–1290 (2011).
16. Besant, J. D., Das, J., Sargent, E. H. & Kelley, S. O. Proximal bacterial lysis and detection in nanoliter wells using electrochemistry. *ACS Nano* **7,** 8183–8189 (2013).
17. Soleymani, L., Fang, Z., Sargent, E. H. & Kelley, S. O. Programming the detection limits of biosensors through controlled nanostructuring. *Nat. Nano* **4,** 844–848 (2009).
18. Lam, B. *et al.* Optimized templates for bottom-up growth of high-performance integrated biomolecular detectors. *Lab. Chip.* **13,** 2569–2575 (2013).
19. Liu, H. & Crooks, R. M. Paper-based electrochemical sensing platform with integral battery and electrochromic read-out. *Anal. Chem.* **84,** 2528–2532 (2012).
20. Wu, C.-G., Lu, M.-I., Chang, S.-J. & Wei, C.-S. A solution-processable high-coloration-efficiency low-switching-voltage electrochromic polymer based on polycyclopentadithiophene. *Adv. Funct. Mater.* **17,** 1063–1070 (2007).
21. Cai, G. F. *et al.* Ultra fast electrochromic switching of nanostructured NiO films electrodeposited from choline chloride-based ionic liquid. *Electrochim. Acta* **87,** 341–347 (2013).
22. Cihanera, A. & Algib, F. A processable rainbow mimic fluorescent polymer and its unprecedented coloration efficiency in electrochromic device. *Electrochim. Acta* **53,** 2574–2578 (2008).
23. Soleymani, L. *et al.* Hierarchical nanotextured microelectrodes overcome the molecular transport barrier to achieve rapid, direct bacterial detection. *ACS Nano* **5,** 3360–3366 (2011).
24. Bin, X., Sargent, E. H. & Kelley, S. O. Nanostructuring of sensors determines the efficiency of biomolecular capture. *Anal. Chem.* **82,** 5928–5931 (2010).
25. Soleymani, L. *et al.* Nanostructuring of patterned microelectrodes to enhance the sensitivity of electrochemical nucleic acids detection. *Angew. Chem. Int. Ed.* **48,** 8457–8460 (2009).
26. Fang, Z. *et al.* Direct profiling of cancer biomarkers in tumor tissue using a multiplexed nanostructured microelectrode integrated circuit. *ACS Nano* **3,** 3207–3213 (2009).
27. Das, J. *et al.* An ultrasensitive universal detector based on neutralizer displacement. *Nat. Chem.* **4,** 642–648 (2012).
28. Das, J. & Kelley, S. O. Protein detection using arrayed microsensor chips: Tuning sensor footprint to achieve ultrasensitive readout of CA-125 in serum and whole blood. *Anal. Chem.* **83,** 1167–1172 (2011).
29. Das, J., Lee, J.-A. & Yang, H. Ultrasensitive detection of DNA in diluted serum using $NaBH_4$ electrooxidation mediated by $[Ru(NH_3)_6]^{3+}$ at indium-tin oxide electrodes. *Langmuir* **26,** 6804–6808 (2010).
30. Lapierre, M. A., O'Keefe, M., Taft, B. J. & Kelley, S. O. Electrocatalytic detection of pathogenic DNA sequences and antibiotic resistance markers. *Anal. Chem.* **75,** 6327–6333 (2003).
31. Jeong, J. *et al.* Arsenic (III) detection using electrochemical-chemical-chemical redox cycling at bare indium-tin oxide electrodes. *Analyst* **139,** 5813–5817 (2014).
32. Brodie, I. & Murray, J. *The Physics of Micro/Nano-Fabrication* 549 (Springer, 1992).
33. Wilson, D. A., Nolte, R. J. M. & van Hest, J. C. M. Autonomous movement of platinum-loaded stomatocytes. *Nat. Chem.* **4,** 268–274 (2012).
34. Vukusic, P. & Sambles, J. R. Photonic structures in biology. *Nature* **424,** 852–855 (2003).
35. Burgess, I. B. *et al.* Wetting in Color: Colorimetric differentiation of organic liquids with high selectivity. *ACS Nano* **6,** 1427–1437 (2012).
36. Kokalj, T., Park, Y., Vencelj, M., Jenko, M. & Lee, L. P. Self-powered imbibing microfluidic pump by liquid encapsulation: SIMPLE. *Lab. Chip.* **14,** 4329–4333 (2014).
37. Laksanasopin, T. *et al.* A smartphone dongle for diagnosis of infectious diseases at the point of care. *Sci Transl Med* **7,** 273re1 (2015).

Acknowledgements

This research was sponsored by the Genome Canada Genomics Applied Partnerships Program grant co-funded by the Ontario Research Fund through the Ministry of Research and Innovation. We also acknowledge support from the Natural Sciences and Engineering Research Council for a Discovery grant that supports work in the Kelley laboratories. We thank the ECTI facility at University of Toronto for their cleanroom facilities. I.B.B. acknowledges support from a Banting Fellowship from the Natural Sciences and Engineering Research Council of Canada.

Author contributions

J.D.B, J.D, I.B.B., W.L., E.H.S and S.O.K. designed the experiments. J.D.B, J.D, I.B.B. and W.L. performed the experiments and interpreted results with assistance from E.H.S. and S.O.K.; J.D.B., J.D., I.B.B., W.L., E.H.S. and S.O.K. composed and refined the manuscript.

Additional information

10

On-the-fly decoding luminescence lifetimes in the microsecond region for lanthanide-encoded suspension arrays

Yiqing Lu[1], Jie Lu[1], Jiangbo Zhao[1], Janet Cusido[2], Françisco M. Raymo[2], Jingli Yuan[3], Sean Yang[4], Robert C. Leif[4], Yujing Huo[5], James A. Piper[1], J. Paul Robinson[6], Ewa M. Goldys[1] & Dayong Jin[1,6]

Significant multiplexing capacity of optical time-domain coding has been recently demonstrated by tuning luminescence lifetimes of the upconversion nanoparticles called 'τ-Dots'. It provides a large dynamic range of lifetimes from microseconds to milliseconds, which allows creating large libraries of nanotags/microcarriers. However, a robust approach is required to rapidly and accurately measure the luminescence lifetimes from the relatively slow-decaying signals. Here we show a fast algorithm suitable for the microsecond region with precision closely approaching the theoretical limit and compatible with the rapid scanning cytometry technique. We exploit this approach to further extend optical time-domain multiplexing to the downconversion luminescence, using luminescence microspheres wherein lifetimes are tuned through luminescence resonance energy transfer. We demonstrate real-time discrimination of these microspheres in the rapid scanning cytometry, and apply them to the multiplexed probing of pathogen DNA strands. Our results indicate that tunable luminescence lifetimes have considerable potential in high-throughput analytical sciences.

[1] Advanced Cytometry Laboratories, ARC Centre of Excellence for Nanoscale BioPhotonics (CNBP), Macquarie University, Sydney, New South Wales 2109, Australia. [2] Laboratory for Molecular Photonics, University of Miami, 1301 Memorial Drive, Coral Gables, Florida 33146-0431, USA. [3] State Key Laboratory of Fine Chemicals, School of Chemistry, Dalian University of Technology, Dalian 116024, China. [4] Newport Instruments, 3345 Hopi Place, San Diego, California 92117-3516, USA. [5] Department of Electronic Engineering, Tsinghua University, Beijing 100084, China. [6] Purdue University Cytometry Laboratories, Bindley Bioscience Center, Purdue University, West Lafayette, Indiana 47907, USA. Correspondence and requests for materials should be addressed to D.J. (email: dayong.jin@mq.edu.au).

Fluorescence lifetimes are typically less than tens of nanoseconds; however, materials containing trivalent lanthanide ions usually exhibit longer lifetimes from microseconds to even milliseconds[1-3]. Such materials have been used as luminescent probes and they show improved sensitivity over conventional fluorescence-based methods when time-gated detection is used to remove the autofluorescence background[4-9]. Moreover, recently it was demonstrated that their microsecond-region lifetimes can be fine-tuned to create multiplexing optical codes and assigned to single upconversion nanoparticles[10,11]. These tunable lifetime values constitute a new temporal dimension of optical codes for data storage, document security and multiplexing assays, which can simultaneously assess a large number of biomolecular species. The next challenge is centred on the decoding issue—how to read and distinguish the slow-decaying lifetimes of each species at high speed.

On the other hand, contemporary life sciences demand new high-throughput analytical technologies, capable of simultaneous identification and quantification of large number of biomolecular species. Suspension arrays are emerging as a future leading technology for multiplexed molecular detection[12,13]. They are based on ensembles of microspheres that have been specially coded, most frequently by varying combinations of fluorescent dyes. The microspheres are thus endowed with a range of identifiable colour codes individually assigned to a specific analyte[14-16]. The major advantages of suspension arrays include rapid reaction kinetics, the absence of washing steps, higher sample throughput, as well as reproducible manufacture of microsphere families. Moreover, suspension arrays have potential as quantitative assays owing to the uniform surface of microspheres, the simplicity of use and their reduced expense compared with alternatives. The test panels are flexible to design by simply selecting a different combination of microspheres.

Towards these desired advances over the planar array biochips, significant efforts have been devoted to create the diverse signatures (barcodes) on microspheres and to expand multi-plexing capacities of suspension arrays[17-19]. In general, although fluorescence encoding and decoding appear to be more practical than non-fluorescence schemes in terms of assay design, cost per test, simplicity and throughput, producing a very large number of independent fluorescence codes by varying the intensity ratios of different colours has been restrained in the crowded spectral domain.

In the present work, we identify a fast fitting algorithm based on the method of successive integration (MSI)[20,21], and evaluate its practicability in realising the on-the-fly measurement of luminescence lifetimes with sensitivity down to single microspheres. This enables a powerful lifetime-decoding capability for an on-the-fly scanning cytometry[22,23]. In our parallel investigation of this work, we adopt the scheme of luminescence resonance energy transfer (LRET)[24-27] to continuously tune the luminescence lifetimes of down-conversion lanthanide complexes within microspheres for encoding. This complementary lifetime coding and decoding by the downconversion lanthanide luminescence within micro-spheres immediately enables a new library of optical identities for high-content screening of multiple pathogen DNA strands in high speed.

Results

Theoretical analysis. The techniques for determining fluorescence lifetimes have been previously developed[28-32], and implemented in the fluorescence-lifetime imaging microscopy[33-35] and the phase-sensitive flow cytometry[36-38]. However, developed for the nano-second region, they are unsuitable for real-time discrimination of

luminescence lifetimes, due to the facts associated with the microsecond region that: (i) the computation time of fitting algorithms increases nonlinearly with the data volume; (ii) an improper detection configuration may lead to significant errors in lifetime measurement; and (iii) various noise sources have to be taken into account in such a long timescale.

Consider an ensemble of luminescence probes excited by a pulse where, for simplicity, we assume that the luminescence decay profiles are monoexponential. From a statistical point of view, these probes emit individual photons with the total number (as a function of time) distributed according to a nonhomogeneous Poisson process with the exponential function as its intensity parameter (for explanation see Supplementary Notes 1–3). The process of lifetime measurement is, in essence, to calculate the lifetime parameter τ in the exponential function from the luminescence decay by using some kind of statistical estimation. Assuming a decay profile is recorded as the counts of luminescence photons individually collected at M temporal intervals (channels) of the same length T, our aim in this section is to determine the optimal M and T.

The Cramér–Rao inequality of the estimation theory indicates that the lowest-possible variance of any unbiased estimator (called Cramér–Rao lower bound, CRLB) is given by the inverse of the Fisher information matrix[39]. In the absence of dark noise, the CRLB for lifetime τ can be explicitly derived (see Supplementary Note 4). Figure 1 illustrates the relative CRLB normalized by the signal intensity (the expected value of the total number of luminescence photons N) as a function of MT/τ for different channel widths T. Note that it approaches unity under ideal conditions, which is a feature of the Poisson process. On one hand, since MT is the entire length of the detection window, these curves indicate an essential condition for accurate measurement of lifetime: the detection window should be sufficiently long compared with the lifetime under test. Otherwise, if the window duration is only about the lifetime under test, for example, the variance in the final results will be at least tenfold larger than its actual value, since no algorithm can achieve a lower variance than the CRLB. In practice, eight times the largest-possible lifetime value as the length of detection window is a rule of thumb suggested by Fig. 1. On the other hand, the decay curves with channel width T smaller than half of the lifetime τ give almost identical results. This indicates that the optimum T for rapid lifetime computation should be around half of the lifetime value

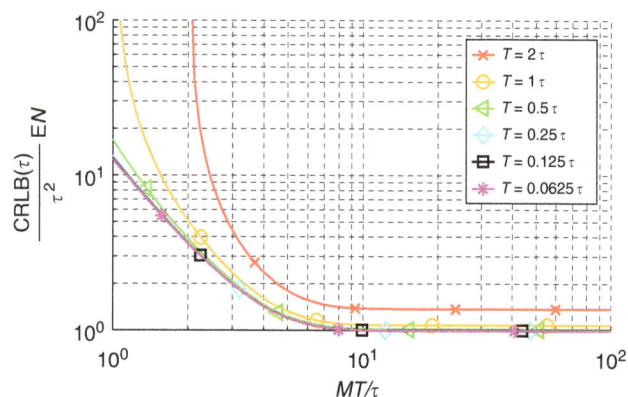

Figure 1 | Cramér-Rao lower bound for lifetime estimation under different detection channel configurations. CRLB(τ) is normalized by the lifetime (τ^2, since CRLB(τ) represents variance) as well as the average number of photons (EN), and plotted as a function of MT/τ. T is the width of every detection channel and M is the total number of detection channels, so that MT indicates the entire length of the detection window.

under test. Alternatively, if sampling is applied at a high frequency, the lifetime computation can be accelerated simply through a pretreatment of data binning to reduce the channel number, without sacrificing the precision. This is particularly useful when different lifetimes need to be measured and T is set about half of the smallest-possible lifetime value.

Numerical simulation. Having identified the optimum detection configuration for the lowest variance estimation of the decay lifetime, we then compared four leading lifetime fitting algorithms through numerical simulations. The aim of this comparison is to determine which algorithm has the best accuracy and speed. The first method (MLE-NF) is nonlinear fitting based on maximum-likelihood estimation (MLE-NF), which is reported in the literature as the most accurate algorithm to estimate parameters from exponential decays[40]. The second method (MLE-PR) is a pattern recognition technique comparing Kullback–Leibler minimum discrimination information, which is effectively MLE in the discrete regime[41,42]. The third is the rapid lifetime determination (RLD) method, which has been widely applied due to its great simplicity[43,44]. The fourth is the MSI, which has attracted recent interest after it was proved to be superior to another popular fast algorithm based on Fourier transform[45]. A detailed description of the numerical simulation can be found in the Methods section, as well as Supplementary Note 5.

Figure 2 summarizes the simulation results. Referring to Fig. 2a, when the data curves are free of background noise (consequently no baseline in the exponential function), the variances of the computed lifetimes through both MLE-NF and MLE-PR asymptotically converge to the CRLB—the minimum value they could ever reach statistically. Although not as efficient as MLE, RLD and MSI are similarly good for shorter detection windows; however, with a longer detection window, the MSI obtains better precision while the RLD becomes inefficient. If a background baseline is added to the model (in this case no close form can be derived for the CRLB), MLE-NF is still the most efficient method. The precision of MLE-PR is very close to that of MLE-NF, only slightly inferior. The performance of RLD suggests its inappropriateness for long lifetime measurement, despite the fact that it has been used in a few purpose-built instruments[35,46,47]. To our surprise, MSI still has the same performance regardless of the background noise. In fact, the

difference of precision between MLE-PR and MSI becomes very small as the signal-to-background ratio further decreases (see Supplementary Fig. 1). Figure 2b compares the lifetime computation by MLE-PR and MSI with different numbers of channels (true lifetime value $\tau = 320\,\mu s$; $MT/\tau = 8$). It was obvious that MSI is much faster than MLE-PR in about 46 times under identical conditions, to obtain the same unbiased results with variance only about 1.3 times larger. Figure 2b also verifies that a channel width around half of the lifetime under test (the case of 16 channels here) ensures the best balance between accuracy and speed.

On the basis of this analysis, we selected the MSI to determine the microsecond-region lifetimes of individual luminescent targets, which offers almost the highest precision but in a much more rapid fashion than alternative algorithms. In fact, since only requiring about 0.078 ms to calculate each lifetime value, it offers the potential for high-throughput lifetime measurement with $> 10^4$ targets per second.

Experimental prototype system. The fast lifetime-decoding algorithm based on MSI has been applied to our scanning cytometry. Figure 3 shows a schematic diagram of the new time-resolved orthogonal scanning automated microscopy (TR-OSAM) system. The initial scan following the OSAM principle[22] localizes any randomly distributed targets labelled with long-lived luminescence on a slide (Fig. 3a). By doing this rapid continuous scan, the wide-field optics can determine the exact positions of each luminescent target. This guides the orthogonal scan along the other direction to be efficient by only following scan lines leading to the preselected targets (Fig. 3b), with real-time recording of their luminescence decay profiles during transition. The OSAM operation efficiently extracts the useful data only from target analytes, thus significantly accelerating the process. This further allows real-time computation for lifetimes of individual targets based on their respective waveforms, consisting of luminescence photon counts in M channels of the same length T. The optimized detection configuration of M and T, and the lifetime fitting algorithm implemented in the system, are explained above, while other details can be found in the Methods section.

LRET suspension arrays. Lifetime-tunable luminescent microspheres were produced through the LRET scheme. A trivalent

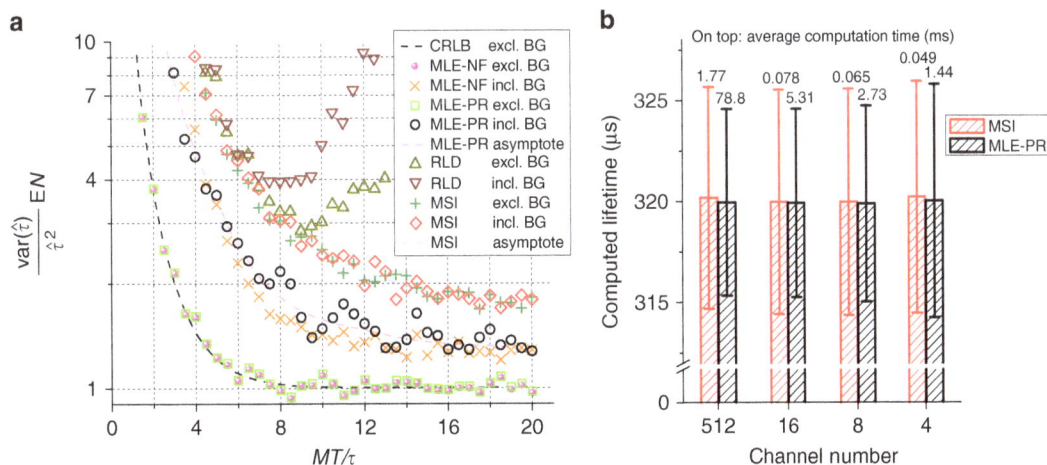

Figure 2 | Numeric simulation results for lifetime fitting algorithms. (a) The relative variances of lifetime estimators, $\mathrm{var}(\hat{\tau})/\tau^2$, normalized by the average number of photons, EN, are shown as a function of MT/τ for different fitting algorithms with and without background taken into account, alongside the CRLB. **(b)** Comparison between MSI and MLE-PR methods in terms of accuracy (error bars stand for ± 1 s.d.) and computation speed, at different channel numbers as a result of data binning pretreatment. Excl, excluding; incl, including.

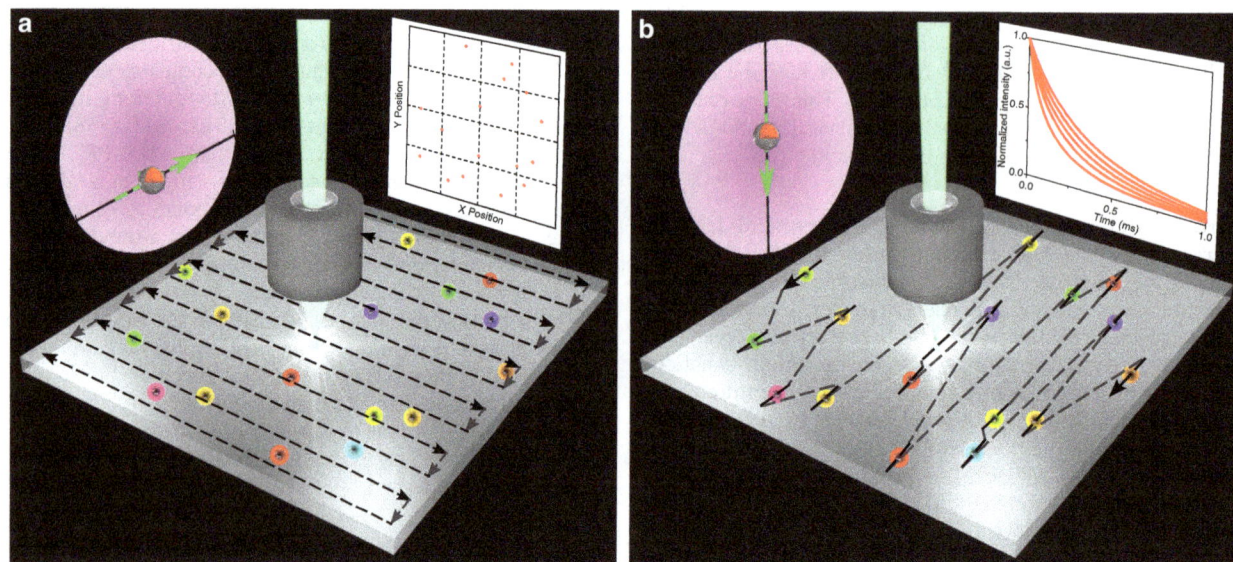

Figure 3 | Schematics of the time-resolved orthogonal scanning automated microscopy (TR-OSAM). The TR-OSAM can identify micron-sized targets randomly distributed on a microscopic slide, and distinguish them by individuals' luminescent lifetimes. (**a**) It typically takes 3 min to scan and map these targets in background-free condition via pulsed excitation and time-gated luminescence detection. The signal trains of luminescence intensity recorded from the detection field-of-view during the transit of the targets are used to obtain their precise locations along the continuous scanning direction. (**b**) The positional coordinates guide sequential orthogonal scans for spot-by-spot inspection of targets at the centre of the field-of-view, and the luminescence lifetime identity of each target is decoded in real time.

europium complex of thenoyltrifluoroacetonate $Eu(TTFA)_3$[48,49] and a hexafluorophosphate salt of cationic coumarin[50] were encapsulated into porous polystyrene beads as donor and acceptor dyes, using a simple method of solvent swelling (see Methods section). When in close proximity, a proportion of excited Eu^{3+} complexes in the beads non-radiatively transferred energy to the acceptor dyes[51,52], as a result of the spectral overlap between the donor emission and the acceptor excitation (see Supplementary Fig. 2). This led to the reduction of the luminescence lifetime of the Eu^{3+} complex at the red emission band around 612 nm. By stepwise varying their concentrations, we manipulated the average donor-to-acceptor distance, making it possible to fine-tune the lifetimes of the microspheres in corresponding steps. Figure 4a illustrates that as the acceptor concentration increases (at identical donor concentration), the mean value of donor lifetimes is effectively tuned down from 359 to 188 µs, as measured from individual microspheres by our TR-OSAM. Each lifetime population has a coefficient of variation in the range of 1.9–5.7%. This allows five populations of microspheres to be separated solely based on their lifetimes, with almost no population overlap, as shown in Fig. 4b.

These five populations of Eu-LRET microspheres were further mixed together and spiked on a slide area of $15 \times 15\, mm^2$. After 3 min of scanning by our TR-OSAM, all microspheres were located, with the varying lifetimes of single microspheres computed in real time (Fig. 5a). The lifetime histogram (Fig. 5b) recovered these five populations with clearly identified boundaries in between, consistent with the results shown in Fig. 4b. Consequently, the TR-OSAM distinguished every type of the Eu-LRET microspheres (Fig. 5c). This suggests that these microspheres can create a new type of suspension arrays for multiplexed assays. Through assigning variable lifetime values to the microcarriers, the spectral crosstalk problem can be avoided and the multiplexing levels may be increased further[10].

To further demonstrate the usefulness of this work underpinning the time-domain multiplexing, we carried out a biological experiment for high-throughput simultaneous detection of different pathogen DNAs (single strands). We designed our time-domain suspension arrays (bead assays) to detect four kinds of pathogens—human immunodeficiency virus (HIV), Ebola virus (EV), hepatitis B virus (HBV) and human papillomavirus 16 (HPV-16). The four types of the Eu-LRET microspheres (R-315, R-276, R-237 and R-188; 'R' stands for red colour and the numbers refer to the mean lifetime values) were conjugated to the capture probes, respectively, which were the complementary strands for the target DNAs. The other type (R-359) was conjugated to a control DNA probe to verify the specificity of the multiplexed assay. In the detection test, the pathogen DNAs were mixed in one sample and added into the test panel consisting of the five types of conjugated microspheres, followed by adding Qdot 565 as a universal reporter dye (see Fig. 6a for the multiplexing scheme and Methods section for experimental details). The reacted suspension was spread onto a glass slide and analysed by the TR-OSAM. It recovered all the lanthanide-encoded microspheres on the slide by detecting the time-gated luminescence from Eu^{3+} complex, and identified their types by decoding the lifetimes in the same colour band (Fig. 6b). The reporter fluorescence from Qdot 565 was also measured to confer the presence and amount of the particular target DNAs, indicated by the types of Eu-LRET microspheres (Fig. 6c). This result shows that the lifetime-encoded Eu-LRET microspheres can successfully recognize multiple pathogen DNAs in a single test, using the standard format of beads assay. In particular, it is important to note that although the reporter dye is excited by the same light source for the Eu^{3+} complex, its fluorescence does not interfere with the encoding luminescence lifetime, due to their complete separation in the temporal domain; while traditional multiplexing techniques based on colour and intensity have to be carefully designed to avoid serious crosstalk issues.

Discussion

In principle, the method for lifetime determination presented here is applicable to both the microseconds to millisecond region

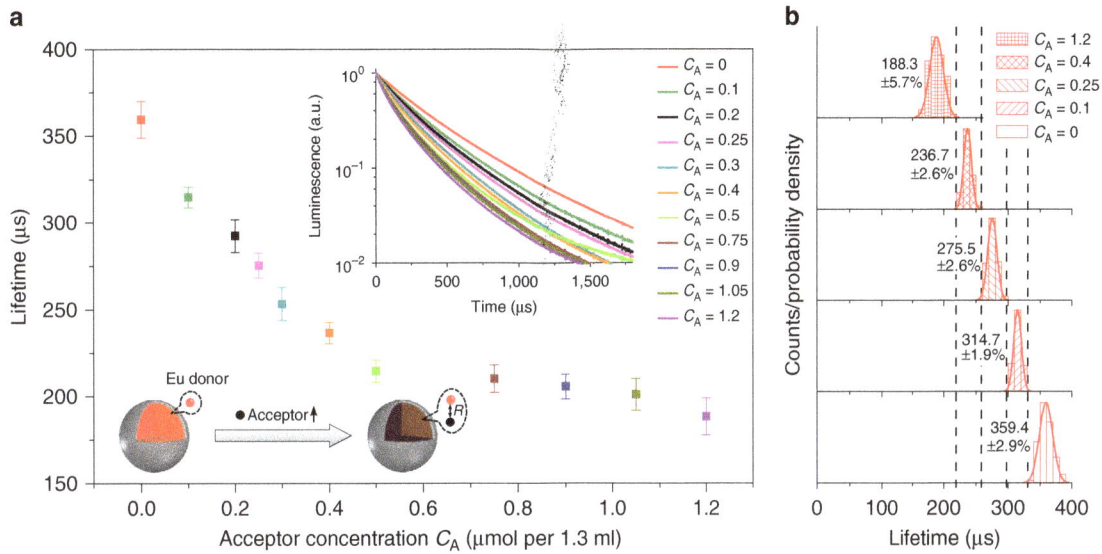

Figure 4 | Lifetime measurements from individual Eu-containing microspheres engineered by LRET scheme. Different solutions, each having a total volume of 1.3 ml and containing identical amount of Eu complexes (2.5 μmol) as donor but incremental amount of acceptor dyes, were used to produce individual groups of encoded polymer microspheres, followed by TR-OSAM analysis. (**a**) Luminescence lifetime measured at Eu^{3+} red emission band shortened as the acceptor concentration in the original dye solution increased, as a result of stronger LRET effect. The error bars stand for ± 1 s.d. The inset curves are the luminescence decay signals measured from single Eu-LRET microspheres. (**b**) Among all samples, five types of microspheres gave separate lifetime histograms, so that they can be clearly differentiated in multiplexing assays by the TR-OSAM. The numerals at the left to each histogram are the mean lifetime ± the lifetime coefficient of variation for its Gaussian fitting.

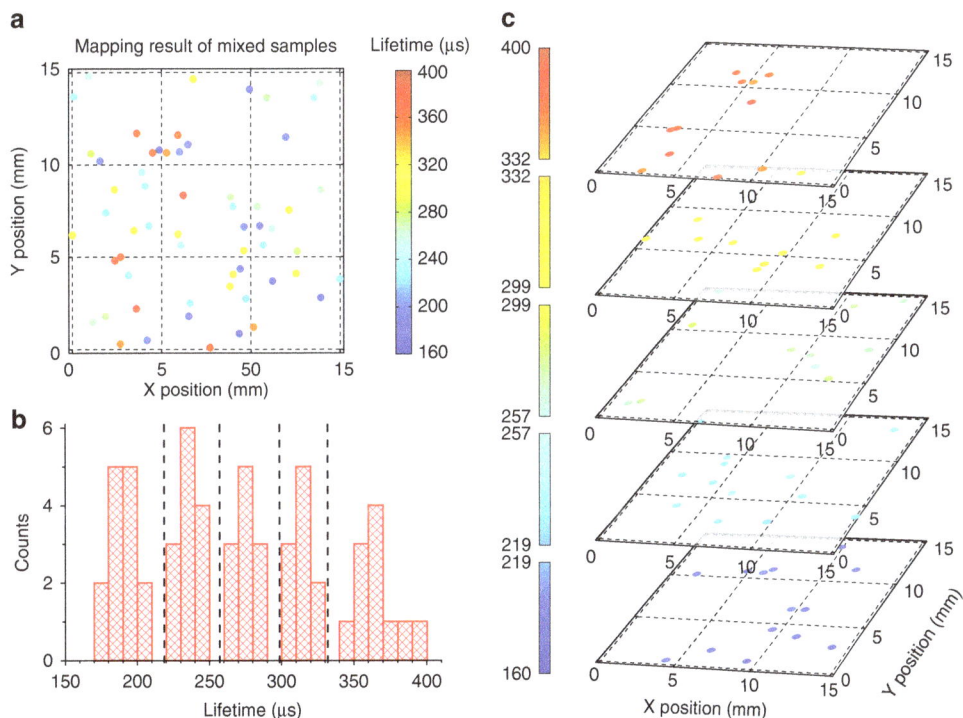

Figure 5 | Multiplexed detection results of lifetime-encoded microspheres using TR-OSAM decoding. (**a**) The mapping result of a mixture of five selected types of Eu-LRET microspheres (refer to Fig. 4b) showed the locations of individual microspheres on a microscopic slide. Pseudo-colours are used to represent their lifetime values. (**b**) Each types of microspheres in the mixed sample were recognized based on the separation of lifetime populations using well-defined boundaries in between. (**c**) The initial mapping result was thus decomposed into five different lifetime ranges for each types of Eu-LRET microspheres carrying lifetime identities.

and the nanosecond region; however, it is in the long lifetime region that this algorithm becomes superior to others for practical circumstances where both high accuracy and high speed are guaranteed. One the other hand, luminescent probes with even

longer lifetimes above 1 ms will necessitate extended recording window, thereby less favourable for high-throughput analysis. The microsecond region (that is, 1–1,000 μs) is probably the most advantageous timescale for time-domain multiplexing, since it is

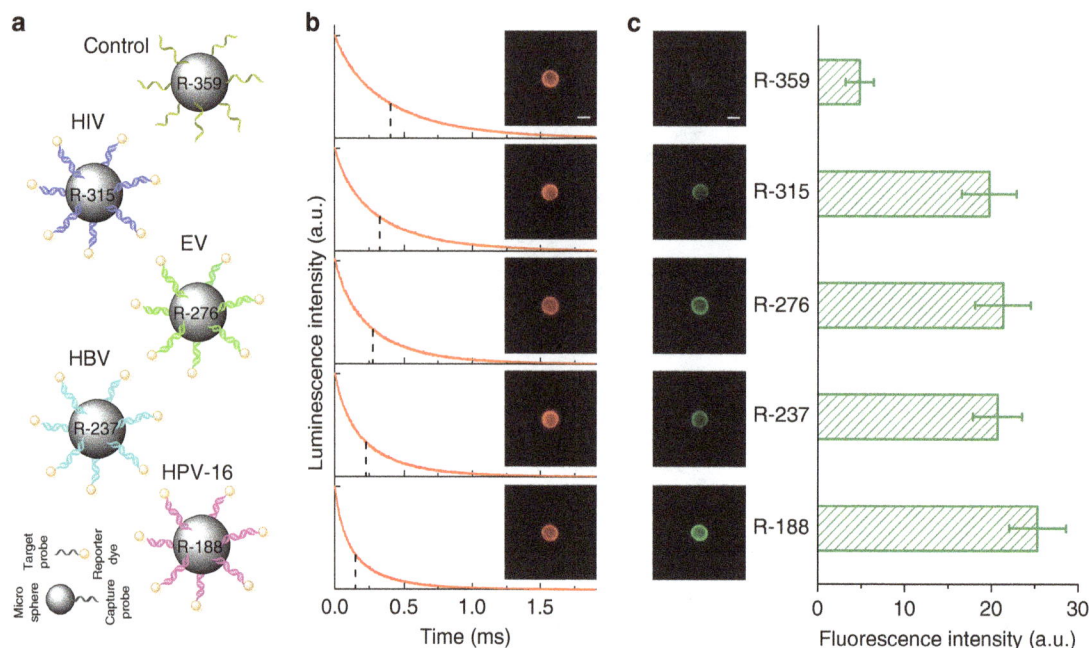

Figure 6 | Multiplexed DNA detection using lifetime-encoded Eu-LRET microspheres and TR-OSAM decoding. (a) The five selected types of Eu-LRET microspheres (refer to Fig. 4b) were conjugated to five different DNA sequences (four of them were complementary to pathogen DNAs from HIV, EV, HBV and HPV-16, and the other was for control purpose). Qdot 565 was used as the universal reporter dye in the beads assay. **(b)**The microspheres were identified based on their lifetimes measured from Eu^{3+} luminescence decays. **(c)**The amounts of target pathogen DNAs in the test sample were determined by the reporter fluorescence intensities (error bars stand for ±1 s.d.). Images were taken with different emission filters at 600 ms exposure time under continuous excitation (scale bar, 10 μm).

sufficiently broad to accommodate a number of lifetime channels while the background interference from autofluorescence and scattering can be completely removed. The feasible time domain also relies on the tunable range of luminescence lifetimes via material engineering, which is currently from tens of microseconds to hundreds of microseconds for both lanthanide-based up- and downconversion probes.

It should be noted that the luminescence decays from the LRET microspheres do not exactly follow monoexponential curves, which can be seen from Fig. 4a inset. However, our TR-OSAM based on monoexponential fitting algorithm is sufficient to discriminate different microsphere populations. Also, note that the MSI is applicable to multiexponential cases as well (see Supplementary Note 6), which can be integrated for more advanced analytical needs.

Moreover, the lifetime codes are independent of either colour in the spectral domain or intensity, and therefore this work also suggests that other LRET microspheres with different emission bands can be simply engineered to further expand the multiplexing capacity of LRET suspension arrays. For example, there are several more spectral bands from the downconversion luminescent or phosphorescent materials[53,54] as the potential donors, such as the terbium (Tb) complexes emitting at 490 and 545 nm, dysprosium (Dy) at 580 nm, samarium (Sm) at 600 and 650 nm, and the platinum (Pt) complexes emitting at 700 nm, which can create more lifetime-coding digits through the LRET route. We envisage that this work will lay the foundation to build a library of specific microspheres carrying more than 10,000 codes, to give the users far greater scope and higher flexibility with assay design. Thus, this work unlocks the real potential of suspension arrays as a particularly powerful analytical technique able to cope with the complexity challenges in life sciences and medicine.

In summary, we demonstrate both theoretically and experimentally that luminescence lifetimes of microscopic targets in the

microsecond region can be measured rapidly and precisely in a simple fashion. This is achieved by the MSI fitting algorithm with an optimized detection configuration. To ensure sufficient accuracy, the key requirements for the detection configuration must be satisfied, that is, the detection window has to be at least eight times of the longest lifetime values under test and the channel width smaller than half of the shortest lifetime under test. Our TR-OSAM system is rapid, sensitive and accurate in resolving the luminescence lifetimes from individual microspheres prepared by the LRET scheme, which can be used to simultaneously probe multiple analytes, such as DNA strands. These results make it possible to take full advantage of the exceptionally long lifetimes offered by luminescent probes, such as lanthanide complexes and upconversion nanocrystals, for multiplexed sensing and imaging in life sciences and photonic applications.

Methods

Time-resolved orthogonal scanning automated microscopy. The TR-OSAM shares the same optical and electronic configurations as the original OSAM reported previously[22]. Briefly, it was modified on an epi-fluorescence inverted microscope (Olympus IX71) to comprise an ultraviolet light-emitting diode (NCCU033A, Nichia; peak wavelength 365 nm, excitation power density ~2.7 W cm^{-2} in the field-of-view) and a dichroic filter (400DCLP, Chroma) to excite the sample slide placed on a motorized stage (H117, Prior Scientific) through an ×60 objective lens (NT38-340, Edmund Optics). The luminescence signals from the sample were collected by the same objective, split from the excitation optical path by the dichroic mirror, transmit through a band-pass filter (either FF01-607/36 for Eu^{3+} luminescence or FF01-560/14 for the reporter fluorescence, both from Semrock) and finally collected by an electronically gateable photon-counting avalanche photodiode (SPCM-AQRH-13-FC, PerkinElmer). The converted electronic photon counts were acquired by a computer at a sampling rate of 500 kHz through a multifunctional data acquisition device (PCI-6251, National Instruments), which also generates synchronised control sequences for time-gated detection as well as stage scanning.

On top of the OSAM scheme, the TR-OSAM integrated the lifetime measurement function based on the MSI algorithm and data binning, which were verified by experiments to be appropriate for rapid and robust lifetime computation, especially with data curves superimposed by noise. The period of time-gating cycles was set to 4 ms, consisting of 90 µs excitation pulse, 10 µs time delay and 3,900 µs detection window. The recorded luminescence decay curves were processed by a purposely built LabVIEW program in real time to calculate the lifetimes for individual luminescent targets identified on the slide sample.

Numeric simulation to compare lifetime fitting algorithms. The mathematic models and computation principles for luminescence lifetime are explained in Supplementary Notes 4 and 5. The Poisson processes for the time series of a total number 10,000 of the luminescence photon counts were generated from 10,000 random numbers chosen from the exponential distribution with mean parameter of a true lifetime value of 320 µs. The time series were counted within M adjacent channels with all widths equal to T, to obtain a simulated decay curve. A random number chosen from the Poisson distribution with parameter BT ($B = 10^5$ for Fig. 3 and $B = 10^6$ for Supplementary Fig. 1) was superimposed for each channel to simulate the dark counts. For every set of (M, T), computation via each fitting algorithm was repeated 1,000 times, each time with a new decay curve. All simulations were executed by a MATLAB program running on a PC with 2.83 GHz quad-core CPU and 4 GB memory.

Preparation of Eu-LRET microspheres and slide samples. A series of 400 µl dye solutions were prepared, each containing a fixed amount of 2.5 µmol Eu(TTFA)$_3$ complex (europium tris(thenoyltrifluoroacetonate), 99.9% (rare-earth oxides), Rare Earth Products) and a varied amount of the hexafluorophosphate salt of cationic coumarin ranging from 0 to 1.2 µmol in dichloromethane. Encapsulation of dyes into microspheres was performed through dropwise addition of one dye solution into a 900-µl suspension of polystyrene beads (PC07N/8783, 15.14 µm in diameter, Bangs Laboratories; containing 2.6×10^5 beads in ethanol, ultrasonically treated before use). The suspension was stirred for 48 h, and then the dye-containing swollen beads were shrunk in absolute ethanol solvent. After that, the impregnated Eu-LRET microspheres were quickly isolated by centrifugation and washed three times in ethanol:water (vol/vol 1:1), and then stored in 500 µl deionised water. To prepare a test sample for TR-OSAM analysis, a certain amount of suspension typically containing 300 ~ 500 Eu-LRET microspheres were drop-cast onto the surface of a glass slide and sealed by a glass coverslip.

Multiplexed DNA detection. The target DNA sequences of HIV, EV, HBV and HPV-16 were identified from the National Center for Biotechnology Information database. Amino-modified capture DNA oligos and 5'-biotin-conjugated target DNA oligos were purchased from Integrated DNA Technologies (see Supplementary Table 1 for detailed sequences). Each 50 µl of as-prepared Eu-LRET microspheres was used to conjugate each type of capture probes activated by 1-ethyl-3-(3-dimethylaminopropyl)carbodiimide hydrochloride (purchased from Sigma Aldrich) in 2-(N-morpholino)ethanesulphonic acid buffer (50 nM, pH 6.0, purchased from Sigma Aldrich) for 2 h at room temperature. The functionalised microspheres were collected by centrifugation and then mixed to form a test panel. Then, a test sample containing 5 nmol of each pathogen DNA oligos was added into the test panel, and 2 × saline-sodium citrate (purchased from Invitrogen) was used as the hybridization buffer. After 1 h of incubation, the microspheres were washed and resuspended into 100 µl solution containing 10 nM streptavidin-conjugated Qdot 565 (Invitrogen), and then incubated for another 1 h. Finally, the suspension of microspheres was purified and prepared onto a glass slide for analysis by the TR-OSAM.

References

1. Eliseeva, S. V. & Bunzli, J. C. G. Lanthanide luminescence for functional materials and bio-sciences. *Chem. Soc. Rev.* **39**, 189–227 (2010).
2. Leif, R. C., Vallarino, L. M., Becker, M. C. & Yang, S. Increasing the luminescence of lanthanide complexes. *Cytometry A* **69A**, 767–778 (2006).
3. Gnach, A. & Bednarkiewicz, A. Lanthanide-doped up-converting nanoparticles: merits and challenges. *Nano Today* **7**, 532–563 (2012).
4. Soini, E. & Kojola, H. Time resolved fluorometer for lanthanide chelates–a new generation of nonisotopic immunoassays. *Clin. Chem.* **29**, 65–68 (1983).
5. Hemmila, I. & Webb, S. Time-resolved fluorometry: an overview of the labels and core technologies for drug screening applications. *Drug Discov. Today* **2**, 373–381 (1997).
6. Hanaoka, K., Kikuchi, K., Kobayashi, S. & Nagano, T. Time-resolved long-lived luminescence imaging method employing luminescent lanthanide probes with a new microscopy system. *J. Am. Chem. Soc.* **129**, 13502–13509 (2007).
7. Song, B., Vandevyver, C. D. B., Chauvin, A.-S. & Buenzli, J.-C.G. Time-resolved luminescence microscopy of bimetallic lanthanide helicates in living cells. *Org. Biomol. Chem.* **6**, 4125–4133 (2008).
8. Gahlaut, N. & Miller, L. W. Time-resolved microscopy for imaging lanthanide luminescence in living cells. *Cytometry A* **77A**, 1113–1125 (2010).
9. Jin, D. Y. & Piper, J. A. Time-gated luminescence microscopy allowing direct visual inspection of lanthanide-stained microorganisms in background-free condition. *Anal. Chem.* **83**, 2294–2300 (2011).
10. Lu, Y. *et al.* Tunable lifetime multiplexing using luminescent nanocrystals. *Nat. Photonics* **8**, 32–36 (2014).
11. Zhao, J. *et al.* Upconversion luminescence with tunable lifetime in NaYF4:Yb,Er nanocrystals: role of nanocrystal size. *Nanoscale* **5**, 944–952 (2013).
12. Braeckmans, K., De Smedt, S. C., Leblans, M., Pauwels, R. & Demeester, J. Encoding microcarriers: present and future technologies. *Nat. Rev. Drug Discov.* **1**, 447–456 (2002).
13. Wilson, R., Cossins, A. R. & Spiller, D. G. Encoded microcarriers for high-throughput multiplexed detection. *Angew. Chem. Int. Ed.* **45**, 6104–6117 (2006).
14. Fulton, R. J., McDade, R. L., Smith, P. L., Kienker, L. J. & Kettman, J. R. Advanced multiplexed analysis with the FlowMetrix(TM) system. *Clin. Chem.* **43**, 1749–1756 (1997).
15. Han, M. Y., Gao, X. H., Su, J. Z. & Nie, S. Quantum-dot-tagged microbeads for multiplexed optical coding of biomolecules. *Nat. Biotechnol.* **19**, 631–635 (2001).
16. Zhang, F. *et al.* Fluorescence upconversion microbarcodes for multiplexed biological detection: nucleic acid encoding. *Adv. Mater.* **23**, 3775–3779 (2011).
17. Cao, Y. W. C., Jin, R. C. & Mirkin, C. A. Nanoparticles with Raman spectroscopic fingerprints for DNA and RNA detection. *Science* **297**, 1536–1540 (2002).
18. Lee, H., Kim, J., Kim, H. & Kwon, S. Colour-barcoded magnetic microparticles for multiplexed bioassays. *Nat. Mater.* **9**, 745–749 (2010).
19. Bendall, S. C. *et al.* Single-cell mass cytometry of differential immune and drug responses across a human hematopoietic continuum. *Science* **332**, 687–696 (2011).
20. Tittelbachhelmrich, K. An integration method for the analysis of multiexponential transient signals. *Meas. Sci. Technol.* **4**, 1323–1329 (1993).
21. Halmer, D., von Basum, G., Hering, P. & Murtz, M. Fast exponential fitting algorithm for real-time instrumental use. *Rev. Sci. Instrum.* **75**, 2187–2191 (2004).
22. Lu, Y., Xi, P., Piper, J. A., Huo, Y. & Jin, D. Time-gated orthogonal scanning automated microscopy (OSAM) for high-speed cell detection and analysis. *Sci. Rep.* **2**, 837 (2012).
23. Lu, J. *et al.* Resolving low-expression cell surface antigens by time-gated orthogonal scanning automated microscopy. *Anal. Chem.* **84**, 9674–9678 (2012).
24. Kupcho, K. R. *et al.* Simultaneous monitoring of discrete binding events using dual-acceptor terbium-based LRET. *J. Am. Chem. Soc.* **129**, 13372–13373 (2007).
25. Rajapakse, H. E. *et al.* Time-resolved luminescence resonance energy transfer imaging of protein-protein interactions in living cells. *Proc. Natl Acad. Sci. USA* **107**, 13582–13587 (2010).
26. Kim, S. H., Gunther, J. R. & Katzenellenbogen, J. A. Monitoring a coordinated exchange process in a four-component biological interaction system: development of a time-resolved terbium-based one-donor/three-acceptor multicolor FRET system. *J. Am. Chem. Soc.* **132**, 4685–4692 (2010).
27. Geißler, D., Stufler, S., Löhmannsröben, H.-G. & Hildebrandt, N. Six-color time-resolved Förster resonance energy transfer for ultrasensitive multiplexed biosensing. *J. Am. Chem. Soc.* **135**, 1102–1109 (2012).
28. O'Connor, D. V. & Phillips, D. *Time-correlated Single Photon Counting* (Academic Press, 1984).
29. Istratov, A. A. & Vyvenko, O. F. Exponential analysis in physical phenomena. *Rev. Sci. Instrum.* **70**, 1233–1257 (1999).
30. Barber, P. R., Ameer-Beg, S. M., Pathmananthan, S., Rowley, M. & Coolen, A. C. C. A Bayesian method for single molecule, fluorescence burst analysis. *Biomed. Opt. Express* **1**, 1148–1158 (2010).
31. Li, D. D. U. *et al.* Video-rate fluorescence lifetime imaging camera with CMOS single-photon avalanche diode arrays and high-speed imaging algorithm. *J. Biomed. Opt.* **16**, 096012 (2011).
32. Arlt, J. *et al.* A study of pile-up in integrated time-correlated single photon counting systems. *Rev. Sci. Instrum.* **84**, 103105 (2013).
33. Bastiaens, P. I. H. & Squire, A. Fluorescence lifetime imaging microscopy: spatial resolution of biochemical processes in the cell. *Trends Cell Biol.* **9**, 48–52 (1999).
34. Suhling, K., French, P. M. W. & Phillips, D. Time-resolved fluorescence microscopy. *Photochem. Photobiol. Sci.* **4**, 13–22 (2005).
35. Gerritsen, H. C., Asselbergs, M. A. H., Agronskaia, A. V. & Van Sark, W. Fluorescence lifetime imaging in scanning microscopes: acquisition speed, photon economy and lifetime resolution. *J. Microsc.* **206**, 218–224 (2002).
36. Pinsky, B. G., Ladasky, J. J., Lakowicz, J. R., Berndt, K. & Hoffman, R. A. Phase-resolved fluorescence lifetime measurements for flow cytometry. *Cytometry* **14**, 123–135 (1993).
37. Cui, H. H., Valdez, J. G., Steinkamp, J. A. & Crissman, H. A. Fluorescence lifetime-based discrimination and quantification of cellular DNA and RNA with phase-sensitive flow cytometry. *Cytometry A* **52A**, 46–55 (2003).

38. Cao, R., Pankayatselvan, V. & Houston, J. P. Cytometric sorting based on the fluorescence lifetime of spectrally overlapping signals. *Opt. Express* **21**, 14816–14831 (2013).

39. Kollner, M. & Wolfrum, J. How many photons are necessary for fluorescence-lifetime measurements. *Chem. Phys. Lett.* **200**, 199–204 (1992).

40. Hall, P. & Selinger, B. Better estimates of exponential decay parameters. *J. Phys. Chem.* **85**, 2941–2946 (1981).

41. Kollner, M. *et al.* Fluorescence pattern recognition for ultrasensitive molecule identification: comparison of experimental data and theoretical approximations. *Chem. Phys. Lett.* **250**, 355–360 (1996).

42. Eggeling, C., Fries, J. R., Brand, L., Gunther, R. & Seidel, C. A. M. Monitoring conformational dynamics of a single molecule by selective fluorescence spectroscopy. *Proc. Natl Acad. Sci. USA* **95**, 1556–1561 (1998).

43. Ballew, R. M. & Demas, J. N. An error analysis of the rapid lifetime determination method for the evaluation of single exponential decays. *Anal. Chem.* **61**, 30–33 (1989).

44. Moore, C., Chan, S. P., Demas, J. N. & DeGraff, B. A. Comparison of methods for rapid evaluation of lifetimes of exponential decays. *Appl. Spectrosc.* **58**, 603–607 (2004).

45. Everest, M. A. & Atkinson, D. B. Discrete sums for the rapid determination of exponential decay constants. *Rev. Sci. Instrum.* **79**, 023108 (2008).

46. Marriott, G., Clegg, R. M., Arndt-Jovin, D. J. & Jovin, T. M. Time resolved imaging microscopy phosphorescence and delayed fluorescence imaging. *Biophys. J.* **60**, 1374–1387 (1991).

47. Ramshesh, V. K. Pinhole shifting lifetime imaging microscopy. *J. Biomed. Opt.* **13**, 064001 (2008).

48. Leif, R. C. *et al.* In: *The Automation of Uterine Cancer Cytology.* (eds Wied, G. L., Bahr, G. F. & Bartels, P. H.) 313–344 (Tutorials of Cytology, 1976).

49. Leif, R. C. *et al.* Calibration beads containing luminescent lanthanide ion complexes. *J. Biomed. Opt.* **14**, 024022 (2009).

50. Deniz, E., Sortino, S. & Raymo, F. M. Fluorescence switching with a photochromic auxochrome. *J. Phys. Chem. Lett.* **1**, 3506–3509 (2010).

51. Heyduk, T. & Heyduk, E. Luminescence energy transfer with lanthanide chelates: interpretation of sensitized acceptor decay amplitudes. *Anal. Biochem.* **289**, 60–67 (2001).

52. Selvin, P. R. Principles and biophysical applications of lanthanide-based probes. *Annu. Rev. Biophys. Biomol. Struct.* **31**, 275–302 (2002).

53. Petoud, S., Cohen, S. M., Bünzli, J.-C. G & Raymond, K. N. Stable lanthanide luminescence agents highly emissive in aqueous solution: multidentate 2-hydroxyisophthalamide complexes of Sm^{3+}, Eu^{3+}, Tb^{3+}, Dy^{3+}. *J. Am. Chem. Soc.* **125**, 13324–13325 (2003).

54. Zhao, Q., Huang, C. H. & Li, F. Y. Phosphorescent heavy-metal complexes for bioimaging. *Chem. Soc. Rev.* **40**, 2508–2524 (2011).

Acknowledgements

We thank Olympus Australia for providing the scanning stage and microscopy accessories, and Deming Liu, Arun Dass, Run Zhang, Lixin Zhang, Xianlin Zheng and Bingyang Shi for extensive technical discussions. Y.L. thanks Long Zhao (UT Austin) for discussions about stochastic processes. This project is financially supported by the Australian Research Council Discovery Project (DP1095465) and Future Fellowship (D.J., FT130100517), Macquarie University Research Fellowship (Y.L.) and Macquarie University Postgraduate Research Scholarships. Y.L. and D.J. acknowledge the International Society for Advancement of Cytometry for support as ISAC Scholars.

Author contributions

D.J., J.A.P., R.C.L. and J.P.R. conceived the project. D.J. and Y.L. designed the experiments and supervised the research. Y.L., J.L., J.Z. and D.J. were primarily responsible for data collection and analysis. Y.L. and D.J. prepared figures and wrote the main manuscript text. All authors contributed to the data analysis, discussions and manuscript preparation.

Additional information

11

Electrografting of calix[4]arenediazonium salts to form versatile robust platforms for spatially controlled surface functionalization

Alice Mattiuzzi[1], Ivan Jabin[1], Claire Mangeney[2], Clément Roux[3], Olivia Reinaud[4], Luis Santos[5], Jean-François Bergamini[5], Philippe Hapiot[5] & Corinne Lagrost[5]

An essential issue in the development of materials presenting an accurately functionalized surface is to achieve control of layer structuring. Whereas the very popular method based on the spontaneous adsorption of alkanethiols on metal faces stability problems, the reductive electrografting of aryldiazonium salts yielding stable interface, struggles with the control of the formation and organization of monolayers. Here we report a general strategy for patterning surfaces using aryldiazonium surface chemistry. Calix[4]tetra-diazonium cations generated *in situ* from the corresponding tetra-anilines were electrografted on gold and carbon substrates. The well-preorganized macrocyclic structure of the calix[4]arene molecules allows the formation of densely packed monolayers. Through adequate decoration of the small rim of the calixarenes, functional molecules can then be introduced on the immobilized calixarene subunits, paving the way for an accurate spatial control of the chemical composition of a surface at molecular level.

[1] Laboratoire de Chimie Organique, Université Libre de Bruxelles (U.L.B.), CP 160/06, 50 avenue F.D. Roosevelt, 1050 Brussels, Belgium. [2] ITODYS, Université Paris Diderot-Paris 7 and CNRS, UMR n°7086, 15 rue Jean de Baïf, 75013 Paris, France. [3] Department of Chemistry, University of Canterbury, MacDiarmid Institute for advanced materials and nanotechnology, Christchurch, New Zealand. [4] Laboratoire de Chimie et Biochimie Pharmacologiques et Toxicologiques, PRES Sorbonne Paris Cité, Université Paris Descartes and CNRS, UMR n°8601, 45 rue des Saints-Pères, 75006 Paris, France. [5] Sciences Chimiques de Rennes, Equipe MaCSE, Université de Rennes 1 and CNRS, UMR n° 6226, Campus de Beaulieu, 35042 Rennes cedex, France. Correspondence and requests for materials should be addressed to C.L. (email: corinne.lagrost@univ-rennes1.fr).

Surface functionalization has attracted considerable interest in recent years, with the aim to design surfaces with tailored properties (such as wettability, corrosion resistance, adhesion, lubricant, biocompatibility) or with operating functions ('smart' surfaces bearing stimuli-responsive architectures)[1–4]. When developed for modifying electrode surfaces, it finds applications in molecular electronics,[5] bioelectronics[6,7], and in the development of charge transport devices[8,9] or sensors[10]. Particularly crucial for sensor studies, the controlled grafting of monolayers allows the design of complex recognition interfaces and provides responses akin to molecular level over the associated transduction reactions[11].

Three major issues for constructing such interfaces can thus be pointed out: control of the thickness and density of the deposited layer, which ideally, as a first step, requires the formation of a monolayer; spatial control of the connected functions and robustness versus temperature, solvents, ageing.

The most popular strategy for fabricating monolayers is the spontaneous two-dimensional adsorption of thiol derivatives, especially alkanethiols, onto coinage metal (Pt, Au, Ag, Cu) forming self-assembled monolayers (SAMs) (Fig. 1a)[12]. Although highly versatile, this strategy could present important limitations related to their lack of thermal and long-term stability, narrow electrochemical window and stochastic behaviour of the gold–thiolate bond[12–14]. Besides, the spatial control of covalently bound functions is particularly difficult to achieve. Indeed, direct self-assembly of alkylthiols covalently connected at their extremity to functional objects generally leads to 'bouquet'-like repartition of the objects on the surface, with empty spaces in between (Fig. 1b)[12]. Dilution strategies with unfunctionalized alkylthiols allowing strong packing of the alkyl chain at the surface and simultaneous connection of large functional objects at the surface often lead to small 'islands of objects', not to regular distribution at the surface[12,15–19]. An alternative strategy for connecting organic moieties onto electrodes is the reductive grafting of aryldiazonium salts that yields covalent attachment of phenyl derivatives onto the surface. Electrografting in particular has gained tremendous interest over the last decade and is now a well-recognized method for surface functionalization[20–24]. Contrary to alkylthiol SAMs, the organic layers obtained from this method are generally highly stable, being strongly resistant to heat,

chemical degradation and ultrasonication[20,25]. Furthermore, the method is easy to process and fast (deposition time on the order of 10 s instead of 10–18 h for well-organized thiol-Au SAMs) and can be applied to a wide range of materials: carbon (graphite, glassy carbon (GC), nanotubes, diamond), metals (Fe, Co, Ni, Cu, Zn, Pt, Au), semiconductor (SiH, SiO_2, SiOC…), indium tin oxide and even organic polymers and dyes[22]. In spite of all these desirable properties, this strategy suffers from a major disadvantage, which is the poor control of layers thickness and regularity. Indeed, the reactive aryl radicals that are transiently produced attack not only the electrode surface but also already-grafted aryl layers, generally yielding multilayers and 'cauliflower'-like surfaces (Fig. 1c). This may result in the loss or decrease of efficiency of desirable properties, as observed for instance for the inferior charge transport properties of ferrocenyl redox moieties bound to such polyaryl layers[26].

Although the diazonium grafting could be empirically controlled to some extent through the appropriate choice of experimental conditions (nature of solvent, aryldiazonium salt concentration, electrolysis time and applied potential), an ultrathin and homogeneous coating is generally difficult to obtain, thus limiting the capability of the method in designing complex surface patterning. Introduction of sterically encumbered substituents on the aryl rings has been successfully used to prevent the uncontrolled polymerization process, thus allowing the formation of a near monolayer[27]. However, an undesirable counterback is the chemical inertness of these molecules, which precludes further functionalization. To remedy this problem, aryldiazonium salts bearing a pendant function masked by a protecting group exhibiting steric hindrance[28–31] and electronic shielding[29] properties have been used (Fig. 1d). After removal of the protecting group, a postfunctionalization step allows for the attachment of functional molecules onto the newly activated monolayer. However, because of the leftover void spaces between two adjacent grafted aryl molecules after the deprotection step, this two-step strategy could be disadvantageous when the direct formation of compact, pinhole-free layers is desired. A very recent article describes a more straightforward one-step strategy based on the reductive electrografting of a benzene(p-bisdiazonium) salt, leaving a pendant diazonium protecting group for further chemical coupling[32]. However, this route suffers from a lack of long-term stability of the diazonium-terminated layers.

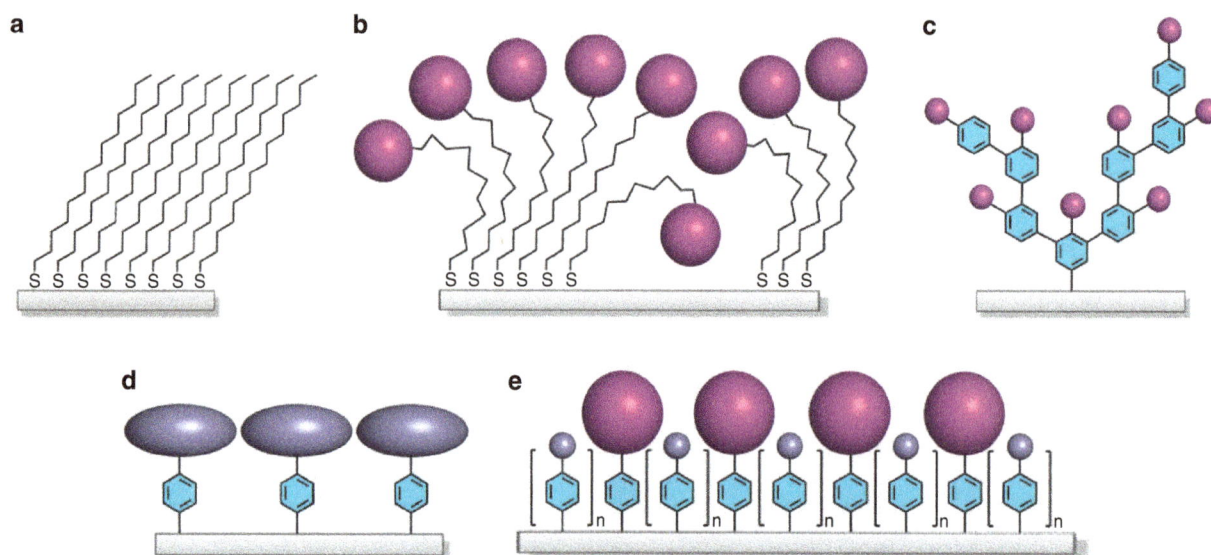

Figure 1 | Schematized strategies based on thiols and diazoniums for grafting organic layers onto surfaces. (**a,b**) SAMs (**a**) with simple alkylthiols, (**b**) bouquet-like layers with functionalized alkylthiols. (**c–e**) Diazonium methodology (**c**) cauliflower-like layers, (**d**) protection-deprotection strategy, (**e**) 'ideal' monolayers (see text for further explanation).

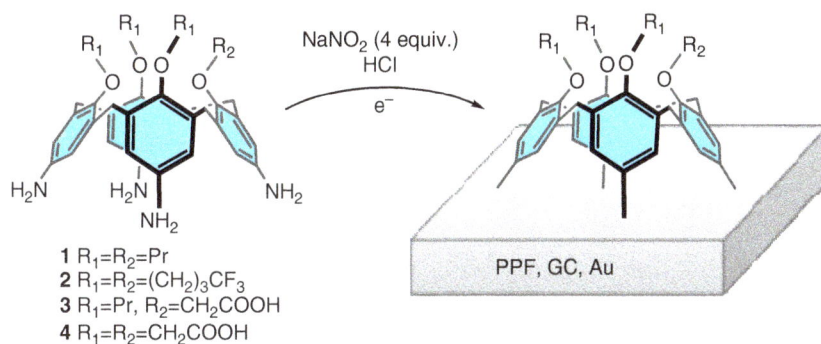

1 $R_1=R_2=Pr$
2 $R_1=R_2=(CH_2)_3CF_3$
3 $R_1=Pr, R_2=CH_2COOH$
4 $R_1=R_2=CH_2COOH$

Figure 2 | Representation of the four calixarenes compounds undergoing *in situ* transformation to diazonium salts for electrografting. Structures of the calix[4]tetra-anilines (**1-4**) used in this work. The modification procedure of the gold, GC or PPF substrates relies on the *in situ* diazotation of the calix[4]tetraanilines in the presence of $NaNO_2$ in acidic aqueous medium followed by the electrochemical reduction of the corresponding diazonium salts.

Here, we propose a different strategy based on surface pavement with versatile platforms that exhibit preorganization properties. This original concept uses rigid tetrapodant calix[4]arenes[33] as building blocks to form robust, closely packed and stable monolayers that can be postfunctionalized. Immobilization of calixarenes onto a surface by self-assembly techniques[34] is classically achieved by anchoring the small rim of the calixarenes onto the substrate through their phenol or thiophenol functions. In contrast, our strategy relies on the large rim grafting of calix[4]arenes, thanks to the diazonium coupling route, whereas the small rim is used for the introduction of spatially controlled postfunctionalizable groups. Adequate decorations of the small rim with appending arms make it possible to direct the introduction of functional molecules with a fine spatial control imposed by the geometry of the small rim, especially when two (or more) molecules are linked to the same calixarene. Close packing of the calix molecules and chemical control of the grafting process are ensured by the constrained cone conformation of the macrocycle that orients all four diazonium groups in the same direction and prevents polymerization of the transiently produced aryl radicals, thanks to the methylene bridges at the ortho-positions of the phenoxy substituents. It is noteworthy to mention that the preparation and the reductive grafting of calix[n]arenes fully substituted at the large rim by diazonium groups have never been reported.

Results

Synthesis and grafting procedures of calix[4]tetra-anilines. *p*tBu-calix[4]arene was first *O*-alkylated by four alkyl or acyl groups following classical procedures. This allows to constrain the macrocycle in a cone conformation and to introduce at the small rim a carboxylic function that can easily be postfunctionalized. A sequence of *ipso*-nitration–reduction then yielded compounds **1–4** displayed in Fig. 2 (see the Supplementary Methods)[35]. Non-postfunctionalizable calix[4]tetra-anilines **1** and **2** were first employed as model compounds for the surface modification and characterization, whereas compounds **3** and **4** were used to explore the postfunctionalization feasibility.

The grafting procedure was similarly conducted whatever the calix[4]tetra-aniline or the nature of substrate used. The corresponding diazonium salts **1–4** were prepared *in situ* in 0.5 M aqueous HCl in the presence of $NaNO_2$ (4 eq.)[36]. Gold, pyrolized photoresist film (PPF) or glassy carbon (GC) (disk electrodes) substrates were functionalized through the reduction of the *in situ* generated diazonium cations from calix[4]tetra-anilines **1–4** either by six voltammetric cycles or by applying a constant potential. Figure 3a shows a typical cyclic voltammogram for the electrografting of **1** on a GC

electrode taken as an example. It exhibits the characteristic behaviour of the reductive electrografting of aryldiazonium salts[37]. The potential was swept between +0.6 V and −0.6 V versus saturated calomel electrode (SCE). On the initial potential cycle, a broad irreversible peak located at about −0.25 V versus SCE was observed then, a dramatic decrease of current intensity was recorded upon the following cycles. The peak stems from the reduction of diazonium cations into phenyl radicals that subsequently react at the electrode surface, forming a covalently linked blocking layer at the electrode interface. Alternatively, the reduction of the *in situ* generated diazonium cations was performed by applying a constant potential at –0.5 V versus SCE for 5 min.

Characterization of the modified surfaces. After copious rinsing with a series of solvents (water, ethanol, dichloromethane, toluene) and drying under a stream of argon, the grafted layers were characterized by analysing their blocking properties toward the electrochemical response of redox probes such as $Fe(CN)_6^{3-/4-}$ (Supplementary Fig. S1), $Ru(NH_3)_6^{3+/2+}$ and dopamine.

Figure 3b–d shows typical voltammograms obtained with $Ru(NH_3)_6^{3+/2+}$ and dopamine from a GC electrode before and after electroreduction of the *in situ* generated diazonium cations from **2 or 4**. As compared with the reversible redox processes at the bare electrode, a complete inhibition of the electrochemical response of the grafted surface was evidenced for all the redox probes. This indicates that an organic layer was formed at the electrode surface, and that this layer acts as a barrier for redox processes. Particularly informative regarding the compactness of the layer is the behaviour of dopamine. Oxidation of dopamine to the corresponding o-quinone has been demonstrated to require adsorption onto the GC electrode surface for exhibiting fast charge transfer kinetics and it was established that the coverage of the electrode surface with a dense monolayer is enough to inhibit the oxidation of dopamine[38]. Thus, observation of an electrochemical activity implies that the molecular probe could reach the carbon substrate that makes dopamine oxidation a specific test for detecting pinholes[38,39]. In our case, we observed the complete inhibition of the dopamine redox activity after the grafting of calixarenes **2** and **4** layers. This observation demonstrates that the possible pinholes/defects/empty spaces within the grafted layers have size smaller than that of dopamine.

To further characterize the grafted organic layer, contact angles of gold-modified substrates were measured using the sessile drop method (Fig. 4). The static contact angle with water was found to be 78±3° for the surface modified with **1** and 89±2° for the surface modified with **2**, in comparison with the value of 62±2° for a bare gold surface.

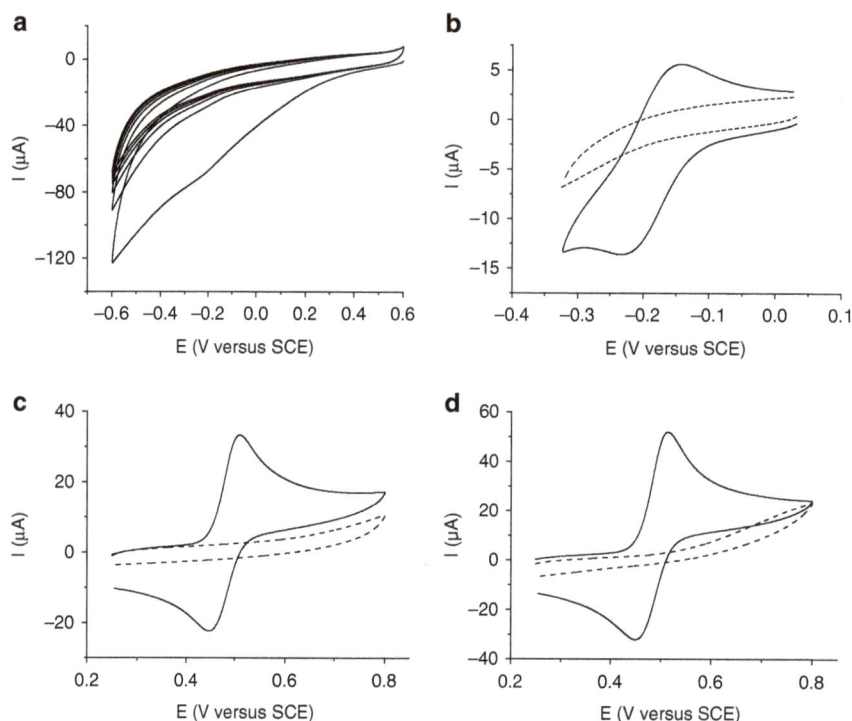

Figure 3 | Electrochemical answers at a GC electrode. (**a**) Cyclic voltammetry of calix[4]arene diazonium *in situ* generated from 5 mM calix[4]tetra-aniline 1 in 0.5 M aqueous HCl in the presence of 4 eq. NaNO₂, scan rate 0.1 V s⁻¹. The first cycle displays the reduction of diazonium cations; subsequent cycles (2–6) exhibit a clear decrease of the current. (**b**) Cyclic voltammetry of 1 mM Ru(NH₃)₆³⁺/²⁺ in 0.1 M KCl at a bare electrode (full line) and at a modified electrode after electrografting of **2** (dashed line). (**c,d**) Cyclic voltammetry of 1 mM Dopamine in 0.1 M H₂SO₄ at a bare electrode (full line) (**c,d**) and at a modified electrode after electrografting of **2** (dashed line) (**c**) and after electrografting of **4** (dashed line) (**d**).

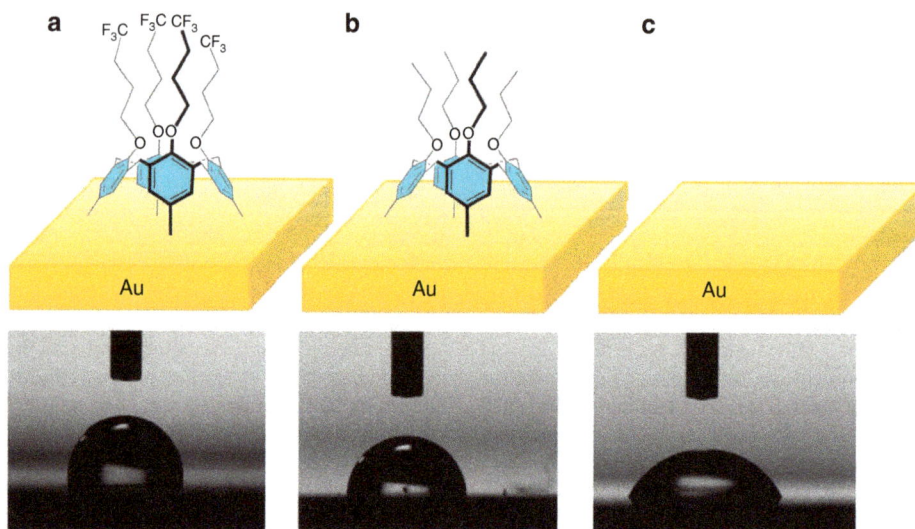

Figure 4 | Characterization of the modified gold substrates by contact angle measurements. Images of a 2 µl water droplet in contact with a gold substrate modified (**a**) with **2**, (**b**) with **1** and (**c**) on a bare substrate. Bare gold substrate is the most hydrophilic sample, showing that the droplet spreads out while the coated substrates, much less hydrophilic because of the grafted organic layer, show a different behaviour of the water droplet.

These data indicate that hydrophobic films were formed on the gold surfaces due to the immobilization of the calixarenes. The larger contact angle for the surface modified with **2** is fully consistent with the stronger hydrophobic nature of the calix[4]arene **2** as compared with **1**. Remarkably, wetting studies for SAMs on gold formed with thio-calix[4]arenes yielded similar contact angle values[40,41].

X-ray photoelectron spectroscopy (XPS) analyses were performed on both PPF and gold substrates modified by the CF₃-containing calix[4]tetra-aniline **2**, the CF₃ groups providing a convenient chemical tag for XPS studies. The survey spectra showed main peaks corresponding to the elemental species C, N and O, whereas peaks due to Au were observed on gold substrates (Supplementary Fig. S2).

In the case of gold substrates, an overall attenuation of the Au photoelectron peaks and a significant increase in the C1s and O1s peak heights, compared with the bare substrate, confirmed the coating of the gold substrate by an organic layer. A characteristic peak of the F1s photoelectrons appeared at 688.5 eV after surface functionalization by the calix[4]arene **2**. The presence of the CF_3 group was further confirmed by the emergence of a new component at 293 eV in the C1s core level spectrum (Fig. 5). The C1s core level showed a main peak centred at 285 eV assigned to the aromatic carbons of the calix[4]arene phenyl groups (and to the carbon atoms of the substrate when analysing the PPF substrates). A component at 286.2 eV was also evidenced, which corresponds to the C–O environment present in the calixarene frame. The small components at the high binding energy side (at ~287.3 and 289 eV) accounts for oxidized carbon from contamination species already detected on the bare substrates.

The absence of a N1s peak attributed to $-N_2^+$ (at 403.8 eV) indicates a complete transformation of the diazonium cations, whereas the very low intensity signal at 400 eV reveals the formation of a very small proportion of azo groups (–N=N–) derived from the diazonium salts as already reported in the literature[36,42–46].

Atomic force microscopy (AFM) analysis was performed by using preferentially PPF substrates because of their low surface roughness in comparison with GC or gold substrates employed in this work. PPF plates grafted with the reduction of diazonium cations issued from **1** or **2** showed surface structures almost identical and of same sizes as the bare PPF, indicating a uniform deposit on the whole surface (Fig. 6). The surface roughness increased only slightly upon the electrografting of the calix[4]arenes (root-mean-square roughness, r.m.s. < 1 nm) as compared with a bare PPF substrate (r.m.s. < 0.8 nm)[47]. These observations fall in line with the formation of very thin layers. Remarkably, the surface roughness was not affected by electrolysis time (30 s, 1 min, 5 min, 10 min) employed for the electrografting of the *in situ* generated diazonium cations, suggesting that the system is self-inhibiting.

AFM technique was further used for estimating the thicknesses of the organic layers. The AFM tip was used in contact mode to scratch a rectangular area on the PPF functionalized samples by exercising a force sufficient to remove the organic part without damaging the PPF sample. Profiling depth measurements of the scratched area in an intermittent contact mode gave the average thickness of the layer, as described by Mc Creery and co-workers[48]. The thickness was thus estimated to be 1.3 ± 0.1 nm (Fig. 6c). This result agrees well with the height of calix[4]arenes **1** and **2** (*c.a.* 1.0 and 1.1 nm, respectively, estimated from MM2 energy minimizations with Chem-Bio3D software), and indicates that the electrografting of the corresponding diazonium cations led to the formation of monolayers. This result is further confirmed by ellipsometric measurements of gold substrates modified with calix[4]arene **2**. Thickness value of 1.09 ± 0.20 nm was found, in very good agreement with the height of calix[4]arene **2**. Both techniques unambiguously demonstrate the formation of monolayers. From the strong blocking effect toward redox probe responses, namely regarding the answer of the monolayer towards dopamine, it can be further concluded that the formed monolayers are rather compact.

Taking account the formation of monolayers, XPS data could be used to give a reliable estimation of surface coverage for the

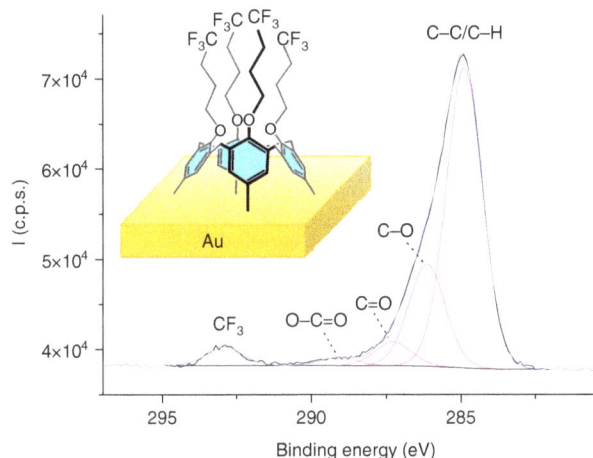

Figure 5 | High-resolution peak-fitted C1s region for gold substrate modified by calix[4]arene 2. Surface composition is indicated in the integrated areas.

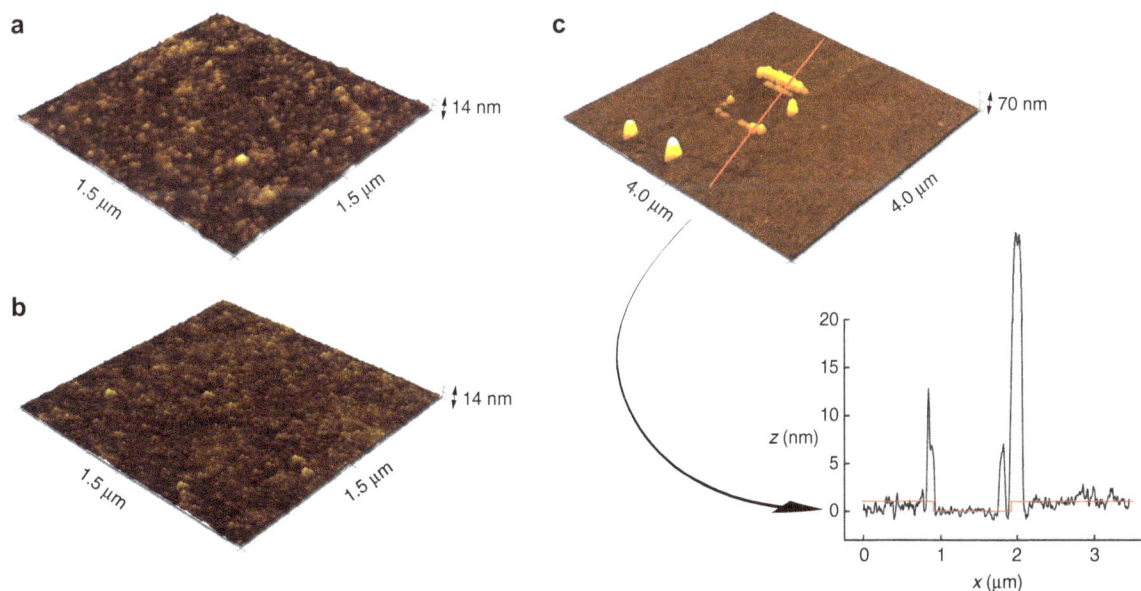

Figure 6 | AFM characterization of PPF sample modified with 2. (a) AFM image (surface scale 1.5×1.5 μm^2) of the bare PPF sample. **(b)** AFM image (surface scale 1.5×1.5 μm^2) of the PPF sample after *in situ* generation and electrografting of diazonium cation from **2**. **(c)** Scratching of the PPF sample as imaged in **b** (surface scale 4×4 μm^2). Depth-profile of the scratched area.

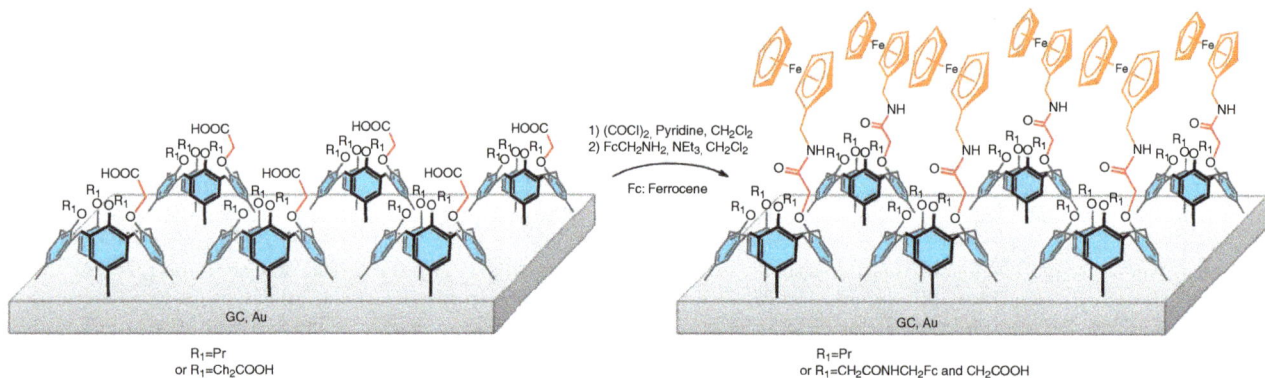

Figure 7 | Chemical postfunctionalization with ferrocenemethylamine of the calix[4]arene-modified substrates. The substrates were initially modified with calix[4]arenes **3** or **4** exhibiting one or four appending COOH groups at the small rim. The easy postfunctionalization is exemplified with the covalent binding of ferrocene redox centres.

non-electroactive calixarene **2**. On gold samples, the average calix[4]arene coverage α on the gold surface could be expressed from F_{1s} and Au_{4f} peak intensities related to calix[4]arenes layers and the gold substrate, respectively[49]. The calculation (details provided in the Supplementary Methods) leads to a surface coverage α of the gold substrate by the calix[4]arene **2** molecules of 79%. Calix[4]arenes present an external diameter of 14 Å (large rim)[50]. Considering a hexagonal compact arrangement of disks (or spheres) of 14 Å diameter, the surface concentration for the closest packing of calixarenes corresponds to $\Gamma_{CPML} = 9.8 \times 10^{-11}\,mol\,cm^{-2}$ and the closest-packing monolayer covers 90.7% of the gold surface. It comes that the surface concentration of calixarene **2** onto gold substrate estimated from XPS data is equal to $8.5 \times 10^{-11}\,mol\,cm^{-2}$ ($0.5\,calix\,nm^{-2}$).

All these results evidence the formation of a dense and compact monolayer of calixarenes.

Postfunctionalization of the modified surfaces. Having demonstrated the straightforward formation of monolayers from the electroreduction of diazonium cations derived from compounds **1** and **2**, calix[4]tetra-anilines **3** and **4** (Fig. 7) were then used to modify substrates as a proof-of-concept for postfunctionalization purposes. The presence of one (for **3**) or four (for **4**) appending carboxyl group(s) at the calixarenes small rim allows a further chemical coupling using acyl chloride activation. Accordingly, ferrocene moieties were covalently introduced through an amide bond formation with ferrocenemethylamine ($FcCH_2NH_2$) following a procedure previously described (Fig. 7)[51].

After the chemical postfunctionalization, the modified substrates were thoroughly rinsed with CH_2Cl_2, then transferred to a pure CH_2Cl_2 + 0.2 M NBu_4PF_6 electrolyte solution. Well-defined oxidation and reduction peaks corresponding to the introduced ferrocene/ferrocenium couple were detected at *c.a.* 0.4 V versus SCE, as illustrated in Fig. 8. It was also observed that the current linearly increased with scan rate in the $(0.05-10\,V\,s^{-1})$ range as expected for surface-confined electroactive species.

For experiments using calixarene **3** (Fig. 8a), the surface density of attached ferrocene centres was derived from integration of the voltammetric peaks at low scan rates. This allowed an estimation of the surface coverage of calixarenes themselves onto the substrate since each calixarene frame bears one functionalizable carboxyl substituent. Because the postfunctionalization reaction may not take place with 100% yield, this value could be regarded as a lower estimate of the calixarenes surface concentration. Surface concentration was found to be $\Gamma_1 = 6.9 \pm 0.6 \times 10^{-11}\,mol\,cm^{-2}$, corresponding to a density of $0.42\,calix\,nm^{-2}$. This result is fully consistent with the surface concentration value estimated from XPS data and

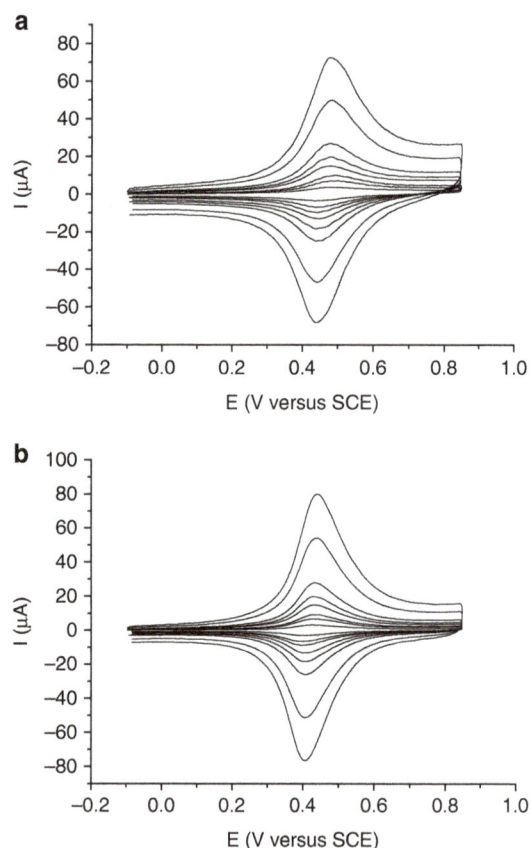

Figure 8 | Cyclic voltammetry of GC electrodes modified with calixarenes and postfunctionalized with ferrocene. The modified surfaces were obtained from the electrografting of diazonium cations of **3** (**a**) and **4** (**b**). CH_2Cl_2 + 0.2 M NBu_4PF_6. Ferrocene redox centres were covalently introduced on the calixarenes platforms from a chemical coupling using the COOH appending groups present at the small rim of calixarenes **3** and **4**. Cyclic voltammetry show the typical signal of surface-confined ferrocene centres. Scan rate varied from 0.1 to $3\,V\,s^{-1}$.

confirms the formation of rather densely packed monolayers, in line with the good blocking properties of the layers towards the electronic transfer. Note that the surface concentration here found is in good agreement with surface concentration reported for gold–thiols SAMs of calix[4]arenes[52,53], or compares favourably with the density reported for the covalent and rigid grafting of calix[4]arenes onto silica surfaces[54], the molecular footprint of the described calixarenes being similar to the calixarenes **1–4**.

The surface concentration of electroactive ferrocene was similarly determined for experiments using the tetra-carboxylic acid decorated calix[4]tetra-aniline **4**. A surface concentration $\Gamma_2 = 1.6 \pm 0.4 \times 10^{-10}$ mol cm^{-2} was calculated. This surface concentration is one third the value reported for the closest packing of ferrocene[55] and corresponds to *c.a.* twice the Γ_1 value. This result indicates that two ferrocenyl moieties could be attached onto each calixarene subunit. This allows the introduction of functional groups on the surface with a precise spatial control, which is imposed by the calixarene small rim geometry.

Discussion

Calix[4]tetra-anilines have been successfully electrografted to various substrates by generating *in situ* their corresponding diazonium cations. Thanks to its well-preorganized and unique 3D conic structure, the calix[4]arene core prevents the formation of multilayers and leads to densely packed monolayers[56]. Characterization of the modified surfaces points out to an efficient packing of the calix molecules, probably ensured by the constrained cone conformation of the macrocycle that orients all four diazonium groups in the same direction. However, this does not necessarily imply that four aryl–substrate bonds are formed. This would depend on the availability of gold or carbon atoms sites and the possible degree of distortion of the calixarene molecule (Supplementary Fig. S3). The important point is that the calix grafts with its cone axis perpendicular to the surface, whatever the number of aryl–substrate bonds formed, and the thicknesses reported in this work are in good agreement with this point. Providing that the calixarene platforms possess chemically reactive appending arms at the small rim, this strategy allows for the straightforward preparation of versatile robust monolayers that can be directed at the spatially controlled introduction of functional entities. This approach can be made even more sophisticated by using calix[4]arenes that exhibit two different anchoring termini on the same frame, thus opening new possibilities for the design of multicomponent films with a spatial control at the molecular level. Work is currently under way to implement this last approach.

Methods

Synthesis. The synthesis of the various calix[4]arenes compounds and their characterization are described in the Supplementary Methods.

Surface modifications procedures. The reagents and solvents used for surface chemistry were of high purity grade (Sigma-Aldrich, SDS, Alfa Aesar). The GC substrates were home-made disk electrodes of 3 mm diameter, thoroughly polished (successively with 5 μm SiC paper (Struers), 1 μm DP–Nap paper (Struers) with 0.3 μm Al$_2$O$_3$ slurry (Struers)) and rinsed with ultrapure water, and absolute ethanol under sonication. Gold substrates (gold layer onto Si substrates) were purchased from Aldrich and thoroughly cleaned before surface functionalization (piranha solution). PPF samples were prepared according to previously published procedure[57]. Surface modification by electrochemical reduction of *in situ* generated diazonium cations was carried out in an ice bath. NaNO$_2$ (4 eq.) was added to the aqueous electrolytic solution containing 5 mmol^{-1} calix[4]tetra-aniline and 0.5 mol^{-1} HCl under stirring. After 5 min, the modification was achieved potentiostatically or by potential cycling between 0.6 and −0.6 V/SCE. After modification, the substrates were rinsed (sonication) with ultrapure water, abs EtOH, freshly distiled CH$_2$Cl$_2$ and toluene (high-performance liquid chromatography grade). The subsequent ferrocenyl postfunctionalization was carried out according to ref. 51.

Electrochemistry. Cyclic voltammograms were recorded with an Autolab electrochemical analyzer (PGSTAT 30 potentiostat/galvanostat from Eco Chemie B.V.) in a three-electrode setup with an SCE reference electrode and a platinum foil as counter electrode.

The surface coverage of electroactive ferrocene is obtained from Faraday's Law, $\Gamma = Q/nFA$ where Q is the charge passed, n the number of exchanged electrons, F the Faraday constant and A the area of electrode. Q is obtained from the integration of the area under the voltammetric peaks at low scan rates using the

integration function of the Origin software. Polynomial baselines can be solved and subtracted before the curve integration.

AFM microscopy analyses. AFM experiments used a Picoscan II microscope (Agilent) coupled to a pico-SCAN2500 controller. The images were acquired in acoustic mode using silicon nitride tips (resonance frequency *c.a* 350 kHz). Scratching technique was used to determine the thickness of the organic layer grafted on the PPF samples. Square-shaped scratches were obtained in contact mode with a force set-point voltage around 1 V. Several depth-profiles were measured for each scratched region. Height distributions were also investigated on images to confirm the depth-profiles measurements.

XPS measurements. XPS photoelectron spectra were collected from modified GC and gold plates on a Thermo VG Scientific Escalab 250 system fitted with a monochromatic Al Kα X-ray source (hν = 1486.6 eV, spot size = 650 μm, power 15 kV×200 W). The pass energy was set at 150 and 40 eV for the survey and the narrow regions, respectively. Spectral calibration was determined by setting the main C1s component at 285 eV. The surface composition was estimated using the integrated peak areas and the corresponding Scofield sensitivity factors corrected for the analyzer transmission function.

Contact angle measurements. The static contact angles of 2 μl ultrapure water drops were measured on four different spots on the surface with an easy drop goniometer (Krüss). The contact angles were determined using a tangent 2 or circle fitting models.

Ellipsometry. Thicknesses of the layers on Au were measured with a mono wavelength ellipsometer Sentech SE400. The following values were taken for gold: $n_s = 0.203$, $k_s = 3.431$. These values were measured on clean surfaces before grafting and the layers thicknesses were determined from the same plates after modification, taking $n_s = 1.5$, $k_s = 0$ for the layer. Measurements reported are the average of ten data points across each substrate (two substrates were analysed).

References

1. Simao, C. *et al.* A robust molecular platform for non-volatile memory devices with optical and magnetic responses. *Nat. Chem.* **3**, 359–364 (2011).
2. Areephong, J., Browne, W. R., Katsonis, N. & Feringa, B. L. Photo- and electro-chromism of diarylethene modified ITO electrodes - towards molecular based read-write-erase information storage. *Chem. Commun.* 3930–3932 (2006).
3. Amatore, C., Genovese, D., Maisonhaute, E., Raouafi, N. & Schollhorn, B. Electrochemically driven release of picomole amounts of calcium ions with temporal and spatial resolution. *Angewandte Chem. Int. Ed.* **47**, 5211–5214 (2008).
4. Lahann, J. *et al.* A reversibly switching surface. *Science* **299**, 371–374 (2003).
5. Kim, B. S. *et al.* Temperature and length dependence of charge transport in redox-active molecular wires incorporating ruthenium(II) bis (σ-arylacetylide) complexes. *J. Phys. Chem. C* **111**, 7521–7526 (2007).
6. Willner, I. Biomaterials for sensors,fuel cells, and circuitry. *Science* **298**, 2407–2408 (2002).
7. Choi, H. J., Kim, N. H., Chung, B. H. & Seong, G. H. Micropatterning of biomolecules on glass surfaces modified with various functional groups using photoactivatable biotin. *Anal. Biochem.* **347**, 60–66 (2005).
8. Adams, D. M. *et al.* Charge Transfer on the nanoscale: current status,. *J. Phys. Chem. B* **107**, 6668–6697 (2003).
9. Finklea, H. O. Electrochemistry of organized monolayers of thiols and related molecules on electrodes. in *Electroanalytical chemistry*, Vol 19 (eds Bard A.J. and Rubinstein) 109–337 (Marcel Dekker, Inc. New York, 1996).
10. Gooding, J. J. Advances in interfacial design for electrochemical biosensors and sensors: aryl diazonium salts for modifying carbon and metal electrodes. *Electroanalysis* **20**, 573–582 (2008).
11. Gooding, J. J. & Ciampi, S. The molecular level modification of surfaces : from self-assembled monolayers to complex molecular assemblies. *Chem. Soc. Rev.* **40**, 2704–2718 (2011).
12. Love, C. J., Estroff, L. A., Kriebel, J. K., Nuzzo, R. G. & Whitesides, G. M. Self-assembeld monolayers of thiolates on metal as a form of nanotechnology. *Chem. Rev.* **105**, 1103–1169 (2005).
13. Vericat, C., Vela, M. E., Benitez, G., Carro, P. & Salvarezza, R. C. Self-assembled monolayers of thiols and dithiols on gold : new challenges for a well-known system. *Chem. Soc. Rev.* **39**, 1805–1834 (2010).
14. Kind, M. & Wöll, C. Organic surfaces exposed by self-assembled organothiol monolayers: preparation, characterization, and application. *Progr. Surf. Sci.* **84**, 230–278 (2009).
15. Sullivan, T. P. & Huck, W. T. S. Reactions on monolayers: organic synthesis in two dimensions. *Eur. J. Org. Chem.* 17–23 (2003).
16. Bain, C. D. & Whitesides, G. M. Formation of monolayers by the coadsorption of thiols on gold: variation in the length of the alkylchain. *J. Am. Chem. Soc.* **111**, 7164–7175 (1989).

17. Folkers, J. P., Laibinis, P. E. & Whitesides, G. M. Self-assembled monolayers of alkanethiols on gold: comparisons of monolayers containing mixture of short- and long-chain constituents with CH$_3$ and CH$_2$OH terminal groups. *Langmuir* **8**, 1330–1341 (1992).

18. Stranick, S. J., Parikh, A. N., Tao, Y. T., Allara, D. L. & Weiss, P. S. Phase separation of mixed-composition self assembled monolayers into nanometer scale molecular domains. *J. Phys. Chem.* **98**, 7636–7646 (1994).

19. Tamada, K., Akiyama, H., Wei, T. X. & Kim, S. A. Photoisomerization reaction of unsymmetrical azobenzene disulfide self-assembled monolayers: modification of azobenzene dyes to improve thermal endurance for photoreaction. *Langmuir* **19**, 2306–2312 (2003).

20. Downard, A. J. Electrochemically assisted covalent modification of carbon electrodes. *Electroanalysis.* **11**, 1085–1096 (2000).

21. Pinson, J. & Podvorica, F. Attachment of organic layers to conductive or semiconductive surfaces by reduction of diazonium salts. *Chem. Soc. Rev.* **34**, 429–439 (2005).

22. Pinson, J. & Bélanger, D. Electrografting: a powerful method for surface modification. *Chem. Soc. Rev.* **40**, 3995–4048 (2011).

23. Mahouche-Chergui, S., Gam-Derouich, S., Mangeney, C. & Chehimi, M. H. Aryl diazonium salts: a new class of coupling agent for bonding polymers, biomacromolecules and nanoparticles to surfaces. *Chem. Soc. Rev.* **40**, 4143–4166 (2011).

24. Aryl Diazonium Salts. *New Coupling Agents in Polymer and Surface Science* (ed. Chehimi, M.M.), (Wiley-VCH, Weinheim, 2012).

25. Gui, A. L. *et al.* A comparative study of modifying gold and carbon electrode with 4-sulfophenyl diazonium salt. *Electroanalysis* **22**, 1283–1289 (2010).

26. Zigah, D., Noel, J. M., Lagrost, C. & Hapiot, P. Charge transfer electroactive species immobilized on carbon surfaces by aryl diazonium reduction. SECM investigations. *J. Phys. Chem. C* **114**, 3075–3081 (2010).

27. Combellas, C., Kanoufi, F., Pinson, J. & Podvorica, F. I. Sterically hindered diazonium salts for the grafting of a monolayer on metals. *J. Am. Chem. Soc.* **130**, 8576–8577 (2008).

28. Leroux, Y. R., Fei, H., Noel, J.- M., Roux, C. & Hapiot, P. Efficient covalent modification of a carbon Surface: Use of a silyl protecting group to form an active monolayer. *J. Am. Chem. Soc.* **132**, 14039–14041 (2010).

29. Malmos, K. *et al.* Using a hydrazone-protected benzenediazonium salt to introduce a near-monolayer of benzaldehyde on glassy carbon surfaces. *J. Am. Chem. Soc.* **131**, 4928–4938 (2009).

30. Nielsen, L. T. *et al.* Electrochemical approach for constructing a monolayer of thiophenolates from grafted multilayers of diaryl disulfides. *J. Am. Chem. Soc.* **129**, 1888–1889 (2007).

31. Peng, Z., Holm, A. H., Nielsen, L. T., Pedersen, S. U. & Daasberg, K. Covalent sidewall functionalization of carbon nanotubes by a "formation-degradation" approach. *Chem. Mater.* **20**, 6068–6075 (2008).

32. Marshall, N. & Locklin, J. Reductive electrografting of benzene (*p*-bisdiazonium hexafluorophosphate): a simple and effective protocol for creating diazonium-functionalized thin films. *Langmuir* **27**, 13367–13373 (2011).

33. Gutsche, C. D. in *Calixarenes Revisited, Monographs in Supramolecular Chemistry* (ed. Stoddart, J.F.), (The Royal Society of Chemistry, Cambridge, 1998).

34. Kim, H. J., Lee, M. H., Mutihac, L., Vicens, J. & Kim, J. S. Host-guest sensing by calixarenes on the surface. *Chem. Soc. Rev.* **41**, 1173–1190 (2012).

35. Sansone, F., Chierici, E., Casnati, A. & Ungaro, R. Thiourea-linked upper rim calix[4]arene neoglycoconjugates: synthesis, conformations and binding properties. *Org. Biomol. Chem.* **1**, 1802–1809 (2003).

36. Baranton, S. & Bélanger, D. Electrochemical derivatization of carbon surface by reduction of *in situ* generated diazonium cations. *J. Phys. Chem. B* **109**, 24401–24410 (2005).

37. Delamar, M., Hitmi, R., Pinson, J. & Savéant, J. -M. Covalent modification of carbon surfaces by grafting of functionalized aryl radicals produced from electrochemical reduction of diazonium salts. *J. Am. Chem. Soc.* **114**, 5883–5884 (1992).

38. DuVall, S. H. & McCreery, R. L. Control of catechol and hydroquinone electron-transfer kinetics on native and modified glassy carbon electrodes. *Anal. Chem.* **71**, 4594–4602 (1999).

39. DuVall, S. H. & McCreery, R. L. Self-catalysis by catechols and Quinones during heterogeneous electron transfer at carbon electrodes. *J. Am. Chem. Soc.* **122**, 6759–6764 (2000).

40. Cygan, M. T. *et al.* Calixarene monolayers as quartz crystal microbalance sensing element in aqueous solution. *Anal. Chem.* **71**, 142–148 (1999).

41. Zhang, S., Song, F. & Echegoyen, L. Synthesis, self-assembled monolayers and alkaline earth metal ion recognition of *p-tert*-butylcalix[4]arene derivatives. *Eur. J. Org. Chem.* 2936–2943 (2004).

42. Toupin, M. & Bélanger, D. Thermal stability study of aryl modified carbon black *in situ* generated diazonium salt. *J. Phys. Chem. C* **111**, 5394–5401 (2007).

43. Saby, C., Ortiz, B., Champagne, G. Y. & Bélanger, D. Electrochemical modification of glassy carbon electrode using aromatic diazonium salt. 1. Blocking effect of 4-nitrophenyl and 4-carboxyphenyl groups. *Langmuir* **13**, 6805–6813 (1997).

44. Lyskawa, J. & Bélanger, D. Direct modification of a gold electrode with aminophenyl groups by electrochemical reduction of *in situ* generated aminophenyl monodiazonium cations. *Chem. Mater.* **18**, 4755–4763 (2006).

45. Hurley, B. L. & McCreery, R. L. Covalent bonding of organic molecules to Cu and Al alloy 2024 T3 surfaces via diazonium ion reduction. *J. Electrochem. Soc.* **151**, B252–B259 (2004).

46. Doppelt, P., Hallais, G., Pinson, J., Podvorica, F. & Verneyre, S. Surface modification of conducting substrates. Existence of azo bonds in the structure of organic layers obtained from diazonium salts. *Chem. Mater.* **19**, 4570–4575 (2007).

47. Ranganathan, S. & McCreery, R. L. Electroanalytical performance of carbon films with near-atomic flatness. *Anal. Chem.* **73**, 893–900 (2001).

48. Anariba, F., DuVall, S. H. & McCreery, R. L. Mono- and multilayer formation by diazonium reduction on carbon surfaces monitored with atomic force microscopy 'scratching'. *Anal. Chem.* **75**, 3837–3844 (2003).

49. Dahoumane, S. A *et al.* Protein-functionalized hairy diamond nanoparticles. *Langmuir* **25**, 9633–9638 (2009).

50. Kim, J. -H., Kim, Y. -G., Lee, K. -H., Kang, S. -W. & Koh, K. -N. Size selective molecular recognition of calix[4]arenes in Langmuir-Blodgett monolayers. *Synth. Met.* **117**, 145–148 (2001).

51. Noel, J. -M., *et al.* Flexible strategy for immobilizing redox-active compounds using *in situ* generation of diazonium salts. Investigation of the blocking and catalytic properties of the layers. *Langmuir* **25**, 12742–12749 (2009).

52. Chen, H. *et al.* Comparative study of protein immobilization properties on calixarenes monolayers. *Sensors* **7**, 1091–1107 (2007).

53. Cormode, D. P., Evans, A. J., Davis, J. J. & Beer, P. D. Amplification of anion sensing by disulfide functionalized ferrocene and ferrocene-calixarene receptors adsorbed onto gold surfaces. *Dalton Trans.* **39**, 6532–6541 (2010).

54. Notestein, J. M., Katz, A. & Iglesia, E. Energetics of small molecule and water complexation in hydrophobic calixarene cavities. *Langmuir* **22**, 4004–4014 (2006).

55. Seo, K., Jeon, I. C. & Yoo, D. J. Electrochemical characteristics of ferrocenecarboxylate-coupled aminoundecylthiol self-assembled monolayers. *Langmuir* **20**, 4147–4154 (2004).

56. Mattiuzzi, A., Jabin, I., Reinaud, O., Hapiot, P. & Lagrost, C. European patent EP 12164038 (2012).

57. Brooksby, P. A. & Downard, A. J. Electrochemical and atomic force microscopy study of carbon surface modification via diazonium reduction in aqueous and acetonitrile solutions. *Langmuir* **20**, 5038–5045 (2004).

Acknowledgements

This work was supported by the Agence Nationale de la Recherche (ANR10-BLAN-714 Cavity-zyme(Cu) project) and by the FNRS (A.M. PhD grant). We thank Pr. A.J. Downard (University of Canterbury, N.Z.) for fruitful discussions and advices concerning PPF samples.

Author contributions

O.R and I.J. conceived and designed the synthesis of calixarenes. P.H and C.L conceived and designed the surface functionalization experiments. A.M. performed the calixarenes syntheses. A.M., C.L and L.S. contributed to surface functionalization. J.F.B performed AFM investigations, C.M. performed XPS analyses. C.R prepared the PPF sample and performed preliminary experiments on PPF samples. O.R., I.J., A.M., C.L. and P.H analysed the data. C.L., I.J. and O.R. wrote the paper. All authors discussed and commented on the manuscript.

Additional information

12

Hydride bridge in [NiFe]-hydrogenase observed by nuclear resonance vibrational spectroscopy

Hideaki Ogata[1,*], Tobias Krämer[1,*], Hongxin Wang[2,3,*], David Schilter[4,*,†], Vladimir Pelmenschikov[5,*], Maurice van Gastel[1], Frank Neese[1], Thomas B. Rauchfuss[4], Leland B. Gee[2], Aubrey D. Scott[2], Yoshitaka Yoda[6], Yoshihito Tanaka[7,†], Wolfgang Lubitz[1] & Stephen P. Cramer[2,3]

The metabolism of many anaerobes relies on [NiFe]-hydrogenases, whose characterization when bound to substrates has proven non-trivial. Presented here is direct evidence for a hydride bridge in the active site of the ^{57}Fe-labelled fully reduced Ni-R form of *Desulfovibrio vulgaris* Miyazaki F [NiFe]-hydrogenase. A unique 'wagging' mode involving H^- motion perpendicular to the Ni(μ-H)^{57}Fe plane was studied using ^{57}Fe-specific nuclear resonance vibrational spectroscopy and density functional theory (DFT) calculations. On Ni(μ-D)^{57}Fe deuteride substitution, this wagging causes a characteristic perturbation of Fe–CO/CN bands. Spectra have been interpreted by comparison with Ni(μ-H/D)^{57}Fe enzyme mimics [(dppe)Ni(μ-pdt)(μ-H/D)^{57}Fe(CO)$_3$]$^+$ and DFT calculations, which collectively indicate a low-spin Ni(II)(μ-H)Fe(II) core for Ni-R, with H^- binding Ni more tightly than Fe. The present methodology is also relevant to characterizing Fe–H moieties in other important natural and synthetic catalysts.

[1] Max Planck Institute for Chemical Energy Conversion, Mülheim an der Ruhr 45470, Germany. [2] Department of Chemistry, University of California, Davis, California 95616, USA. [3] Division of Physical Biosciences, Lawrence Berkeley National Laboratory, Berkeley, California 94720, USA. [4] Department of Chemistry, University of Illinois, Urbana, Illinois 61801, USA. [5] Institut für Chemie, Technische Universität Berlin, Berlin 10623, Germany. [6] Division of Research and Utilization, SPring-8/JASRI, Hyogo 679-5198, Japan. [7] Materials Dynamics Laboratory, RIKEN SPring-8, Hyogo 679-5148, Japan. * These authors contributed equally to this work. † Present address: IBS Center for Multidimensional Carbon Materials, Ulsan National Institute of Science and Technology, Ulsan 689-798, Republic of Korea (D.S.); Graduate School of Material Science, University of Hyogo, Hyogo 678-1297, Japan (Y.T.). Correspondence and requests for materials should be addressed to S.P.C. (email: spjcramer@ucdavis.edu).

Acentral goal of the hydrogen economy is to forestall a continual buildup of CO_2 and the threat of global climate change[1]. However, achieving independence of carbon-based fuels necessitates the development of better H_2-processing catalysts from earth-abundant materials. In this regard, bidirectional hydrogenase enzymes, which catalyse both production and consumption of H_2 (refs 2,3), have attracted interest either for direct utilization[4] or as aspirational targets for biomimetic catalysts[5–11].

The redox-active hydrogenases are classified as either [FeFe]- or [NiFe]-hydrogenase, according to the metals present at their active sites. While both types possess high catalytic activity[12–14], the latter are attractive practically in that they are more O_2-tolerant[12,15]. The [NiFe]-hydrogenase active site features Ni and Fe centres bridged by two Cys residues (Fig. 1a), with two further Cys ligands binding Ni terminally, and the Fe coordination sphere being completed by one CO and two CN^- groups. The [NiFe] site functions in concert with an electron transport chain comprising three Fe–S clusters. The catalytic cycle is generally thought to involve three key redox states of the bimetallic centre: electron spin (paramagnetic) resonance (EPR)-silent Ni-SI$_a$, EPR-active Ni-C and another EPR-silent species known as Ni-R[16–18]. Despite progress in characterizing hydrogenases by crystallography, spectroscopy, and theory[16], questions remain about the molecular and electronic structure of various intermediates and inhibited species. This particularly applies to the electronic structure of Ni-R.

Up to three isoelectronic Ni(II)Fe(II) forms of Ni-R are characterized by their pH-dependent Fourier transform infrared (FT-IR) signatures[18]. The structures proposed for these Ni-R subspecies (Fig. 1b–d) most commonly have a bridging hydride at the active site (Ni(μ-H)Fe)[19–22], with some even featuring an additional (terminal) hydride at Ni (HNi(μ-H)Fe)[23,24]. Another suggested form has an Fe-bound dihydrogen ligand (NiFe(η^2-H_2))[25,26]. A recent high-resolution crystallographic analysis of Ni-R1, a subspecies of Ni-R, indicates a Ni(μ-H)Fe core with a protonated terminal cysteinate[27]. Among proposals supporting the bridging hydride, there is further debate as to whether H^- is bound more strongly to Fe or Ni (ref. 28) and whether Ni(II) is high[29] or low spin[30] or whether both spin configurations coexist in the bulk[31,32].

The Ni-R state represents a special challenge for spectroscopy in that it is EPR-silent, photolabile (and thus difficult to study with Raman spectroscopy) and expected to feature an active site M–H moiety (being notoriously difficult to observe using infrared methods)[33]. The present study instead employs a synchrotron radiation technique called nuclear resonance vibrational spectroscopy (NRVS), which involves X-ray excitation of a Mössbauer-active nuclide[34–37]. The raw NRVS data are commonly translated into partial vibrational density of states (PVDOS) spectra[38], which show vibrational energy contributions specifically from the Mössbauer-active nuclei, such as ^{57}Fe. PVDOS can also be predicted using density functional theory (DFT) or empirical force field calculations, assisting in confident spectral assignments.

NRVS is uniquely suited to detailed investigations of ^{57}Fe-labelled enzyme active sites, avoiding interference from the thousands of protein modes present in a typical infrared or Raman spectrum. NRVS has enabled the observation of Fe–CN and Fe–CO bending and stretching modes for the active sites in [NiFe]-hydrogenases[39,40], despite the presence of 11 (or more) Fe centres in clusters of the electron transport chain. This is because Fe–CN and Fe–CO modes are strongest in the region from 440 to $640\,cm^{-1}$, while Fe–S cluster modes only make significant contributions below $440\,cm^{-1}$ (refs 41–43). Another recent application of NRVS to a trans-H/D-^{57}Fe-(CO) compound shed light on coupling of Fe–H/D and Fe–CO bending modes[33].

The present study discloses the first spectroscopic evidence for the bridging hydride in the Ni-R active site and the unprecedented NRVS observation of a Fe–H stretching mode in a synthetic Ni–H–Fe system. A combined experimental/theoretical analysis of both Ni-R and its synthetic mimics is presented here, an approach that we anticipate to be of broad utility for the characterization of (bio)inorganic Fe-hydride catalysts.

Results

NRVS of model complexes. The Ni(II)(μ-thiolate)$_2$(μ-H)Fe(II) core proposed for Ni-R is reproduced by the diamagnetic H_2-evolving catalyst [(dppe)Ni(μ-pdt)(μ-H)Fe(CO)$_3$]$^+$ ([**1H**]$^+$, dppe = 1,2-bis(diphenylphosphino)ethane = 1,2-Ph$_2$PCH$_2$CH$_2$PPh$_2$, pdt^{2-} = 1,3-propanedithiolate = $^-$SCH$_2$CH$_2$CH$_2$S$^-$)[44,45] shown in Fig. 2a, which has recently been studied using resonance Raman spectroscopy[46]. Synthetic methodology allowed the preparation of the labelled analogue [(dppe)Ni(μ-pdt)(μ-H/D) ^{57}Fe(CO)$_3$]$^+$ in both hydride [**1′H**]$^+$ and deuteride [**1′D**]$^+$ forms for ^{57}Fe NRVS analysis (see Supplementary Figs 1–16 and Supplementary Note 1). Complementary ^{13}CO-labelled species [**1″H/D**]$^+$ were prepared as well.

The NRVS spectra for [**1′H/D**]$^+$ are presented in Fig. 2b. Apart from low-frequency (180–340 cm^{-1}) Fe–S modes, there are strong bands in the 440–630 cm^{-1} region assigned to $\nu_{Fe–CO}$ stretches and $\delta_{Fe–CO}$ bends (Supplementary Table 1). On the high-energy side, NRVS analysis of [**1′H**]$^+$ also reveals peaks at 1,532 and 1,468 cm^{-1} that have previously been assigned to $\nu_{Fe–H}$ modes of two different conformations of the pdt^{2-} ligand ('Ni/Fe-flippamers'[47–50], see Fig. 2a and Supplementary Fig. 17 using Raman spectroscopy on [**1H**]$^+$ (ref. 46, Supplementary Table 2). As expected, these features red-shift to ~1,101 cm^{-1} on D substitution in the bridge. While NRVS has uncovered $\delta_{Fe–H/D}$ (ref. 33) and $\nu_{Fe–D}$ modes[51] for other species, the 1,532 and 1,468 cm^{-1} bands for [**1′H**]$^+$ are the first Fe–H stretching modes detected using this technique, as well as being the highest-frequency bands observed using NRVS to date. The band at 954 cm^{-1} for [**1′H**]$^+$, previously assigned to a $\nu_{Ni–H}$ mode[46], is evidently also Fe-coupled, given its detection using NRVS. In [**1′D**]$^+$, the corresponding Ni–D stretch is at 708 cm^{-1}. As

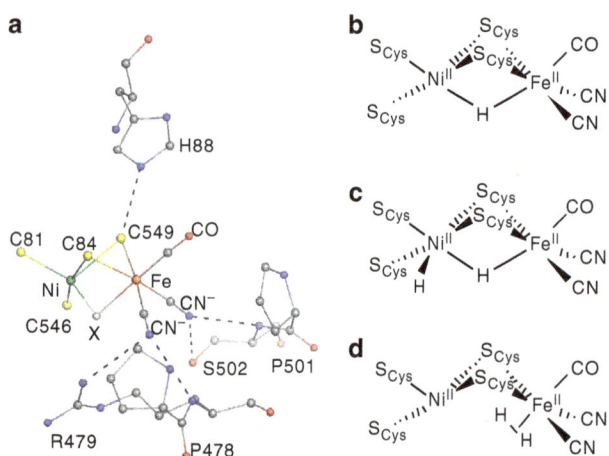

Figure 1 | The Ni-R reduced state of [NiFe]-hydrogenase. (a) X-ray structure of the *Desulfovibrio vulgaris* Miyazaki F (*Dv*MF) [NiFe]-hydrogenase active site from PDB entry 1WUI (ref. 62). The bridging ligand X is oxygenic for deactivated states, with catalytically active states having either a hydride or a vacant site. **(b–d)** Some of the structures proposed for isoelectronic Ni-R forms[19–26].

a

Ni-flippamer   Fe-flippamer

d 774 cm^{-1}

e 1,022 cm^{-1}

f 1,479 cm^{-1}

b Experiment

Ni-H/D-Fe wag   Ni-H/D stretch

608
608 708 758
495 617
587
131 146 275 558
146 275
131 495 580 752
580

Fe-H/D stretch

954 1,101 1,468 1,532
1,023

c DFT

506 612
508 615
743 767 774* ×4
558 616
556 580 617
137 580
137 766
138
138 264 283
264 283
505 574
508 575

1,030 1,053
1,022* 1,061

1,447 1,479*

[1'D]$^+$
[1'H]$^+$

0   500   1,000   1,500

Energy (cm^{-1})

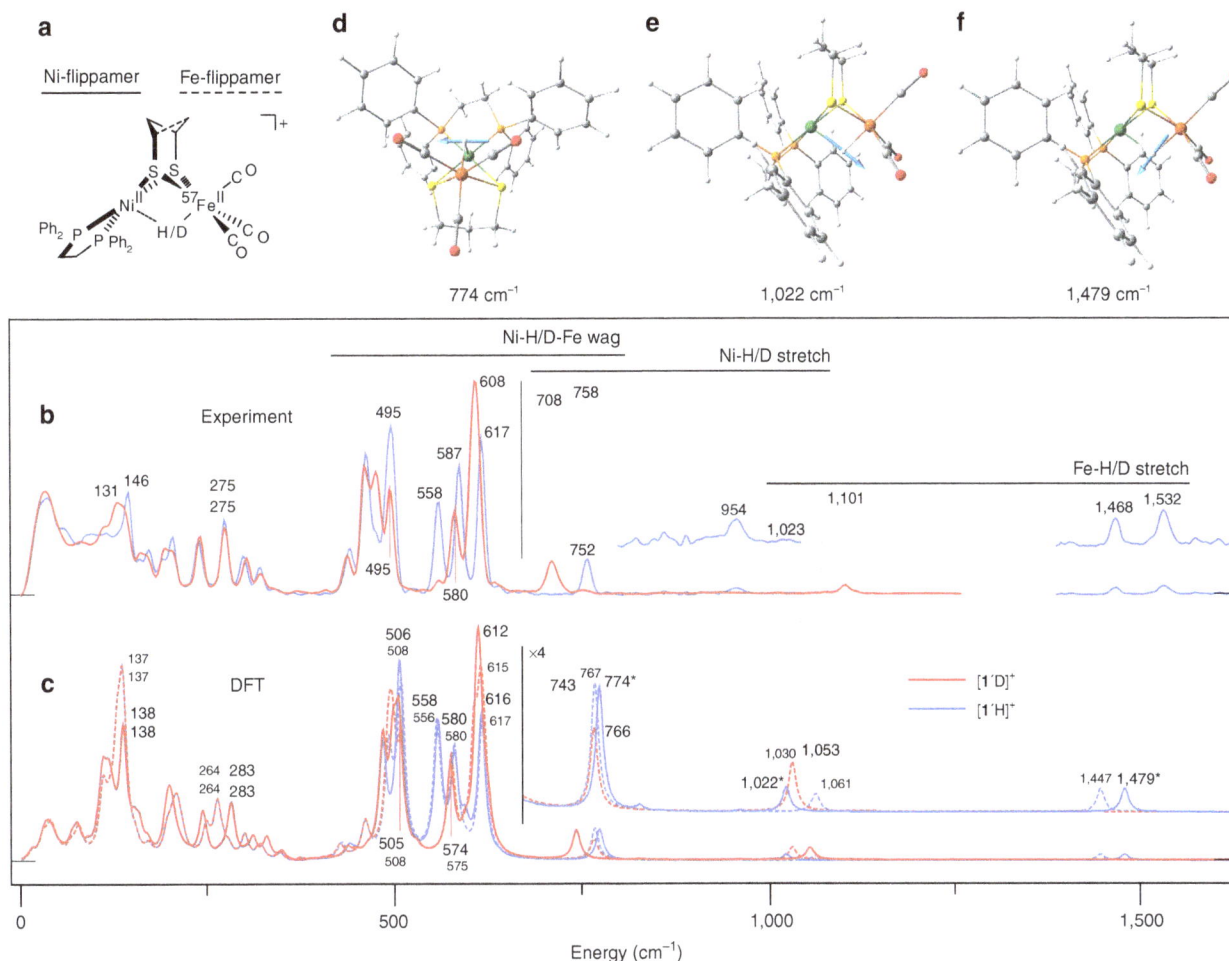

Figure 2 | Metal-hydride bands for complexes [1'H/D]$^+$. (a) Structure of [1'H]$^+$ showing 'flippamer' conformations of the pdt^{2-} ligand. **(b,c)** Full-range ^{57}Fe PVDOS for [1'H]$^+$ (blue trace) and [1'D]$^+$ (red trace) from NRVS experiments (**b**) and DFT calculations (**c**). In **c**, the plain/broken traces are spectra calculated for the dominant/alternative Ni-/Fe-flippamer, respectively. Spectra are repeated in the region >700 cm^{-1} with their intensities × 4 amplified. The colour bars highlight specific M–H/D bands, as well as their shifts on isotopic substitution. **(d-f)** Scaled-arrow representations of the M–H normal modes calculated for the Ni-flippamer of [1'H]$^+$ are shown, with the corresponding bands indicated (*) in **c**. Unscaled-arrow and animated representations of these M–H modes can be found in Supplementary Fig. 22 and Supplementary Movies 1-3, respectively.

expected from a harmonic oscillator in which H/D binds a much heavier nucleus, the ν_{Fe-H}/ν_{Fe-D} and ν_{Ni-H}/ν_{Ni-D} frequency ratios are $\sim 2^{1/2}$. This difference in the H/D and ^{57}Fe nuclear masses also results in pure hydride bands having low NRVS intensities, as such vibrations involve only small displacements of the ^{57}Fe centre. In addition, our successful use of NRVS to detect the well-defined ν_{Ni-D} stretch in ^{57}Fe-labelled [1'D]$^+$ contrasts Raman studies of natural Fe abundance [1D]$^+$, in which the Ni-D band was obscured by solvent modes[46].

On the basis of previous studies, one would expect $\delta_{OC-Fe-H/D}$ modes to appear in the 530–750/410–640 cm^{-1} regions, respectively[33,46]. The hydride [1'H]$^+$ exhibits a well-defined NRVS feature at 758 cm^{-1} consistent with a δ_{Fe-H} mode, this region being obscured by solvent bands in Raman data[46]. As will become clear, this 758 cm^{-1} mode observed for [1'H]$^+$ is of particular relevance to the interpretation of NRVS data for [NiFe]-hydrogenase. A distinct assignment for the corresponding δ_{Fe-D} mode in [1'D]$^+$ is prevented by its mixing with the Fe–CO bending modes, such that the NRVS intensity is redistributed throughout the 440–630 cm^{-1} region.

Analysis of [1'H]$^+$ revealed several intense Fe–CO bands in the 440–630 cm^{-1} region (Fig. 2b), the energies of which are

almost identical to those of the conjugate base (dppe)Ni(μ-pdt)^{57}Fe(CO)$_3$ (1')[47], which lacks the hydride bridge. This is exemplified by the δ_{Fe-CO} NRVS triplets for [1'H]$^+$ (558, 587 and 617 cm^{-1}) and 1' (557, 588 and 613 cm^{-1}) being virtually coincident (Supplementary Table 1), suggesting that the presence of H$^-$ does not significantly perturb the δ_{Fe-CO} dynamics. In contrast, the δ_{Fe-CO} region for [1'D]$^+$ collapses to a pair of bands at 580 and 608 cm^{-1}, consistent with significant coupling to the δ_{Fe-D} bending motion. Thus, although Fe–D modes have intrinsically low NRVS intensity, the Fe–D/Fe–CO coupling allows for the high-intensity δ_{Fe-CO} region to serve as an indicator of whether or not a Fe–D moiety is present[33].

DFT of model complexes. DFT calculations were undertaken to better understand the dynamics of H/D-coupled motions in the model complexes. The DFT-simulated ^{57}Fe PVDOS of [1'H/D]$^+$ Ni/Fe-flippamers were compared with NRVS data over the range 0–1,600 cm^{-1} (Fig. 2c, see also Supplementary Fig. 18 and Supplementary Tables 1 and 2). As with our recently reported spectra for [1']$^{0/+}$ (ref. 52), the observed and calculated band positions and intensities are in good agreement below 700 cm^{-1}.

Given that the broad $>700 \, cm^{-1}$ region is solely populated by Ni/Fe–H/D bands, DFT also allows for a confident assignment of these NRVS features (Fig. 2b,c) despite their low intensities and the difficulties recognized in the accurate theoretical prediction of M–H vibrational frequencies[46].

In line with previous DFT calculations and Raman analyses[46], the key distinction between calculated NRVS data for the two $[\mathbf{1'H}]^+$ Ni/Fe-flippamers results from splitting of the $v_{Fe-H} = 1,479/1,447 \, cm^{-1}$ and $v_{Ni-H} = 1,022/1,061 \, cm^{-1}$ modes, respectively, as indicated in Fig. 2c. For the Ni-flippamer, the calculated $v_{Fe/Ni-H}$ modes are shown correspondingly in Fig. 2f,e and Supplementary Movies 1 and 2. As expounded in the 'Further Discussion on Model Complex' section of the Supplementary Discussion, a fine yet noticeable interplay of the optimized Ni/Fe–H distances (Supplementary Table 3) gives rise to the *inverted* character of these ~ 30–$40 \, cm^{-1}$ flippamer-dependent splittings.

One of the most interesting and useful results from the present calculations on $[\mathbf{1'H}]^+$ is the prediction of a ^{57}Fe PVDOS band at $774/767 \, cm^{-1}$ for the Ni/Fe-flippamers, respectively. With only a small flippamer-dependent splitting of $7 \, cm^{-1}$, this mode gives rise to the most intense feature above $700 \, cm^{-1}$ and aligns well with the NRVS band observed at $758 \, cm^{-1}$ (Fig. 2b,c). Inspection of the DFT-calculated nuclear displacements (see Fig. 2d and Supplementary Movie 3 for the mode animation) shows a H nucleus motion normal to the $Ni(\mu$-H)Fe plane, in what is a unique 'wagging' mode. While the Ni–H–Fe wag is relatively isolated in $[\mathbf{1'H}]^+$ (H motion accounts for $\sim 80\%$ of the kinetic energy), the corresponding Ni-D-Fe motion in $[\mathbf{1'D}]^+$ is predicted to be heavily mixed with Fe–CO modes (Supplementary Table 1). The results of such mixing are evident in the intense 440–$630 \, cm^{-1}$ Fe–CO region (Fig. 2b,c). Thus, the NRVS signatures of the Ni–H–Fe moiety in the enzyme mimic are the weak wag band observed for $[\mathbf{1'H}]^+$ and the change in the amplified features in the Fe–CO region when comparing spectra of $[\mathbf{1'H}]^+$ and $[\mathbf{1'D}]^+$.

Hydrogenase NRVS results. NRVS data for Ni-R in H_2/H_2O and in D_2/D_2O are compared in Fig. 3a. While vibrations of the three electron transport Fe–S clusters exclusively populate the $<420 \, cm^{-1}$ region[39,40,43], this work instead focuses on the Fe–CO/CN region and higher-energy NRVS features to assign spectroscopic markers characteristic of the bridging hydride. Analysis of Ni-R in H_2O revealed a sharp band at $549 \, cm^{-1}$ previously assigned to a v_{Fe-CO} mode[39], with additional features at 454, 475 and $502 \, cm^{-1}$ arising from a mixture of Fe–CO and Fe–CN modes. Compared with our previous results[39], the absence of shoulders and additional features around the v_{Fe-CO} band indicates a higher level of sample purity. The NRVS band positions are similar to but nevertheless distinct from Raman peaks for the Ni-L Ni(I)Fe(II) state of [NiFe]-hydrogenase, for which v_{Fe-CO} was observed at $559 \, cm^{-1}$ (ref. 53). Samples of Ni-R in H_2O exhibit bands at 590 and $609 \, cm^{-1}$ that collapse to a single peak at $609 \, cm^{-1}$ when D_2O is instead used. Qualitatively, one can attribute the differences in this Fe–CO/CN region to a different coupling to Fe–H and Fe–D motion in the respective samples, as discussed above for $[\mathbf{1'H/D}]^+$. Other details about the active site, such as whether cysteine ligands are unprotonated or protonated, cannot be addressed on the basis of the NRVS data alone.

NRVS analysis of [NiFe]-hydrogenase in H_2O also revealed a weak but well-resolved band at $675 \, cm^{-1}$ not observed for other samples. This band is presumably related to the Ni–H–Fe wag exposed for $[\mathbf{1'H}]^+$ at $758 \, cm^{-1}$ (Fig. 2b,d), making this the first assignment of a Fe–H-related mode in any enzyme by NRVS and the first direct spectroscopic evidence for a $Ni(\mu$-H)Fe core in

Ni-R. Deuteration of Ni-R is expected to red-shift this mode into the 420–$620 \, cm^{-1}$ Fe–CO/CN region, in agreement with the changes observed and calculated for $[\mathbf{1'D}]^+$. Analogous to the model complex, the Ni-D-Fe wag is strongly mixed with Fe–CO/CN modes, which, in the case of Ni-R, makes its unique assignment very difficult.

Hydrogenase DFT results. To interpret NRVS measurements in terms of suitable structural candidates for Ni-R, we performed DFT calculations on a series of active site models featuring different binding modes of the H_2 substrate or its heterolysis products (see Supplementary Fig. 19). Limiting our models to the [NiFe] site is appropriate in that Fe–S clusters do not feature NRVS bands in the $>420 \, cm^{-1}$ region of interest[39,40]. Two main structures were considered: one in which substrate is present in the form of a dihydrogen ligand $((\eta^2$-H_2)NiFe, **I** or $NiFe(\eta^2$-H_2)$, **II**) and another where a bridging hydride is present $(Ni(\mu$-H)Fe, **III** or $HNi(\mu$-H)Fe, **IV**). In addition, variants of **III**, in which a terminal Cys ligand is protonated $((Cys546)SHNi(\mu$-H)Fe, **V** and $(Cys81)SHNi(\mu$-H)Fe, **VI**), were also studied. Taking into account the Ni(II)Fe(II) Ni-R active site[16,18], and assuming Fe(II) remains low-spin, each model may exist in electronic singlet $(S = 0)$ or triplet $(S = 1)$ Ni(II) states, both of which were evaluated computationally. The DFT-calculated ^{57}Fe PVDOS for selected models were compared with the NRVS data for Ni-R over the range 400–$750 \, cm^{-1}$ (Fig. 3a–c, see 'Further Discussion of Enzyme Cluster Models' of the Supplementary Discussion and Supplementary Figs 20–30). DFT-calculated NRVS for models $\mathbf{V^S}$ and $\mathbf{VI^S}$ (Fig. 3d,e, superscript S denotes singlet Ni(II)) match experimental data remarkably well, with the number and positions of absorption bands being in accordance. Relative intensities of the calculated peaks are also in good agreement with our measurements.

According to the normal mode analysis of model $\mathbf{V^S}$ (Supplementary Figs 28 and 29; animated representations of vibrational modes for models $\mathbf{V^S}$ and $\mathbf{VI^S}$ are provided in Supplementary Movies 4–34), vibrations in the 440–$504 \, cm^{-1}$ region predominantly involve Fe–CN bending and stretching, while higher-energy bands $(543$–$613 \, cm^{-1})$ are derived from Fe–CO vibrations. The feature calculated at $543 \, cm^{-1}$ has significant v_{Fe-CO} character, while that at $613 \, cm^{-1}$ is assigned to a δ_{Fe-CO} mode. Likewise, the $588 \, cm^{-1}$ band can be assigned to an H–Fe–CO bend in which H, Fe and C remain nearly collinear. The above normal modes are highly mixed, which make the assignment of individual fragments complicated.

Both bridging hydride models $\mathbf{V^S}$ and $\mathbf{VI^S}$ are predicted to exhibit a weak Ni–H–Fe out-of-plane wagging band (at 727 and $692 \, cm^{-1}$, respectively, see Fig. 3; Supplementary Movies 11 and 28 for mode animations) whose intensity is comparable to that of the $675 \, cm^{-1}$ feature observed for Ni-R. When compared with the Fe–CO/CN region, both theory and experiment predict lower NRVS intensity of the wag in Ni-R than that observed and calculated for $[\mathbf{1'H}]^+$ (at 758 and $774/767 \, cm^{-1}$, respectively, see Fig. 2). Moreover, simulations for $\mathbf{V^S}$ and $\mathbf{VI^S}$ accurately reproduce the disappearance of the $675 \, cm^{-1}$ band in data for Ni-R in D_2O. The difference between the experimental and calculated frequencies of the wagging mode are likely due to limitations in our model, which does not take into account direct contacts between the protonated cysteines and surrounding residues, as well as anharmonicity effects. Moreover, one would expect an intrinsic error of the chosen functional/basis set combination. While hydride bands are extremely sensitive to (electronic) structure, we note that the observed error is still well within normal limits[54–58] expected for this methodology. However, since the full spectral information is considered in

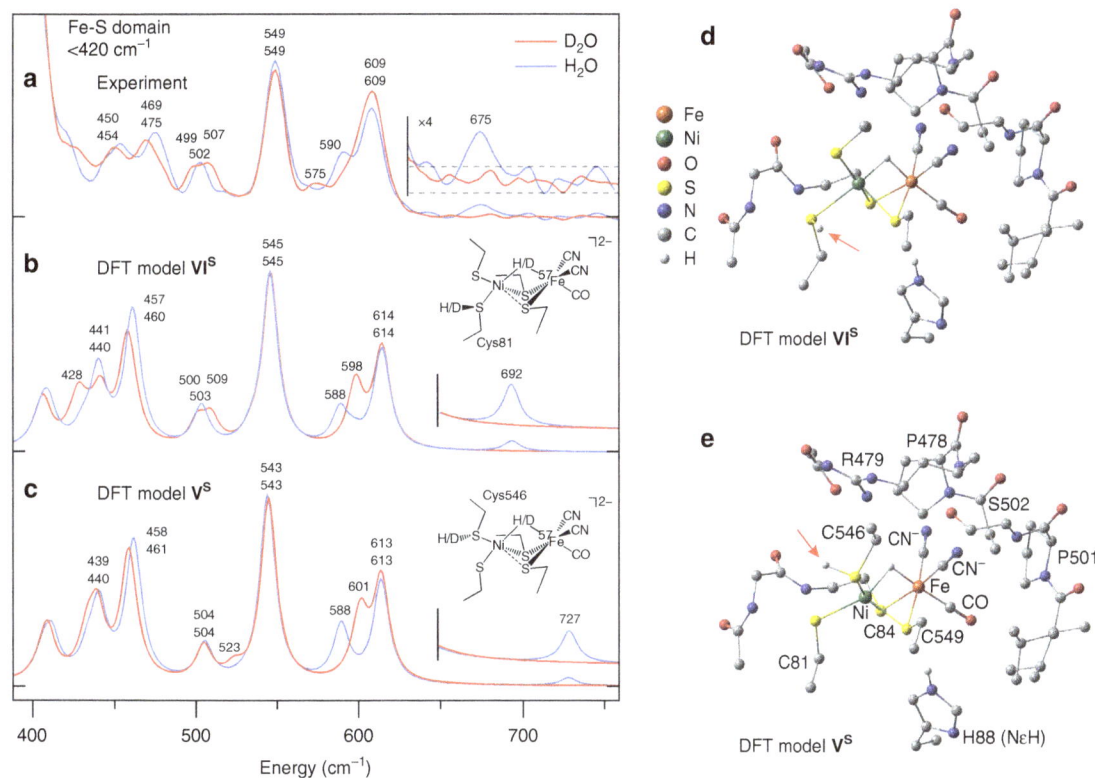

Figure 3 | Ni–H–Fe-hydride wag exposed in the reduced state Ni-R of [NiFe]-hydrogenase. (**a**–**c**) High-frequency NRVS for [NiFe]-hydrogenase reduced in H_2O (blue trace) and D_2O (red trace; **a**) and the corresponding ^{57}Fe PVDOS simulations given for the representative DFT models **VI^S** (**b**) and **V^S** (**c**). The higher regions of spectra containing the Ni–H–Fe wag band (in H_2O samples) are repeated with their intensities $\times 4$ amplified. The low-energy region of the Ni-R spectrum in H_2O reveals a triplet of bands (454, 475 and 502 cm^{-1}) that correspond to those located at 440, 461 and 504 cm^{-1} in the calculated spectrum of the model **V^S**. Further, two intense bands seen at 549 and 609 cm^{-1} in Ni-R map on calculated bands at 543 and 613 cm^{-1}, with an additional weak band observed at 590 cm^{-1} that can be correlated with the calculated band appearing at 588 cm^{-1}. (**d,e**) Representative DFT-optimized models **VI^S** (**d**) and **V^S** (**e**) for the Ni-R active site. Arrows indicate the position of CysSH. Non-substrate H atoms have been omitted for clarity (excluding $HN\varepsilon$ of His88).

the interpretation, the present conclusions can be made with confidence. Finally, calculated H/D isotope shifts for the two representative models (Supplementary Tables 4 and 5) in the low-energy region are also fully consistent with the observed data. The overall analysis here identifies the 675 cm^{-1} feature in the Ni-R NRVS spectrum as the Ni–H–Fe wag mode.

Discussion

Our NRVS measurements on synthetic bridging hydrides and Ni-R, combined with DFT calculations, provide new constraints on the structure of this key catalytic state of [NiFe]-hydrogenase. Spectra of [1'H/D]$^+$ feature characteristic Fe–H/D stretches whose energies (1,532/1,468 cm^{-1} for the Ni/Fe-flippamer, respectively) are comparable to those for bridging hydrides in other structures, including another recently reported Ni(μ-thiolate)$_2$(μ-H)Fe species for which infrared spectroscopy revealed a ν_{Fe-H} band at 1,687 cm^{-1} (ref. 28). Symmetrical μ_2-H$^-$ bridges, such as those in Fe$_4$H$_4^+$ clusters, give rise to symmetric stretches at around 1,400 cm^{-1} (ref. 59), while purely terminal hydrides have $\nu_{Fe-H} \sim 1,700$–2,300 cm^{-1} (refs 33,60). The sensitivity of hydride vibrations to structural perturbations underscores their enormous diagnostic value in understanding catalyst structure and function.

Unfortunately, even the strongest of these relatively pure stretches, ν_{Fe-D}, is predicted to have much lower NRVS intensity than the ν_{Fe-CO} modes. Thus, while [1'H/D]$^+$ allowed for direct observation of $\nu_{Fe-H/D}$ and $\nu_{Ni-H/D}$ modes, the resolution of

similar bands for [NiFe]-hydrogenase is beyond our current capabilities.

Of special significance is the DFT prediction of a Ni–H–Fe wag (Fig. 2d), this vibration being assigned to an observed NRVS band at 758 cm^{-1} for [1'H]$^+$, a mode likely obscured by solvent bands in Raman spectra[46]. A key advantage of NRVS is thus demonstrated in that its sole detection of modes coupled to the Mössbauer-active ^{57}Fe nucleus makes it unaffected by solvent or matrix modes. The NRVS intensity of the Ni–H–Fe wag is at least four times greater than those of the Ni–H/Fe-H stretches. The Ni–H–Fe wag is a valuable diagnostic probe of the Ni-R structure, with NRVS data for Ni-R in H_2O featuring a weak but reproducible band at 675 cm^{-1} that is absent when a D_2O medium was used (Fig. 3). Given that ^{57}Fe NRVS-active modes necessarily involve motion of this metal centre, observation of an H/D isotopically sensitive band at 675 cm^{-1} is strong evidence for the presence of a Fe–H moiety in Ni-R.

A second observable indicating the presence of a bridging H$^-$/D$^-$ in both Ni-R and its mimics [1'H/D]$^+$ stems from the coupling of Fe–CO and Fe–CN stretches and bends with the Ni–D–Fe wag. While coupling to the Fe–CO/CN modes makes resolution of the Ni-D-Fe wag impossible, it also results in marked changes in band position and intensity in the 440–630 cm^{-1} region on H/D substitution (Figs 2 and 3). In line with the marked isotope effects described above for [1'H/D]$^+$, NRVS spectra of Ni-R display similar shifts in position and intensity around 450–480 cm^{-1}, splitting of a band at 502 cm^{-1}, and disappearance of a shoulder peak at $590\,cm^{-1}$. This coupling provides an indirect but powerful method for characterizing the Fe–H/D-binding geometry. The NRVS experiments were complemented by DFT simulations, which further pointed to the presence of an active-site bridging hydride. Spectra predicted for the singlet ($S = 0$) models $(Cys81)SHNi(\mu\text{-H})Fe$ (**VI^S**) and $(Cys546)SHNi(\mu\text{-H})Fe$ (**V^S**) reproduce the data exceptionally well, in particular with respect to the Ni–H–Fe wag in each H/D isotopologue. The model **V^S** is also supported by the recent high-resolution crystallographic analysis of Ni-R1 (ref. 27). As detailed in the 'Further Discussion of Enzyme Cluster Models' section of the Supplementary Discussion, the NRVS data are best reproduced assuming a low-spin Ni(II). DFT also indicated asymmetric binding of H^- to the Ni(II)Fe(II) core, with the ligand more strongly bound to Ni than to Fe (see 'Electronic Structure of Model V' of the Supplementary Discussion).

Taken together, NRVS analysis of Ni-R, in combination with NRVS and DFT data for $[\mathbf{1}'\text{H/D}]^+$, indicates that models **VI^S** and **V^S** provide a consistent and detailed picture of Ni-R. This cohesive study thus represents the first evidence from vibrational spectroscopy for the presence of a bridging H^- in the active site of Ni-R, a state central to the function of [NiFe]-hydrogenase. The combined experimental and theoretical approach described here has applicability far beyond the Ni-R and [NiFe]-hydrogenase. Ideal for the detailed study of Fe–H fragments, we envisage that such methods will also be of importance for unravelling the mechanisms of [FeFe]-hydrogenase and nitrogenase[61], as well as for the development of synthetic catalysts inspired by these metalloenzymes.

Methods

General. Protocols employed for chemical and biochemical synthesis, as well as NRVS measurements and DFT calculations, are outlined here. Full procedures, including associated references, are given in the Supplementary Methods.

Model complex preparation. Metallic ^{57}Fe was converted to the complex $(dppe)Ni(\mu\text{-pdt})^{57}Fe(CO)_3$ (**1$'$**) via the intermediates $^{57}Fe_2I_4(2\text{-propanol})_4$, $(dppe)Ni(\mu\text{-pdt})^{57}FeI_2$ and $[(dppe)Ni(\mu\text{-pdt})(\mu\text{-I})^{57}Fe(CO)_3]BF_4$ ($[\mathbf{1}'I]BF_4$) (ref. 52). Species **1$'$** underwent facile exchange with ^{13}CO (1 atm), allowing access to isotopologue $(dppe)Ni(\mu\text{-pdt})^{57}Fe(^{13}CO)_3$ (**1$''$**). The two Ni(I)Fe(I) derivatives **1$'$** and **1$''$** were then subjected to protonation and deuteronation (effected with excess HBF_4 and HBF_4/CD_3OD, respectively), affording the Ni(II)Fe(II) salts $[\mathbf{1}'H]BF_4$, $[\mathbf{1}'H]BF_4$, $[\mathbf{1}'H]BF_4$ and $[\mathbf{1}''D]BF_4$ as crystalline solids. The isotopic purity of the hydrides and deuterides was confirmed using multinuclear (^1H, ^2H, ^{13}C and ^{31}P) NMR spectroscopy, infrared spectroscopy and ESI mass spectrometry. NRVS analysis was conducted on a solid sample of each of the four model complexes (*vide infra*).

D. vulgaris Miyazaki F [NiFe]-hydrogenase preparation. [NiFe]-hydrogenase expressed in *D. vulgaris* was isolated and purified as described earlier[62]. The as-isolated protein was transferred from 25 mM Tris-HCl (pH = 7.4) buffer to 100 mM MES (pH = 5.0). The solution was placed in a tube, which was sealed, degassed and then purged with H_2 (1.2 bar) for 8 h. In the case of the sample in D_2O, the buffer was replaced by 100 mM MES (pD = 5.0) in D_2O and the mixture placed under D_2 (1.3 bar) for 8 h. Samples were transferred to an anaerobic chamber and loaded into NRVS cells. FT-IR spectra were recorded on a Bruker IFS66v/S FT-IR spectrometer with a $2\,cm^{-1}$ spectral resolution at 293 K (Supplementary Fig. 31).

NRVS measurements and data analysis. The NRVS data were collected according to a published procedure[39] at SPring-8 BL09XU (with flux $\sim 1.4 \times 10^9$ photons s^{-1}) and BL19LXU ($\sim 6 \times 10^9$ photons s^{-1}) using 14.4 keV radiation at 0.8 meV resolution. The spectral maximum counts/second (cts s^{-1}) at BL19 is ~ 2.6–3 times of that at BL09. To compare the data from different beamlines, we rescale the BL19 counting time on the basis of its max cts s^{-1} versus the max cts s^{-1} at BL09 and create BL09 equivalent seconds, for example, the 10 (s) per point (s/pt) at BL19 is corresponding to 26 or 30 equivalent s/pt at BL09. Delayed nuclear and Fe K fluorescence (from internal conversion) were recorded with a 2×2 APD (avalanche photodiode) array in either beamline, and raw NRVS data were converted to single-phonon ^{57}Fe PVDOS using the PHOENIX software[39]. Sample temperatures were maintained at 30–50 K during analysis.

The average cts s^{-1} is 0.6–0.8 at the Fe–CO peak ($609\,cm^{-1}$) and 0.10–0.12 at X–Fe–H peak ($675\,cm^{-1}$), while the measured dark background cts are ~ 0.03 cts s^{-1}. To improve weak features, we use 1–3 s/pt from -240 to $400\,cm^{-1}$, 5–10 s/pt for the Fe–CN and Fe–CO regions, and 10–30 s/pt for the candidate X–Fe–H bending region (at 620–$770\,cm^{-1}$).

Model complex calculations. Initial coordinates for the DFT calculations on $[\mathbf{1}'\text{H/D}]^+$ were extracted from the X-ray structure of $[\mathbf{1}H]BF_4 \cdot 3THF$ (ref. 44). The methodology applied was mostly equivalent to our earlier set-up on $[\mathbf{1}']^{0/+}$ (ref. 52). Structural optimizations and subsequent normal mode analyses were performed using GAUSSIAN 09 on the basis of the densities exported from single point calculations using JAGUAR 7.9. The BP86 functional and the LACV3P** basis set were employed. The environment was considered using a self-consistent reaction field model. ^{57}Fe PVDOS spectra were generated using Q-SPECTOR, successfully applied earlier[33]. Simulated spectra were broadened by convolution with a full-width at half-maximum = $12\,cm^{-1}$ Lorentzian.

Enzyme cluster model calculations. DFT calculations on active-site cluster models were performed using ORCA 3.0. The initial geometry was prepared from the crystal structure of reduced [NiFe]-hydrogenase from *Desulfovibrio vulgaris* Miyazaki F (PDB 1H2R)[63]. Constraint geometry optimizations and vibrational frequency calculations at the scalar relativistic level (ZORA) employed B3LYP(-D3) hybrid-GGA with RIJCOSx and the COSMO model ($\varepsilon = 4$). Segmented all-electron relativistically contracted basis sets with corresponding auxiliary basis sets were used (def2-TZVPP: Ni, Fe, CN^-, CO, $S\gamma$, H^-, H^+, H_2; def2-SV(P): remaining atoms). Vibrational frequencies and normal mode compositions were utilized to simulate NRVS data (Lorentzian fitting, linewidth $12\,cm^{-1}$).

References

1. Bockris, J. O. M. The hydrogen economy: its history. *Int. J. Hydrogen Energy* **38**, 2579–2588 (2013).
2. Jugder, B. E., Welch, J., Aguey-Zinsou, K. F. & Marquis, C. P. Fundamentals and electrochemical applications of Ni-Fe uptake hydrogenases. *RSC Adv.* **3**, 8142–8159 (2013).
3. Kim, J. Y. H. & Cha, H. J. Recent progress in hydrogenase and its biotechnological application for viable hydrogen technology. *Korean J. Chem. Eng.* **30**, 1–10 (2013).
4. Mertens, R. & Liese, A. Biotechnological applications of hydrogenases. *Curr. Opin. Biotechnol.* **15**, 343–348 (2004).
5. Cammack, R., Frey, M. & Robson, R. (Eds.) *Hydrogen as a Fuel: Learning from Nature* (Taylor & Francis, 2001).
6. Heinekey, D. M. Hydrogenase enzymes: recent structural studies and active site models. *J. Organomet. Chem.* **694**, 2671–2680 (2009).
7. Tard, C. & Pickett, C. J. Structural and functional analogues of the active sites of the [Fe]-, [NiFe]-, and [FeFe]-hydrogenases. *Chem. Rev.* **109**, 2245–2274 (2009).
8. Yang, J. Y., Bullock, M., Rakowski DuBois, M. & DuBois, D. L. Fast and efficient molecular electrocatalysts for H_2 production: Using hydrogenase enzymes as guides. *MRS Bull.* **36**, 39–47 (2011).
9. Fritsch, J., Lenz, O. & Friedrich, B. Structure, function and biosynthesis of O_2-tolerant hydrogenases. *Nat. Rev. Microbiol.* **11**, 106–114 (2013).
10. Simmons, T. R. & Artero, V. Catalytic hydrogen oxidation: Dawn of a new iron age. *Angew. Chem. Int. Ed.* **52**, 6143–6145 (2013).
11. Matsumoto, T., Kim, K., Nakai, H., Hibino, T. & Ogo, S. Organometallic catalysts for use in a fuel cell. *ChemCatChem* **5**, 1368–1373 (2013).
12. Evans, R. M. *et al.* Principles of sustained enzymatic hydrogen oxidation in the presence of oxygen—the crucial influence of high potential Fe-S clusters in the electron relay of [NiFe]-hydrogenases. *J. Am. Chem. Soc.* **135**, 2694–2707 (2013).
13. Armstrong, F. A. Dynamic electrochemical experiments on hydrogenases. *Photosynth. Res.* **102**, 541–550 (2009).
14. Shafaat, H. S., Rüdiger, O., Ogata, H. & Lubitz, W. [NiFe] hydrogenases: a common active site for hydrogen metabolism under diverse conditions. *Biochim. Biophys. Acta* **1827**, 986–1002 (2013).
15. Guiral, M. *et al.* Hyperthermostable and oxygen resistant hydrogenases from a hyperthermophilic bacterium *Aquifex aeolicus*: physicochemical properties. *Int. J. Hydrogen Energy* **31**, 1424–1431 (2006).
16. Lubitz, W., Ogata, H., Rüdiger, O. & Reijerse, E. Hydrogenases. *Chem. Rev.* **114**, 4081–4148 (2014).
17. Ogata, H., Lubitz, W. & Higuchi, Y. [NiFe] hydrogenases: structural and spectroscopic studies of the reaction mechanism. *Dalton Trans.* 7577–7587 (2009).
18. Pandelia, M.-E., Ogata, H. & Lubitz, W. Intermediates in the catalytic cycle of [NiFe] hydrogenase: functional spectroscopy of the active site. *ChemPhysChem* **11**, 1127–1140 (2010).
19. Dole, F. *et al.* Nature and electronic structure of the Ni-X dinuclear center of *Desulfovibrio gigas* hydrogenase. Implications for the enzymatic mechanism. *Biochemistry* **36**, 7847–7854 (1997).

20. Bruschi, M. *et al.* A theoretical study of spin states in Ni-S$_4$ complexes and models of the [NiFe] hydrogenase active site. *J. Biol. Inorg. Chem.* **9**, 873–884 (2004).

21. Lill, S. O. N. & Siegbahn, P. E. M. An autocatalytic mechanism for NiFe-hydrogenase: reduction to Ni(I) followed by oxidative addition. *Biochemistry* **48**, 1056–1066 (2009).

22. Zampella, G., Bruschi, M., Fantucci, P. & De Gioia, L. DFT investigation of H$_2$ activation by [M(NHPnPr$_3$)('S3')] (M = Ni, Pd). Insight into key factors relevant to the design of hydrogenase functional models. *J. Am. Chem. Soc.* **127**, 13180–13189 (2005).

23. Amara, P., Volbeda, A., Fontecilla-Camps, J. C. & Field, M. J. A hybrid density functional theory/molecular mechanics study of nickel-iron hydrogenase: investigation of the active site redox states. *J. Am. Chem. Soc.* **121**, 4468–4477 (1999).

24. Stein, M. & Lubitz, W. Relativistic DFT calculation of the reaction cycle intermediates of [NiFe] hydrogenase: a contribution to understanding the enzymatic mechanism. *J. Inorg. Biochem.* **98**, 862–877 (2004).

25. Siegbahn, P. E. M., Blomberg, M. R. A., Pavlov, M. W. N. & Crabtree, R. H. The mechanism of the Ni-Fe hydrogenases: a quantum chemical perspective. *J. Biol. Inorg. Chem.* **6**, 460–466 (2001).

26. Niu, S., Thomson, L. M. & Hall, M. B. Theoretical characterization of the reaction intermediates in a model of the nickel-iron hydrogenase of *Desulfovibrio gigas*. *J. Am. Chem. Soc.* **121**, 4000–4007 (1999).

27. Ogata, H., Nishikawa, K. & Lubitz, W. Hydrogens detected by subatomic resolution protein crystallography in a [NiFe] hydrogenase. *Nature* **520**, 571–574 (2015).

28. Ogo, S. *et al.* A functional [NiFe] hydrogenase mimic that catalyzes electron and hydride transfer from H$_2$. *Science* **339**, 682–684 (2013).

29. Fan, H. J. & Hall, M. B. High-spin Ni(II), a surprisingly good structural model for [NiFe] hydrogenase. *J. Am. Chem. Soc.* **124**, 394–395 (2002).

30. Jayapal, P., Robinson, D., Sundararajan, M., Hillier, I. H. & McDouall, J. J. W. High level *ab initio* and DFT calculations of models of the catalytically active Ni-Fe hydrogenases. *Phys. Chem. Chem. Phys.* **10**, 1734–1738 (2008).

31. Bruschi, M., Zampella, G., Fantucci, P. & De Gioia, L. DFT investigations of models related to the active site of [NiFe] and [Fe] hydrogenases. *Coord. Chem. Rev.* **249**, 1620–1640 (2005).

32. Yson, R. L., Gilgor, J. L., Guberman, B. A. & Varganov, S. A. Protein induced singlet-triplet quasidegeneracy in the active site of [NiFe]-hydrogenase. *Chem. Phys. Lett.* **577**, 138–141 (2013).

33. Pelmenschikov, V., Guo, Y., Wang, H., Cramer, S. P. & Case, D. A. Fe-H/D stretching and bending modes in nuclear resonant vibrational, Raman and infrared spectroscopies: comparisons of density functional theory and experiment. *Faraday Discuss.* **148**, 409–420 (2011).

34. Champeney, D. C. The scattering of Mössbauer radiation by condensed matter. *J. Rep. Prog. Phys.* **42**, 1017–1054 (1979).

35. Seto, M., Yoda, Y., Kikuta, S., Zhang, X. W. & Ando, M. Observation of nuclear resonant scattering accompanied by phonon excitation using synchrotron radiation. *Phys. Rev. Lett.* **74**, 3828–2831 (1995).

36. Alp, E. E. *et al.* Vibrational dynamics studies by nuclear resonant inelastic X-ray scattering. *Hyperfine Interact.* **144/145**, 3–20 (2002).

37. Sturhahn, W. Nuclear resonant spectroscopy. *J. Phys. Condens. Matter* **16**, S497–S530 (2004).

38. Sturhahn, W. *et al.* Phonon density of states measured by inelastic nuclear resonant scattering. *Phys. Rev. Lett.* **74**, 3832–3835 (1995).

39. Kamali, S. *et al.* Observation of the Fe-CN and Fe-CO vibrations in the active site of [NiFe] hydrogenase by nuclear resonance vibrational spectroscopy. *Angew. Chem. Int. Ed.* **52**, 724–728 (2013).

40. Lauterbach, L. *et al.* Nuclear resonance vibrational spectroscopy reveals the FeS cluster composition and active site vibrational properties of an O$_2$-tolerant NAD(+)-reducing [NiFe] hydrogenase. *Chem. Sci.* **6**, 1055–1060 (2015).

41. Xiao, Y. *et al.* Dynamics of *Rhodobacter capsulatus* [2Fe-2S] ferredoxin VI and *Aquifex aeolicus* ferredoxin 5 *via* nuclear resonance vibrational spectroscopy (NRVS) and resonance raman spectroscopy. *Biochemistry* **47**, 6612–6627 (2008).

42. Mitra, D. *et al.* Dynamics of the [4Fe-4S] cluster in *Pyrococcus furiosus* D14C ferredoxin *via* nuclear resonance vibrational and resonance raman spectroscopies, force field simulations, and density functional theory calculations. *Biochemistry* **50**, 5220–5235 (2011).

43. Mitra, D. *et al.* Characterization of [4Fe-4S] cluster dynamics and structure in nitrogenase Fe Protein at three oxidation levels *via* combined NRVS, EXAFS and DFT analyses. *J. Am. Chem. Soc.* **135**, 2530–2543 (2013).

44. Barton, B. E., Whaley, C. M., Rauchfuss, T. B. & Gray, D. L. Nickel-iron dithiolato hydrides relevant to the [NiFe]-hydrogenase active site. *J. Am. Chem. Soc.* **131**, 6942–6943 (2009).

45. Barton, B. E. & Rauchfuss, T. B. Hydride-containing models for the active site of the nickel-iron hydrogenases. *J. Am. Chem. Soc.* **132**, 14877–14885 (2010).

46. Shafaat, H. S., Weber, K., Petrenko, T., Neese, F. & Lubitz, W. Key hydride vibrational modes in [NiFe] hydrogenase model compounds studied by resonance Raman spectroscopy and density functional calculations. *Inorg. Chem.* **51**, 11787–11797 (2012).

47. Bertini, L., Greco, C., Bruschi, M., Fantucci, P. & De Gioia, L. CO affinity and bonding properties of [FeFe] hydrogenase active site models. A DFT study. *Organometallics* **29**, 2013–2025 (2010).

48. Liu, T. B., Li, B., Singleton, M. L., Hall, M. B. & Darensbourg, M. Y. Sulfur oxygenates of biomimetics of the diiron subsite of the [FeFe]-hydrogenase active site: Properties and oxygen damage repair possibilities. *J. Am. Chem. Soc.* **131**, 8296–8307 (2009).

49. Schilter, D. *et al.* Mixed-valence nickel-iron dithiolate models of the [NiFe]-hydrogenase active site. *Inorg. Chem.* **51**, 2338–2348 (2012).

50. Justice, A. K. *et al.* Redox and structural properties of mixed-valence models for the active site of the [FeFe]-hydrogenase: progress and challenges. *Inorg. Chem.* **47**, 7405–7414 (2008).

51. Bergmann, U. *et al.* Observation of Fe-H/D modes by nuclear resonant vibrational spectroscopy. *J. Am. Chem. Soc.* **125**, 4016–4017 (2003).

52. Schilter, D. *et al.* Synthesis and vibrational spectroscopy of ^{57}Fe-labeled models of [NiFe] hydrogenase: first direct observation of a nickel-iron interaction. *Chem. Commun.* **50**, 13469–13472 (2014).

53. Siebert, E. *et al.* Resonance Raman spectroscopy as a tool to monitor the active site of hydrogenases. *Angew. Chem. Int. Ed.* **52**, 5162–5165 (2013).

54. Petrenko, T. *et al.* Characterization of a genuine iron(V)-nitrido species by nuclear resonant vibrational spectroscopy coupled to density functional calculations. *J. Am. Chem. Soc.* **129**, 11053–11060 (2007).

55. Peng, Q., Pavlik, J. W., Scheidt, W. R. & Wiest, O. Predicting nuclear resonance vibrational spectra of [Fe(OEP)(NO)]. *J. Chem. Theory Comput.* **8**, 214–223 (2012).

56. Wong, S. D. *et al.* Elucidation of the Fe(IV) = O intermediate in the catalytic cycle of the halogenase SyrB2. *Nature* **499**, 320–323 (2013).

57. Park, K. *et al.* Nuclear resonance vibrational spectroscopic and computational study of high-valent diiron complexes relevant to enzyme intermediates. *Proc. Natl Acad. Sci. USA* **110**, 6275–6280 (2013).

58. Li, J. F. *et al.* Comprehensive Fe-ligand vibration identification in {FeNO}^{6} hemes. *J. Am. Chem. Soc.* **136**, 18100–18110 (2014).

59. Swart, I. *et al.* H$_2$ adsorption on 3d transition metal clusters: a combined infrared spectroscopy and density functional study. *J. Phys. Chem. A* **112**, 1139–1149 (2008).

60. Kubas, G. J. Fundamentals of H$_2$ binding and reactivity on transition metals underlying hydrogenase function and H$_2$ production and storage. *Chem. Rev.* **107**, 4152–4205 (2007).

61. Hoffman, B. M., Lukoyanov, D., Yang, Z. Y., Dean, D. R. & Seefeldt, L. C. Mechanism of nitrogen fixation by nitrogenase: the next stage. *Chem. Rev.* **114**, 4041–4062 (2014).

62. Ogata, H. *et al.* Activation process of [NiFe] hydrogenase elucidated by high-resolution X-ray analyses: conversion of the ready to the unready state. *Structure* **13**, 1–8 (2005).

63. Higuchi, Y., Ogata, H., Miki, K., Yasuoka, N. & Yagi, T. Removal of the bridging ligand atom at the Ni-Fe active site of [NiFe] hydrogenase upon reduction with H$_2$, as revealed by X-ray structure analysis at 1.4 Å resolution. *Structure* **7**, 549–556 (1999).

Acknowledgements

We thank Patricia Malkowski (MPI-CEC) for her assistance with the ^{57}Fe [NiFe]-hydrogenase sample preparation. This work was supported by the DOE Office of Biological and Environmental Research (S.P.C.), NIH grant GM-65440 (S.P.C.), DOE grant DEFG02-90ER14146 (T.B.R.), BMBF (03SF0355C), EU/Energy Network project SOLAR-H2 (FP7 contract 212508), DFG-funded Cluster of Excellence RESOLV (EXC1069), Max Planck Society (W.L., T.K., M.v.G., F.N. and H.O.) and the DFG-funded 'Unifying Concepts in Catalysis' (UniCat) initiative (V.P.). Use of SPring-8 is supported by JASRI (proposals 2012A0032-2014B1032) and RIKEN (proposals 20120107, 20130022 and 20140033).

Author contributions

H.O. prepared the [57]Fe-labelled [NiFe]-hydrogenase samples. H.W., L.B.G., A.D.S., Y.Y., Y.T. and S.P.C. measured and analysed the NRVS spectra. T.B.R. and D.S. prepared and characterized the model complexes. V.P. performed the DFT calculations on the model complexes. T.K., M.v.G., F.N. performed the DFT calculations on [NiFe]-hydrogenase. H.O., T.K., H.W., D.S., V.P., M.v.G., W.L. and S.P.C. wrote the manuscript. S.P.C. designed and coordinated the project.

Additional information

Competing financial interests: The authors declare no competing financial interests.

13

Supersensitive fingerprinting of explosives by chemically modified nanosensors arrays

Amir Lichtenstein[1,*], Ehud Havivi[1,*], Ronen Shacham[1], Ehud Hahamy[1,2], Ronit Leibovich[1,2], Alexander Pevzner[3], Vadim Krivitsky[3], Guy Davivi[3], Igor Presman[1], Roey Elnathan[3], Yoni Engel[3], Eli Flaxer[3,4] & Fernando Patolsky[3,5,6]

The capability to detect traces of explosives sensitively, selectively and rapidly could be of great benefit for applications relating to civilian national security and military needs. Here, we show that, when chemically modified in a multiplexed mode, nanoelectrical devices arrays enable the supersensitive discriminative detection of explosive species. The fingerprinting of explosives is achieved by pattern recognizing the inherent kinetics, and thermodynamics, of interaction between the chemically modified nanosensors array and the molecular analytes under test. This platform allows for the rapid detection of explosives, from air collected samples, down to the parts-per-quadrillion concentration range, and represents the first nanotechnology-inspired demonstration on the selective supersensitive detection of explosives, including the nitro- and peroxide-derivatives, on a single electronic platform. Furthermore, the ultrahigh sensitivity displayed by our platform may allow the remote detection of various explosives, a task unachieved by existing detection technologies.

[1] Tracense Ltd., Hanadiv 71 Street, Hertzelia 46485, Israel. [2] School of Mathematics, The Raymond and Beverly Sackler Faculty of Exact Sciences Tel-Aviv University, Tel Aviv 69978, Israel. [3] School of Chemistry, The Raymond and Beverly Sackler Faculty of Exact Sciences Tel-Aviv University, Tel Aviv 69978, Israel. [4] Afeka Tel Aviv Academic College of Engineering, Tel Aviv 69978, Israel. [5] The Center for Nanoscience and Nanotechnology, Tel-Aviv University, Tel Aviv 69978, Israel. [6] Department of Materials Science and Engineering, The Iby and Aladar Fleischman Faculty of Engineering, Tel-Aviv University, Tel Aviv 69978, Israel. * These authors contributed equally to this work. Correspondence and requests for materials should be addressed to F.P. (email: fernando@post.tau.ac.il).

Numerous recent adverse events have led to an increased requirement for improving our ability to detect highly energetic chemicals, explosives, making their detection an issue of burning immediacy, and a major current challenge in terms of research and development. The capability to detect traces of these hazardous chemical species sensitively, selectively, accurately and rapidly could be of great benefit for applications relating to civilian national security and environmental monitoring. Unfortunately, the reliable detection of explosives still represents a challenging, and a largely unmet, arduous problem. This is mainly due to the requirement to sensitively and selectively detect, and quantify, traces of a broad variety of explosive species in any given encountered environment. This challenge is further exacerbated by the inherently low vapour pressures of most of these compounds[1]. For instance, trinitrotoluene, TNT, exhibits a vapour pressure of a few parts-per-billion at room temperature (1 p.p.b. corresponds to only one TNT molecule in 1 billion molecules of air), while the more powerful, and less volatile, explosives RDX (1,3,5-Trinitro-1,3,5-triazacyclohexane) and HMX (Octahydro-1,3,5,7-tetranitro-1,3,5,7-tetrazocine) stands at a few parts-per-trillion (p.p.t.) and part-per-quadrillion (p.p.q.), respectively. This directly means that the detection system must be highly sensitive, while simultaneously exhibiting rapid response and high selectivity. Furthermore, the detection of improvised peroxide-based explosive species, that is, TATP (Triacetone triperoxide) and HMDT (Hexamethylene triperoxide diamine), is strongly challenged by the inherent absence of nitro groups and their poor spectroscopic features. Even though multiple strategies have been developed for the detection of explosive species[2–4], these methods are usually time consuming, require bulky equipment, tedious sample preparation and a trained operator. Furthermore, these systems cannot be miniaturized and lack the ability to perform automated real-time high-throughput analysis, strongly handicapping their mass deployment.

Due to their reduced dimensions, nanomaterials offer the ability of incorporating multiple sensors capable of detecting numerous chemical threats simultaneously on a single miniature array platform[5–7]. Thus, nanoscale device arrays have the potential to generate sensors with ultrahigh sensitivity, exceptional specificity towards a wide-range expandable 'library' of detectable threats, real-time unlabelled detection, samples versatility (gas, liquid and particles), and portable easily-operable devices. In general, 1D-nanowire building blocks, when configured as nanoFET (nano Field Effect Transistor) devices, are extremely sensitive to the absorption of chemical species on their surface, due to their intrinsically high surface-to-volume ratio, leading to significant changes in their resulting electrical conductance[8–12]. Traditionally, analyte detection relies on the specific chemical interactions between a ligand and its probe, entailing complexity, high costs and time consuming fabrication processes. Besides, a one probe-one ligand approach excludes the identification of not yet identified analytes. The growing interest to make use of such intriguing nanoscale building blocks for the detection of explosive species, has led to the recent application of chemically modified nanowire FETs as ultrasensitive label-free trace sensors for TNT molecules[13], attaining an unprecedented limit of detection in the low femtomolar (p.p.q.) concentration range. Despite the exquisite achieved sensitivity, the use of nanodevices chemically modified with a single 'unspecific' molecular binding agent led to a strongly limited selectivity, and the consequent absolute inability to discriminate between the binding of explosive molecules and that of different, similar and dissimilar, chemical species[2]. Clearly, in order to make full advantage of nanosensing devices in the field of explosives detection, and other small molecular species, existing challenges related to sensitivity and selectivity must be jointly overcome.

Inspired by nature, artificial sensing systems mimic the mammalian olfactory system, in which a finite number of not highly selective cross-reactive receptors generate a response pattern that is perceived by the brain as a particular odour. This general approach make use of a library of sensitive materials, the artificial cross-reactive sensors that differentially interact with the target molecules, and generates a fingerprinting pattern that is recognized by pattern-recognition chemometric tools[14–18].

Here, we present the development of an ultrasensitive and highly selective platform for the detection and discriminative identification of explosive chemical species, based on the use of large-scale arrays of chemically modified electrical nanosensors, defining a plurality of multiple functional reaction sites enabling a high-throughput assay. When chemically modified in a multiplexed mode, by the surface immobilization of multiple non-specific small chemical receptors, nanowire-based field-effect-transistor arrays (NW-FETs) enable the supersensitive discriminative detection, fingerprinting, of multiple explosive molecules, down to the parts-per-quadrillion concentration range[19]. The differential identification between explosives is achieved by pattern recognizing the naturally inherent interaction, both kinetically and thermodynamically[20–27], between the chemically modified nanosensors array and the chemical analytes under test. Thus, a unique explosives fingerprinting database can be created, enabling to set apart similar chemical entities, and providing a fast and reliable method to identify individual chemical agents. This allows the direct detection of explosives species displaying the lowest volatilities, as well as the remote detection of various explosive species.

Results

The chemically modified nanosensors array.
The multiplexed high-throughput detection and identification of explosives by a single chip is achieved by fabricating a SiNW-FETs multiarray (144 FET devices per chip, functioning devices yield of 60–70%), containing eight subarrays in total, each subarray consisting of 18 FET nanodevices, being individually modified by a unique surface binding agent. All subarrays are fed by a common integrated microfluidic channel that enables the flow and interaction of all analytes with the modified array, Fig. 1a,b. p-type boron-doped silicon nanowires (p-type SiNWs) were grown by the vapour-liquid-solid chemical vapour deposition method, followed by dry transferring to their destined locations on chip, and the formation of electrical contacts by common lithography and metal deposition steps[13,28,29]. Metal source and drain contacts were passivated by a thin 60 nm silicon nitride layer, deposited using PECVD (plasma-enhanced chemical vapour deposition), which electrically isolates the metal electrodes from the sensing solution (Supplementary Fig. 1A,B). The resulting 144 nano-sensor devices were electrically characterized and effectively normalized by their respective transconductance values, thus minimizing the device-to-device signal variability[30], Fig. 2. Using SiNWs as the sensing elements, various selected surface-bound electron-rich aminosilane derivatives chemically pair with the explosive molecules in a sample, by charge-transfer (e.g., Meisenheimer complexes) and additional chemical inter-actions[31], thus leading to the formation of charges or surface dipole alteration, in close vicinity to the sensing surfaces. Since the analyte molecules are not of biological nature, salt-free baseline solution is used in all our sensing experiments (0.1% dimethyl sulfoxide (DMSO) in H_2O), enabling electrolyte screening-free conditions, resulting in a large Debye length ($\sim 1\,\mu m$[21,32]) that confers extremely high detection sensitivity.

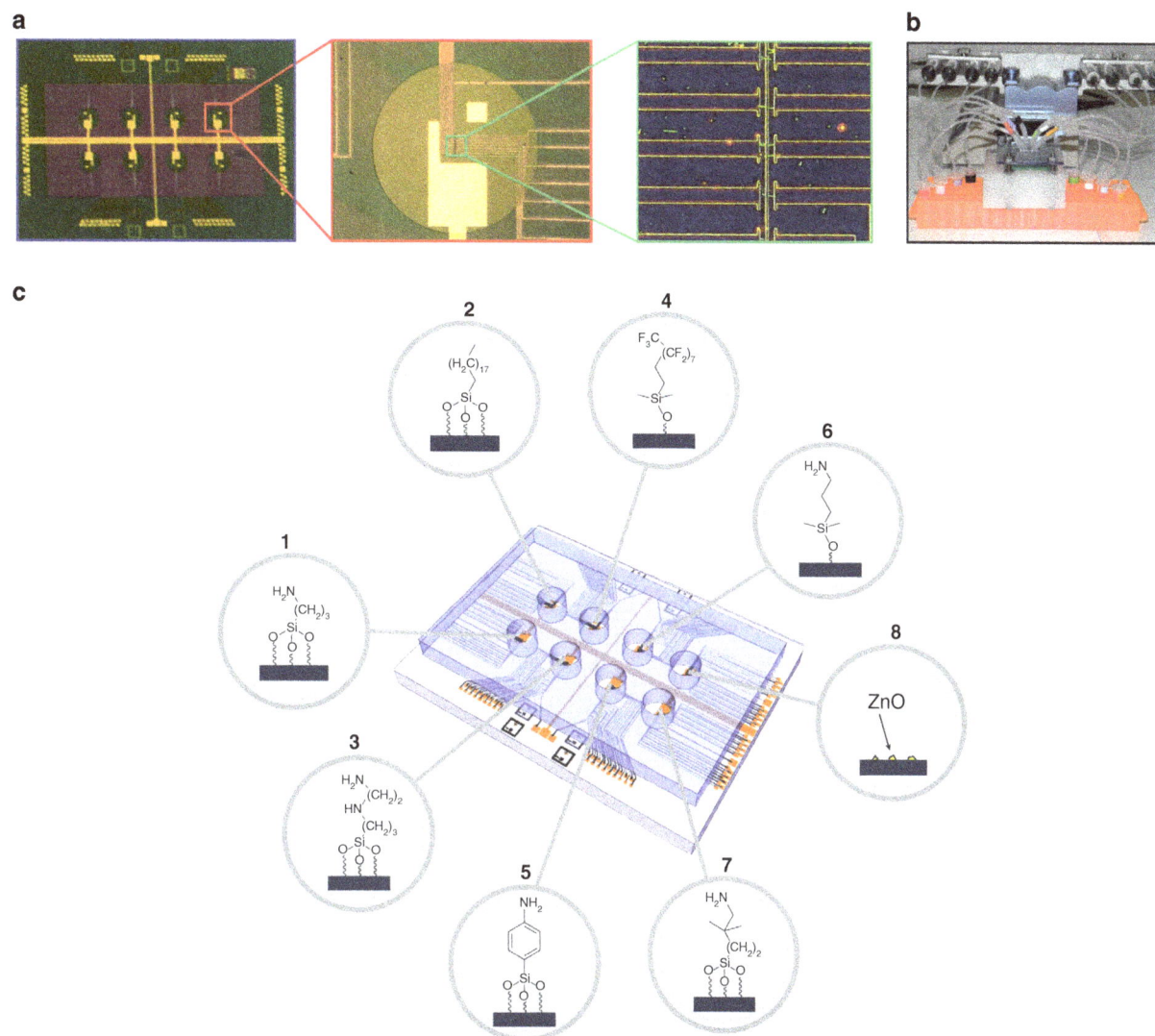

Figure 1 | View of the multiplex sensing chip. 144 NanoFET devices sharing a common liquid gate, arrayed into eight separate subregions, chemically modified with eight different chemical receptors. (**a**) Progressive magnifications of a sensing chip, wherein p-doped SiNWs channels, bridge a common source (middle electrode) with its surrounding drain electrodes. (**b**) Eight chemically differentiated subregions are created by simultaneously flowing different silane derivatives onto their designated pads, using a dedicated eight-channel fluidic device. (**c**) Scheme of a chip bearing eight chemically differentiated subregions. (**1**) Aminopropyltriethoxysilane (APTES), (**2**) N-Octadecyltriethoxysilane (OTS), (**3**) n-(2-aminoethyl)-3-3aminopropyltrimethoxysilane, (en-APTAS), (**4**) (Heptadecafluoro-1,1,2,2-tetrahydrodecyl)dimethylchlorosilane (Fluorosilane), (**5**) p-aminophenyltrimethoxysilane (p-APhS), (**6**) Aminopropyldimethylmethoxysilane (APDMES), (**7**) 4-amino-3,3-dimethylbutyl triethoxysilane (tBu), (**8**) silicon oxide, non-modified SiNW, or NPs-decorated devices.

Up to eight different surface-confined chemical receptors, molecules **1** to **8** (Fig. 1c), were selectively anchored to separate nanosensors subarray spots, each subarray containing 18 devices, by the use of area-selected silane-coupling procedures[33–35], as schematically depicted in Fig. 1 (Supplementary Figs 2 and 3). The successful modification of the different silane derivatives was characterized by XPS measurements (X-ray photoelectron spectroscopy; see Methods and Supplementary Methods Sections for experimental details and data). The surface coverage of the different silane derivatives was measured by means of QCM (quartz crystal microbalance) measurements (see Methods for details), and strongly depends on the chemical nature of the silane derivative molecule. Polar silanes display lower surface coverage, in the order of ~ 3–6×10^{13} molecules·cm^{-2}, while non-polar derivatives display higher surface coverage of ~ 1–4×10^{14} molecules·cm^{-2}. The silanes layers were demonstrated to be highly stable for periods of weeks of continuous operation under

our experimental conditions. Generally, the interaction of electron-deficient nitro-containing explosives with amine groups of silane receptor molecules leads to the formation of surface-confined charges[36,37], consequently altering the nanodevices electrical response. During the measurements, differential gating can be applied at the subthreshold or threshold regimes[38], while the applied V_{SD} bias (source-drain voltage) allows operation at a particular region of the device's transconductance.

The detection of explosives by nanosensors array. All our sensing experiments using nanoFETs arrays are performed in liquid phase, in a solution of 0.1% DMSO in H$_2$O, as stated in the manuscript body and methods section. All sensor calibrations, detection limits extraction, kinetic and thermodynamic parameters measurements and real-world tests, are done in the liquid phase using explosive standard solutions (purchased from

AccuStandard Inc.). When collecting explosives' vapours off the air, the analyte molecules are partitioned onto a solid surface whence they are washed off with the sensing solution collected and tested.

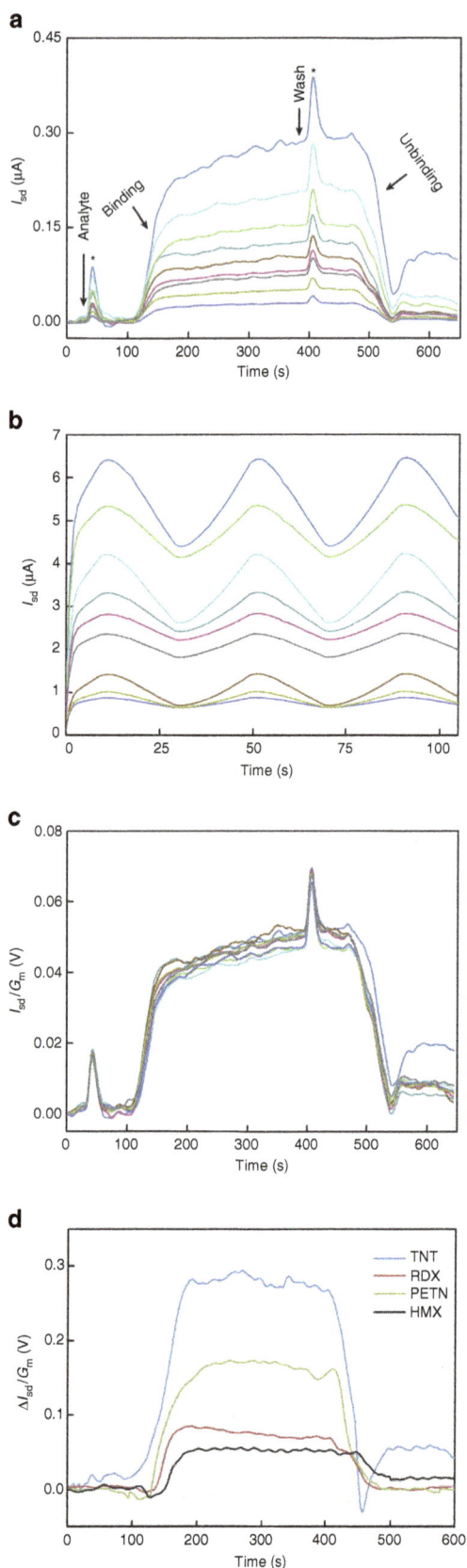

In practice, the detection of real-world samples is separated into two sequential steps, performed automatically by our sensing prototype, where the air sample is collected for 5 s through a micro-porous filter support by means of an air sampling device (sampling of 50–100 l min^{-1}), then the adsorbed analytes are flushed off the collecting membrane by the sensing solution, and finally channelled into the sensing area for detection.

The resulting change in the source-drain current (I_{SD}) that flows through the chemically modified nanodevices is a function of the change in the chemical and electronic environment. Thus, when target molecules in the sample attach to the nanodevices' surface, a change occurs in their conductance–resulting in a change in the measured current and indicating the detection of the target molecule. Figure 2a displays the electrical shifts observed upon the interaction of several nanodevices on a single chip, modified with the amine-containing receptor APTES (aminopropyl triethoxy silane) (1), with TNT-containing samples. The devices are real-time sensors, and quickly regenerate after each sensing cycle by flushing a pulse of washing solution that restores the devices to their initial baseline. Hundreds of sensing cycles can be performed using a single sensing platform, for periods of weeks, without considerable degradation of the detection performance. Notably, the whole detection cycle, from explosive sample injection till final detection (binding, plateau, unbinding), can be speeded up down to ∼30 s, Supplementary Fig. 4. Clearly, the calibrated responses of the devices to the analyte (ΔI_{sd}), when normalized by their respective transconductance values (device transconductance ($G_m = \frac{dIds}{dVg}$)), Fig. 2b, will result in relatively small device-to-device variability, even for originally largely dissimilar devices[30], Fig. 2c. Furthermore, a large number of sensors allows redundancy, thus improving the signal-to-noise ratio, and offers the ability to reject the output of poorly performing sensing devices[15]. Moreover, the reproducibility displayed by a single FET sensor to the repeated interaction with a certain explosive is remarkable, Supplementary Fig. 4A–D. Clearly, the single-sensor sequential responses display a remarkable level of reproducibility, both regarding their signal plateau (thermodynamically, values of ΔG from baseline to peak) as well as regarding the complex kinetics responses. Of, course, the same behaviour is observed for all devices in a single array, regardless their chemical modifier and location on the array. These results show the ability of our platform, to identify different molecular species with a high level of reliability and reproducibility.

Importantly, all devices modified with molecular receptor APTES (1), also show considerable electrical shifts and cross-reactivity, upon their interaction with different explosive species at concentration of 10 p.p.t.: TNT, RDX, HMX and PETN ([3-Nitroxy-2,2-bis(nitromethyl)propyl]nitrate), Fig. 2d, as well as upon their interaction with non-explosive interferents (for example, Ammonium Nitrate, Musk ketone, Musk moskene). Although the interaction with the highly electron-deficient TNT molecules leads to the largest measured electrical shifts, with

Figure 2 | Signal processing steps. (**a**) The raw sensing response to TNT over time of nine APTES-modified active sensing devices. (*) two peak rises at 50 s and 400 s are caused by a pinch valve pressure artifact. (**b**) The transconductance response of the same sensing devices as a function of the gate voltage (time at a given scan rate) being periodically modulated between − 0.3 V and + 0.3 V. (**c**) Calibrated responses of devices in **a**, after normalizing each device raw response by its gate dependence ($\frac{dIds}{dVg}$). After calibration, all nine devices display almost identical electrical behaviour. (**d**) Calibrated responses of a representative APTES-modified nano FET device to different explosive species at the same concentration, 10 p.p.t., displaying the inherent lack of selectivity.

detection limits down to the p.p.q. (parts-per-quadrillion) concentration range, these results show the complete incapability to selectively detect different explosive species by nanodevices chemically modified with a single molecular receptor, regardless the chemical receptor used for detection, as per previously reported cases[13].

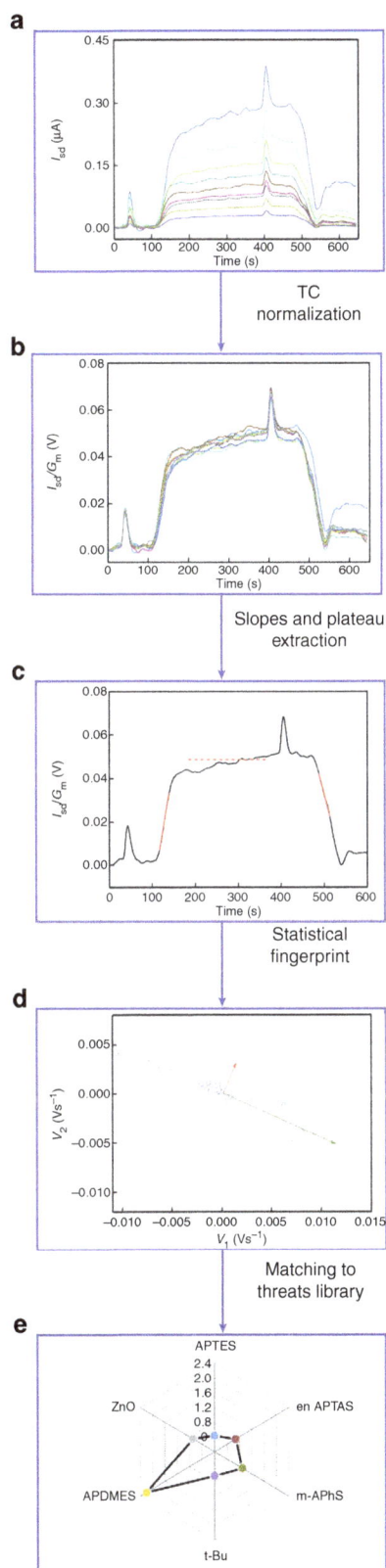

The discriminative identification of explosive species. A potential advantageous approach for the selective detection of explosives, and their discriminative identification, is based on the real-time mathematical analysis of their interaction, both kinetically and thermodynamically, with a single array bearing multiple molecular binding agents. To that end we have tested a broad library of aminosilane derivatives for their interaction with multiple explosive species of relevance. Our selected surface-confined binding agents (Fig. 1c) consist of primary amines (APTES, en-APTAS 3-(2-Aminoethylamino)propyltrimethoxysilane), tBu (4-amino-3,3-dimethylbutyl triethoxysilane), APDMES (3-aminopropyl)-dimethyl-ethoxysilane), secondary amines (en-APTAS), aromatic amines (aniline derivatives m-APhS (aminophenyl triethoxy silane) and p-APhS), as well as amine-free silane derivatives (OTS (octadecyltrichloro silane), fluorosilane derivatives), serving as control explosives-non-binding compounds (Supplementary Table 1). Clearly, each type of aminosilane modifier has different chemical, steric and electronic characteristics. Our nanoarray chip, with integrated multichannel microfluidics, allows for the differential yet simultaneous chemical modification of all eight subarrays with multiple surface binding agents, as illustrated in Fig. 1b. Generally, 1% ethanolic solutions of each silane derivative are simultaneously and separately flowed along their respective wells, by means of individual inlet and outlet ports, followed by washing the unbound silanes excess before the final curing step (see Methods for detailed information and Supplementary Figs 2 and 3). In this case, each explosive species is expected to exhibit a distinctive pattern of interaction, both kinetically and thermodynamically, with the chemically modified nanodevice array.

For that purpose, Fig. 3c, we chose to make use of two parameters inherent to the interaction between any explosive chemical species and the surface-confined chemical receptors: (1) the signal amplitude, or electrical shift, measured after reaching the asymptotic saturation value of the device (I_{sd} after equilibrium is achieved) at any concentration of the given explosive, when the ΔI_{sd} was calibrated by its gate dependence, the transconductance value (G_m), and (2) the initial rates of association, and dissociation, measured after the interaction of the array with a certain explosive species, and the sequential flushing of the sensors surface with a clean baseline solution leading to the unbinding of the explosive species, V_1 and V_2 respectively. To achieve this goal, we developed a five-tier algorithm (for more information see Supplementary Methods) able to extract, and analyse, in real time all the above mentioned parameters, Fig. 3a–e, followed by the final identification of the molecular species under test based on a formerly built fingerprints database Fig. 3e (see Supplementary Methods).

Figure 4a,b displays the results obtained from the interaction of several common explosives with a multireceptor array of nanosensors modified with molecular receptors (**1**, **3**, **5**, **6**, **7** and **8** of Fig. 1c). Explosives samples were obtained by collecting air samples of explosives-containing solid sources, at room

Figure 3 | Signal processing algorithm steps. Raw response (**a**), followed by calibrating each device's absolute raw electrical response against its gate dependence ($\frac{d/ds}{dVg}$) (**b**). Mathematically extracting the following parameters: the kinetic ratio (V_1/V_2) and the calibrated absolute response under steady state conditions, at plateau (**c**). Data processing ensues, whereby the statistical dependence of the kinetic constants is globally estimated by the slope of a straight line in the V_1-V_2 plane obtained by regression using R^2 as a target function, taking into account all accumulated V_1-V_2 pairs (**d**). Chemical identification, wherein a matching is done between the pattern generated by the given ensemble response and a library of previously calculated responses (**e**).

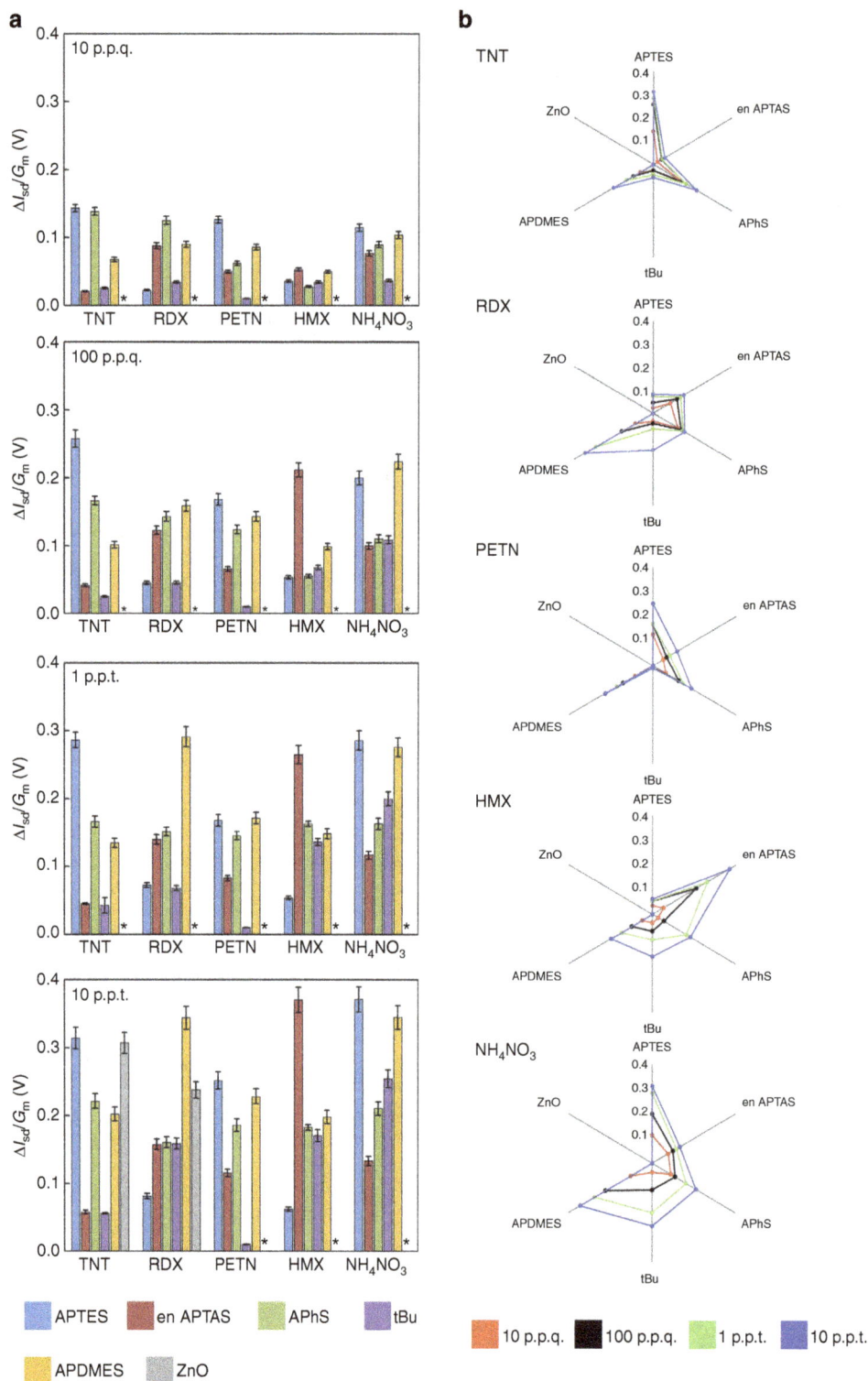

Figure 4 | Thermodynamics-based analysis. (a) At a steady state, the calibrated absolute response of different silanes to a certain explosive species is discriminative, such that their relative responses create a concentration-independent pattern to make an explosive's unique fingerprint. (**1**) Aminopropyl triethoxysilane (APTES), (**3**) n-(2-aminoethyl)-3-3aminopropyltrimethoxysilane, (en-APTAS), (**5**) p-aminophenyltrimethoxysilane (p-APhS), (**6**) Aminopropyldimethylmethoxysilane (APDMES), (**7**) 4-amino-3,3-dimethylbutyl triethoxysilane (tBu), (**8**) ZnO NPs-decorated devices, against different explosives (TNT, RDX, HMX, PETN and NH_4NO_3). Errors bars represent the measured values range for $n > 50$ measurements. Asterisks denote the absence of electrical response of the nanodevice upon interaction with the respective explosive. (**b**) The 'radar plot' two-dimensional presentation for the interaction of each explosive species with six different surface receptors, at different concentrations. The radar plots of each explosive at different concentrations are overlapped in order to demonstrate the concentration independency of our fingerprinting approach.

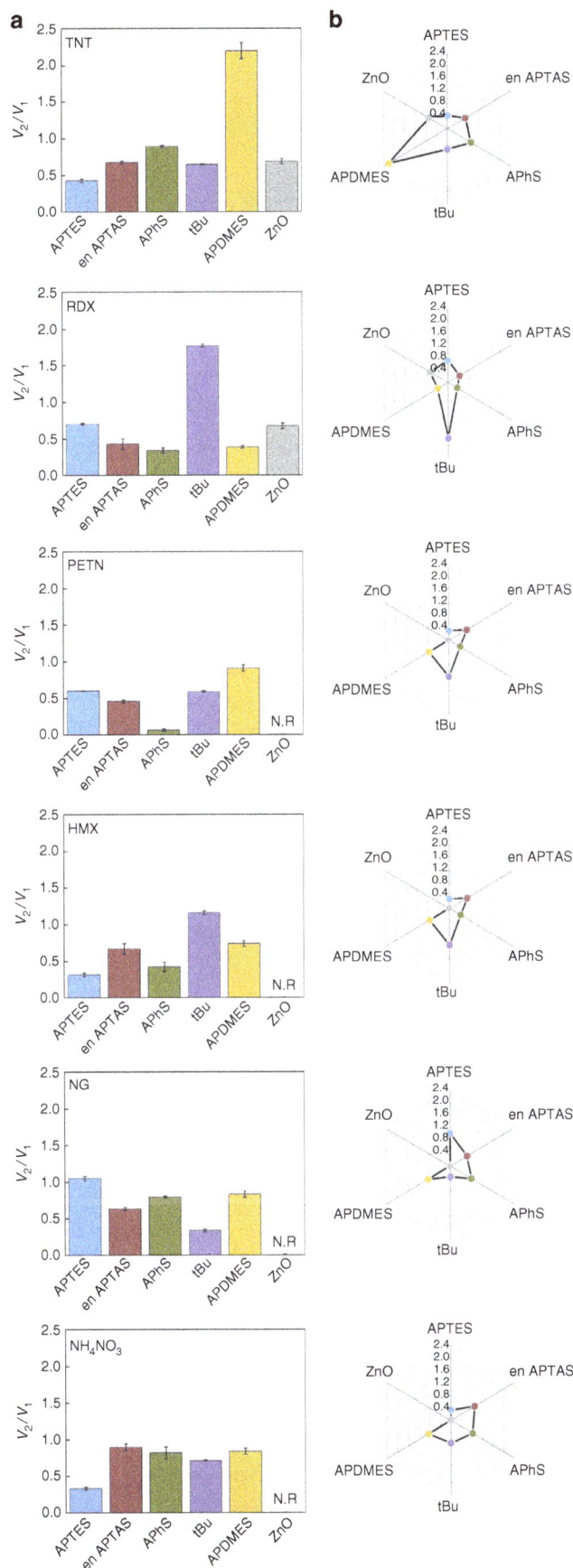

temperature, on filter supports (by short sampling pulses of 5 s). Alternatively, calibrated standard solutions containing known concentrations of the respective explosives were prepared and flowed into the sensing system. Figure 4a,b presents the interaction of each explosive species, in terms of $\Delta I_{sd}/G_m$ values measured under steady state conditions, (constant analyte concentration under constant flow) with various molecular binding receptors of the chemically modified nanoarray. Clearly, different explosives species display a distinctive pattern of interaction with the nanosensing array, thus allowing for a simple and straightforward identification of the molecule under test. The interaction pattern of each explosive species is kept constant regardless its concentration in the tested sample, Fig. 4b. Notably, all devices display a clear concentration-dependent behaviour when exposed to different concentrations of a certain explosive species, Supplementary Fig. 5a,b. Also, this distinctive fingerprinting pattern of interaction is common to all devices within a single nanoarray platform (for all the eight subarrays of 18 nanoFETs each), and across different nanoarray chips ($n > 100$ chips). Furthermore, control unmodified nanodevices on the same chip, as well as devices modified with non-amino silanes (for example, alkyl or fluoroalkyl silanes) do not produce any observable signal upon interacting with even relatively high concentrations of explosives (in the high μM range, hundreds of p.p.m. level, confirming previous reports[13,21]). Importantly, these devices demonstrate the need for electron-donating functional groups, for example, amine, thiol groups[9,39] and possibly calyx[4]pyrrole receptors[40], required for the binding interaction of the electron-deficient explosive species to the nanosensors surface, and thus will be applied as the internal control sensors throughout our studies.

In addition, kinetic parameters inherent to the chemical interaction between different explosives species and the chemically modified nanosensors array can be simultaneously extracted and analysed, together with thermodynamic parameters, to enhance the selectivity of the sensing process, and the accurate identification of the molecular species under test. The 'kinetic parameters' referred in this work are the initial slopes, the very first seconds, of the reactions that occur between the explosive species and surface-confined binding receptors. In other words, after each explosive binds and unbinds, the initial slopes of I_{SD} as a function of time indicate the binding (V_1) and unbinding (V_2) processes, respectively. After measuring V_1 and V_2, the kinetic factors of six silane-based receptors against different explosive species, as per Figs 4 and 5, are calculated. The kinetic data is summarized in Fig. 5a,b. Significantly, every silane-to-explosive couple has its own and unique kinetic V_1/V_2 ratio, creating together a kinetic data matrix. This matrix is actually a library–or 'fingerprint'–of each one of the explosives under measurement; of course, measuring additional materials (both explosives and silanes) will expand this database and will enable us to update it,

Figure 5 | Kinetics-based analysis. (a) The algorithm-derived kinetic ratios of the real-time measured transient electrical signals (V_1–V_2) map. Multiple explosive species are tested against the silanes-modified sensing array, to create a unique fingerprint pattern that enables the discrimination of all tested explosives (see 'radar' two-dimensional presentation on panel (**b**)). The whole set of mathematically derived kinetics produces a profile library (left panel, bar graph), where each result is calculated over >50 experiments performed with at least five different chips. (APTES, Aminopropyltriethoxysilane; en-APTAS, n-(2-aminoethyl)-3-3aminopropyltrimethoxysilane; p-APhS, m-aminophenyltrimethoxysilane; APDMES, Aminopropyldimethylmethoxysilane; tBu, 4-amino-3,3-dimethylbuthyltriethoxysilane; ZnO, zinc oxide-decorated SiNWs). Errors bars represent the measured values range for $n > 50$ measurements.

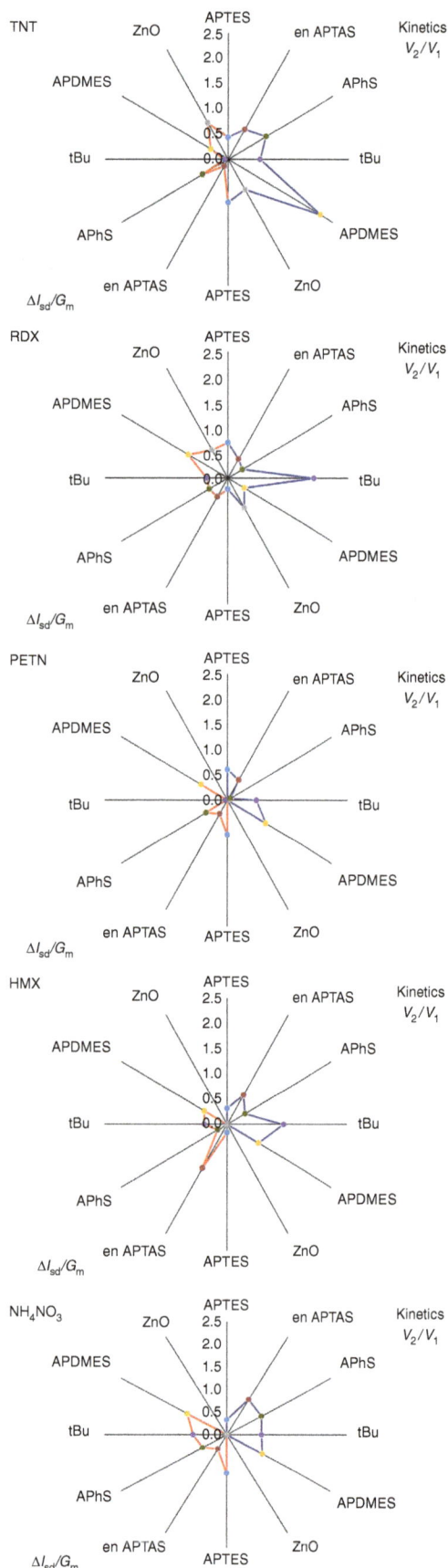

and enable the detection and identification of novel types of explosives. Note that our measurements of the kinetics complex ratios of any given explosive species with the chemically modified array do not show a concentration-dependence behaviour (in the range of 1 p.p.q.-1 p.p.b.), but only affected by the nature of the explosive-to-receptor pair under investigation. Notably, potential interfering molecules will give rise to considerably different interaction patterns with the chemically modified nanosensors array, and may be easily discriminated in real-time from known explosive molecules of interest. In addition, by virtue of our developed algorithm, and the multiplexed nature of the sensing array, we are capable of detecting and identifying explosives mixtures (Supplementary Fig. 6), interfering species and taggant species (Supplementary Fig. 7). Furthermore, the multiparameter combined use of kinetics- and thermodynamics-based analyses,- Figs 4 and 5, allows for a more accurate and clear-cut identification of the molecular species under test, as described in Fig. 6. Figure 6 practically exemplifies the complete set of results obtained when exposing the multireceptor nanosensor array to a certain explosive sample. Furthermore, in order to show the selectivity and relative insensitivity of our sensing platform to potentially interfering environments, we have carried out experiments for the detection of explosives under highly contaminated conditions, for example, cigarette-smoke heavily polluted air samples. Supplementary Fig. 8 demonstrates the capability to detect a sample of 10 p.p.t. TNT on a series of APTES-modified nanoFET devices away from smoke-highly contaminated samples. The direct 30 s collection of air samples from a lighted cigarette, by the air sampling device through a filter support, does not cause any electrical responses on the sensor devices, while a 5 s air collection of TNT vapour samples (from above a solid TNT source) leads to clear-cut electrical responses from all FET devices.

Detecting peroxide-based explosives on a single platform. Lastly, but not of lesser importance, is our capability to directly and sensitively detect and identify peroxide-based explosives, such as Triacetone trioxide (TATP) and HMDT, infamous for their clandestine production due to the availability of the ingredients and the ease of production[41]. Conventional explosives sensors rely on the presence of nitro groups. Unlike conventional nitro-containing explosives, TATP displays a significant vapour pressure at ambient temperature (6.95 Pa at 25 °C[42]), but because of the fact that it does not contain nitro groups, its detection poses a serious challenge[43]. HMTD further exacerbates this limitation by displaying an extremely low vapour pressure. Thus, the development of novel approaches for the sensitive and rapid detection of these hazardous molecules is of great importance in the field of homeland security. Unfortunately, peroxide-based explosives do not show any appreciable interaction with the chemically modified nanosensors array, modified with silanes **1–8** (Figs 7c and 1c), and even at the highest tested concentrations of tens of p.p.m. (part-per-millions). In order to overcome this limitation, we developed a strategy for the area-selected decoration of nanowire devices with ZnO nanoparticles, by a low-temperature ALD (atomic layer deposition) process[44].

Figure 6 | Fingerprinting explosives by combining kinetics and thermodynamics. The algorithm-derived analysis results of the interaction of the multireceptor sensor platform (consisting of **1**, **3**, **5**, **6**, **7** and **8**) after its exposure against various explosive species, combining the results separately described in Figs 4 and 5. Left side $\Delta G/G_{max}$ represents the thermodynamic-derived results for each explosive (red colour), and the right side represents the kinetics-derived results for each explosive species (blue colour).

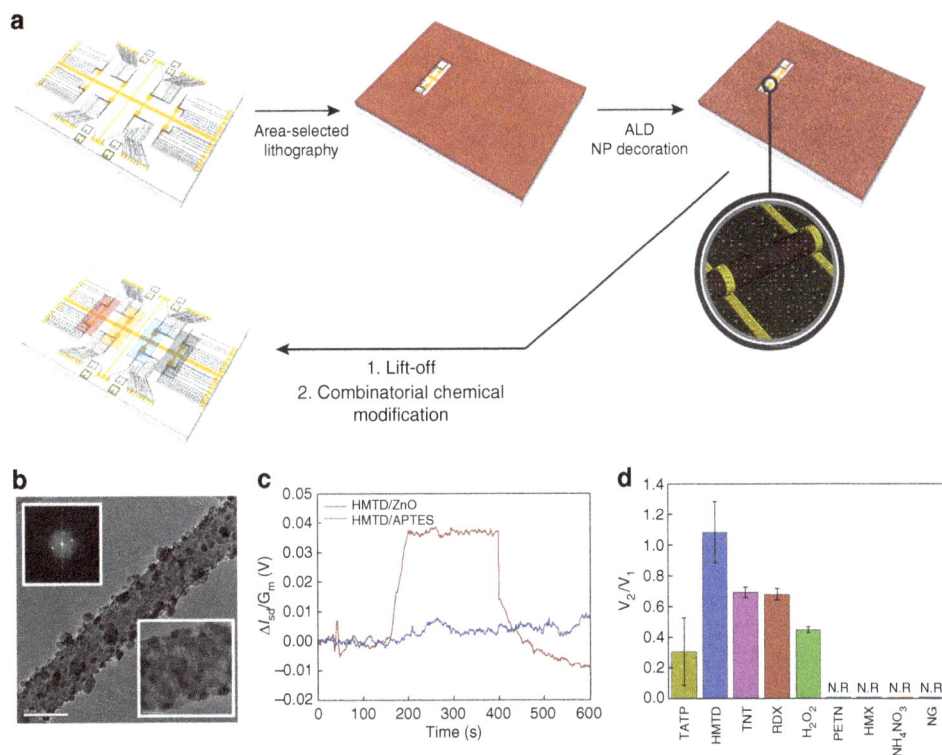

Figure 7 | Peroxides-based explosives detection. Peroxide-based explosives (TATP, HMTD) are detected by their unique complexation with zinc oxide nanoparticles decorating silicon nanowires (ZnO-SiNW). (**a**) Area-selected sensing regions are created on a multiplex chip by the ALD-assisted decoration of Si nanoFETs with ZnO nanoparticles, followed by resist removal to lay bare other subregions, to be in turn, chemically modified selectively with silane derivative receptors, as described in Fig. 1. (**b**) Representative transmission electron microscopy image of ALD-deposited zinc oxide nanoparticles (70 °C, 10 cycles) decorating p-type SiNWs ($\sim$20 nm diameter). (**c**) Receptors that are sensitive to the nitro-carrying explosives family are mostly insensitive to the peroxide-based explosives family. The HMTD concentration is 1 p.p.t. (**d**) Besides peroxide-based explosives, ZnO complexes with TNT and RDX, yet is inert to HMX and the linear structured nitro-carrying explosives (PETN, NG). ZnO is known to donate electrons to H_2O_2, but hydrogen peroxide is detected only at high concentrations above 250 p.p.m. Irrespective of their concentration, all explosives are discriminated through their respective kinetic ratios. Errors bars represent the measured values range for $n > 50$ measurements.

The area-selected decoration of ZnO NPs is allowed by an additional photolithography masking step, before the subsequent chemical modification of the undecorated array areas with the silane derivatives can be performed, Fig. 7a. The decoration process leads to the formation of a dense layer of ZnO NPs on the NWs sensing elements, with the capability of controlling over the nanoparticles dimensions, between 4–10 nm, Fig. 7b. In addition, the same ALD-based process can be readily applied for the decoration of nanowire sensing elements with TiO_2 and In_2O_3 nanoparticles. The deposited metal oxide nanoparticles will serve as a basis for the complexation and further detection of this challenging explosives family. Past theoretical studies have shown a plausible approach based on the formation of complexes between the molecular ring structures of these explosives and a central metal moiety, analogous to clatherates and crown ethers that selectively bind to ionic species in solution. These studies have predicted that TATP molecules can bind to several ions of different valency, with In^{3+}, Zn^{2+} and Ti^{4+} showing the highest binding energy[45]. On the basis of these studies, we confidently assumed that surface-decorated metal oxide nanoparticles will serve as effective complexating centres for the binding and detection of TATP and HMDT. Accordingly, Fig. 7d shows the interaction of ZnO NPs-decorated nanosensing devices in the presence of TATP, HMTD (Hexamethylene triperoxidediamine), TNT, RDX, PETN and HMX. These devices can sensitively detect TATP and HMTD down to the low p.p.t. concentration range, and effectively discriminate and identify these molecular species against all the nitro-containing explosives species tested before, Figs 4–6, by the simple use of a single chemically modified sensing platform. Although TNT and RDX are sensed by the ZnO-decorated devices, they can only be detected at high concentrations of >100 p.p.t., and can be still easily discriminated by the complete thermodynamic and kinetic fingerprinting pattern extracted from their interaction with the multireceptor platform. As well, a clear discrimination, kinetically and thermodynamically, against simple peroxide species, that is, hydrogen peroxide, can be readily attained.

Discussion

Last but not least, the ultrahigh sensitivity attained by our nanosensing devices, in the low p.p.q. level, may allow the remote detection of various explosives species, without the need for physically contacting and collecting explosives' trace particles, as commonly practiced by current sensitivity-challenged devices. In this context, we have performed a broad series of real-world sensing experiments for the detection of ultra-low vapour traces of explosives, done by the collection of short pulses of air samples above solid sources of different explosives, as well as at different distances from these explosives sources (see Supplementary Note 1 and Supplementary Table 2). Our preliminary tests, have clearly shown the capability to detect the volatile TATP explosive at distances of at least 5 m from the solid open source. Also, the less volatile TNT explosive can be readily detected and discriminated from distances of up to 4 m from the explosive solid source. These results are in accordance with the remarkable sensitivity of our sensors, as well as with the calculated concentration-to-distance curves extracted for these explosives (see Supplementary Notes Section for detailed information on discussed results and calculations)[46]. These promising results demonstrate the potential capability of our sensing platform for the remote detection of explosive species.

In conclusion, we have demonstrated the development of a chemically modified nanosensors array sensing platform for the supersensitive detection and identification of a broad range of chemically dissimilar explosive species. Besides its intrinsically unprecedented sensitivity, down to the low p.p.q. concentration range in gas and aqueous samples, the presented approach allows for the clear-cut discriminative identification of multiple explosives species based on the novel application of a real-time mathematical analysis, kinetically and thermodynamically, of the measured chemical interaction curves between the molecules under test and the individual surface-confined molecular receptors on the nanosensing elements. This analysis results in a clear fingerprinting of individual explosives molecules, and allows their identification and discrimination from non-explosives potential interferents. Furthermore, the sensitive rapid detection of peroxide-based improvised explosive species, TATP and HMTD can be performed by the decoration of nanosensing elements with metal oxide nanoparticles, e.g., ZnO, instead of the conventional surface modification by molecular silane derivatives. By means of a simple area-selected lithography procedure, we have demonstrated for the first time the fabrication of multiplexed sensing arrays consisting of molecularly-modified, as well as NP-decorated, nanosensing devices altogether in a single detection platform. In addition, the capability of this platform to remotely detect various explosive species was preliminarily demonstrated.

We believe that these solid practical results represents the first generation of supersensitive, rapid and real-time analytical platform for the ultra-trace detection and identification of a broad range of explosives species. Undoubtedly, this approach can be naturally extended to the detection of additional small molecular and biomolecular species of interest.

Methods

Materials and chemicals. All silanes are products of Gelest Inc. (Morrisville, PA). All explosives are products of AccuStandard (New Haven, CT), NH_4NO_3 from Sigma-Aldrich, Israel, while Musk moskene, Musk ketone and Musk xylene are from Fluka (Germany). Standard solutions of the explosives TNT, RDX, HMX PETN (1 part per thousand) are provided in methanol–acetonitrile 1:1mixtures, while HMTD, TATP (0.1 p.p.t.h) are solubilized in acetonitrile alone. All explosives, ammonium nitrate and the musk derivatives are solubilized and diluted in 0.1% DMSO (Sigma, Israel) solution in pure deionized water, 18.4 MΩ. Dimethyl Zinc and titanium isopropoxide ALD precursors, for the decoration of nanowire devices, were purchased from Sigma-Aldrich and used as received. AuNPs (20 nm) from BBI International (UK), and poly-L-lysine from Ted Pella Inc. (Redding Ca). SiH_4 gas 99.9999%, high-purity Argon 99.9999%, and diborane were purchased from Linde Germany. Three inch 600 nm thermal oxide-covered p-type <100> silicon wafers were supplied by Silicon Quest International (Reno, NV). LOR-5A copolymer and S-1805 photoresist were purchased from MicroChem Corp. Analytical-grade toluene was purchased from Frutarom, Israel.

Experimental procedures. SiNW growth is described in detail elsewhere[29]. Briefly, all NWs used in this work were grown by the VLS technique, using high-purity silane (SiH_4), high-purity argon as a carrier gas and 20 nm AuNPs serving as catalysts adhered to the growth substrate by a layer of poly-L-lysine. Nanowires were doped *in situ* with diborane, leading to 1:4,000 (B:Si ratio) p-type nanowire elements. The resultant crystalline NWs, with a final diameter of *ca.* 20 nm and *ca.* 10–20 microns long, exhibit a smooth surface with a thin native silicon oxide layer (*ca.* 2 nm), as confirmed by transmission electron microscopy.

Nanowire-based FETs array fabrication. The Si wafers were cleaned by washing with acetone, isopropyl alcohol (IPA), DIW, incubation in hot Piranha for 5 min, rinsing thoroughly with deionized water (DIW), blowing with dry N_2 followed by oxygen plasma treatment (100 W, 50 sccm O_2 for 200 s (Axic HP-8 USA)).

Source and drain electrodes were defined by mask exposure of a multilayer resist structure consisting of 500 nm LOR-5A copolymer and 500 nm S-1805 photoresist (purchased from MicroChem Corp.). After exposure, development and gold metallization of the gate, drain and source electrodes pattern (VST Israel), SiNWs were arrayed on their respective eight pads by a dry transfer method, as published in past[47]. Briefly, a lab-built contact printer brought together the opposing surfaces of the SiNWs donor and the chip aligned in such a way that the SiNWs would be transferred onto their destined 8 pads, with the NWs aligned perpendicularly to the surface of each source and its respective drain. A pressure of 10 g cm^{-2} was applied, at a constant velocity of 10 mm min^{-1}, for a distance of 10 mm.

To eliminate all redundant NWs, except those destined to be channels at the active eight pads, a protecting mask of a resist pattern was created and the chip was treated by DRIE (PlasmaTherm, USA) under the following conditions: C_4F_8 (10 sccm), SF_6 (50 sccm), RF1 (bias) 100 W, $P = 5$ mTorr and $t = 13.5$ s followed by incubation at PG remover (15 min, 70 °C), wash with acetone and IPA.

The drain electrodes were defined, developed and metalized by e-beam evaporation of Ti/Pd/Ti (5/60/10 nm), respectively (VST Israel), and then passivated with an insulating dielectric layer of Si_3N_4 (50 nm-thick) deposited by Inductively Coupled Plasma-Enhanced Chemical Vapour Deposition (ICP-PECVD) (Axic, USA).

To eliminate the creation of analytes depots on the chip that might affect the measured results, the chip was protected from the environment by SU8, except the active pads areas. Open pads were defined by photolithography using negative photoresist, SU8 2000.5, spin coated at 500 r.p.m. for 10 s and at 3,000 r.p.m. for 30 s, to achieve a layer of ~500 nm thickness, followed by a pre-baking step (95 °C for 1 min). Photolithography is done at 350–430 nm, with an exposure time of 7 s, using a lithography mask (Supplementary Fig. 1), followed by post-baking step (95 °C; 1 min) and development (1 min) and finally, IPA wash (10 s), and drying with N_2. The wafer was hard baked at 150 °C for 15 min.

Atomic layer deposition (Savannah 200 Cambridge Nanotech, Waltham, MA) was used to grow Zinc oxide nanoparticles (ZnO or TiO_2 NPs) over SiNWs located on their destined exposed pads (70 °C, 10 cycles, each cycle having the following sequence: H_2O 0.015 s, wait 5 s, dimethyl zinc (or titanium isopropoxide) 0.015 s, wait 10 s.

Devices electrical characterization. IV measurements were taken for each device using a probe station (Janis Research Co., Washington) at room temperature. (1) The drain current (I_{ds}) response to the applied V_{ds}, varying between -0.3 V to $+0.3$ V at 10 mV s^{-1} rate, at constant gate bias (taken from $+0.4$ V to -0.4 V). The gate dependence of the device was established by taking the drain current I_{DS} response in time for varying V_g from -0.3 V to $+0.3$ V at 10 mV s^{-1} rate, at a constant V_{DS} bias (0.1 V). The transconductance (g_m) is given by the slope of $I_{ds} - V_g$ at the linear section. Since the separation between the source and drain electrodes dictates a channel length of 3 μm and because the dry transfer aligns the SiNWs perpendicular to the electrode's surface, we confidently conclude that the g_m need not be corrected since the length of the active channel is constant.

Surface modification of a multiplex chip. Silane surface modification is performed as follows: The cleaned chip is dried over a hotplate (115 °C, 1 h) and fitted over with a multiple ports flow through PDMS microfluidic system (plasma treated first (30 W, 50sccm O_2, 3 min)) that enables the surface modification of the chip with several silanes simultaneously (Supplementary Fig. 3). All silane solutions are made of 1% silane (v/v) in 95%: 5% mixture of ethanol with water adjusted to pH = 5 (with glacial acetic acid), while heptadecafluoro-1,1,2,2, is modified in 2% silane solutions of 95% ethanol/DIW and 1:1dichloromethane/heptane. All mixtures are incubated for 20 min at RT and filtered through a 0.2 μm-cutoff. Silanes are infused through the microfluidic device over their destined pads at 3 μl min^{-1} flow for 2 h, after ethanol is being flowed first (20 μl min^{-1} for 15 min) and after the modification is finished. The PDMS is removed, the chip is washed again with ethanol for 10 s. The chip is placed on a hot plate at 115 °C for 3 h.

OTS monolayers where assembled over dedicated pads beforehand following Sagiv's method[48], whereby the cleaned chip is immersed for *ca.* 30 s in a 5 mM solution of OTS in cyclohexane under controlled environment (25 °C, vapour pressure 50%), followed by sonication for 2 min in toluene. This procedure is repeated once followed by exposing the chip for 12 h to a water saturated atmosphere (100% RH at 40 °C), followed by an additional immersion in the OTS/BCH solution for *ca.* 30 s, sonication in toluene, and final annealing for 10 min in a preheated oven at 115 °C. This sequence of operations is repeated three times, ending with the thermal annealing at 115 °C.

Sensing measurements. Qualified devices that ensure good sensor performance, as determined by electrical characterization, were wire-bonded (West-Bond 7476E, Anaheim, CA) to the outer pads of the chip carrier (PCB technologies Ltd Israel. A commercial DC source measure unit (SM32P, FES Israel) enables the simultaneous reading of 64 devices under test. PDMS-molded microfluidic heads supply the pads with alternating control and sensing solutions. The dimensions of the channels dictate a laminar flow with a Pe ~200. The fluid is infused or withdrawn by a syringe pump (Mitos XS Dolomite, UK) at a rate of 10–1,000 μl min^{-1}.

XPS measurements. XPS studies were performed on SiNWs substrates, using 5,600 Multi-Technique System (PHI, USA) with a base pressure of 2.5×10^{-10} torr. Samples were irradiated with an Al Kα monochromated source (1486.6 eV) and the out coming electrons were analysed by a Spherical Capacitor Analyzer using a slit aperture of 0.8 mm. Sample charging was compensated (if required) with a charge neutralizer (C1s at 285 eV was taken as an energy reference). The SAM samples were analysed at a shallow take-off angle of 23°. High-resolution XPS measurements were taken at pass energy of 11.75 eV with 0.05 eV per step interval. SiNWs wafers with 600 nm thermal oxide, SSP prime grade, were cut for the XPS studies to 5×5 mm^2 pieces. Before the chemical modification, samples were cleaned by an oxygen plasma step, and dehydration on a hot plate for 1–2 h. The different modifications steps were done as described above in the surface functionalization section, except that the chips were immersed in the silane solution for 30 min, washed with ethanol, IPA, ethanol, blown with dry N$_2$ prior to drying on a hot plate at 115 °C for 3 h (for additional details, see Supplementary Figs 9–15).

QCM measurements. A home-built QCM analyzer equipped with a Fluke 164T multifunction counter was used for the microgravimetric quartz crystal micro-balance experiments. Quartz crystals (AT cut, 9 MHz) sandwiched between two Au electrodes (roughness factor ca. 3.5 with an area of 0.196 cm^2) were used in microgravimetric experiments, after deposition of a thin silicon oxide layer of 5 nm by PECVD deposition. Quartz electrodes were cleaned with a piranha solution (70% H$_2$SO$_4$:30% H$_2$O$_2$) for 15 min, then rinsed thoroughly with DDW and dried with a stream of argon before chemical modification steps.

Thickness measurements by ellipsometry. Ellipsometric measurements were carried out on a M-2000DUV Spectroscopic Ellipsometer (J.A.Woollam Co., Inc.). The angles of incidence were 65°, 70° and 75° with a spot size of 2–3 mm. The data were analysed using WVASE32 software installed with the ellipsometer. The film thicknesses of the examined layers were calculated by using the Cauchy model.

References

1. Brady, J. E., Smith, J. L., Hart, C. E. & Oxley, J. Estimating ambient vapor pressures of low volatility explosives by rising-temperature thermogravimetry. *Propell. Explos. Pyrot.* **37**, 215–222 (2012).
2. Caygill, J. S., Davis, F. & Higson, S. P. Current trends in explosive detection techniques. *Talanta* **88**, 14–29 (2012).
3. Beardah, M. S., Doyle, S. P. & Hendey, C. E. Effectiveness of contamination prevention procedures in a trace explosives laboratory. *Sci. Justice* **47**, 120–124 (2007).
4. Nambayah, M. & Quickenden, T. I. A quantitative assessment of chemical techniques for detecting traces of explosives at counter-terrorist portals. *Talanta* **63**, 461–467 (2004).
5. McAlpine, M. C., Ahmad, H., Wang, D. & Heath, J. R. Highly ordered nanowire arrays on plastic substrates for ultrasensitive flexible chemical sensors. *Nat. Mater.* **6**, 379–384 (2007).
6. Timko, B. P., Cohen-Karni, T., Qing, Q., Tian, B. & Lieber, C. M. Design and implementation of functional nanoelectronic interfaces with biomolecules, cells, and tissue using nanowire device arrays. *IEEE Trans. Nanotechnol.* **9**, 269–280 (2010).
7. Patolsky, F., Zheng, G. & Lieber, C. M. Nanowire sensors for medicine and the life sciences. *Nanomedicine (Lond)* **1**, 51–65 (2006).
8. Cui, Y. & Lieber, C. M. Functional nanoscale electronic devices assembled using silicon nanowire building blocks. *Science* **291**, 851–853 (2001).
9. Gao, X. P., Zheng, G. & Lieber, C. M. Subthreshold regime has the optimal sensitivity for nanowire FET biosensors. *Nano Lett.* **10**, 547–552 (2009).
10. Hahm, J.-i. & Lieber, C. M. Direct ultrasensitive electrical detection of DNA and DNA sequence variations using nanowire nanosensors. *Nano Lett.* **4**, 51–54 (2004).
11. Zheng, G., Patolsky, F., Cui, Y., Wang, W. U. & Lieber, C. M. Multiplexed electrical detection of cancer markers with nanowire sensor arrays. *Nat. Biotechnol.* **23**, 1294–1301 (2005).
12. Cohen-Karni, T. *et al.* Synthetically encoded ultrashort-channel nanowire transistors for fast, pointlike cellular signal detection. *Nano Lett.* **12**, 2639–2644 (2012).
13. Engel, Y. *et al.* Supersensitive detection of explosives by silicon nanowire arrays. *Angew. Chem. Int. Ed.* **49**, 6830–6835 (2010).
14. Persaud, K. & Dodd, G. Analysis of discrimination mechanisms in the mammalian olfactory system using a model nose. *Nature* **299**, 352–355 (1982).
15. Albert, K. J. *et al.* Cross-reactive chemical sensor arrays. *Chem. Rev.* **100**, 2595–2626 (2000).
16. Anslyn, E. V. Supramolecular analytical chemistry. *J. Org. Chem.* **72**, 687–699 (2007).
17. Legin, A., Rudnitskaya, A., Vlasov, Y. G., Di Natale, C. & d'Amico, A. The features of the electronic tongue in comparison with the characteristics of the discrete ion-selective sensors. *Sensor. Actuat. B-Chem.* **58**, 464–468 (1999).
18. Ciosek, P. & Wróblewski, W. Sensor arrays for liquid sensing—electronic tongue systems. *Analyst* **132**, 963–978 (2007).
19. Regonda, S. *et al.* Silicon multi-nanochannel FETs to improve device uniformity/stability and femtomolar detection of insulin in serum. *Biosens. Bioelectron.* **45**, 245–251 (2013).
20. Davidson, III G. & Peppas, N. A. Solute and penetrant diffusion in swellable polymers: VI. The Deborah and swelling interface numbers as indicators of the order of biomolecular release. *J. Control. Release* **3**, 259–271 (1986).
21. Senesac, L. & Thundat, T. G. Nanosensors for trace explosive detection. *Mater. Today* **11**, 28–36 (2008).
22. Cetó, X., O'Mahony, A. M., Wang, J. & del Valle, M. Simultaneous identification and quantification of nitro-containing explosives by advanced chemometric data treatment of cyclic voltammetry at screen-printed electrodes. *Talanta* **107**, 270–276 (2013).
23. Qi, P. *et al.* Toward large arrays of multiplex functionalized carbon nanotube sensors for highly sensitive and selective molecular detection. *Nano Lett.* **3**, 347–351 (2003).
24. Wongwiriyapan, W., Inoue, S., Honda, S.-i. & Katayama, M. Adsorption Kinetics of NO2 on Single-Walled Carbon Nanotube Thin-Film Sensor. *Jpn J. Appl. Phys.* **47**, 8145 (2008).
25. Hassibi, A., Vikalo, H. & Hajimiri, A. On noise processes and limits of performance in biosensors. *J. Appl. Phys.* **102**, 014909–014912 (2007).
26. Potyrailo, R. A. & Mirsky, V. M. Combinatorial and high-throughput development of sensing materials: the first 10 years. *Chem. Rev.* **108**, 770–813 (2008).
27. Tsitron, J., Ault, A. D., Broach, J. R. & Morozov, A. V. Decoding complex chemical mixtures with a physical model of a sensor array. *PLoS Comput. Biol.* **7**, e1002224 (2011).
28. Fan, Z. *et al.* Wafer-scale assembly of highly ordered semiconductor nanowire arrays by contact printing. *Nano Lett.* **8**, 20–25 (2007).
29. Patolsky, F., Zheng, G. & Lieber, C. M. Fabrication of silicon nanowire devices for ultrasensitive, label-free, real-time detection of biological and chemical species. *Nat. Protoc.* **1**, 1711–1724 (2006).
30. Ishikawa, F. N. *et al.* A calibration method for nanowire biosensors to suppress device-to-device variation. *ACS Nano* **3**, 3969–3976 (2009).
31. Terrier, F. Rate and equilibrium studies in Jackson-Meisenheimer complexes. *Chem. Rev.* **82**, 77–152 (1982).
32. Hozumi, A. *et al.* Amino-terminated self-assembled monolayer on a SiO surface formed by chemical vapor deposition. *J. Vac. Sci. Technol. A* **19**, 1812 (2001).
33. Plueddemann, E. P. *Silane coupling agents* (Springer, 1982).
34. He, T. *et al.* Controlled modulation of conductance in silicon devices by molecular monolayers. *J. Am. Chem. Soc.* **128**, 14537–14541 (2006).
35. Metwalli, E., Haines, D., Becker, O., Conzone, S. & Pantano, C. Surface characterizations of mono-, di-, and tri-aminosilane treated glass substrates. *J. Colloid Interface Sci.* **298**, 825–831 (2006).
36. Buncel, E. & Webb, J. An Aromatic Amine-1, 3, 5-trinitrobenzene σ-complex. *Can. J. Chem.* **50**, 129–131 (1972).
37. Foster, R. & Fyfe, C. The interaction of electron acceptors with bases—XIX: The interaction of 1, 3, 5-trinitrobenzene with secondary amines in acetone solution. *Tetrahedron* **22**, 1831–1842 (1966).
38. Yang, X., Frensley, W. R., Zhou, D. & Hu, W. Performance Analysis of Si Nanowire Biosensor by Numerical Modeling for Charge Sensing. *IEEE Trans. Nanotechnol.* **11**, 501–512 (2012).
39. Kabalka, G. W. & Varma, R. S. *Reduction of nitro and nitroso compounds.* vol. 8 (Pergamon Press Oxford, 1991).
40. Nielsen, K. A. *et al.* Tetra-TTF calix [4] pyrrole: a rationally designed receptor for electron-deficient neutral guests. *J. Am. Chem. Soc.* **126**, 16296–16297 (2004).
41. Fitzgerald, M. & Bilusich, D. Sulfuric, hydrochloric, and nitric acid-catalyzed triacetone triperoxide (TATP) reaction mixtures: an aging study. *J. Forensic Sci.* **56**, 1143–1149 (2011).
42. Oxley, J. C., Smith, J. L., Luo, W. & Brady, J. Determining the vapor pressures of diacetone diperoxide (DADP) and hexamethylene triperoxide diamine (HMTD). *Propell. Explos. Pyrot.* **34**, 539–543 (2009).
43. Önnerud, H., Wallin, S. & Östmark, H. in Intelligence and Security Informatics Conference (EISIC), 2011 European. 238–243 (IEEE).
44. Devika, M., Koteeswara Reddy, N., Pevzner, A. & Patolsky, F. Heteroepitaxial Si/ZnO hierarchical nanostructures for future optoelectronic devices. *Chemphyschem* **11**, 809–814 (2010).
45. Dubnikova, F., Kosloff, R., Zeiri, Y. & Karpas, Z. Novel approach to the detection of triacetone triperoxide (TATP): Its structure and its complexes with ions. *J. Phys. Chem. A* **106**, 4951–4956 (2002).
46. Turner, D. B. *Workbook of atmospheric dispersion estimates: an introduction to dispersion modeling* (CRC press, 1994).
47. Fan, Z. *et al.* Wafer-scale assembly of highly ordered semiconductor nanowire arrays by contact printing. *Nano Lett.* **8**, 20–25 (2008).
48. Zeira, A., Chowdhury, D., Maoz, R. & Sagiv, J. Contact electrochemical replication of hydrophilic—hydrophobic monolayer patterns. *ACS Nano* **2**, 2554–2568 (2008).

Acknowledgements

We sincerely thank Tracense Ltd. for its financial support. We also thank the Legacy Foundation and ISF (Israel Science Foundation) for partial support of this research.

Author contributions

F.P. conceived the experiments, supervised the research work and wrote the paper. A.L., E.H. (Ehud Havivi), R.S., A.P., V.K. and G.D. carried out sensing experiments and helped analysing the data and writing manuscript. E.H. (Ehud Hahami) and R.L. developed mathematical analysis algorithm and analysed the experimental data. I.P. designed and fabricated sensing devices. E.F. developed detection system electronics and software, and helped analysing experimental data. R.E. and Y.E. helped designing and analysing experimental data during manuscript revision and revising the manuscript.

Additional information

14

A protective layer approach to solvatochromic sensors

Jung Lee[1], Hyun Taek Chang[1], Hyosung An[1], Sora Ahn[1], Jina Shim[1] & Jong-Man Kim[1,2]

As they have been designed to undergo colorimetric changes that are dependent on the polarity of solvents, the majority of conventional solvatochromic molecule based sensor systems inevitably display broad overlaps in their absorption and emission bands. As a result, colorimetric differentiation of solvents of similar polarity has been extremely difficult. Here we present a tailor-made colorimetric and fluorescence turn-on type solvatochromic sensor that enables facile identification of a specific solvent. The sensor system displays a colorimetric transition only when a thin protective layer, which protects the solvatochromic materials, is destroyed or disrupted by a specific solvent. The versatility of the strategy is demonstrated by designing a sensor that differentiates chloroform and dichloromethane colorimetrically and one that performs sequence selective colorimetric sensing. In addition, the approach is employed to construct a solvatochromic molecular AND logic gate. The new strategy could open new avenues for the development of novel solvatochromic sensors.

[1]Department of Chemical Engineering, Hanyang University, Seoul 133-791, Korea. [2]Institute of Nano Science and Technology, Hanyang University, Seoul 133-791, Korea. Correspondence and requests for materials should be addressed to J.-M.K. (email: jmk@hanyang.ac.kr).

A challenging task in chemistry has been the development of a solvatochromic sensor that is responsive to a specific solvent. Various organic[1-10], organometallic[11,12], metal organic framework[13,14] and hybrid[15,16] materials have been investigated to determine their solvatochromic properties in diverse solvents. Conventional colorimetric sensors, however, inevitably display changes in absorption and emission peaks that are in indiscriminant in their response to organic solvents. This phenomenon is a consequence of the fact that the probe molecules are designed to undergo spectral shifts that depend solely on the polarity of surrounding medium. Because of this limitation, visual differentiation of solvents that have similar polarities has been very difficult.

In this study, we devise a new approach to designing a system for colorimetric differentiation of common organic solvents. The new tailor-made colorimetric and fluorescence turn-on type solvent sensor system enables facile naked eye identification of one among several solvents. The key strategy employed for the sensor system is schematically described in Fig. 1a. A solvatochromic material is first coated on a solid substrate and then covered with a thin protective layer. As a result, the solvatochromic sensor molecules are protected from direct exposure to organic solvents unless the solvent disrupts the protective layer by either dissolution or swelling. In the latter event, the solvatochromic molecules are exposed to the solvent and undergo an observable colorimetric transition. As the colorimetric transition of the sensor system is dependent on the properties of the protective layer and the solvent, it does not require that the solvatochromic substance respond in a specific manner to a certain solvent. By using the new approach, we devise a system that is able to distinguish between dichloromethane and chloroform, two solvents that are very difficult to differentiate

colorimetrically. In addition, the new solvatochromic strategy is used to fabricate a sequence selective solvatochromic sensor as well as a colorimetric AND logic gate[17-24]. The significant features of the solvatochromic sensor system developed in this study are as follows. First, the colorimetric signal generated upon exposure of the system to a specific target solvent is easily recognized by using the naked eye. Second, a single solvatochromic dye can be employed in systems that differentiate several different solvents. Third, commercially available and inexpensive polymers can be used as the protective layers. Fourth, the sensor film can be readily fabricated by utilizing simple spin-coating or drop-casting techniques. Fifth, colorimetric changes of the sensor film occur in most cases within 1 min of exposure to the solvent. Finally, the strategy can be employed in the preparation of a variety of tailor-made sensors that are comprised of properly selected dyes and protective layers.

Results

Colorimetric and fluorescence turn-on sensor. In order to determine the feasibility of the turn-on solvatochromic sensor strategy described above, studies were carried out using the conjugated polydiacetylene (PDA) polymer[25-39] derived from 10,12-pentacosadiynoic acid (PCDA, $CH_3(CH_2)_{11}C\equiv C - C \equiv C(CH_2)_8COOH$), which is a well-known solvatochromic material (Supplementary Fig. S1). A thin film (*ca.* $1.0 \mu m$) was prepared on a glass substrate by first spin-coating a viscous solution PCDA ($40 \, mg \, ml^{-1}$) and polystyrene (PS, Mw: $280,000 \, g \, mol^{-1}$) (Fig. 1b) followed by irradiation with UV light ($254 \, nm$, $1 \, mW \, cm^{-2}$, $3 \, min$) to induce polymerization. As a photomask was used in the irradiation step, blue-phase PDAs are generated only in UV-exposed areas. Finally, the generated PDA

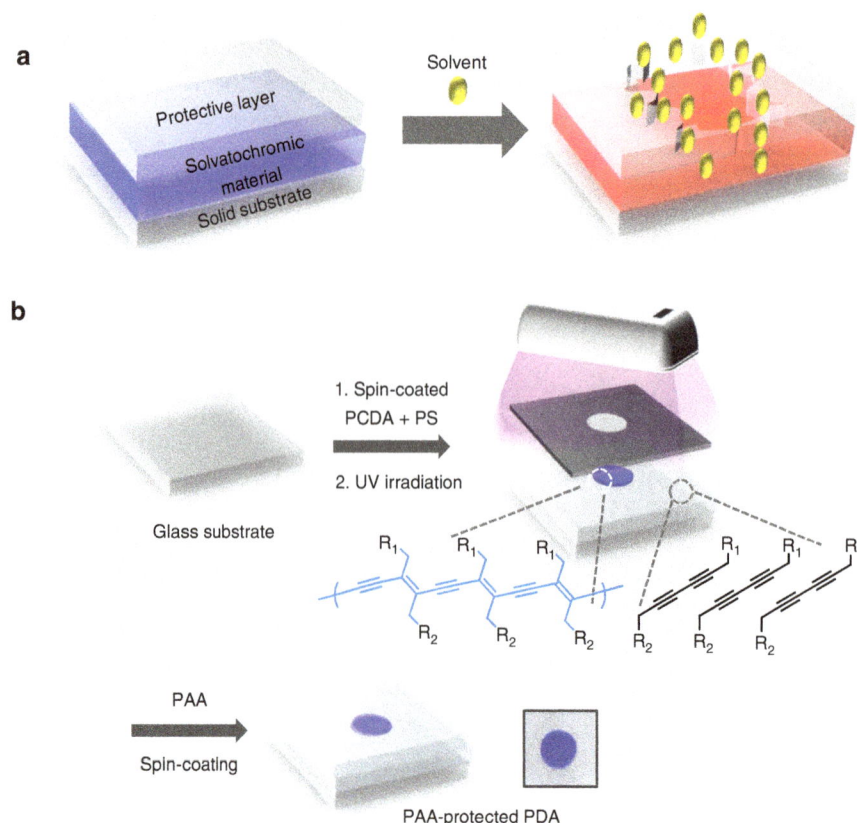

Figure 1 | Fabrication of the solvatochromic sensor system. (**a**) Schematic representation of the colorimetric sensor system. (**b**) Fabrication of polyacrylic acid (PAA)-protected polydiacetylene (PDA) film on a glass substrate.

film was coated to a thickness of *ca.* 1.5 μm using a methanol solution of poly(acrylic acid) (PAA, Mw: 450,000 g mol^{-1}). As neither PS nor PDA is soluble in water, the PS film containing PDA is stable during the PAA coating process. Implementation of this simple procedure led to fabrication of a PAA-protected blue-phase PDA film on a glass substrate.

Very interesting observations were made when pipette drops (*ca.* 100 μl) of common organic solvents were applied to the tops of unprotected and PAA-protected PDA films (Fig. 2a). As expected, unprotected PS films containing PDAs undergo an observable colour changes when exposed to most of the tested solvents, except for methyl alcohol (MeOH), isopropyl alcohol (IPA), hexane and acetonitrile (ACN) (Fig. 2a, top). In contrast, when the PAA-protected PDA films were exposed to the solvents, only the one treated with tetrahydrofuran (THF) undergoes a blue-to-red colorimetric transition (Fig. 2a, middle) (see also Supplementary Movie 1). As the red coloured form of the PDA is fluorescent while the blue counterpart is virtually nonfluorescent[40,41], only the THF-exposed film emits red fluorescence (Fig. 2a, bottom).

Visible absorption spectra of the PAA-coated PDA films were also recorded after exposure to the solvents. A significant spectral shift associated with the blue-to-red transition was observed to take place only with the film that was treated with THF (Fig. 2b). The chemical nature of the colour change process was also probed by using Raman spectroscopy (Supplementary Fig. S2). The conjugated alkyne–alkene groups in the Raman spectrum of the blue-phase PDA appear at 2081 (C≡C) and 1,451 cm^{-1} (C=C)[42]. Inspection of the Raman spectrum of the red-phase PDA, obtained by exposure of the film to THF, shows that the alkyne–alkene bands at 2,081 and 1,451 cm^{-1} are shifted to higher frequencies (2,121 and 1,515 cm^{-1}, respectively). This finding demonstrates that most of the blue-phase PDAs are transformed to red-phase counterparts upon THF treatment.

Two critical factors must be considered when deciphering the solvatochromic behaviour of the PAA-protected PDA film described above. The film disrupting power of the solvent caused by dissolution and/or swelling is one important parameter, because these processes must occur in order for solvent molecules

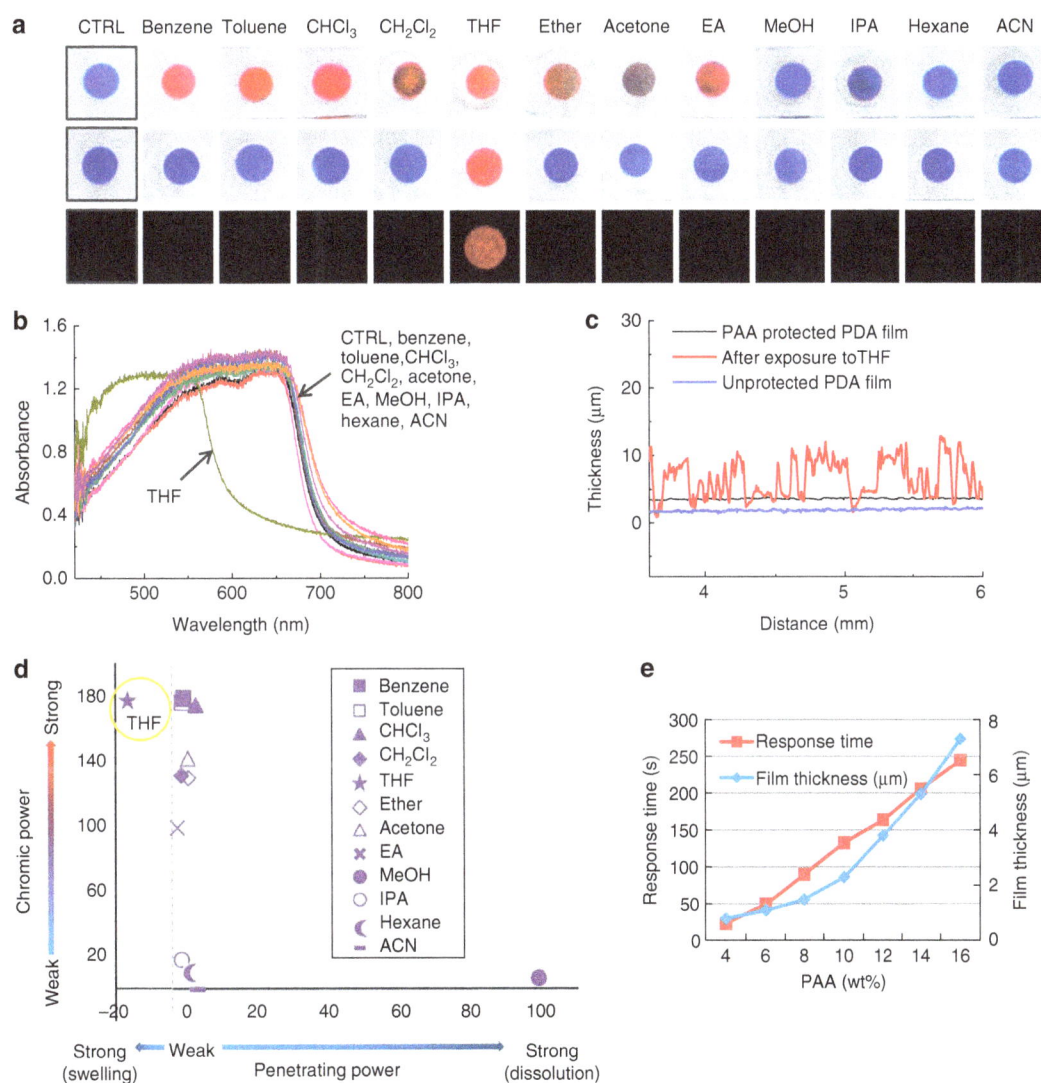

Figure 2 | THF-specific solvatochromism. (a) Photographs of solvent-exposed unprotected (top) and PAA-protected (middle) PCDA-derived PDA films. Fluorescence microscope images of the solvent-treated PAA-protected PDA film are also displayed (bottom). Photographs were taken 1 min after adding a drop (*ca.* 100 μl) of each solvent to the top of the film. **(b)** Absorption spectra of PAA-protected PDA film upon exposure to various solvents. **(c)** Film thickness as a function of distance before (blue line) and after (black line) PAA protection. The fluctuating red line is obtained after exposure of the PAA-protected PDA film to THF. **(d)** Plots of PDA chromic power versus PAA penetrating power for various solvents tested. **(e)** Plots of colorimetric response time and film thickness as a function of concentration of protective PAA.

to reach and interact with the PDAs. PAA layer disruption was demonstrated by determining the thickness profile of the film before and after exposure to solvents. The results show that a significant disruption of the sensor film occurs when it is placed in contact with THF (Fig. 2c) but that this phenomenon does not take place when other solvents, except MeOH, are used (Supplementary Fig. S3).

The layer disrupting power of a solvent can also be evaluated by measuring the weights of a PAA-coated film before and after incubation in a solvent (Supplementary Fig. S4). The results of these measurements show that no significant change in weight takes place for films treated with all tested solvents except MeOH and THF. In the presence of MeOH, PAA layer is nearly completely removed while the weight of THF-treated PAA film increases by 16.4% after incubation in THF. The latter observation indicates that immobilization of THF molecules occurs in the film.

The other critical factor responsible for promoting the colorimetric transition is the ability of the solvent to induce the blue-to-red associated phase transition of the PDA molecules (Supplementary Fig. S5). The colorimetric change inducing ability of each solvent, displayed in Supplementary Fig. S5, was determined by measuring the intensity of the red colour developed when the unprotected PDA film is exposed to each solvent. The results show that not all solvents, which penetrate the protective PAA layer, are capable of inducing the blue-to-red colour transition. For instance, MeOH readily dissolves the hydrophilic PAA layer but it is ineffective in promoting the phase change of the PDA supramolecules in the film state. It should be noted that incubation of the PDA powder alone in MeOH causes a blue-to-purple colour change (Supplementary Fig. S1), a finding that indicates that the hydrophobic PS matrix used for the fabrication of PDA film serves as a protective layer repelling hydrophilic MeOH. Inspection of Fig. 2d, in which a plot correlating the two important parameters related to colorimetric response and penetrating ability is given, demonstrates that these parameters are well related to the solvent specific colorimetric changes depicted in Fig. 2a. Specifically, because THF has both strong colorimetric and film penetrating (by swelling) properties in contrast to other solvents, it can be selectively identified by using a properly designed colorimetric and fluorescence turn-on type THF selective solvatochromic sensor system.

The protective layer strategy also enables manipulation of the colorimetric response time of the sensor film. For instance, the thickness of the protective layer can be readily controlled by varying either the concentration of the protective polymer solution or the number of spin-coating using a fixed concentration of the polymer solution (Supplementary Figs S6 and S7). This expectation is confirmed by the results displayed in Fig. 2e, which show that the thickness of the protective layer increases as the concentration of PAA increases. Importantly, as time is required for the THF molecule to disrupt the PAA layer and reach the solvatochromic PDA layer, the response time for promotion of the blue-to-red colour transition increases as the thickness of the protective layer increases.

In order to address important sensitivity related issues with regard to the solvatochromic sensor system, additional experiments were carried out. It is obvious that if the diameter of the sensor spot exposed to the organic solvent is smaller, a lower amount of the solvent would be required to promote the colorimetric transition. Interestingly, additional studies showed that only one microliter of THF is sufficient to cause the blue-to-red colour change of the polymer when a sensor spot of 2 mm diameter is employed (Supplementary Fig. S8). In addition, the solvatochromic sensor system is also applicable to systems in which the solvent of interest is diluted with other solvent. To demonstrate this feature, solvatochromic tests were carried out on mixtures of THF and ACN. ACN was selected as a diluting solvent because it does not induce the colorimetric transition of the PDA and it is completely miscible with THF. We observed that the sensor system functions well for solutions up to a 50 vol% THF-ACN (Supplementary Fig. S9). Regarding the detection limit in terms of time, we observed that the time required for the colorimetric response decreases as the thickness of the protective layer decreases (Fig. 2e). In addition, if the protective layer is too thin (below a micrometre), no colorimetric selectivity is achieved for the THF sensor due to the loss of the function of the protective layer. Thus, minimum contact time required for good solvatochromic discrimination is *ca.* 20 s.

Differentiation between chloroform and dichloromethane. We next investigated the design of a more challenging sensor system that is capable of distinguishing between the very closely related solvents, chloroform and dichloromethane. Supplementary Fig. S1 shows that both chloroform and dichloromethane bring about an indistinguishable blue-to-red colour change when they are individually applied to the PCDA-derived PDA powder. Thus, we anticipated that the protective layer approach could be used to carry out the challenging visual differentiation between these two solvents. Among various commercially available polymers, poly(vinylchloride) (PVC) has a strikingly different solubility in the two solvents, being highly soluble in dichloromethane and only poorly soluble in chloroform (Fig. 3d, Supplementary Fig. S10). Consequently, these two solvents should be colorimetrically distinguishable when PVC is used as the protective layer. In order to test this proposal, a sensor system, created by using a modified double-layer protection approach, was prepared (Fig. 3a). First, a thin PS film containing the PDA was applied to a glass substrate (as described in Fig. 1b). Second, spin-coating a MeOH solution containing polyvinylpyrrolidone (PVP, Mw: 360,000 g mol^{-1}, 12 wt%) afforded a thin PVP layer on the top of the PDA film. The coating of PVP, which is highly soluble in both chloroform and dichloromethane (Supplementary Fig. S11), is required as a 'dummy' layer in this case because direct coating of PVC on the top of the PDA causes disruption the PDA supramolecules and a premature blue-to-red colorimetric change of the polymer film. Finally, the solvent distinguishable PVC layer was introduced on top of the PVP film by spin-coating a THF solution (12 wt%) containing this polymer (Mw: 620,000 g mol^{-1}).

When individually applied, both chloroform and dichloromethane induce an immediate blue-to-red colour change of a polymer film comprised of the upper 'dummy' PVP and lower sensing PDA layer (Fig. 3b, top). In contrast, individual application of the two solvents to the PVC-protected film results in completely different outcomes. The dichloromethane-treated film undergoes a colour transition to red while the chloroform-treated sample does not experience a colour change (Fig. 3b, bottom) (see also Supplementary Movie 2). Absorption spectroscopic monitoring of these processes (Fig. 3c) has also been used to follow the changes occurring in the solvent and protective layer dependent processes. Consequently, because both chloroform and dichloromethane have a strong colorimetric transition power for PDA, their different abilities to solubilize PVC can be utilized as the basis for a dichloromethane-selective colorimetric sensor system.

IPA selective solvatochromic sensor. During the course of this study, we observed that an unprotected PDA film derived from 10,12-tricosadiynoic acid (TCDA, $CH_3(CH_2)_9C\equiv C-C\equiv C(CH_2)_8$ COOH) undergoes a blue-to-red colorimetric transition when treated with alcoholic solvents such as MeOH and IPA

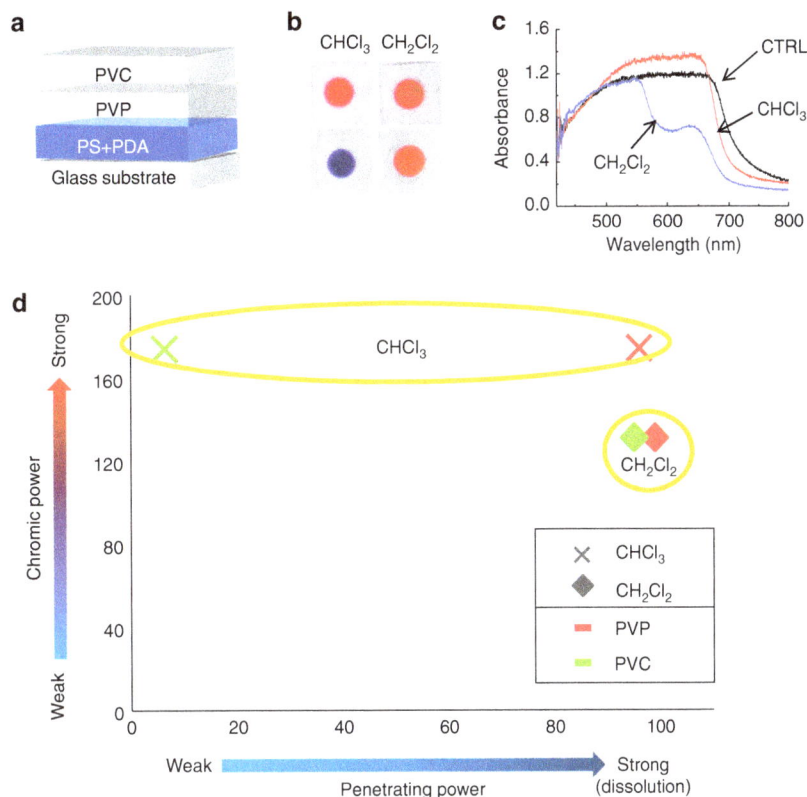

Figure 3 | Colorimetric differentiation between dichloromethane and chloroform. (a) Schematic of a double-layer protected sensor system. **(b)** Photographs of solvent-exposed single- (top, PVC), and double- (bottom, PVC-on-PVP) layer protected PCDA-derived PDA films. Photographs were taken 1 min after adding a drop (*ca.* 100 μl) of each solvent to the top of the film. **(c)** Absorption spectra of double-layer protected PDA films upon exposure to chloroform and dichloromethane. **(d)** Plots of PDA chromic power versus layer penetrating power for chloroform and dichloromethane.

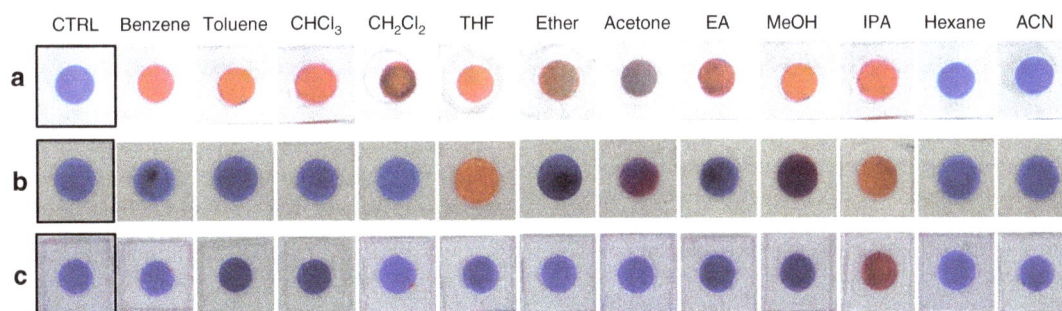

Figure 4 | IPA selective solvatochromic sensor. (a–c) Photographs of the solvent-exposed unprotected **(a)**, single-layer protected (PAA/PDA) **(b)** and double-layer protected (PVP/PAA/PDA) **(c)** TCDA-derived PDA films. Photographs were taken 1 min after dropping each solvent on the top of the film.

(Supplementary Fig. S12). TCDA contains a two carbon shorter alkyl chain than PCDA, which makes the colour transition of the TCDA-derived polymer more sensitive to alcoholic solvents than the polymer derived from PCDA, which does not undergo a MeOH or IPA promoted colorimetric transition when a incorporated in a PS film (Fig. 2a, top).

These observations led to the design an IPA selective solvatochromic sensor. A double-layer protected sensor film, comprised of (PVP/PAA/PDA), was constructed for this purpose. As expected, the unprotected TCDA-derived PDA film undergoes a colour transition when treated with all of the tested solvents except hexane and ACN (Fig. 4a). In contrast, the single-layer (PAA) protected film is only highly colorimetrically responsive to IPA and THF and to a lesser extent MeOH and acetone, owing to the polar nature of the PAA matrix (Fig. 4b). Finally, only IPA is able to

penetrate the double-layer protected sensor film and cause a blue-to-red colour change of the buried PDA supramolecules (Fig. 4c).

Solvatochtomic logic circuits. One unique advantage of the protective layer approach to solvent identification not found in conventional film or solution-based sensors is that it enables construction of a sequence selective sensor system. We observed that a poly(methyl methacrylate) (PMMA) and PVP-protected PCDA-derived PDA film (Fig. 5a) responds to two solvents in a sequence selective manner. Accordingly, a blue-to-red colour transition occurs when this film is exposed sequentially to IPA and toluene (Fig. 5b) but a colour change does not take place when the sensor film is treated in a reverse sequential manner (that is, toluene first) with these solvents. The solubility

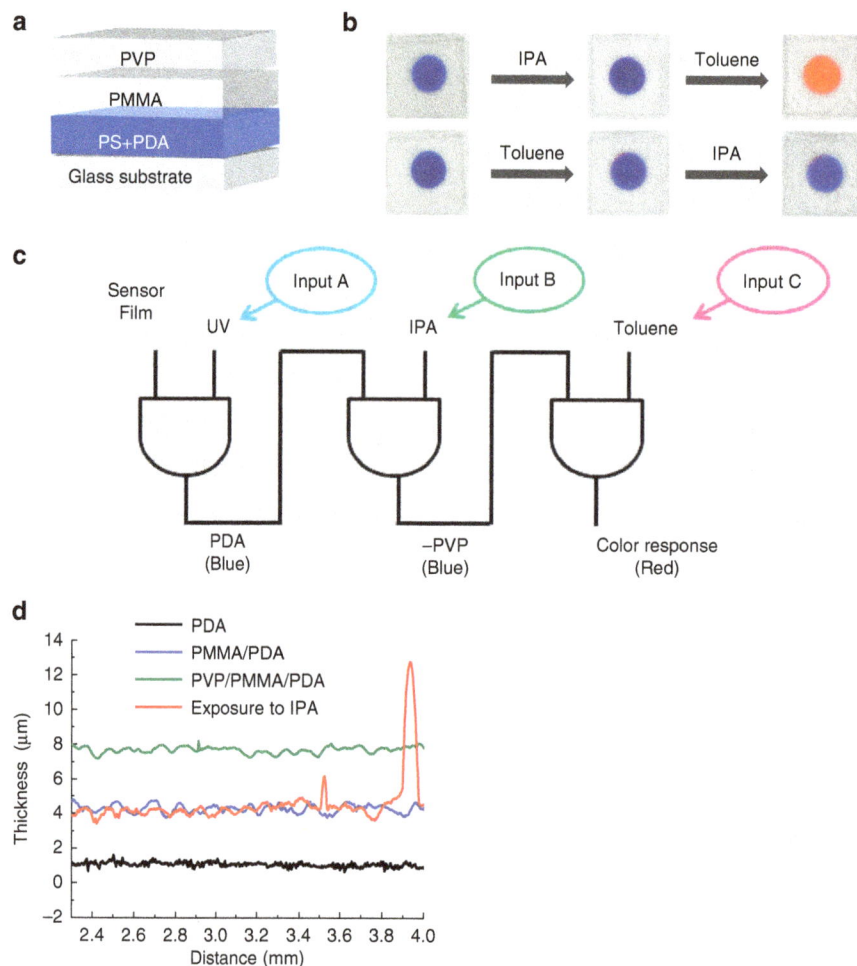

Figure 5 | Sequence selective logic gates. (**a**) Schematic of a double-layer protected PCDA-derived PDA film for a sequence-specific colorimetric sensor. (**b**) Photographs of the sensor films after exposure to IPA and toluene in different order. (**c**) A solvatochromic keypad lock logic circuit. (**d**) Film thickness as a function of distance before (black) and after protection with PMMA (blue) and PVP (green). The red line is obtained after exposure of the sensor film to IPA.

differences of the two protective polymers, PVP and PMMA, in IPA versus toluene is responsible for this sequence selective solvatochromic behaviour. Application of IPA to the sensor film results in removal of the PVP layer owing to the solubility of this polymer in alcoholic solvents (Supplementary Fig. S11). The thickness profile data presented in Fig. 5d demonstrate this property. In the initial constructing stage, the thickness of the sensor film increases sequentially as PMMA (1st layer) (blue line) and PVP (2nd layer) (green line) layers are coated on the top of the PDA polymer film (black line). Exposure of the resulting film to IPA causes a decrease in the thickness of the PMMA layer (red line), but toluene is ineffective in disrupting/dissolving the PVP layer. However, owing to its solubilizing properties toluene easily penetrates the PMMA layer so that it can induce colorimetric transition of the polymeric PDA molecules (Supplementary Fig. S13).

These observations demonstrate that the protective layer approach can be used to fabricate a sensor system that displays the general mechanistic features of a keypad lock logic circuit[43] (Fig. 5c), in which UV irradiation, IPA and toluene inputs need to be applied in an ordered sequence in order to turn on the system (Table 1). Another salient feature of the protective layer strategy is that it enables the design of solvatochromic molecular AND logic gates. For example, the sensor system displayed in Fig. 6a functions only when water and THF are present as input signals.

Table 1 | Truth table for the solvatochromic keypad lock system upon varying the order of the input signals.

Input 1	Input 2	Input 3	Output
A	B	C	1
A	C	B	0
B	A	C	0
B	C	A	0
C	A	B	0
C	B	A	0

No colour change occurs if the sensor film is exposed to water alone because the hydrophobic PVC layer is not soluble in this solvent (Fig. 6b, top). In addition, even though it disrupts the top PVC layer, THF does not reach the sensor layer owing to poor penetrability through the poly(vinyl alcohol) (PVA) layer. As a result, no apparent colour change of the PCDA-derived PDA occurs (Fig. 6b, middle). However, a blue-to-red colorimetric transition does take place when the sensor system is exposed to aqueous THF (Fig. 6b, bottom) and, consequently, the sensor system serves as a two-input (H_2O and THF) solvatochromic AND logic gate (Fig. 6c, Table 2). It is significant to note that this sensor system functions well even when a 2 vol% H_2O-THF solution is used (Fig. 6d), suggesting that it can be applied as

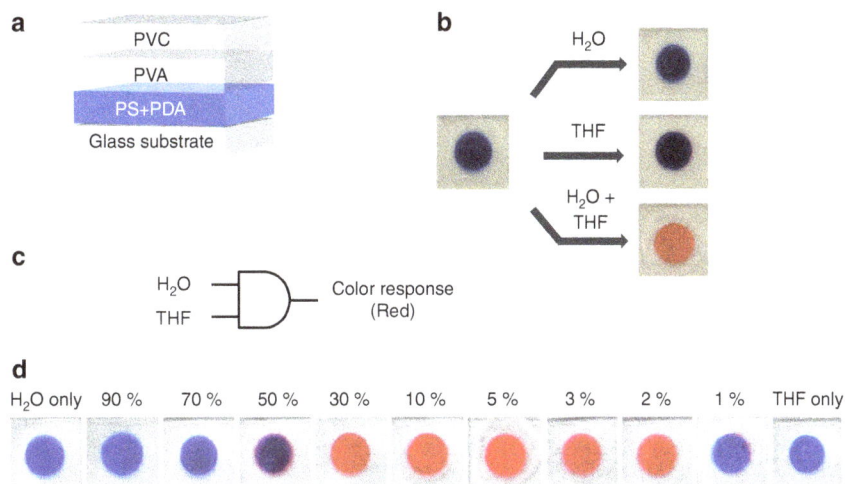

Figure 6 | Solvatochromic AND logic gates. (a) Schematic of a double-layer protected PCDA-derived PDA film for an AND logic gate. **(b)** Photographs of the sensor film after exposure to H_2O, THF and 20% aqueous THF. **(c)** A two-input solvatochromic AND logic gate. **(d)** Colorimetric response of the sensor film after exposure to various volume compositions of water in THF.

Table 2 | Truth table for the two-input (H_2O and THF) AND gate.

2 Input AND gate		
Input 1 (H_2O)	Input 2 (THF)	Output (Color response)
0	0	0
0	1	0
1	0	0
1	1	1

sensitive colorimetric system to determine the anhydrous nature of solvents if more water sensitive protective layers are employed.

Fluorescence turn-on sensor system. The flexible nature of the protective layer strategy enables it to be applied to the design of turn-on sensors that contain target and response guided sensor matrixes. For example, the formation of a fluorescamine-primary amine adduct can be utilized for the fabrication of a solvent responsive fluorescence turn-on sensor (Fig. 7). Fluorescamine is virtually nonfluorescent but it reacts with primary amines to generate strongly fluorescent adducts[44]. By taking advantage of this property, we have designed a novel layered fluorescence sensor system (Fig. 7a). A thin PS film containing fluorescamine, prepared on a glass substrate, was sequentially coated with a PVP layer and then a PMMA layer containing the primary amine, octadecylamine. Octadecylamine was selected for this purpose because its low volatility avoids losses that could occur during sensor preparation. In this device, the primary amine and fluorescamine are separated by the intervening PVP layer. We expected that disruption of the PVP layer by solvents would remove the barrier between two reactants and, thus, allow free diffusion and reaction between fluorescamine and amine. In Fig. 7b are shown vials containing fluorescamine and octadecylamine in the selected organic solvents toluene, chloroform, THF, ethyl acetate and methanol. Each vial emits blue adduct derived fluorescence upon irradiation with 365 nm UV light. To demonstrate the importance of the protective layer, a polymer film containing a PS-fluorescamine and PMMA-octadecylamine layer and not possessing an intervening PVP layer was prepared by using spin-coating. As can be seen in Fig. 7c, this film emits blue fluorescence under UV light (see also Supplementary Fig.

S14). In contrast, the corresponding polymer film containing the intervening PVP protective layer displays no fluorescence emission (see also Supplementary Fig. S14). Exposure to chloroform results in the generation of a circled fluorescence pattern owing to the coffee ring effect that is generally observed with a liquid drop (Fig. 7e)[45]. The results obtained in this study of the fluorescamine- and octadecylamine-based film demonstrate that unique solvatochromic turn-on fluorescence sensor systems can be devised by simply choosing appropriate stimuli responsive molecules and protective layers.

Discussion

PDAs undergo a distinct colour change (typically blue-to-red) when their arrayed p-orbitals are distorted under the influence of environmental perturbations and the observed change in colour is dependent on the degree of the distortion. As PDAs are formed from self-assembled diacetylene monomers, individual PDA chain interacts strongly with neighbouring PDAs, a feature that makes most PDAs insoluble in most solvents. When solvent molecules disrupt the densely packed PDA chains that cause an increase in the interchain distance, a colorimetric transition occurs as a result of the partial distortion of the p-orbital overlap. If individual PDA chains are separated by their individual interactions with a good solvent, the resulting severe distortion of the conjugated p-orbital array results in a blue-to-yellow colour transition. But the poor solubility of PDAs in most organic solvents leads to a lower degree of distortion and, consequently, a blue-to-red (or purple) colour change.

PDA supramolecules derived from a single chain diacetylene that contains a terminal carboxylic acid (for example, PCDA) display a solvent nonspecific colour change when they are exposed to common organic solvents. The study described above has demonstrated the viability and generality of a new solvatochromic sensor strategy. By taking advantage of the protective layer approach, we were able to devise unprecedented colorimetric and fluorescence turn-on type solvatochromic film sensors that carry out the demanding task of visual differentiation of solvents. The source of the selectivity of this type of sensor resides in a colorimetric transition of solvatochromic molecules that only occur when a target solvent disrupts a properly selected protecting layer. The PAA-protected PDA system clearly demonstrates that a tailor-made THF-specific colorimetric sensor can be fabricated. In

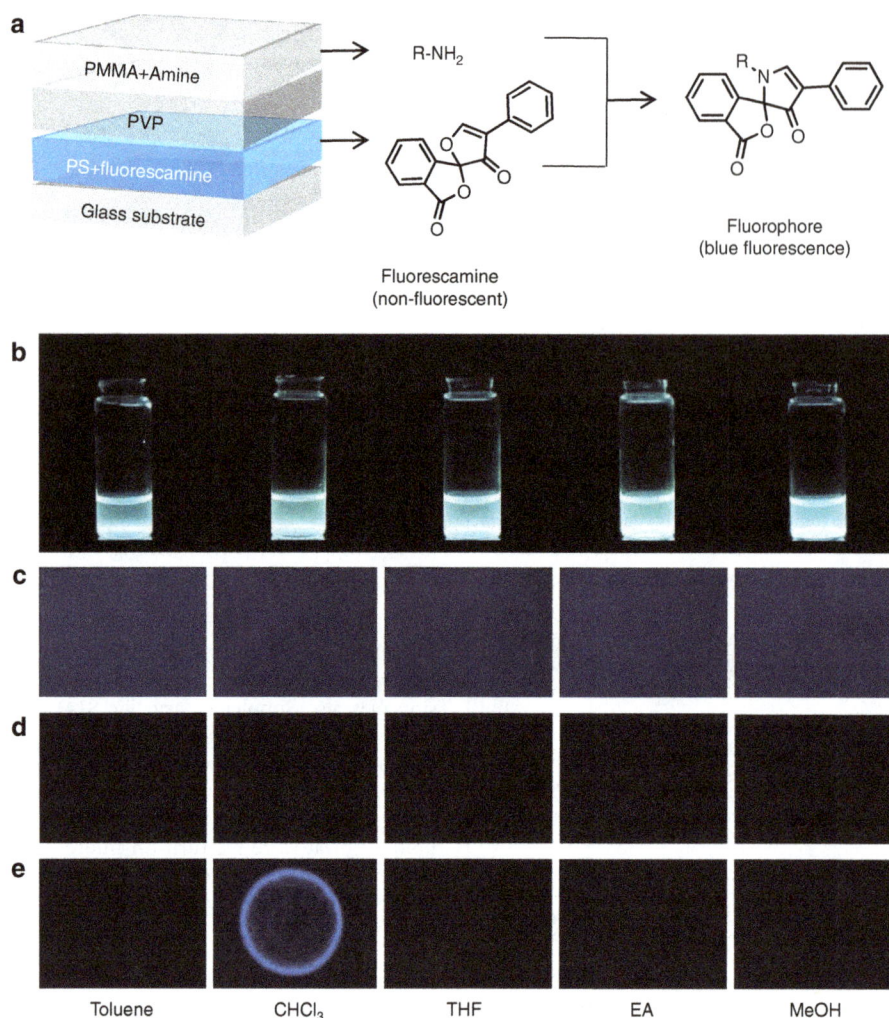

Figure 7 | A fluorescence turn-on solvatochromic sensor system. (**a**) Schematic representation of a fluorogenic sensor system based on fluorescamine-amine adduct formation. (**b**) Photographs of vials containing solutions of fluorescamine ($2\,mg\,ml^{-1}$) and amine ($2\,mg\,ml^{-1}$) under 365 nm UV light. (**c**) Photographs of unprotected films (without PVP layer) under 365 nm UV light. (**d**) Photographs of PVP-protected films under 365 nm UV light. (**e**) Photographs of PVP-protected films under 365 nm UV light after exposure to various solvents.

addition, by taking advantage of the protective layer approach, we were able to devise a sensor that carries out the demanding task of colorimtrically distinguishing between dichloromethane and chloroform. Facile fabrication of a sensor system that has solvent sequence selectivity as well as ready construction of a solvatochromic molecular AND logic gate are additional meritorious features of employing the protective layer-based sensor strategy. Finally, the flexibility of the protective layer approach was further demonstrated by constructing a system, which relies on a fluorescamine-amine adduct forming reaction, for solvent-selective fluorescence turn-on sensing. It is believed that the new strategy developed in this effort will find wide application in the design of tailor-made solvatochromic chemosensors.

Methods

Materials and instruments. 10,12-Pentacosadiynoic acid (PCDA) and TCDA were purchased from GFS Chemicals, Ohio, USA. PAA (Mw: $450,000\,g\,mol^{-1}$), PVP (Mw: $360,000\,g\,mol^{-1}$), PVC (Mw: $620,000\,g\,mol^{-1}$), PMMA (Mw: $120,000\,g\,mol^{-1}$), poly(ethylene oxide) (Mw: $200,000\,g\,mol^{-1}$) and poly(vinyl alcohol) (PVA, Mw: $89,000–98,000\,g\,mol^{-1}$), used as protective layers, were obtained from Aldrich Co. Fluorescamine and octadecylamine were purchased from Aldrich Co. Spectroscopic and HPLC grade solvents (Burdick and Jackson) were used for the solvatochromic studies. Film thicknesses were measured using an

alpha step instrument. Raman spectra were obtained using excitation at 785 nm laser and a Raman microscope (Kaiser Optical Systems).

Fabrication of a polymer-layer protected PDA film. A typical procedure for the preparation of a protected sensor film is as follows. A thin polymer film (*ca.* $1.0\,\mu m$) was prepared on a glass substrate by spin-coating a viscous polymer solution containing 10,12-pentacosadiynoic acid (PCDA, $40\,mg\,ml^{-1}$) and PS (Mw: $280,000\,g\,mol^{-1}$, 5 wt%). The polymer film was then irradiated with a hand-held UV lamp (254 nm, $1\,mW\,cm^{-2}$, 3 min) to induce photopolymerization of PCDA molecules. A readily available photomask, printed on a transparent polymer film, was used to generate PDAs only in the UV-exposed areas. On to the photoirradiated PS film was spin-coated a methanol solution of PAA (Mw: $450,000\,g\,mol^{-1}$, 8 wt%). The thickness of the PAA-protected film was found to be *ca.* $2.5\,\mu m$.

Solubility test of a polymer film in an organic solvent. A thin polymer film used as a protective layer was prepared on a glass substrate ($2.5 \times 2.5\,cm^2$) by spin-coating a polymer (PAA, PVC, PVP, PMMA and so on) solution. The polymer-coated glass substrate was incubated in the test solvent for 10 min and placed in a fume hood for 5 min. The weight of the glass substrate was determined before and after incubation in the solvent. Three independent measurements were made and the average value was used.

Measurement of colorimetric transition power. A thin PS film (*ca.* $1.0\,\mu m$ thickness) containing PCDA ($40\,mg\,ml^{-1}$) was prepared and UV irradiated (254 nm, $1\,mW\,cm^{-2}$, 3 min) to generate a blue coloured PDA film. The film was exposed to a test solvent and the red intensity values of 10 different spots on the solvent-treated film were extracted using the Adobe Photoshop program.

References

1. Benedetti, E., Kocsis, L. S. & Brummond, K. M. Synthesis and photophysical properties of a series of cyclopenta[*b*]naphthalene solvatochromic fluorophores. *J. Am. Chem. Soc.* **134**, 12418–12421 (2012).

2. Kucherk, O. A., Didier, P., Mély, Y. & Klymchenko, A. S. Fluorene analogues of prodan with superior fluorescence and solvatochromism. *J. Phys. Chem. Lett.* **1**, 616–620 (2010).

3. Giordano, L., Shvadchak, V. V., Fauerbach, J. A., Jares-Erijman, E. A. & Jovin, T. M. Highly solvatochromic 7-aryl-3-hydroxychromones. *J. Phys. Chem. Lett.* **3**, 1011–1016 (2012).

4. Achelle, S., Barsella, A., Baudequin, C., Caro, B. & Robin-le Guen, F. Synthesis and photophysical investigation of a series of push-pull arylvinyldiazine chromophores. *J. Org. Chem.* **77**, 4087–4096 (2012).

5. Chiu, M. *et al.* N,N′-Dicyanoquinone diimide-derived donor-acceptor chromophores: conformational analysis and optoelectronic properties. *Org. Lett.* **14**, 54–57 (2012).

6. Ooyama, Y. *et al.* Solvatochromism of novel donor-π-acceptor type pyridinium dyes in halogenated and non-halogenated solvents. *N. J. Chem.* **33**, 2311–2316 (2009).

7. Kucherak, O. A., Richert, L., Mély, Y. & Klymchenko, A. S. Dipolar 3-methoxychromones as bright and highly solvatochromic fluorescent dyes. *Phys. Chem. Chem. Phys.* **14**, 2292–2300 (2012).

8. Fakhari, M. A. & Rokita, S. E. A new solvatochromic fluorophore for exploring nonpolar environments created by biopolymers. *Chem. Commun.* **47**, 4222–4224 (2011).

9. Do, J., Huh, J. & Kim, E. Solvatochromic fluorescence of piperazine-modified bipyridazines for an organized solvent-sensitive film. *Langmuir* **25**, 9405–9412 (2009).

10. Li, H. & Jäkle, F. Donor-π-acceptor polymer with alternating triarylborane and triphenylamine moieties. *Macromol. Rapid Commun.* **31**, 915–920 (2010).

11. Lo, L. T.-S., Lai, S.-W., Yiu, S.-M. & Ko, C.-C. A new class of highly solvatochromic dicyano rhenate(I) diamine complexes – synthesis, photophysics and photocatalysis. *Chem. Commun.* **49**, 2311–2313 (2013).

12. Sevakumar, P. M., Nadella, S., Fröhlich, R., Albrecht, M. & Subramanian, P. S. A new class of solvatochromic material: geometrically unsaturated Ni (II) complexes. *Dyes Pigments* **95**, 563–571 (2012).

13. Lu, Z.-Z., Zhang, R., Li, Y.-Z., Guo, Z.-J. & Zheng, H.-G. Solvatochromic behavior of a nanotubular metal-organic framework for sensing small molecules. *J. Am. Chem. Soc.* **133**, 4172–4174 (2011).

14. Cui, J., Li, Y., Guo, Z. & Zheng, H. A porous metal-organic framework based on Zn_6O_2 clusters: chemical stability, gas adsorption properties and solvatochromic behavior. *Chem. Commun.* **49**, 555–557 (2013).

15. Takagi, S. *et al.* Unique solvatochromism of a membrane composed of a cationic porphyrin-clay complex. *Langmuir* **26**, 4639–4641 (2010).

16. Schreiter, K. *et al.* Novel periphery-functionalized solvatochromic nitrostilbenes as precursors for class II hybrid materials. *Chem. Mater.* **22**, 2720–2729 (2010).

17. de Silva, A. P. & Uchiyama, S. Molecular logic and computing. *Nat. Nanotechnol.* **2**, 399–410 (2007).

18. Bozdemir, O. A. *et al.* Selective manipulation of ICT and PET processes in styryl-bodipy derivatives: applications in molecular logic and fluorescence sensing of metal ions. *J. Am. Chem. Soc.* **132**, 8029–8036 (2010).

19. de Ruiter, G. & van der Boom, M. E. Surface-confined assemblies and polymers for molecular logic. *Acc. Chem. Res.* **44**, 563–573 (2011).

20. Andréasson, J. *et al.* All-photonic multifunctional molecular logic device. *J. Am. Chem. Soc.* **133**, 11641–11648 (2011).

21. Kumar, M., Kumar, N. & Bhalla, V. A naphthalimide based chemosensor for Zn^{2+}, pyrophosphate and H_2O_2: sequential logic operations at the molecular level. *Chem. Commun.* **49**, 877–879 (2013).

22. Chen, S., Yang, Y., Tian, H. & Zhu, W. Multi-addressable photochromic terarylene containing benzo[*b*]thiophene-1,1-dioxide unit as ethane bridge: multifunctional molecular logic gates on unimolecular platform. *J. Mater. Chem* **22**, 5486–5494 (2012).

23. Park, K. S., Seo, M. W., Jung, C., Lee, J. Y. & Park, H. G. Simple and universal platform for logic gate operations based on molecular beacon probes. *Small* **14**, 2203–2212 (2012).

24. Elstner, M., Weisshart, K., Müllen, K. & Schiller, A. Molecular logic with a saccharide probe on the few-molecules level. *J. Am. Chem. Soc.* **134**, 8098–8100 (2012).

25. Wegner, G. Topochemical polymerization of monomers with conjugated triple bonds. *Makromol. Chem.* **154**, 35–48 (1972).

26. Sun, A., Lauher, J. W. & Goroff, N. S. Preparation of poly(diiododiacetylene), an ordered conjugated polymer of carbon and iodine. *Science* **312**, 1030–1034 (2006).

27. Lu, Y. *et al.* Self-assembly of mesoscopically ordered chromatic polydiacetylene/silica nanocomposites. *Nature* **410**, 913–917 (2001).

28. Yarimaga, O., Jaworski, J., Yoon, B. & Kim, J.-M. Polydiacetylenes: supramolecular smart materials with a structural hierarchy for sensing, imaging and display application. *Chem. Commun.* **48**, 2469–2485 (2012).

29. Yoon, B. *et al.* Inkjet printing of conjugated polymer precursors on paper substrates for colorimetric sensing and flexible electrothermochromic display. *Adv. Mater.* **23**, 5492–5497 (2011).

30. Peng, H. *et al.* Electrochromatic carbon nanotube/polydiacetylene nanocomposite fibres. *Nat. Nanotechnol* **4**, 738–741 (2009).

31. Chen, X. *et al.* Magnetochromatic polydiacetylene by incorporation of Fe_3O_4 nanoparticles. *Angew. Chem. Int. Ed.* **50**, 5486–5489 (2011).

32. Chen, X., Zhou, G., Peng, X. & Yoon, J. Biosensors and chemosensors based on the optical responses of polydiacetylenes. *Chem. Soc. Rev.* **41**, 4610–4630 (2012).

33. Hsu, T. J., Fowler, F. W. & Lauher, J. W. Preparation and structure of a tubular addition polymer: a true synthetic nanotube. *J. Am. Chem. Soc.* **134**, 142–145 (2012).

34. Lee, J. & Kim, J. Multiphasic sensory alginate particle having polydiacetylene liposome for selective and more sensitive multitargeting detection. *Chem. Mater.* **24**, 2817–2822 (2012).

35. Bai, F., Sun, Z., Lu, P. & Fan, H. Smart polydiacetylene nanowire paper with tunable colorimetric response. *J. Mater. Chem.* **22**, 14839–14842 (2012).

36. Shimogaki, T. & Matsumoto, A. Structural and chromatic changes of host polydiacetylene crystals during interaction with guest alkylamines. *Macromolecules* **44**, 3323–3327 (2011).

37. Schott, M. The colors of polydiacetylenes: a commentary. *J. Phys. Chem. B* **110**, 15864–15868 (2006).

38. Pang, J. *et al.* Thermochromatism and structural evolution of metastable polydiacetylene crystals. *J. Phys. Chem. B* **110**, 7221–7225 (2006).

39. Bloor, D. Dissolution and spectroscopic properties of the polydiacetylene poly(10,12-docosadiyne-1,12-diol-bisethylurethane). *Macromol. Chem. Phys.* **201**, 1410–1423 (2001).

40. Baughman, R. H. & Chance, R. R. Comments on the optical properties of fully conjugated polymers: analogy between polyenes and polydiacetylenes. *J. Polym. Sci. Polym. Phys. Ed.* **14**, 2037–2045 (1976).

41. Ahn, D. J. & Kim, J.-M. Fluorogenic polydiacetylene supramolecules: Immobilization, micropatterning, and application to label-free chemosensors. *Acc. Chem. Res.* **41**, 805–816 (2008).

42. Giorgetti, E. *et al.* UV polymerization of self-assembled monolayers of a novel diacetylene on silver: a spectroscopic analysis by surface plasmon resonance and surface enhanced Raman scattering. *Langmuir* **22**, 1129–1134 (2006).

43. Strack, G., Ornatska, M., Pita, M. & Katz, E. Biocomputing security system: concatenated enzyme-based logic gates operating as a biomolecular keypad lock. *J. Am. Chem. Soc.* **130**, 4234–4235 (2008).

44. Griesser, T., Kuhlmann, J.-C., Wieser, M., kern, W. & Trimmel, G. UV-induced modulation of the refractive index and the surface properties of photoreactive polymers bearing N-phenylamide groups. *Macromolecule* **42**, 725–731 (2009).

45. Cui, L. *et al.* Suppression of the coffee ring effect by hydrosoluble polymer additives. *ACS Appl. Mater. Interfaces* **4**, 2775–2780 (2012).

Acknowledgements

We thank the National Research Foundation of Korea (NRF) for financial support through Basic Science Research Program (20120006251 and 2012R1A6A1029029), Nano-Material Technology Development Program (2012M3A7B4035286) and Centre for Next Generation Dye-Sensitized Solar Cells (2013004800).

Author contributions

J.L. designed and carried out the experiments and interpreted the data. H.T.C. developed the fluorescence turn-on sensor system. H.A. and J.S. measured polymer film thicknesses using an alpha step instrument. S.A. developed the solvatochromic molecular AND logic gate. J.-M.K. interpreted the data and wrote the manuscript.

Additional information

15

Nanoscale visualization of redox activity at lithium-ion battery cathodes

Yasufumi Takahashi[1,2], Akichika Kumatani[1], Hirokazu Munakata[3], Hirotaka Inomata[2], Komachi Ito[2], Kosuke Ino[2], Hitoshi Shiku[2], Patrick R. Unwin[4], Yuri E. Korchev[5], Kiyoshi Kanamura[3] & Tomokazu Matsue[1,2]

Intercalation and deintercalation of lithium ions at electrode surfaces are central to the operation of lithium-ion batteries. Yet, on the most important composite cathode surfaces, this is a rather complex process involving spatially heterogeneous reactions that have proved difficult to resolve with existing techniques. Here we report a scanning electrochemical cell microscope based approach to define a mobile electrochemical cell that is used to quantitatively visualize electrochemical phenomena at the battery cathode material $LiFePO_4$, with resolution of $\sim 100\,nm$. The technique measures electrode topography and different electrochemical properties simultaneously, and the information can be combined with complementary microscopic techniques to reveal new perspectives on structure and activity. These electrodes exhibit highly spatially heterogeneous electrochemistry at the nanoscale, both within secondary particles and at individual primary nanoparticles, which is highly dependent on the local structure and composition.

[1] WPI-Advanced Institute for Materials Research, Tohoku University, Sendai 980-8577, Japan. [2] Graduate School of Environmental Studies, Tohoku University, Sendai 980-8579, Japan. [3] Graduate School of Urban Environmental Sciences, Tokyo Metropolitan University, Tokyo 192-0397, Japan. [4] Department of Chemistry, University of Warwick, Coventry CV4 7AL, UK. [5] Division of Medicine, Imperial College London, London W12 0NN, UK. Correspondence and requests for materials should be addressed to Y.T. (email: takahashi@bioinfo.che.tohoku.ac.jp).

Lithium-ion batteries have a remarkably wide range of applications from portable electronics to hybrid electric vehicles, where high energy, power and long-term cycling stability are in demand[1]. However, the rational development and improvement of battery technologies requires a better view of fundamental properties of redox activity at battery electrode surfaces. Although several techniques have been developed to visualize physicochemical processes in batteries and battery materials[2-7], mapping redox activity of battery electrodes remains challenging owing to a lack of effective analytical tools, not least because battery electrodes are typically very rough on the microscale but show activity variations on the nanoscale. To address this issue, in this paper, we describe a powerful approach for visualizing redox activity at complex composite electrodes at high spatial resolution. Although electrochemical imaging via scanning electrochemical microscopy (SECM) can achieve high resolution in certain applications, such measurements are rare and tend to be applied to flat (topographically uninteresting) surfaces if activity is to be measured quantitatively[8,9].

Hitherto, microscopic understanding of the redox activity of the battery electrodes has relied on just a few techniques. A microelectrode–particle contact method has been used to evaluate the electrochemical processes at single active cathode electrode particles[10-12], but the process was either measured for the whole grain (relatively large secondary particle) or, in special circumstances, information was obtained at a resolution of several microns (or larger) by using in situ optical or Raman visualization techniques[12]. At higher resolution, atomic force microscopy (AFM) and scanning tunneling microscopy (STM) have been used to visualize dynamic structural changes during charge/discharge, for instance, solid electrolyte interface formation[13,14]. More specialized scanned probe microscopes have proven particularly powerful for in situ high-resolution dynamic mapping, as exemplified by electrochemical strain microscopy[15-18] and scanning ion conductance microscopy[19] studies. For local dynamic electrochemical measurements, SECCM[20-25] and related techniques[26] appear particularly promising for direct, local electrochemical measurements, and we thus develop this platform herein to identify key processes underpinning the performance of battery electrodes.

As a model system, but one that is also of practical importance, we study lithium iron phosphate (LiFePO$_4$) cathode materials, which have found application in both aqueous[27-29] and non-aqueous[10,30] electrolyte lithium-ion batteries. LiFePO$_4$ has attracted particular interest owing to its low cost, excellent safety profile and for environmental considerations. However, a drawback is that LiFePO$_4$ suffers from low electronic conductivity[31], requiring that it is blended with a conductive material such as acetylene black (AB). This results in a complex composite material for which the local structure–function properties are largely unknown, and hence is considered as an ideal system to demonstrate the capabilities of SECCM imaging and multi-microscopy[32].

Here, we use single-channel SECCM to map spatial heterogeneities in the electrochemistry of LiFePO$_4$ electrodes, and to link these to local topography and surface chemistry using complementary imaging techniques. We attain a spatial resolution for electrochemical activity mapping that is more than two orders of magnitude better than the best resolution from previous (recent) electrochemical imaging studies of such materials[33]. We are also able to investigate the behaviour of individual pure LiFePO$_4$ primary nanoparticles for the first time. This enables us to identify key factors controlling electrochemistry and (de)intercalation, from elementary nanomaterials to whole device electrodes.

Results

Localized electrochemical measurements. Figure 1a is a schematic of single-channel SECCM that allows us to visualize

and measure various electrochemical properties simultaneously with electrode topography. The principles and operation of the technique are similar to those outlined elsewhere[26], but we use a 50-nm radius nanopipette in a hopping mode to obtain high spatial resolution electrochemical images, even on a very rough electrode surface. Moreover, we explore new modes for meniscus-based imaging, such as zero-current (essentially open circuit) cell potential mapping and local charge/discharge modification.

The technique uses a moveable nanopipette probe (typical radius of 50 nm; Supplementary Note 1 and Supplementary Fig. 1) usually containing 3 M LiCl electrolyte solution and an Ag/AgCl quasi-reference counter electrode (QRCE). The nanopipette was brought into close contact with a sample electrode surface via a liquid meniscus at the nanopipette end. For initial experiments, the sample LiFePO$_4$ composite electrode was connected to a current amplifier and a bias voltage of $+0.65$ V versus Ag/AgCl QRCE was applied with the QRCE connected to ground. The cell current versus probe (meniscus) to sample separation is shown in Fig. 1b for three different nanopipettes approached towards the surface at 20 nm ms^{-1}. When the meniscus just made contact with the surface, a small anodic current flowed and a set point of 2 pA triggered the nanopipette to stop from further approaching. During imaging, the tip height position at this point was recorded, enabling the topography of the surface to be tracked.

Localization of the electrochemical cell to the footprint of the meniscus allowed the LiFePO$_4$ charge/discharge reaction to be probed at the nanoscale without the LiFePO$_4$ electrode being completely immersed in solution. The redox reaction of interest is:

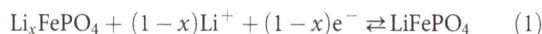

$$Li_xFePO_4 + (1-x)Li^+ + (1-x)e^- \rightleftharpoons LiFePO_4 \qquad (1)$$

with the process left to right corresponding to discharge (Li$^+$ intercalation), which is a main focus of the studies herein. Major advantages of our approach are that working electrode/electrolyte interface can be created at defined locations via a moveable nanopipette, and the nanoscale footprint of the electrochemical

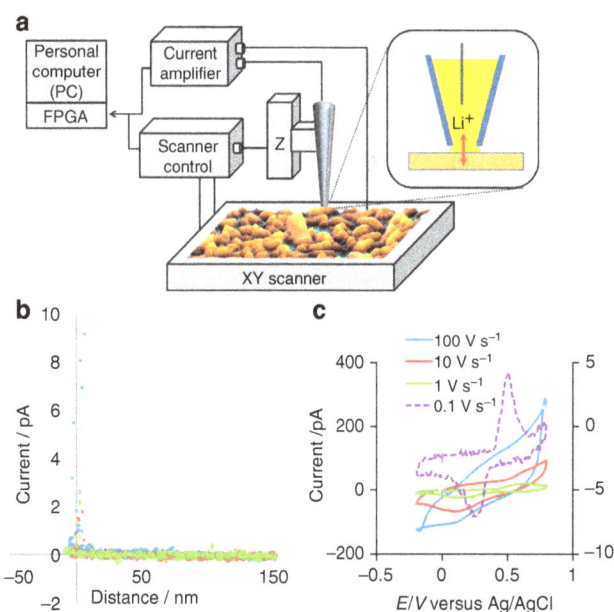

Figure 1 | Experimental overview and local electrochemical measurements. (**a**) Schematic of the SECCM. (**b**) Approach characteristics (current–distance) for three different nanopipettes (3 M LiCl) moved towards a LiFePO$_4$ electrode with a bias voltage of $+0.65$ V versus Ag/AgCl QRCE. (**c**) CVs obtained with nanopipette meniscus contact (3 M LiCl) at different scan rates on a LiFePO$_4$ electrode. CV with 0.1 V s^{-1} uses the right axis and the other CVs use the left axis.

cell diminishes ohmic effects and capacitive currents during electrochemical measurements and imaging. Figure 1c shows examples of cyclic voltammograms (CVs) on a LiFePO$_4$ thin-film electrode. The working electrode potential was linearly ramped up from 0 to 0.8 V, returned to -0.2 V and then scanned back to 0 V versus Ag/AgCl (scan rates in the range 0.1 to 100 V s^{-1}). At 0.1 V s^{-1}, there is a clear anodic peak at ca. 0.49 V (at 0.1 V s^{-1}) corresponding to the oxidation of Fe^{II} to Fe^{III} with deintercalation of Li$^+$. Upon reversing the potential, the reverse process (Fe^{III} reduction to Fe^{II}, with Li$^+$ intercalation) occurs at ca. 0.27 V. The peak-to-peak separation is similar to that measured at the same timescale on LiFePO$_4$ microelectrodes[34], confirming the validity of our new approach, which is most powerfully used for imaging as discussed next. The (de)intercalation process is quasi-reversible[30], as evidenced by the peak-to-peak separation increasing further with increasing scan rate in the range 1–100 V s^{-1}. The upper value is the fastest voltammetric timescale applied to LiFePO$_4$ redox activity and (de)intercalation by two orders of magnitude[34], and it is clear that the transformation is reasonably facile. Note that when the solution in the nanopipette was replaced with one containing 3.0 M KCl, negligible redox activity and (de)intercalation was observed (Supplementary Note 2 and Supplementary Fig. 2).

Electrochemical imaging of composite LiFePO$_4$ electrodes. In this section, we highlight major capabilities of SECCM in visualizing redox activity at high spatial resolution and linking this to topography and surface chemistry at the nanoscale. We also demonstrate the measurement of charge–discharge curves at desired nanoscale locations and carry out surface potential measurements to allow 'chemical' mapping (identification of LiFePO$_4$ domains). We further show how SECCM allows the charge state of a complex electrode to be changed and visualized locally, opening up the manipulation of the electrode surfaces and new measurement possibilities.

First, a secondary electroactive particle ($\varphi = 3\,\mu$m) of LiFePO$_4$, attached with poly(vinylidenedifluoride) (PVdF) binder and AB in a composite electrode, was measured. Figure 2b shows typical topography and interfacial current images obtained simultaneously by SECCM, with the electrode potential at $+0.65$ V (just into the Li$^+$ deintercalation process; Fig. 1c). Notably, a high current response was observed at the location of the mound in the region probed. This topographical feature can reasonably be assigned as a secondary particle based on Scanning Electron Microscope (SEM)–energy dispersive X-ray spectroscopy (EDS) images of a typical electrode area and the fact that there is excellent correspondence of the current and particle position maps, with little current flow elsewhere. Importantly, however, although the secondary particle (comprising primary particles of different orientation) can be identified, it is also clear that within the particle the redox activity is highly heterogeneous. We consider the origin of this further below, when examining individual primary nanoparticles. As shown in Supplementary Note 3 and Supplementary Fig. 3, rates (currents) of the intercalation and deintercalation processes are strongly spatially correlated across a LiFePO$_4$ electrode surface.

After imaging, the nanopipette was positioned to make meniscus contact at locations corresponding to the particle (blue arrow in Fig. 2a) and away from the particle (red arrow) to enable localized CV measurements (Fig. 2b). Significant redox activity and (de)intercalation was observed on the particle, confirming the correlation of the current with the position of LiFePO$_4$. Away from the particle, only a very weak current–voltage response was observed, suggesting that this was an area of AB.

SECCM also allowed charge/discharge characteristics to be determined on secondary particles within a composite electrode

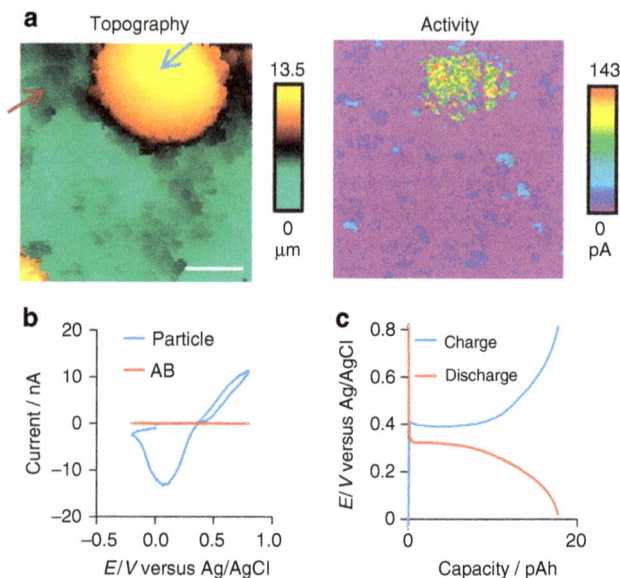

Figure 2 | Topography and current activity of a LiFePO$_4$ electrode. (a) Simultaneous SECCM topography (left) and current (right) images. Scan ranges are $20 \times 20\,\mu$m. The substrate potential was $+0.65$ V versus Ag/AgCl QRCE (Li$^+$ deintercalation; scale bar, 5 μm). **(b)** CVs at different points on a LiFePO$_4$ electrode surface, corresponding to the blue and red arrow of **b**. Scan rate is 0.1 V s^{-1}. **(c)** Local charge (deintercalation) and discharge (intercalation) characteristics applying current magnitudes of 200 pA in each case via SECCM.

(Fig. 2c). Here a current magnitude of 200 pA was applied via nanoscale meniscus contact, and the potential–time curves for charge and discharge were recorded. There is excellent correspondence of the morphology of two curves, from which the capacity was estimated to be $\sim$15 pA h. The reacted volume is $\sim$25 μm^3 using theoretical volumetric capacity (612 mA h cm^{-3}), corresponding to a spherical radius of 1.8 μm. This is about the same size of a small secondary particle, such as that in Fig. 2c. Thus, this analysis indicates that individual secondary particles within a LiFePO$_4$ composite electrode can be targeted and characterized by local electrolyte contact, without influence from neighbouring particles.

Potential imaging of composite LiFePO$_4$ electrodes. We now demonstrate SECCM as a powerful technique for local potential mapping. The condition of this measurement was zero current in an electrochemical cell (applied after landing a meniscus at each point on the surface investigated), so that a potential close to an open circuit value was measured, as described in the experimental section. Figure 3a shows an example of cell potential mapping of a LiFePO$_4$ film with AB and PVdF binder. A highly spatially inhomogeneous potential can be seen. Regions where the value is ca. $+0.4 \sim 0.6$ V can be assigned to LiFePO$_4$, based on the redox potential for Li$^+$ (de)intercalation (for example, Fig. 1c). In other regions, it was difficult to achieve a steady potential. We attribute these to AB regions that would show behaviour close to that of an ideally polarizable electrode[35]. The SECCM data in this two-component system indicate that potential mapping is powerful for highlighting different types of material in a composite electrode.

We were also able to change the local redox state (Li$^+$ concentration) in the LiFePO$_4$ by inducing the deintercalation reaction using nanopipette meniscus contact and by imaging the resulting cell potential distribution. For these measurements, the nanopipette meniscus was positioned at the centre of the sample

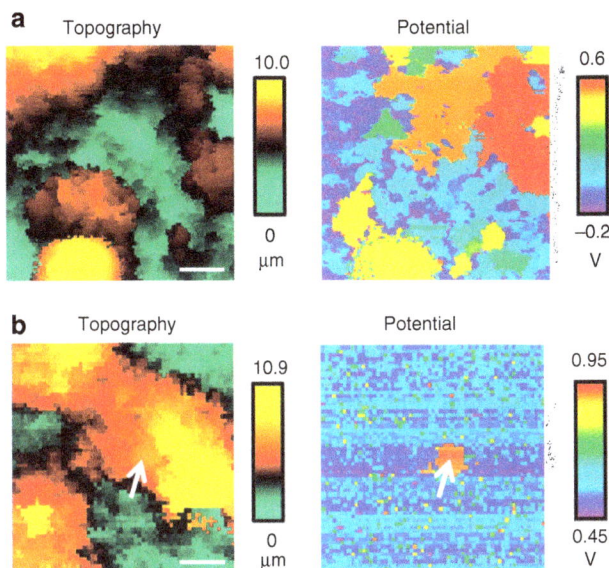

Figure 3 | Topography and potential images of a LiFePO₄ electrode.
(**a**) Simultaneous SECCM topography (left) and potential (right) images
(15 × 15 μm; scale bar, 3 μm). (**b**) Simultaneous SECCM topography (left)
and surface potential (right) images after charging the electrode (Li⁺
deintercalation) locally at a point on the electrode (until the potential
reached + 0.8 V; scale bar, 5 μm). Images are 25 × 25 μm. The white arrow
shows the charging point on the topography and potential map.

surface and localized charging was performed (+ 200 pA) until
the voltage reached 0.8 V (taking ca. 1 h). The cell potential was
then mapped, yielding the image in Fig. 3b. At the centre of the
LiFePO₄ (white arrow), there is a higher potential (∼ 0.8 V) at the
position where the oxidation of Fe^{II} to Fe^{III} (Li⁺ deintercalation)
occurred, highlighting that the local redox state and composition
in LiFePO₄ can be manipulated with high control. This local
perturbation approach might be usable in the future to probe
solid state diffusion coefficients, among other applications.

Local galvanostatic charge/discharge property mapping. To
further demonstrate charge/discharge imaging and to highlight
unequivocally how this linked to the chemical character of a
composite electrode, localized charging was performed
(+ 5.0 pA) for 600 ms (total charge 3 fC) at each measurement
point, with the potential–time characteristic measured simultaneously (10 ms time resolution). The 3 fC charge corresponds to a
spherical radius of ca. 57 nm based on the LiFePO₄ theoretical
volumetric capacity (612 mA h cm⁻³), ensuring that this measurement was surface sensitive. We further visualized the same
region of the LiFePO₄ thin film by SEM–EDS to allow the
unambiguous identification of LiFePO₄ particles from the strong
oxygen signal (Fig. 4a). To enable different microscopies to be
applied to the same region of the sample, we focused on a region
near the edge of the LiFePO₄ thin film; the unique microscale
topography of the sample allowed distinct features to be used for
co-location of the different microscopy techniques. Particles
evidently appear as small (micron scale) mounds in the surface
(as proposed for the analysis of Fig. 2a); compare the morphologies obtained by different techniques in Fig. 4a,b.

By acquiring potential–time data at each pixel, the resulting
data can be the most powerful represented as a potential–time
movie (Supplementary Movie 1), with several snapshots shown in
Fig. 4c. At the secondary LiFePO₄ particle, there is good
correlation between the SECCM potential and LiFePO₄ particle,
which is identified by oxygen (from phosphate) in the EDS image.

At the centre of the secondary LiFePO₄ particle (red arrow), the
potential is essentially fixed at 0.40 V on this timescale (consistent
with the data in Fig. 2c). This is also seen clearly in the potential–
time plot in Fig. 4d from a pixel indicated by the red arrows in
Fig. 4a,c. In contrast, away from the secondary particle, which can
reasonably be considered as an AB area (black arrow), the
potential did not reach steady state because of its polarizable
property (Fig. 4d).

To illustrate the relationship between potentiostatic redox
activity and galvanostatic charge activity, we carried out both
types of measurements in the same area of the LiFePO₄
composite electrode (Supplementary Note 4 and Supplementary
Fig. 4). Evidently, the stable potential-response area corresponds
with the area of high current, which in turn has been shown to be
a LiFePO₄ particle from topographical imaging and EDS
mapping. These results mean that we are able to link the surface
component (LiFePO₄ secondary particle or AB) and function
(charge/discharge) unequivocally and unambiguously via a multi-
microscopy approach in which SECCM is a central technique.

Single LiFePO₄ nanoparticle imaging. Finally, we investigated
individual primary LiFePO₄ nanoparticles dispersed on a Pt
substrate (Fig. 5a). This presented a heterogeneous substrate at
two levels: first, the nanopipette meniscus mainly landed on a Pt
substrate and only a small charging current was measured; second, even when the nanopipette located an LiFePO₄ nanoparticle,
Li⁺ deintercalation current (electrode potential + 0.65 V) would
depend on the orientation of the nanoparticle. Thus, a distinct
inhomogeneity of the current, linked to the topography (nanoparticle height), measured by SECCM in hopping mode (Fig. 1b),
was observed (Fig. 5b). The topographical image reveals the
nanoparticle size distribution (Fig. 5c). The lowest currents
(< 10 pA), which comprise ca. 98% of the sampled surface, are
mainly owing to the Pt substrate, whereas the broad current
distribution at higher values can be attributed to the LiFePO₄
nanoparticles that are reasonably expected to have different
crystal orientations[36] and phases[37]. This result ties closely to the
activity images of secondary particles, discussed above (for
example, Fig. 2b), in which a broad range of reaction rates within
a particle was found, which appears to be owing to different
orientations of the primary nanoparticles, from which the
secondary particles are composed. In the future, it should be
possible to visualize any influence of grain orientation on redox
activity and (de)intercalation kinetics, enabling optimum crystal
structures and orientations to be identified.

Discussion
In this paper, we have proposed a novel single-channel
nanopipette SECCM method for the detailed characterization of
Lithium-ion battery cathode materials. We have been able to map
localized redox activity on LiFePO₄ composite electrode surfaces
at the nanoscale for the first time, identifying significant
variations in reaction rates that depend on local composition.
Furthermore, local charging/discharging at LiFePO₄ particles in
composite electrodes reveals that such processes are largely
limited to the particle where the perturbation is made. This
provides an opportunity to probe single particles and agglomer-
ates *in situ* in complex composite materials, and to develop an
understanding of how structure and function are related in
complex composite battery electrodes. Importantly, where
comparisons can be made with previous measurements, good
agreement has been found, thereby giving confidence in the new
insights obtained from this new local technique.

We have highlighted the versatility of SECCM by mapping the
open circuit potential of LiFePO₄ electrodes at the nanoscale and

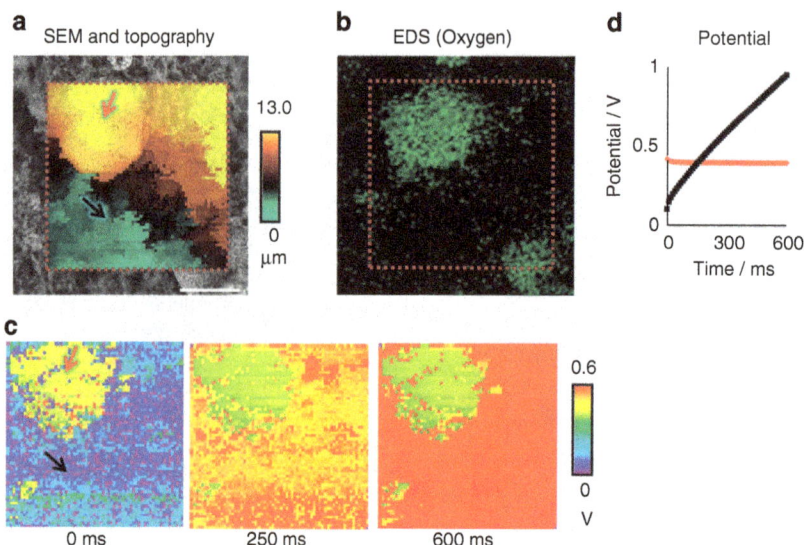

Figure 4 | Galvanostatic charge property mapping. (**a**) SEM and topography overlap image of a LiFePO$_4$ electrode (scale bar, 10 μm). (**b**) EDS (oxygen) image of the same area. (**c**) Potential images at times of 0, 250 and 600 ms (30 × 30 μm) during galvanostatic charging at +5.0 pA. (**d**) Galvanostatic potential–time curves at pixels on (red) and off (black) a secondary LiFePO$_4$ particle.

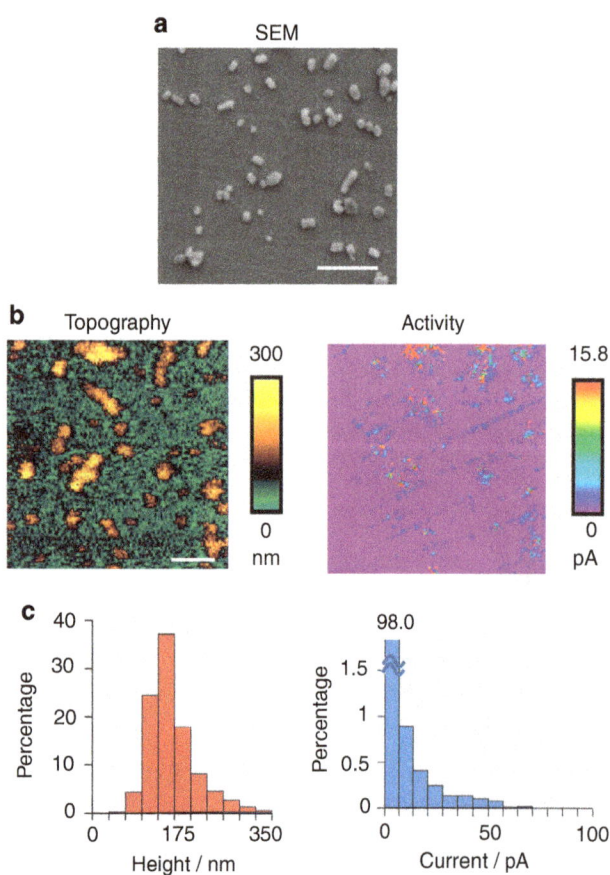

Figure 5 | Characterizing the activity and topography of individual LiFePO$_4$ nanoparticles. (**a**) SEM image of LiFePO$_4$ nanoparticles on a Pt collector electrode (scale bar, 1 μm). (**b**) Simultaneous SECCM topography (left) and current (right) images. Scan range is 5 × 5 μm (scale bar, 1 μm). The substrate potential was 0.65 V versus Ag/AgCl QRCE. (**c**) Data from the SECCM images of height (left), equivalent to particle size, and current (right), equivalent to deintercalation rate, plotted as histograms.

inducing charge/discharge processes locally. We have further been able to chemically map LiFePO$_4$ electrodes (identifying LiFePO$_4$ and AB domains from the cell potential response) and modify nanoscopic portions of macroscopic electrodes, opening up new measurement possibilities. The correlation of potential maps and current activity images with the underlying chemical composition of LiFePO$_4$ composite surfaces highlights the considerable strength of multi-microscopy measurements in understanding the structure and function of complex materials at the nanoscale. Finally, to further understand LiFePO$_4$ activity, we have been able to target primary nanoparticles and determine the distribution of electrochemical activity of such particles. A broad range of activity has been observed that links to the wide range of electrochemical fluxes seen across secondary particles, which comprises an agglomeration of such primary nanoparticles. This part of the study provides a foundation for future work, where it should be possible to determine how the orientation of crystallites influences electrochemistry and ion fluxes.

Methods

Preparation of LiFePO$_4$ thin-film electrodes. Both primary and secondary particles of LiFePO$_4$ were examined in this study. The procedure for synthesizing LiFePO$_4$ particles was described previously[38]. Primary particles were nanocrystals with a characteristic dimension of 100–200 nm that were non-carbon coated and dispersed onto a Pt current collector, which served as a specimen electrode. Test electrodes to examine secondary LiFePO$_4$ particles were prepared by a conventional method[30] in which a slurry of LiFePO$_4$, AB (conductive agent) and PVdF (binder) was coated on an Al current collector using a doctor blade. N-methylpyrrolidone was used as a solvent for the slurry. The mixing ratio of LiFePO$_4$:AB:PVdF was 92:4:4. The resulting secondary particle size was typically in the range of ∼1–5 μm.

Instruments. The design of the instrument was similar to related electrochemical probe microscopes as previously described[39–42]. The current was measured by a dual-channel MultiClamp700B patch-clamp amplifier (Axon Instruments). The current signal during imaging was typically filtered using a low-pass filter at 1 kHz, except for CV at scan rates >1 V s^{-1}, where a 10-kHz low-pass filter was used. The data were digitized and analysed with continuous data acquisition hardware and software (Axon Digidata 1322A, Axon Instruments). The relative sample and probe position was precisely controlled by mounting the sample on an XY piezoelectric translation stage and the nanopipette on a Z stage (Nanocontrol, 621.2CL and 621.ZCL), both controlled by an amplifier module (Nanocontrol,

NC3301). The system was controlled and data were acquired using LabVIEW (National Instruments). As well as being enclosed in a Faraday cage, the instrument was also enclosed in an acoustic isolation box (VIC International, VSD BM-1). The vertical Z positioning of the nanopipette and the movement of the sample in the XY plane were controlled (and recorded) by a Field Programmable Gate Array (FPGA) board (PCIe-7841R, National Instruments). SEM-EDS analysis was carried out with a Field Emission-Scanning Electron Microscope (FE-SEM) (JSM-7800 F, JEOL) and Energy Dispersive X-ray Spectrometer (Oxford Instruments, X-Max Silicon Drift Detector).

Scanning protocol. The following procedure was used to bring the nanopipette towards the sample surface so that the liquid meniscus just made contact (without contact from the nanopipette itself) at a series of predefined positions, with an electrochemical measurement at each point. First, the nanopipette (meniscus) was withdrawn from its existing position by a specified distance, typically 4.0 μm. Next, the vertical position of the probe was maintained for 20 ms, while the nanopositioning stage moved the specimen to a new imaging point in the xy plane. Then, the nanopipette was lowered at constant fall rate of 20 nm ms^{-1} while monitoring the current. Immediately after detecting a current (2 pA threshold) by forming the electrical contact between the nanopipette and the sample through the nanopipette meniscus, the approach was stopped and the vertical position of the nanopipette was saved along with the x,y co-ordinate to form a topography map. Twenty milliseconds then elapsed before the Li$^+$ (de)intercalation current was measured (50 μs duration, averaging 10 samples at a sampling frequency of 200 kHz). After the electrochemical measurement, the nanopipette was quickly withdrawn by the specified distance to start a new measurement cycle. In this way, simultaneous pictures of topography and redox activity were built up. Local cell potential mapping was performed by controlling the applied voltage so as to maintain the current from the sample working electrode to be zero by using the function of MultiClamp700B constant current mode.

Fabrication of nanopipettes. Nanopipettes were fabricated by pulling a borosilicate glass pipette (inner diameter = 1.00 mm and outer diameter = 0.78 mm; GC150F-10, Harvard Apparatus) using a laser puller (Sutter Instruments, model P-2000), with a two-step protocol. For the initial step, the parameters were heat 350, filament 3, velocity 21 and delay 200. The second step, they were heat 350, filament 2, velocity 26, delay 160 and pull 250.

References

1. Tarascon, J. M. & Armand, M. Issues and challenges facing rechargeable lithium batteries. *Nature* **414**, 359–367 (2001).
2. Huang, R. & Ikuhara, Y. STEM characterization for lithium-ion battery cathode materials. *Curr. Opin. Solid. State Mater. Sci.* **16**, 31–38 (2012).
3. Chueh, W. C. *et al.* Intercalation pathway in many-particle LiFePO$_4$ electrode revealed by nanoscale state-of-charge mapping. *Nano Lett.* **13**, 866–872 (2013).
4. Baddour-Hadjean, R. & Pereira-Ramos, J. P. Raman microspectrometry applied to the study of electrode materials for lithium batteries. *Chem. Rev.* **110**, 1278–1319 (2010).
5. Nishimura, S. *et al.* Experimental visualization of lithium diffusion in LixFePO$_4$. *Nat. Mater.* **7**, 707–711 (2008).
6. Gu, L. *et al.* Direct observation of lithium staging in partially delithiated LiFePO$_4$ at atomic resolution. *J. Am. Chem. Soc.* **133**, 4661–4663 (2011).
7. Clark, J. M., Nishimura, S., Yamada, A. & Islam, M. S. High-Voltage pyrophosphate cathode: insights into local structure and lithium-diffusion pathways. *Angew. Chem. Int. Ed.* **51**, 13149–13153 (2012).
8. Shen, M., Ishimatsu, R., Kim, J. & Amemiya, S. Quantitative imaging of ion transport through single nanopores by high-resolution scanning electrochemical microscopy. *J. Am. Chem. Soc.* **134**, 9856–9859 (2012).
9. Laforge, F. O., Velmurugan, J., Wang, Y. X. & Mirkin, M. V. Nanoscale imaging of surface topography and reactivity with the scanning electrochemical microscope. *Anal. Chem.* **81**, 3143–3150 (2009).
10. Munakata, H., Takemura, B., Saito, T. & Kanamura, K. Evaluation of real performance of LiFePO$_4$ by using single particle technique. *J. Power Sources* **217**, 444–448 (2012).
11. Dokko, K., Nakata, N. & Kanamura, K. High rate discharge capability of single particle electrode of LiCoO$_2$. *J. Power Sources* **189**, 783–785 (2009).
12. Jebaraj, A. J. J. & Scherson, D. A. Microparticle electrodes and single particle microbatteries: electrochemical and in situ microRaman spectroscopic studies. *Acc. Chem. Res.* **46**, 1192–1205 (2013).
13. Beaulieu, L. Y., Hatchard, T. D., Bonakdarpour, A., Fleischauer, M. D. & Dahn, J. R. Reaction of Li with alloy thin films studied by in situ AFM. *J. Electrochem. Soc.* **150**, A1457–A1464 (2003).
14. Inaba, M. *et al.* STM study on graphite/electrolyte interface in lithium-ion batteries: solid electrolyte interface formation in trifluoropropylene carbonate solution. *Electrochim. Acta* **45**, 99–105 (1999).
15. Balke, N. *et al.* Nanoscale mapping of ion diffusion in a lithium-ion battery cathode. *Nat. Nanotech.* **l5**, 749–754 (2010).
16. Balke, N. *et al.* Real space mapping of Li-ion transport in amorphous Si anodes with nanometer resolution. *Nano Lett.* **10**, 3420–3425 (2010).
17. Arruda, T. M., Kumar, A., Kalinin, S. V. & Jesse, S. Mapping irreversible electrochemical processes on the nanoscale: ionic phenomena in Li ion conductive glass ceramics. *Nano Lett.* **11**, 4161–4167 (2011).
18. Balke, N. *et al.* Local detection of activation energy for ionic transport in lithium cobalt oxide. *Nano Lett.* **12**, 3399–3403 (2012).
19. Lipson, A. L., Ginder, R. S. & Hersam, M. C. Nanoscale in situ characterization of Li-ion battery electrochemistry via scanning ion conductance microscopy. *Adv. Mater.* **23**, 5613–5617 (2011).
20. Ebejer, N., Schnippering, M., Colburn, A. W., Edwards, M. A. & Unwin, P. R. Localized high resolution electrochemistry and multifunctional imaging: scanning electrochemical cell microscopy. *Anal. Chem.* **82**, 9141–9145 (2010).
21. Lai, S. C. S., Dudin, P. V., Macpherson, J. V. & Unwin, P. R. Visualizing zeptomole (electro)catalysis at single nanoparticles within an ensemble. *J. Am. Chem. Soc.* **133**, 10744–10747 (2011).
22. Güell, A. G., Ebejer, N., Snowden, M. E., Macpherson, J. V. & Unwin, P. R. Structural correlations in heterogeneous electron transfer at monolayer and multilayer graphene electrodes. *J. Am. Chem. Soc.* **134**, 7258–7261 (2012).
23. Güell, A. G. *et al.* Quantitative nanoscale visualization of heterogeneous electron transfer rates in 2D carbon nanotube networks. *Proc. Natl Acad. Sci. USA* **109**, 11487–11492 (2012).
24. Lai, S. C. S., Patel, A. N., McKelvey, K. & Unwin, P. R. Definitive evidence for fast electron transfer at pristine basal plane graphite from high-resolution electrochemical imaging. *Angew. Chem. Int. Ed.* **51**, 5405–5408 (2012).
25. Aaronson, B. D. B. *et al.* Pseudo-single-crystal electrochemistry on polycrystalline electrodes: visualizing activity at grains and grain boundaries on platinum for the Fe^{2+}/Fe^{3+} redox reaction. *J. Am. Chem. Soc.* **135**, 3873–3880 (2013).
26. Williams, C. G., Edwards, M. A., Colley, A. L., Macpherson, J. V. & Unwin, P. R. Scanning micropipet contact method for high-resolution imaging of electrode surface redox activity. *Anal. Chem.* **81**, 2486–2495 (2009).
27. Liu, X. H., Saito, T., Doi, T., Okada, S. & Yamaki, J. Electrochemical properties of rechargeable aqueous lithium ion batteries with an olivine-type cathode and a Nasicon-type anode. *J. Power Sources* **189**, 706–710 (2009).
28. Mi, C. H., Zhang, X. G. & Li, H. L. Electrochemical behaviors of solid LiFePO$_4$ and Li$_{0.99}$Nb$_{0.01}$FePO$_4$ in Li$_2$SO$_4$ aqueous electrolyte. *J. Electroanal. Chem.* **602**, 245–254 (2007).
29. Manickam, M., Singh, P., Thurgate, S. & Prince, K. Redox behavior and surface characterization of LiFePO$_4$ in lithium hydroxide electrolyte. *J. Power Sources.* **158**, 646–649 (2006).
30. Yu, D. Y. W. *et al.* Study of LiFePO$_4$ by cyclic voltammetry. *J. Electrochem. Soc.* **154**, A253–A257 (2007).
31. Gaberscek, M., Dominko, R. & Jamnik, J. Is small particle size more important than carbon coating? An example study on LiFePO$_4$ cathodes. *Electrochem. Commun.* **9**, 2778–2783 (2007).
32. Patten, H. V. *et al.* Electrochemical mapping reveals direct correlation between heterogeneous electron-transfer kinetics and local density of states in diamond electrodes. *Angew. Chem. Int. Ed.* **51**, 7002–7006 (2012).
33. Zampardi, G., Ventosa, E., La Mantia, F. & Schuhmann, W. In situ visualization of Li-ion intercalation and formation of the solid electrolyte interphase on TiO$_2$ based paste electrodes using scanning electrochemical microscopy. *Chem. Commun.* **49**, 9347–9349 (2013).
34. Come, J., Taberna, P. L., Hamelet, S., Masquelier, C. & Simon, P. Electrochemical kinetic study of LiFePO$_4$ using cavity microelectrode. *J. Electrochem. Soc.* **158**, A1090–A1093 (2011).
35. Bard, A. J. & Faulkner, L. R. *Electrochemical Methods: Fundamentals and Applications* (Wiley, 2000).
36. Fisher, C. A. J. & Islam, M. S. Surface structures and crystal morphologies of LiFePO$_4$: relevance to electrochemical behaviour. *J. Mater. Chem.* **18**, 1209–1215 (2008).
37. Cogswell, D. A. & Bazant, M. Z. Theory of coherent nucleation in phase-separating nanoparticles. *Nano Lett.* **13**, 3036–3041 (2013).
38. Nakano, H., Dokko, K., Koizumi, S., Tannai, H. & Kanamura, K. Hydrothermal synthesis of carbon-coated LiFePO$_4$ and its application to lithium polymer battery. *J. Electrochem. Soc.* **155**, A909–A914 (2008).
39. Novak, P. *et al.* Nanoscale live-cell imaging using hopping probe ion conductance microscopy. *Nat. Methods* **6**, 279–281 (2009).
40. Takahashi, Y. *et al.* Simultaneous noncontact topography and electrochemical imaging by SECM/SICM featuring ion current feedback regulation. *J. Am. Chem. Soc.* **132**, 10118–10126 (2010).
41. Takahashi, Y. *et al.* Multifunctional nanoprobes for nanoscale chemical imaging and localized chemical delivery at surfaces and interfaces. *Angew. Chem. Int. Ed.* **50**, 9638–9642 (2011).
42. Takahashi, Y. *et al.* Topographical and electrochemical nanoscale imaging of living cells using voltage-switching mode scanning electrochemical microscopy. *Proc. Natl Acad. Sci. USA* **109**, 11540–11545 (2012).

Acknowledgements

This work was supported by a Grant-in-Aid for Development of Systems and Technology for Advanced Measurement and Analysis, and Advanced Low Carbon Technology Research and Development Program from the Japan Science and Technology Agency, Grants-in-Aid for Scientific Research (A) (no. 25248032) and for Young Scientists (B) (nos. 24710140 and 24750005) from the Japan Society for the Promotion of Science. This work was partly supported by the Cabinet Office, Government of Japan, through its 'Funding Program for Next Generation World-Leading Researchers'. Y.T. was supported by Asahi Glass Foundation and the Murata Science Foundation. P.R.U. acknowledges support from a European Research Council Advanced Investigator Grant (ERC-2009-AdG247143 QUANTIF). Y.E.K.was supported by the Engineering and Physical Sciences Research Council.

Author contributions

Y.T., A.K., H.M., P.R.U., Y.E.K., K.K. and T.M. conceived and designed the project, analysed the data and wrote the manuscript. Y.T., A.K., H.I. and Komachi Ito performed the experiments. Y.T. designed the instrumentation for SECCM. Y.Y., A.K., Kosuke Ino., H.S., P.R.U., K.K. and T.M. conceived the idea of the electrochemical experiments.

Additional information

Competing financial interests: The authors declare no competing financial interests.

16

Probing water micro-solvation in proteins by water catalysed proton-transfer tautomerism

Jiun-Yi Shen[1,*], Wei-Chih Chao[1,*], Chun Liu[1], Hsiao-An Pan[1], Hsiao-Ching Yang[2], Chi-Lin Chen[1,2], Yi-Kang Lan[2], Li-Ju Lin[3], Jinn-Shyan Wang[3], Jyh-Feng Lu[3], Steven Chun-Wei Chou[1], Kuo-Chun Tang[1] & Pi-Tai Chou[1]

Scientists have made tremendous efforts to gain understanding of the water molecules in proteins via indirect measurements such as molecular dynamic simulation and/or probing the polarity of the local environment. Here we present a tryptophan analogue that exhibits remarkable water catalysed proton-transfer properties. The resulting multiple emissions provide unique fingerprints that can be exploited for direct sensing of a site-specific water environment in a protein without disrupting its native structure. Replacing tryptophan with the newly developed tryptophan analogue we sense different water environments surrounding the five tryptophans in human thromboxane A_2 synthase. This development may lead to future research to probe how water molecules affect the folding, structures and activities of proteins.

[1] Department of Chemistry, Center for Emerging Material and Advanced Devices, National Taiwan University, Taipei 10617, Taiwan. [2] Department of Chemistry, Fu-Jen Catholic University, New Taipei City 24205, Taiwan. [3] School of Medicine, Fu-Jen Catholic University, New Taipei City 24205, Taiwan. * These authors contributed equally to this work. Correspondence and requests for materials should be addressed to H.-C.Y. (email: hcyang_chem@mail.fju.edu.tw) or to P.-T.C. (email: chop@ntu.edu.tw).

In proteins, water ubiquitously participates in dictating structure and functionality, including protein secondary and tertiary structures, the spontaneous formation of membranes and the recognition of signaling molecules in signal transduction[1-3]. Recent advances have provided more convincing evidence that the water molecules in proteins are key elements in activating bio-functionality such as enzymatic reactions[4-7] or *vice versa*, the protein (for example, an antifreeze protein) may affect the organization of water molecules[8]. Probing the water environment of a specifically interesting site in proteins thus may pave a way to understanding the underlying mechanisms and functionality. Unfortunately, although enormous efforts have been made in the characterization of fundamental aqueous hydration phenomena on protein surfaces[9-11], little insight has been gained into water micro-solvation in protein[12,13]. Most of the relevant approaches have focused instead on probing the polarity of the local environment in proteins, particular around the active sites, by means of fluorescence solvatochromism[14,15].

To date, **Trp** in bacterial proteins has been popularly substituted by **(7-aza)Trp** or in part with **(2-aza)Trp**, in which the **indole** moiety in **Trp** is replaced by **7-azaindole** or **2-azaindole** (Fig. 1), for either investigating the antimetabolic properties of the **Trp** isosteres[16] or accessing the local polarity of **Trp** in proteins with unknown tertiary structures[17-21]. Catalysed by protic solvent molecules such as alcohols (for example, methanol), **7-azaindole** undergoes solvent-assisted excited-state proton transfer (ESPT), resulting in an N_7-H tautomer form (Fig. 2a) that exhibits green emission (~ 510 nm)[22-24]. In water, however, the N_7-H tautomer green emission is virtually non-observable[25]; this lack has been attributed to the much slower proton-transfer rate constant ($\sim 10^9 \, s^{-1}$) together with dominant radiation-less deactivation pathways[22]. The photophysical properties of **(7-aza)Trp** are very similar to those of **7-azaindole**, exhibiting essentially no N_7-H tautomer form emission[16,26-28] and thus cannot be used for probing any water-associated photophysical phenomena in proteins.

To surpass the limitation of the traditional **Trp** analogues, we strategically design a new type of aza-Trp by overlapping two azaindole parent moieties, that is, **2-azaindole** and **7-azaindole**, yielding a new core moiety **2,7-diazaindole** (Fig. 1) that is capable of sensing a protein water environment. The underlying concept originates from the molecule **3-cyano-7-azaindole**[29] (Fig. 1), in which the cyano electron-withdrawing ability increases hydrogen acidity of the N_1-H form and hence facilitates the overall ESPT rate in water, rendering an intensive N_7-H proton-transfer

tautomer emission in the green (Fig. 2a). For **2,7-diazaindole**, the replacement of pyrrole by the pyrazole moiety, in which the N(2) atom acts as an efficient electron-withdrawing group[30], is expected to increase the N_1-H proton acidity without perturbing the parent geometry. Using **2,7-diazaindole** as the core moiety, we further develop a new **Trp** analogue, **(2,7-aza)Trp** (Fig. 1), that exhibits remarkably different properties from classical aza-Trp analogues in that water-assisted proton-transfer isomerization takes place in both the excited and the ground states (Fig. 2a,b). The proposed schematic diagram of the proton-transfer cycle for **2,7-diazaindole** and **(2,7-aza)Trp** is depicted in Fig. 2c. Figure 2c unveils the water-catalysed isomerization between the N_1-H form and the N_2-H form in the ground state and ESPT for the N_1-H form in the excited state, as supported by the following firm spectroscopy and dynamics evidence. The associated multiple emissions thus serve as a unique fingerprint

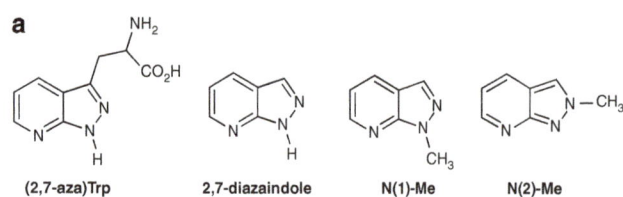

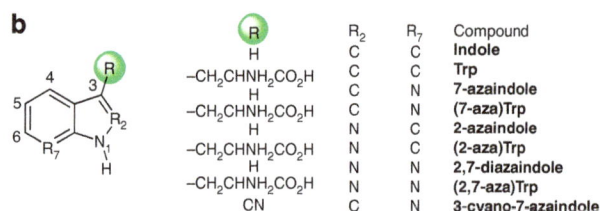

Figure 2 | Proposed ground and electronically excited states mechanism.
(**a**) The proposed protic solvent catalysed ESPT mechanism for **7-azaindole** and its derivatives. The asterisk * indicates the electronically excited state. This mechanism incorporates a fast excited-state equilibrium between polysolvated and 1:1 cyclic hydrogen-bonded N_1-H/H_2O (or methanol) followed by proton tunnelling k_{pt}[22], which has been unambiguously proved by the kinetic deuterium isotope effect[29]. Under the assumption of $k_{-1} > k_1$ and k_{pt}, the overall ESPT rate constant k_{rxn} can be expressed as $k_{rxn} = (k_1/k_{-1})k_{pt}$. (**b**) The ground-state equilibrium for **2,7-diazaindole** and its derivatives between N_1-H form and N_2-H form isomers in water. (**c**) The proposed schematic diagram of a proton-transfer cycle for **2,7-diazaindole** and **(2,7-aza)Trp** in water, in which the barrier and hence potential energy surface are arbitrarily chosen.

Figure 1 | The structural formulae of relevant indole and tryptophan derivatives. (**a**) The chemical structures of **(2,7-aza)Trp**, **2,7-diazaindole**, **N(1)-Me** and **N(2)-Me**. (**b**) The chemical structures of various indole and tryptophan analogues.

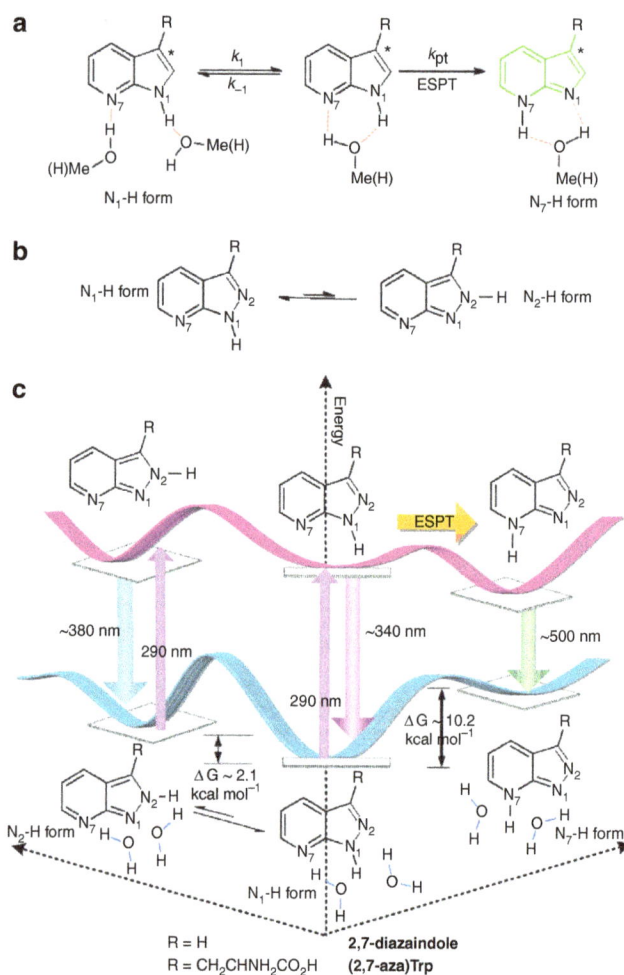

for sensing the water environment. The proof of concept is made by the site-specific substitution **(2,7-aza)Trp** for tryptophan in human thromboxane A$_2$ synthase (TXAS) to resolve a distinct water environment in various **Trp** sites of TXAS.

Results

Proton-transfer tautomerism. Manifested in neutral water (pH = 7.0), either **(2,7-aza)Trp** or **2,7-diazaindole** reveals remarkable triply fluorescent bands (Fig. 3a,b and Table 1), consisting of an apparent shoulder at ∼330 nm, a peak wavelength at ∼370 nm and a much red-shifted emission band maximized at ∼500 nm, corresponding to the existence of N$_1$-H, N$_2$-H and N$_7$-H isomers, respectively, in the excited state. The excitation spectra at 335 nm and at the very red edge (for example, 530 nm) of the emission are identical and they also resemble the absorption spectrum (Supplementary Fig. S1). As **2,7-diazaindole** and **(2,7-aza)Trp** were synthesized independently from different methods (Supplementary Fig. S2 and Supplementary Methods), the possibility of trace impurities contributing to the triple emission bands can be eliminated. Rather, the triple emission can be well rationalized by the proton-transfer tautomerism in both ground and excited states. Support is provided chemically by the methyl derivatives of **2,7-diazaindole**, **N(1)-Me** and **N(2)-Me** (Fig. 1), which exhibit prominent emissions at 342 and 367 nm, respectively, in neutral water (Fig. 3c and Supplementary Fig. S3). The emission of **N(2)-Me**, in view of the peak position (367 nm), is nearly identical to that (370 nm) of **2,7-diazaindole**, so the assignment of

2,7-diazaindole 370 nm emission to the N$_2$-H isomer appears unambiguous. In addition, given the similarity of its emission peak wavelength (342 nm) to that of **N(1)-Me**, the 335 nm band in **2,7-diazaindole** can reasonably be ascribed to the emission of the N$_1$-H isomer. The difference in excitation origins for 335 and 370 nm emission bands (Supplementary Fig. S1) clearly indicates the existence of ground-state equilibrium between N$_1$-H and N$_2$-H isomers for **2,7-diazaindole** (Fig. 2b). In accordance with this assignment, we also propose the occurrence of water-catalysed ESPT in the N$_1$-H form, forming the N$_7$-H isomer, to account for the ∼500 nm green emission. Taking account of the structural and spectral similarity, the same assignment of the triple emission should hold true for **(2,7-aza)Trp**.

Firm support for the tautomerism is given by the correlation of relaxation dynamics among emission bands (Table 1). At the short wavelength shoulder of 320 nm, **(2,7-aza)Trp** exhibits a single decay component of 260 ps in water. At the red edge of the green emission (540 nm), the relaxation dynamics consist of a finite rise component (263 ps) and a decay component with a lifetime of 0.58 ns (Fig. 3d and Table 1). The 263 ps rise dynamic, within experimental uncertainty of ± 30 ps, correlates well with the decay of the 340 nm band, supporting the precursor (340 nm band) and successor (500 nm band) type of reaction kinetics and hence the water-assisted ESPT for the N$_1$-H isomer. On the other hand, the relaxation dynamics at the peak wavelength of 380 nm consists of an instant, system-response-limited rise time (<60 ps), accompanied by two decay components, one with a small amplitude (∼12%), fast decay (265 ps), and a major, long decay component (88%, 10.07 ns, see Fig. 3e). The former 265 ps

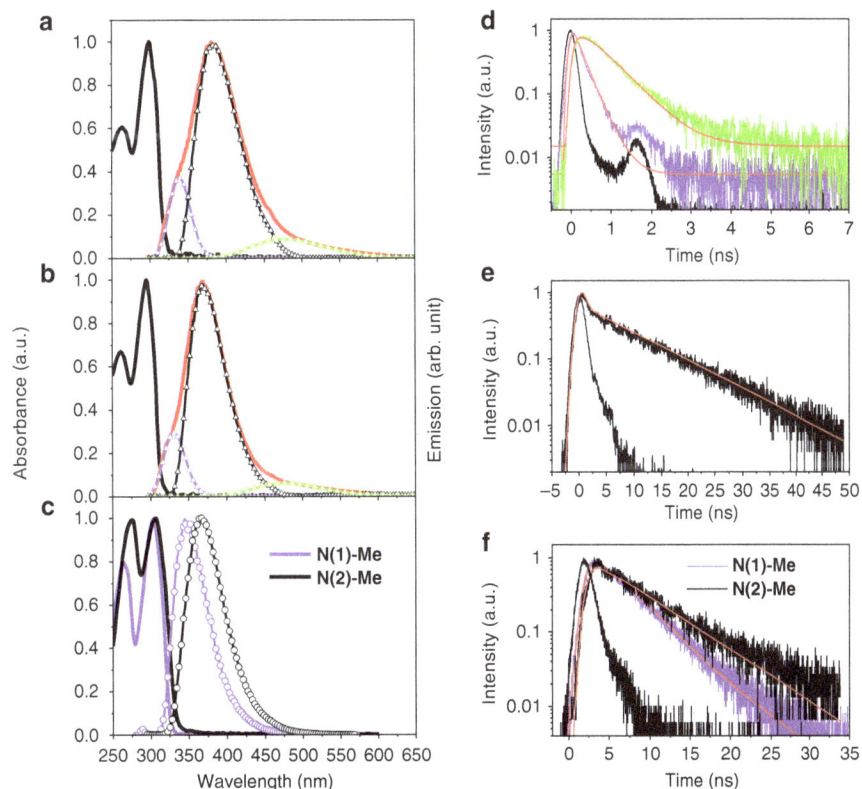

Figure 3 | Identification of the triple emission bands. (a,b) Absorption (black) and emission spectra (red) of **(2,7-aza)Trp** are shown in **a** and **2,7-diazaindole** are shown in (**b**) in neutral water. Further, the decomposed PL spectra are shown (violet, blue and green triangles). $\lambda_{ex} = 290$ nm. (**c**) Absorption and emission spectra of **N(1)-Me** and **N(2)-Me** in neutral water. (**d**) The emission decay dynamics of **(2,7-aza)Trp** monitored at 320 nm (violet) and 540 nm (green). (**e**) The emission decay dynamics of **(2,7-aza)Trp** monitored at 380 nm. (**f**) The emission decay dynamics of **N(1)-Me** and **N(2)-Me**. The instrument response and the fitting curves are marked in black and red lines, respectively. The residuals of the fit of emission decay dynamics as shown in Supplementary Fig. S8.

Table 1 | Photophysical properties.

	λ_{abs} (nm)	λ_{em} (nm)	λ_{mon} (nm)[†]	τ (ns)[‡]
(7-aza)indole[§]	288	386		0.91
(7-aza)Trp	289	400		0.65
2,7-diazaindole	295	335	320	0.22
		370	400	10.10
		495	550	0.21(rise), 1.30
(2,7-aza)Trp	300	340	320	0.26
		380	380	0.27(12%), 10.07(88%)
		500	540	0.26(rise), 0.58
N(1)-Me	304	342	360	3.92
N(2)-Me	306	367	370	5.20
(2,7-aza)Trp^{31,65,133,203,446}-TXAS	300	338	350	0.31(83%), 1.60(17%)
		507	520	0.27(rise), 1.18
(2,7-aza)Trp^{31,133,203,446}Phe^{65}-TXAS	300	340	350	0.28
		503	520	0.31(rise), 1.12
(2,7-aza)Trp^{31}Phe^{65,133,203,446}-TXAS	300	341	350	0.29
		507	520	0.30(rise), 1.23
(2,7-aza)Trp^{65}Phe^{31,133,203,446}-TXAS	300	332	350	1.55
(2,7-aza)Trp^{133}Phe^{31,65,203,446}-TXAS	300	341	350	0.21
		504	520	0.22(rise), 1.32
(2,7-aza)Trp^{203}Phe^{31,65,133,446}-TXAS	300	341	350	0.28
		503	520	0.32(rise), 1.47
(2,7-aza)Trp^{446}Phe^{31,65,133,203}-TXAS	300	341	350	0.30
		503	520	0.30(rise), 1.41

Photophysical properties of various azaindole and azatryptophan analogues in neutral water and mutated TXAS in buffer (20 mM sodium phosphate (pH 7.5), containing 10% glycerol, 0.2% cholic acid and 0.05% lubrol).
†The wavelength at which the measurement of relaxation dynamics was monitored.
‡Values are ± 0.03 ns in uncertainty. Values in parentheses indicate percentages and in brackets denote the rise component.
§Data from the study by Chen et al.[28]

component in the 380 nm emission decay, within the experimental error, is identical to 260 ps monitored at 320 nm and thus is assigned to the residue of the N_1-H emission band. The latter is the population decay of the N_2-H form emission. The same correlation among the triple emission bands is also resolved for 2,7-diazaindole (Table 1), supporting firmly that both 2,7-diazaindole and (2,7-aza)Trp in water undergo proton-transfer tautomerism in the ground state (N_1-H/N_2-H equilibrium) and the excited state (N_1-H $\rightarrow N_7$-H proton transfer).

One fundamental issue lies in the thermodynamics between N_1-H and N_2-H isomers in water (Fig. 2b). Experimentally, the molar absorption coefficient of N(2)-Me, $5.3 \times 10^3\,M^{-1}\,cm^{-1}$ (at 320 nm), was adopted for the N_2-H isomer of 2,7-diazaindole at 320 nm, a wavelength that is free from N_1-H form interference. A known concentration of 2,7-diazaindole was prepared, and the N_2-H/N_1-H equilibrium constant was measured to be 0.024 (detail in Supplementary Methods), corresponding to a difference of free energy of 2.2 kcal mol^{-1} at 298 K. This value is in good agreement with the computational calculations, estimating that the N_2-H form is 2.1 kcal mol^{-1} higher in energy than the N_1-H form for 2,7-diazaindole in water. The same trend is observed but with a rising difference if we consider that less polar solvation by benzene, N_2-H form, has a higher energy by 6.3 kcal mol^{-1} for 2,7-diazaindole. We find that this is true with no detectable N_2-H isomer form emission resolved in organic solvents for 2,7-diazaindole (Supplementary Figs S4 and S5). This proton transfer induced multiple emission in water is remarkable and proves to provide fingerprints for probing a protein water environment.

TXAS homology structure and substrate/water gating channel. The focus of probing water in proteins is on the cytochrome P450 superfamily because recent advances have generated more convincing data indicating that water molecules have pivotal roles in the organism metabolic pathways in P450s[31–33]. We herein selected the structurally pending TXAS[34–36], an endoplasmic reticulum membrane protein containing five tryptophans, as a

prototype to demonstrate the feasibility of (2,7-aza)Trp in probing water environments. The atypical catalysis without reductase or oxygen and the pathogenesis of major diseases make TXAS of prime interest to elucidate the corresponding structure–function relationship[37]. One recent advance indicates that W65 of TXAS is likely to be in a hydrophobic pocket surrounded by the Phe-cluster, whereas the environment of the other four tryptophans is yet to be determined[38]. A prior molecular dynamic (MD) simulation in this study unveils the spatial arrangement of these five tryptophan residues relative to the haem group and gating channels of TXAS (Fig. 4a,b), where W65 residue is on the substrate channel, W133 is located on the aqueduct water channel, W31 lies in the NH_2-terminal loop and W203 and W446 are close to the protein surface (detail in Supplementary Fig. S6).

Expression and analytics of (2,7-aza)Trp-TXAS. To examine the relevant tryptophans positions, TXAS mutants were synthesized in Trp-auxotrophic *Escherichia coli* cells[39], and they proved to be capable of co-translational incorporation by site-specifically replacing Trp residues with (2,7-aza)Trp (see Supplementary Methods). Fig. 4d and e depict the fluorescence spectra of various mutants. As for single (2,7-aza)Trp substitution at W31, W133, W203 and W446, the resulting (2,7-aza)Trp^{31}Phe^{65,133,203,446}-TXAS, (2,7-aza)Trp^{133}Phe^{31,65,203,446}-TXAS, (2,7-aza)Trp^{203}Phe^{31,65,133,446}-TXAS and (2,7-aza)Trp^{446}Phe^{31,65,133,203}-TXAS exhibit prominent, well-resolved dual emissions maximized at ~ 340 nm (N_1-H) and ~ 505 nm (N_7-H), indicating the occurrence of N_1-H to N_7-H proton-transfer in the excited state. By taking advantage of the MD simulations, we were able to examine the proximal environment around various Trp sites of TXAS. We found no relevant amino acids forming direct H-bonds with the mutant (2,7-aza)Trps, also eliminated the possibility of the ESPT catalysed from the protic residues. Therefore, the results ensure the water-catalysed ESPT for the single (2,7-aza)Trp substitution at each of four Trp sites in TXAS. For (2,7-aza)Trp substitution

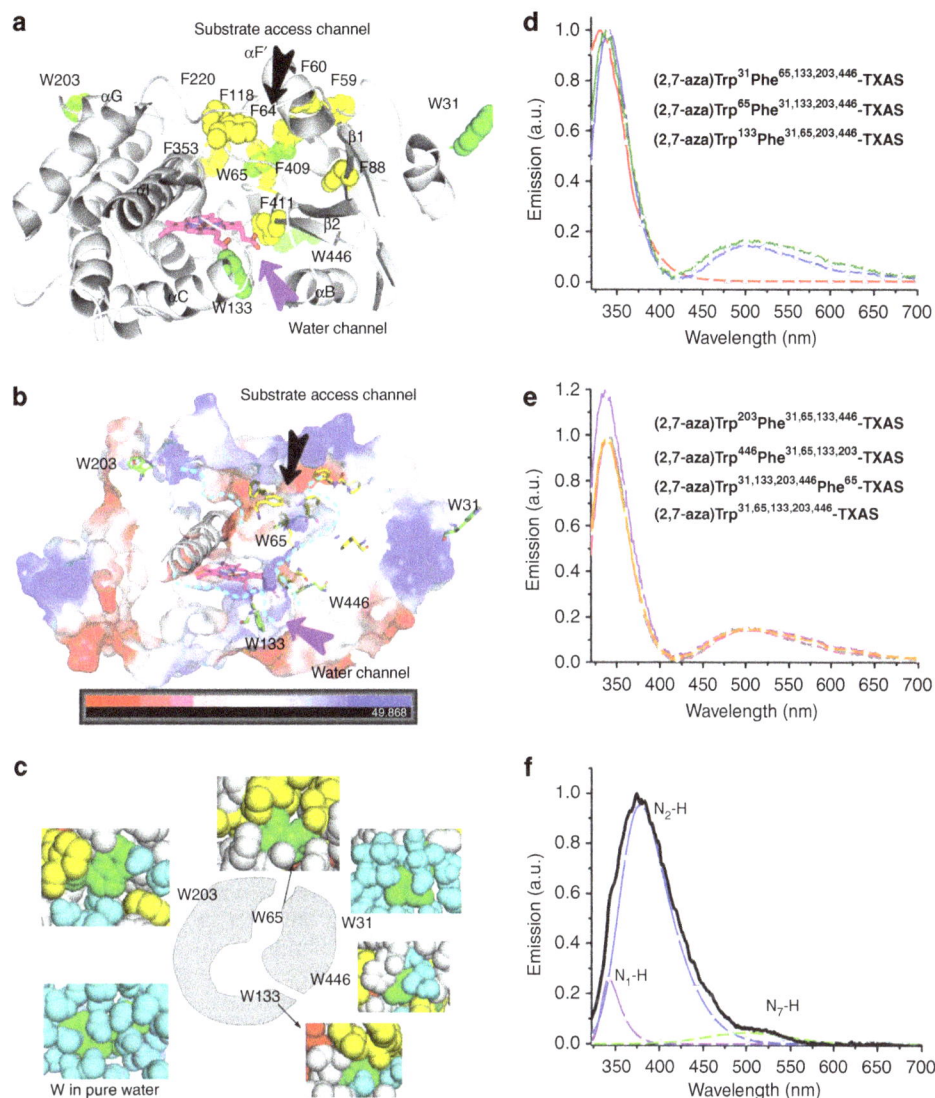

Figure 4 | MD simulation of TXAS protein and expression of TXAS mutants. (**a**) TXAS homology structure shows the characteristic P450 fold, illustrating the phenylalanine residues (yellow), five tryptophan positions (green) and haem (stick figures in violet). (**b**) The solvent-accessible surfaces of the active site cavities calculated using an H probe under an electrostatic potential (ESP) surface with electrophile/nucleophile (blue/red). Proposed gating channels are indicated. (**c**) The local environment of tryptophan (W) in hydrated TXAS (or water), in which W is shown in green, water is blue, acidic residues are red, hydrophobic residues are yellow and other residues are white. (**d,e**) Emission spectra of various TXAS mutants excited at 310 nm. The use of two separated figures is to avoid spectral congestion. (**f**) Emission spectra of denatured (**2,7-aza**)**Trp**^{31,65,133,203,446}**-TXAS** and the decomposed PL spectra.

at W31, W203 and W446, time-resolved measurement monitored at 350 nm clearly reveals a major 0.28–0.30 ns (±0.03) decay component, which correlates well with the rise time of the ∼505 nm emission band of 0.30–0.32 ns (±0.03) (Table 1). Both spectroscopy and dynamics data are reminiscent of the water-catalysed ESPT of the (**2,7-aza**)**Trp** N_1-H isomer in water. As shown in Fig. 4d, the observed dual emission also guarantees the occurrence of water-catalysed ESPT for single (**2,7-aza**)**Trp**-substituted TXAS mutants at W133. However, compared with that of W31, the slight increase of green tautomer emission intensity and the associated faster rate of ESPT (0.21 ± 0.03 ns, Table 1) may imply its difference in a water environment. Supported by water titration experiments for **7-azaindole**[25,40], the breakdown of bulk water clusters may facilitate the rate of ESPT. Accordingly, we tentatively propose that there is a reduced water density

region at the W133 site (*cf.* W31). The results of MD simulation elaborated in a later section seem to support this viewpoint.

In sharp contrast, for the single (**2,7-aza**)**Trp** substitution at W65, the resulting (**2,7-aza**)**Trp**^{65}**Phe**^{31,133,203,446}**-TXAS** exhibits solely an emission band maximized at 332 nm with a long lifetime of ∼1.55 ns (Fig. 4d and Table 1). The apparent lack of a 500 nm emission band speaks to the nature of the prohibition of ESPT at the W65 position in TXAS, concluding the deficiency of water molecules surrounding the proximity of the W65 site. Support is further given by the full (**2,7-aza**)**Trp** substitution of all five tryptophan residues in the expression of (**2,7-aza**)**Trp**^{31,65,133,203,446}**-TXAS** mutant, which exhibits well-resolved dual emissions maximized at ∼340 and 507 nm as well (Fig. 4e). Notice that the intensity ratio for the 340 nm emission band versus the 500 nm emission band is slightly higher than that of quadruple **(2,7-aza)Trp** substitution of **(2,7-aza)Trp**[31,133,203,446]**Phe**[65]**-TXAS** mutant, indicating the increase of 340 nm emission due to the prohibition of ESPT in **(2,7-aza)Trp65**. Statistically, however, the mixture of incomplete substitution of **(2,7-aza)Trp** at any Trp sites is likely to occur in **(2,7-aza)Trp**[31,65,133,203,446]**-TXAS**. Therefore, the corresponding quantitative spectral analysis is not possible. Nevertheless, the selected excitation wavelength at 310 nm should be virtually free of absorption by trace Trp residues. Unambiguously, the results present a unique feature, that the position at W65 is subject to a water-deficient microenvironment wherein the lack of surrounding water molecules prohibits ESPT, resulting in a solely normal N_1-H form 340 nm emission. The other four tryptophans (W31, W133, W203 and W446) in TXAS are in a water-accessible microenvironment that catalyses ESPT, resulting in an N_7-H tautomer green emission.

Importantly, unlike the triple emission of dissolved **(2,7-aza)Trp** in water, all native **(2,7-aza)Trp-TXAS** mutants reveal an absence of N_2-H form 380 nm emission (*cf.* Figs 3a and 4d,e). This seems to describe the vital folding structure of a solvated protein. In a native folding, the associated framework in protein (TXAS) is different from the unfolding residues exposed in the bulk water solution, in which the N_2-H isomer is solvated and stabilized at the sufficient water hydration level. Moreover, the TXAS conformer polarity may not be high enough to stabilize the N_2-H isomer, being similar to the lack of population of N_2-H isomer in organic solvents (*vide supra*, see Supplementary Fig. S4, in methanol). Firm support of this viewpoint is given by the reappearance of the N_2-H form 380 nm emission upon denaturing of **(2,7-aza)Trp**[31,65,133,203,446]**-TXAS** variants (Fig. 4f). The lack of N_2-H isomer emission thus indicates that none of the **(2,7-aza)Trp** in **(2,7-aza)Trp-TXAS** mutants are bulk water-solvated, and so the Trps in wild-type TXAS, due to the structural similarity.

Explicit solvent MD simulations on both wild-type TXAS and **(2,7-aza)Trp**-replaced TXAS have been performed and showed no detectable differences in terms of the solvated homology structure. The results indicate that the TXAS matrices are dynamically coupled to the hydration water by a relay-type of hydrogen-bonding configuration (Fig. 4c). This, together with the close-up inspections of the hydrogen-bonding configurations for each **(2,7-aza)Trp**-replaced tryptophans in TXAS (Supplementary Fig. S6), draws two important remarks. First, there are clearly no water molecules surrounding the W65 site. Second, except for W65, water micro-solvation does exist for all other **(2,7-aza)Trp**. When compared with explicit water MD simulations, a locally fluctuating density profile is revealed (detail in Supplementary Methods and Supplementary Fig. S7), so it seems to be apparent that fewer trapped water molecules associate with the W133 site. Both results are consistent with the experimental observation (*vide supra*). The combination of experimental and MD approaches thus proves the feasibility of **(2,7-aza)Trp** in probing the micro-solvation structures in protein.

As manifested by explicit solvent MD simulations, two distinct channels are identified as the substrate access channel (W65 locus) and the aqueduct water channel (W133 locus) in protein TXAS (Fig. 4b). The associated faster rate of ESPT in **(2,7-aza)Trp133** seems to specify the destination of hydration water behaviours between the protein surface and aqueduct water channel in protein. Also worthy of note is the hydrophobic region located on the substrate channel; this region, with phenylalanine residues (Phe60, Phe64, Phe127, Phe220, Phe409 and Phe411) and W65, forms a 'Phe-cluster'. This distinctive Phe-cluster and hydrophobic residues on the TXAS substrate channel[38] result in a less polar and water-deficient microenvironment at the active site, which is in stark contrast to its counter-enzyme prostacyclin

synthase (PGIS), which has greater water accessibility to the active haem site[39,41]. This prediction is firmly supported by the prohibition of ESPT in (2,7-aza)Trp65. The location of the Phe-cluster results in a gating conformation of the active site of TXAS. MD simulations also reveal that conformational movement of the Phe-cluster residues could open/close the channel and consequently increase/decrease the volume of the active site of TXAS. We hypothesize that this relevant locus of the 'Phe-cluster' is crucial in gating different ligand/water access in an early step of allowance to the active haem site.

Discussion

In summary, we have developed a new tryptophan analogue, **(2,7-aza)Trp**, which undergoes water-catalysed proton-transfer tautomerism in both ground and excited states, giving a multiple emission that is sensitive to the water environment. Incorporating **(2,7-aza)Trp** into human TXAS, we then demonstrate for the first time the feasibility to probe water molecules surrounding in proteins. Manifested by explicit solvent MD simulation, **(2,7-aza)Trp-TXAS** mutant maintains its parent wild-type TXAS structure. Exploiting co-translational **(2,7-aza)Trp** as a probe, correlation between these biocompatible probing spectra and a water micro-solvation environment has been established, thereby identifying specific tryptophan locations on TXAS. The incorporation of **(2,7-aza)Trp** into TXAS proves **(2,7-aza)Trp** to be a novel optical probe that allows transmission of its surrounding water environment into unique fluorescence spectral features. In combination with in-depth MD simulation, future quantitative correlation for the ESPT dynamics versus number/orientation of water molecules is feasible, showing different hydration environments between the substrate access channel, aqueduct water channel and protein surface in, for example, TXAS. The superb water-sensing capability in terms of ESPT and ground-state N_1-H/N_2-H equilibrium thus provides an unprecedented tool for probing the water environment in bio-systems on a structural basis.

Methods

General synthesis and characterization. Full experimental details, synthesis and characterization for all compounds, computational methodology, growth and induction of bacteria are included in the Supplementary Methods.

Steady-state and time-resolved fluorescence spectroscopy. Steady-state absorption and emission spectra were recorded on a Hitachi (U-3310) spectrophotometer and an Edinburgh (FS920) fluorimeter, respectively. Both wavelength-dependent excitation and emission response of the fluorimeter had been calibrated. Time-resolved spectroscopic measurements were carried out by means previously reported elsewhere in detail[42,43]. Briefly, nanosecond (ns) lifetime studies were performed with a time-correlated single-photon counting (TCSPC, Edinburgh FL 900), using a hydrogen-filled lamp as the excitation source. The emission decays were analysed by the sum of exponential functions as below, which allows partial elimination of instrument time broadening and thus renders a temporal resolution of ∼ 300 ps. To achieve a faster time resolution, studies were also performed using a TCSPC system (Edinburgh OB-900L) coupled with an excitation light from the third harmonic generation (THG, at 306 nm) of pulse-selected femtosecond laser pulses at 920 nm (90 fs, Tsunami and Model 3980 pulse picker, Spectra-Physics). The temporal resolution is ∼ 60 ps. Data were analysed by using the nonlinear least-squares procedure in combination with an iterative convolution method.

Tautomer structure calculations. The theoretical approach used the Second-Order Approximate Coupled-Cluster method (CC2)[44] together with a resolution-of-the-identity technique to enhance performance. All atoms were employed using correlation-consistent basis sets of double-ζ quality with the diffuse and polarization functions (aug-cc-pVDZ)[45]. All calculations incorporated the solvation (COSMO model)[46] for ground-state geometry optimization. All the theoretical calculations were performed with the TURBOMOLE programme package (version 6.3)[47-50]. Thermodynamics of various isomers in different solvents were calculated using the DFT approach (B3LYP 6–311 + g(d,p)) incorporating a continuum solvation model (Gaussian 09)[51].

References

1. Ball, P. Water as an active constituent in cell biology. *Chem. Rev.* **108**, 74–108 (2008).
2. Chaplin, M. Do we underestimate the importance of water in cell biology? *Nat. Rev. Mol. Cell Biol.* **7**, 861–866 (2006).
3. Levy, Y. & Onuchic, J. N. Water mediation in protein folding and molecular recognition. *Annu. Rev. Biophys. Biomol. Struct.* **35**, 389–415 (2006).
4. Abel, R. *et al.* Contribution of explicit solvent effects to the binding affinity of small-molecule inhibitors in blood coagulation factor serine proteases. *Chem. Med. Chem.* **6**, 1049–1066 (2011).
5. Robinson, D. D., Sherman, W. & Farid, R. Understanding kinase selectivity through energetic analysis of binding site waters. *Chem. Med. Chem.* **5**, 618–627 (2010).
6. Beuming, T. *et al.* Thermodynamic analysis of water molecules at the surface of proteins and applications to binding site prediction and characterization. *Proteins* **80**, 871–883 (2012).
7. Wang, L., Berne, B. J. & Friesner, R. A. Ligand binding to protein-binding pockets with wet and dry regions. *Proc. Natl Acad. Sci. USA* **108**, 1326–1330 (2011).
8. Meister, K. *et al.* Long-range protein-water dynamics in hyperactive insect antifreeze proteins. *Proc. Natl Acad. Sci. USA* **110**, 1617–1622 (2013).
9. Pal, S. K. & Zewail, A. H. Dynamics of water in biological recognition. *Chem. Rev.* **104**, 2099–2123 (2004).
10. Zhang, L. *et al.* Mapping hydration dynamics around a protein surface. *Proc. Natl Acad. Sci. USA* **104**, 18461–18466 (2007).
11. Lakshmikanth, G. S. & Krishnamoorthy, G. Solvent-exposed tryptophans probe the dynamics at protein surfaces. *Biophys. J.* **77**, 1100–1106 (1999).
12. Ebbinghaus, S. *et al.* An extended dynamical hydration shell around proteins. *Proc. Natl Acad. Sci. USA* **104**, 20749–20752 (2007).
13. Okada, T. *et al.* Functional role of internal water molecules in rhodopsin revealed by X-ray crystallography. *Proc. Natl Acad. Sci. USA* **99**, 5982–5987 (2002).
14. Brand, L. & Gohlke, J. R. Fluorescence probes for structure. *Annu. Rev. Biochem.* **41**, 843–868 (1972).
15. Sackett, D. L. & Wolff, J. Nile red as a polarity-sensitive fluorescent probe of hydrophobic protein surfaces. *Anal. Biochem.* **167**, 228–234 (1987).
16. Ross, J. B., Szabo, A. G. & Hogue, C. W. Enhancement of protein spectra with tryptophan analogs: fluorescence spectroscopy of protein-protein and protein-nucleic acid interactions. *Methods Enzymol.* **278**, 151–190 (1997).
17. Cornish, V. W. *et al.* Site-specific incorporation of biophysical probes into proteins. *Proc. Natl Acad. Sci. USA* **91**, 2910–2914 (1994).
18. Scott, D. J. *et al.* Quaternary re-arrangement analysed by spectral enhancement: the interaction of a sporulation repressor with its antagonist. *J. Mol. Biol.* **293**, 997–1004 (1999).
19. Blouse, G. E. *et al.* A concerted structural transition in the plasminogen activator inhibitor-1 mechanism of inhibition. *Biochemistry* **41**, 11997–12009 (2002).
20. De Filippis, V. *et al.* Incorporation of the fluorescent amino acid 7-azatryptophan into the core domain 1-47 of hirudin as a probe of hirudin folding and thrombin recognition. *Protein Sci.* **13**, 1489–1502 (2004).
21. Richmond, M. H. The effect of amino acid analogues on growth and protein synthesis in microorganisms. *Bacteriol. Rev.* **26**, 398–420 (1962).
22. Mente, S. & Maroncelli, M. Solvation and the excited-state tautomerization of 7-azaindole and 1-azacarbazole: Computer simulations in water and alcohol solvents. *J. Phys. Chem. A* **102**, 3860–3876 (1998).
23. Smirnov, A. V. *et al.* Photophysics and biological applications of 7-azaindole and its analogs. *J. Phys. Chem. B* **101**, 2758–2769 (1997).
24. Waluk, J. Hydrogen-bonding-induced phenomena in bifunctional heteroazaaromatics. *Acc. Chem. Res.* **36**, 832–838 (2003).
25. Chapman, C. F. & Maroncelli, M. Excited-state tautomerization of 7-azaindole in water. *J. Phys. Chem.* **96**, 8430–8441 (1992).
26. Rich, R. L., Smirnov, A. V., Schwabacher, A. W. & Petrich, J. W. Synthesis and photophysics of the optical probe N_1-methyl-7-azatryptophan. *J. Am. Chem. Soc.* **117**, 11850–11853 (1995).
27. Hoesl, M. G., Larregola, M., Cui, H. & Budisa, N. Azatryptophans as tools to study polarity requirements for folding of green fluorescent protein. *J. Pept. Sci.* **16**, 589–595 (2010).
28. Chen, Y., Gai, F. & Petrich, J. W. Single-exponential fluorescence decay of the nonnatural amino acid 7-azatryptophan and the nonexponential fluorescence decay of tryptophan in water. *J. Phys. Chem.* **98**, 2203–2209 (1994).
29. Chou, P. T., Yu, W. S., Wei, C. Y., Cheng, Y. M. & Yang, C. Y. Water-catalysed excited-state double proton transfer in 3-cyano-7-azaindole: the resolution of the proton-transfer mechanism for 7-azaindoles in pure water. *J. Am. Chem. Soc.* **123**, 3599–3600 (2001).
30. Chou, P. T. & Chi, Y. Phosphorescent dyes for organic light-emitting diodes. *Chem. Eur. J.* **13**, 380–395 (2007).
31. Denisov, I. G., Makris, T. M., Sligar, S. G. & Schlichting, I. Structure and chemistry of cytochrome P450. *Chem. Rev.* **105**, 2253–2277 (2005).
32. Zhao, B., Guengerich, F. P., Voehler, M. & Waterman, M. R. Role of active site water molecules and substrate hydroxyl groups in oxygen activation by cytochrome P450 158A2: a new mechanism of proton transfer. *J. Biol. Chem.* **280**, 42188–42197 (2005).
33. Yano, J. K. *et al.* The structure of human microsomal cytochrome P450 3A4 determined by X-ray crystallography to 2.05-A resolution. *J. Biol. Chem.* **279**, 38091–38094 (2004).
34. Samuelsson, B. *et al.* Prostaglandins and thromboxanes. *Annu. Rev. Biochem.* **47**, 997–1029 (1978).
35. Cheng, Y. *et al.* Role of prostacyclin in the cardiovascular response to thromboxane A2. *Science* **296**, 539–541 (2002).
36. Williams, P. A. *et al.* Crystal structures of human cytochrome P450 3A4 bound to metyrapone and progesterone. *Science* **305**, 683–686 (2004).
37. Attwell, D. *et al.* Glial and neuronal control of brain blood flow. *Nature* **468**, 232–243 (2010).
38. Chao, W. C. *et al.* Probing ligand binding to thromboxane synthase. *Biochemistry* **52**, 1113–1121 (2013).
39. Chao, W. C. *et al.* Probing the interaction between prostacyclin synthase and prostaglandin H2 analogues or inhibitors via a combination of resonance Raman spectroscopy and molecular dynamics simulation approaches. *J. Am. Chem. Soc.* **133**, 18870–18879 (2011).
40. Chou, P. T. *et al.* Monohydrate catalysis of excited-state double-proton transfer in 7-azaindole. *J. Phys. Chem.* **96**, 5203–5205 (1992).
41. Li, Y. C. *et al.* Structures of prostacyclin synthase and its complexes with substrate analog and inhibitor reveal a ligand-specific heme conformation change. *J. Biol. Chem.* **283**, 2917–2926 (2008).
42. Chou, P. T. *et al.* Excited-state intramolecular proton transfer in 10-hydroxybenzo(h)quinoline. *J. Phys. Chem. A* **105**, 1731–1740 (2001).
43. Chou, P. T. *et al.* Femtosecond dynamics on excited-state proton/ charge-transfer reaction in 4'-N,N-diethylamino-3-hydroxyflavone. The role of dipolar vectors in constructing a rational mechanism. *J. Phys. Chem. A* **109**, 3777–3787 (2005).
44. Christiansen, O., Koch, H. & Jørgensen, P. The second-order approximate coupled cluster singles and doubles model CC2. *Chem. Phys. Lett.* **243**, 409–418 (1995).
45. Dunning, J. T. H. Gaussian basis sets for use in correlated molecular calculations. I. The atoms boron through neon and hydrogen. *J. Chem. Phys.* **90**, 1007–1023 (1989).
46. Eckert, F. & Klamt, A. Fast solvent screening via quantum chemistry: COSMO-RS approach. *AIChE J.* **48**, 369–385 (2002).
47. Ahlrichs, R., Bär, M., Häser, M., Horn, H. & Kölmel, C. Electronic structure calculations on workstation computers: The program system turbomole. *Chem. Phys. Lett.* **162**, 165–169 (1989).
48. Hattig, C. & Weigend, F. CC2 excitation energy calculations on large molecules using the resolution of the identity approximation. *J. Chem. Phys.* **113**, 5154–5161 (2000).
49. Hattig, C. Geometry optimizations with the coupled-cluster model CC2 using the resolution-of-the-identity approximation. *J. Chem. Phys.* **118**, 7751–7761 (2003).
50. Kohn, A. & Hattig, C. Analytic gradients for excited states in the coupled-cluster model CC2 employing the resolution-of-the-identity approximation. *J. Chem. Phys.* **119**, 5021–5036 (2003).
51. Frisch, M. J. *et al.* *Gaussian 09.* revision A.02 (Gaussian Inc., 2009).

Acknowledgements

This work was supported by Grants NSC 101-2811-M-002-134 and 101-2113-M-030-009-MY3 from the National Science Council of the Republic of China, a Cutting-Edge research grant from the National Taiwan University and a Medical Research Grant from the Medical School, Fu-Jen Catholic University. We thank Dr Ah-Lim Tsai and Lee-Ho Wang for their valuable comments. We also thank the National Center for High-Performance Computing for computational resources.

Author contributions

J.-Y.S. and S.C.-W.C. performed the chemical syntheses. C.L. and K.-C.T. measured and analysed the data. H.-A.P. performed the theoretical calculations. W.-C.C., L.-J.L., J.-S.W. and J.-F.L. performed TXAS mutants, experiments and analysis. C.-L.C. and Y.-K.L. performed the MD simulation. H.-C.Y. and P.-T.C. co-wrote the paper. All the authors discussed the results and commented on the manuscript.

Additional information

17

Polarization-resolved spectroscopy imaging of grain boundaries and optical excitations in crystalline organic thin films

Z. Pan[1,†], N. Rawat[1], I. Cour[1,†], L. Manning[1], R.L. Headrick[1] & M. Furis[1]

Exploration of optical properties of organic crystalline semiconductors thin films is challenging due to submicron grain sizes and the presence of numerous structural defects, disorder and grain boundaries. Here we report on the results of combined linear dichroism (LD)/ polarization-resolved photoluminescence (PL) scanning microscopy experiments that simultaneously probe the excitonic radiative recombination and the molecular ordering in solution-processed metal-free phthalocyanine crystalline thin films with macroscopic grain sizes. LD/PL images reveal the relative orientation of the singlet exciton transition dipoles at the grain boundaries and the presence of a localized electronic state that acts like a barrier for exciton diffusion across the grain boundary. We also show how this energy barrier can be entirely eliminated through the optimization of deposition parameters that results in films with large grain sizes and small-angle boundaries. These studies open an avenue for exploring the influence of long-range order on exciton diffusion and carrier transport.

[1]Department of Physics, Materials Science Program, University of Vermont, Burlington, Vermont 05405, USA. †Present Addresses: CGG, Houston, Texas 77072, USA (Z.P.); Intel Corporation, Hillsboro, Oregon 97124, USA (I.C.). Correspondence and requests for materials should be addressed to M.F. (email: Madalina.Furis@uvm.edu).

Research on organic semiconductors has advanced tremendously for the past two decades as these materials have great potential for the development of novel electronic and photonic devices[1,2]. Among them, small molecules with finite, well-defined π-conjugated systems, such as pentacene, rubrene, perylene or phthalocyanine, exhibit very large charge carrier mobilities[3-5], and represent a cost-effective flexible electronics alternative for certain traditional silicon-based semiconductor applications such as field-effect transistors and photovoltaic devices[5-8]. While it is clear that the electronic properties of such films are highly tunable by chemical methods that simply modify their molecular building blocks or by physical methods through different fabrication techniques (chemical vapour deposition, spin coating, zone casting and molecular beam epitaxy)[5,6], there is a critical need for a deeper fundamental understanding of the influence of long-range ordering on collective phenomena such as exciton diffusion and recombination, or carrier transport. X-ray, electron microscopy and other structural characterization techniques that provide feedback on the molecular packing and crystalline symmetry cannot selectively probe correlations between itinerant electrons.

Optical spectroscopy techniques, such as photoluminescence (PL), differential absorption and circular or linear dichroism (LD), are powerful investigation methods for collective excitations in solids[9,10]. However, in organic films they are severely limited by the large degree of disorder present on the length scale of optical microscopy techniques ($\sim 1\,\mu m$). In many cases, the large concentration of defects effectively quenches the luminescence. For small-molecule polycrystalline films, it is very difficult to study correlations between long-range order and excitons because the typical grain sizes are only in the tens to hundreds of nanometres range and the recorded optical response contains contributions from many randomly oriented grains. Novel solution-based deposition methods[6,11] that produce polycrystalline films with macroscopic millimetre-sized grains offer the opportunity for the first time to employ the wealth of powerful optical spectroscopy techniques to gain insight into the fundamental links between the electronic states, long-range order and the role of grain boundaries in crystalline thin films of organic semiconductors.

In this paper, we report on a spectrally resolved PL/LD polarization microscopy study of excitonic states in individual crystalline grains of metal-free phthalocyanine (H_2Pc) films fabricated with a novel hollow capillary pen-writing deposition technique[11-13]. An in-house built laser scanning polarization-resolved microscopy setup records 90×90-μm-LD images, and probes the symmetry of optically allowed electronic states with a spatial resolution of $\sim 5\,\mu m$. These images reveal how the relative orientation of exciton dipoles at the grain boundaries can be controlled in a marked manner through fine tuning of the deposition parameters (that is, pen-writing speed and solution concentration). The same microscopy setup simultaneously records the polarization-resolved PL from single crystalline grains or individual grain boundaries, which can be correlated with the orientation of optically allowed dipoles. The PL spectra of grain boundaries show the existence of a monomer-like emission exclusively localized at the boundary. Using this dual PL/LD microscopy setup, we conducted a systematic survey of grain boundaries and established a dependence of the localized luminescence intensity on the relative orientation of exciton dipoles at the boundary. These observations indicate the grain boundaries are not molecularly abrupt but rather disordered regions of finite width. We also established that this localized state can be eliminated in films with large long grains where the relative angle between crystalline axes in adjacent grains is $<5°$.

Results

Soluble phthalocyanine: a ubiquitous molecule revisited. Figure 1 displays standard polarized optical microscope images of four thin films of 2,3,9,10,16,17,23,24-Octakis(octyloxy)-29H,31-phthalocyanine (H_2Pc-OC_8), a commercially available peripherally substituted phthalocyanine (Pc) soluble in common organic solvents, our molecule of choice for the present study. The films were deposited at different writing speeds from chloroform or toluene solutions of various concentrations using a 'pen-writing', hollow capillary technique[11,13]. In essence, this technique involves translating a 500-μm-wide glass capillary loaded with solution at a constant speed across a substrate. Optimal results are obtained when crystallization takes place at the contact line between the meniscus and the substrate. In all instances, we used crystalline c plane sapphire as substrate; however, similar results were obtained for other dielectric transparent substrates such as plain microscope glass or indium tin oxide. To produce films with macroscopic grain sizes, the capillary translation speed, the

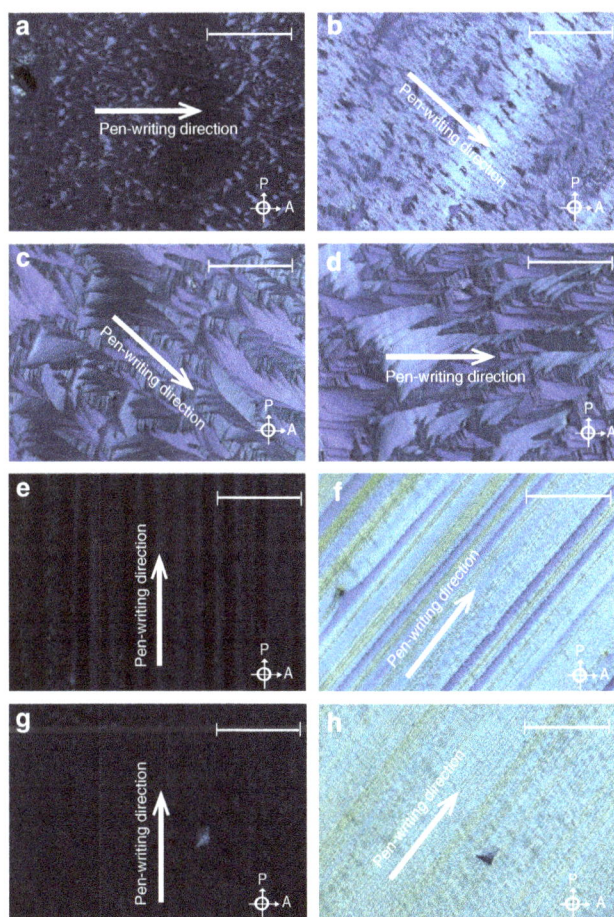

Figure 1 | Long-range ordering in phthalocyanine thin films. Polarized microscopy images of crystalline H_2Pc-OC_8 thin films deposited on c plane sapphire using the solution-processing technique. Optimization of deposition speed, solution concentration and choice of solvent, results in millimetre-sized grains suitable for optical spectroscopy experiments. (**a,b**) 0.5%, 11 $\mu m\,s^{-1}$, toluene; (**c,d**) 1%, 18 $\mu m\,s^{-1}$, toluene; (**e,f**) 1%, 40 $\mu m\,s^{-1}$, chloroform and (**g,h**) 1%, 30 $\mu m\,s^{-1}$, chloroform. During film deposition, the sample stage is moving along the direction marked with a white arrow in each image. The grains with molecular stacking axis oriented along the polarizer direction appear dark in these images. Scale bar, 500 μm.

concentration of molecules in solution, and the choice of solvent must be optimized. The four films in Fig. 1 illustrate this optimization process that can be quite different for each molecule. In the case of $H_2Pc\text{-}OC_8$, the largest grain sizes and most uniform films were obtained when 1% chloroform solution was used (Fig. 1e–h). At the optimum deposition speed of $30\,\mu m\,s^{-1}$ in chloroform (Fig. 1g,h), the grain size exceeds the microscope field of view. For toluene solutions, at lower concentrations (Fig. 1a,b), the optimized deposition results in a disordered film with small grains and poor substrate coverage, whereas for deposition speeds significantly larger than optimal the nucleation happens randomly and simultaneously over the entire area of the sample resulting in a 'fan-like' polycrystalline structure (Fig. 1c,d). This latter case is quite an interesting one since it represents a situation reminiscent of the vapour-deposited films that are polycrystalline structures with randomly oriented grains. The fan-like structure can thus be regarded as a 'scaled-up' model for the vapour-deposited films, with grain sizes large enough for optical spectroscopy investigations. For this reason, our studies of correlations between different grain boundaries and the nature of excited states at the boundary focused precisely on these types of disordered samples rather than the long-grain well-ordered structures that we explored elsewhere[14].

Phthalocyanine, in its non-soluble form, is a well-studied ubiquitous molecule characterized by D_{2h} symmetry that results in optical transition dipoles polarized in the molecular plane. This symmetry and selection rules are preserved for the soluble, substituted Pc, whose absorbance spectrum measured in chloroform exhibits a splitting of the peak associated with the Q-band transitions (Fig. 2c)[15–17]. The optical dipoles associated with these transitions lie along mutually perpendicular directions in the plane of the molecule[18,19] and photons absorbed or emitted are polarized in the molecular plane. The difference in absorbance ($A_x - A_y$) for light polarized parallel and perpendicular to the molecular plane is referred to as LD[20]. In solution or amorphous films, there is no preferential orientation of the molecular plane; therefore, no LD is observed in such cases. It is expected,

however, that in a single crystal the long-range molecular order preserves the selection rules, hence the LD can be observed on a macroscopic scale[21].

Out-of-plane X-ray diffraction (Supplementary Fig. 1) experiments we conducted at the synchrotron facility of Brookhaven National Laboratory confirmed our films crystallize in the orthorhombic phase with the molecules stacked edge-on to the substrate and the stacking axis parallel to the substrate. That means an experimental geometry, such as the one sketched in Fig. 2a,b, where the **k** vector of the incident light is perpendicular to the stacking axis, should render a significant amount of LD (LD $\sim A_x - A_y$) provided the orientation of the molecular stacking remains the same at least on the length scale of the beam diameter. This orientation also explains the polarization-mode contrast observed in Fig. 1: if the stacking axis happens to be oriented along the polarizer axis the grain appears dark in the microscope image.

Typical absorbance (A) and LD spectra from a 0.1% film (Fig. 1c) show a large spectral broadening as a result of molecular interactions[22–24], and exciton–vibrational coupling[17,25–27]. The LD spectrum is of a similar shape to the absorption and the large LD signal (about 30% of that of a commercial polarizer) observed for the bandgap spectral region, implies the majority of the transition dipoles are linearly polarized and oriented perpendicular to the pen-writing direction. Assuming molecular selection rules are preserved in the solid film, the large positive LD confirms the presence of long-range order, where most molecules stack face-to-face, 'edge-on', as illustrated in Fig. 2b. While the orientation of stacking axis can be inferred from the measured LD, it is important to emphasize that, unlike electron microscopy or X-ray diffraction which offer direct structural information but cannot probe extended excitonic states in crystals, optical spectroscopy techniques such as LD or PL microscopy selectively probe the electronic/excitonic band states and the optical transition dipoles orientations. Other absorbance spectra from the samples investigated in this work can be found in Supplementary Fig. 2, which indicates that there are no

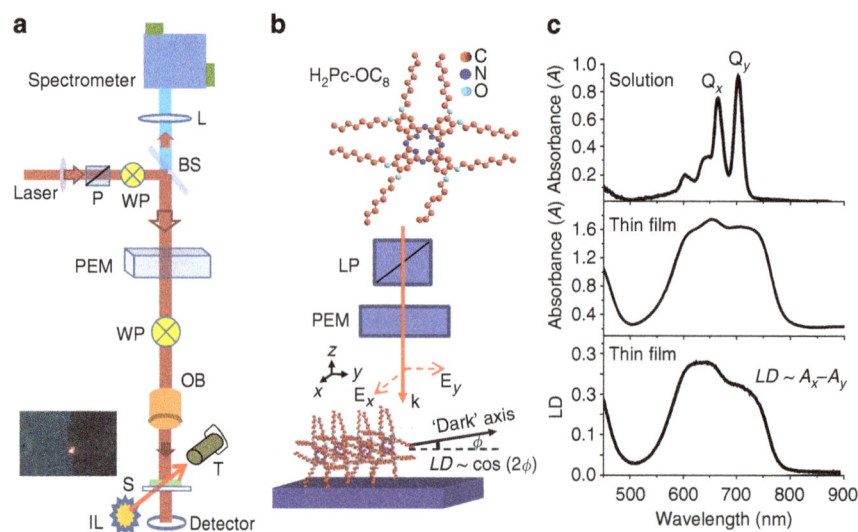

Figure 2 | Linear dichroism (LD) and molecular ordering. (a) Schematics of the LD/PL microscopy experiment employed to study the relative orientation of crystalline grains in the $H_2Pc\text{-}OC_8$ thin films. The LD image is acquired in transmission geometry by scanning a polarization-modulated focused laser beam onto the sample surface while spatially- and polarization-resolved PL spectra are recorded in backscattering geometry, after removing the piezoelectric modulator and waveplate from the setup. BS, beam splitter; IL, illumination lamp; L, singlet lens; OB, objective lens; P, linear polarizer; PEM, piezoelectric modulator; S, sample; T, long working distance observation telescope; WP, half-wave plate. **(b)** Molecular structure of $H_2Pc\text{-}OC_8$ and the orientation of light polarization with respect to the 'edge-on' molecular stacking. Light polarized perpendicular to the stacking ('dark') axis is predominantly absorbed. **(c)** Typical absorbance ($A \sim \alpha d$) and linear dichroism (LD $\sim Ax - Ay$) spectra from an $H_2Pc\text{-}OC_8$ crystalline film deposited using the pen-writing technique (a reference spectrum of the same molecule in solution is shown for comparison).

significant changes in the spectral shape. The overall absorbance does scale with film thickness as expected. All subsequent film thicknesses mentioned in the present work are estimated based on a calibration of the deposition system presented in the Supplementary Information of ref. 12.

Electronic states localized on grain boundaries. The influence of molecular ordering on excitonic states, radiative recombination and states originating at the grain boundaries is probed when spatially resolved PL and LD are simultaneously recorded from a selected area using a focused narrowband laser probe beam. To this end, we combined LD and PL scanning microscopy as illustrated in Fig. 2a to unambiguously establish a particular grain orientation, while selectively probing the bandgap luminescence with a spatial resolution of $\sim 5\,\mu m$. A detailed description of the experimental setup is also available in the Methods section of this paper.

The three samples we explored with our scanning microscopy experiment are presented in Fig. 3a,b and Supplementary Fig. 3. The 30-nm-thick sample A (fabricated from a chloroform solution of 0.1% at a pen-writing speed of $2\,mm\,s^{-1}$) exhibits a stripe-like geometry where all molecules within any individual stripe are oriented in the same direction, with some difference in orientation between two adjacent stripes, which accounts for the image contrast. For larger deposition concentrations, the microscope image of the 200-nm-thick sample B (deposited from a chloroform 1% solution at $20\,\mu m\,s^{-1}$ pen-writing speed) reveals more randomly oriented grains with 'fan-like' structures that are expected in a concentrated solution with many nucleation centres. All films have grain sizes ranging from 10 to $100\,\mu m$, enabling us to spatially resolve the properties of electronic states in the vicinity of the bandgap within a single-grain and probe single-grain boundaries using this dual-microscopy technique.

Figure 3c,d reproduces LD two-dimensional (2D) scans of 90×90-μm areas marked with white squares in the microscopy images (Fig. 3a,b). The pen-writing direction for each sample is also indicated in the microscope image. Sample A exhibits parallel stripes of $\sim 10\,\mu m$ in width, orthogonal to the pen-writing direction. The LD values are positive across the scanned area with a 25% difference between adjacent stripes. Since the LD is proportional to $\cos(2\varphi)^{21}$, where φ is the angle that defines the stacking axis orientation with respect to the polarizer axis (Fig. 2), the LD contrast observed in sample A corresponds to a relative angle $\Delta\varphi \sim 5^0$ between stacking axes from adjacent stripes. In striking contrast, sample B exhibits a distinctly different, 'fan-like' structure with well-defined grain boundaries in polarization-mode microscopy images. The LD scan across an area traversed by a single-grain boundary, revealed a very large contrast between the adjacent grains corresponding to a relative angle $\Delta\varphi \sim 25^0$ between the stacking axes of the two grains. A preliminary LD imaging survey conducted on films deposited at different concentrations ranging between 0.1% and 1% indicated that there is a direct relationship between the relative orientation of the stacking axes at the grain boundaries and solution concentration

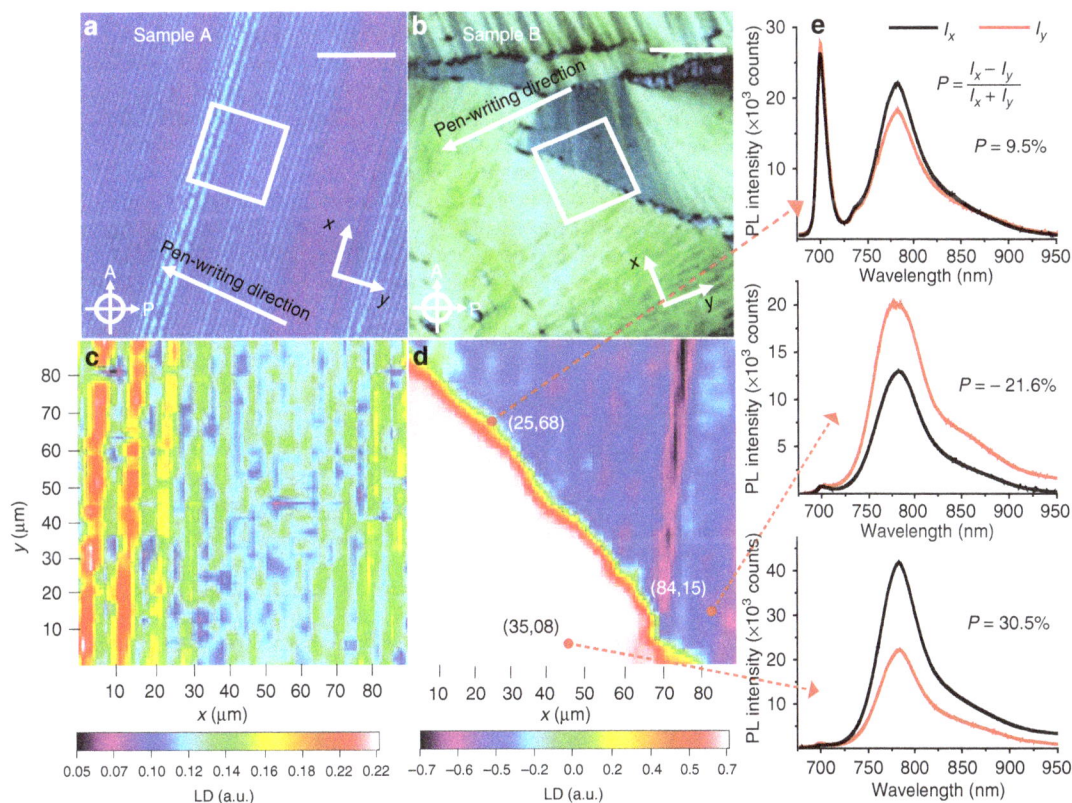

Figure 3 | LD/PL microscopy of electronic states at the grain boundaries. (**a,b**) Standard polarization-mode microscopy images of two distinct polycrystalline H_2Pc-OC_8 films: sample A (0.1% solution, $2\,mm\,s^{-1}$) and sample B (1% solution, $20\,\mu m\,s^{-1}$) that reveal very different grain structures. The orientation of pen-writing direction with respect to the polarizer and analyser (x–y) axes is marked with a white arrow. (**c,d**) High-resolution LD microscopy images of 90×90-μm areas identified with a white square in (**a,b**). (**e**) polarization and spatial resolved PL spectra of sample B recorded at three distinct locations identified through their (x,y) coordinates. The LD contrast indicates the different orientation of the stacking axis in adjacent grains. Depending on the growth conditions, the grains assemble in stripe-like films with low-angle ($\sim 5°$) grain boundaries (sample A (**a,c**)) or fan-like films with high-angle ($\sim 25°$) grain boundaries (sample B (**b,d**)). An additional monomer-like feature is present at the high-angle grain boundary. Similar spectra are recorded from sample A, but no monomer-like feature is observed at the low-angle grain boundaries.

and deposition speed. A film of intermediate concentration and thickness (0.5%, 150 nm) deposited with the same writing speed ($20\,\mu m\,s^{-1}$), sample C, has parallel grains with large positive LD. The sample thicknesses quote here were estimated based on the thickness versus deposition speed curve found by Cour et al.[12] The absorbance measurements performed on our samples (Supplementary Fig. 2) are consistent with this estimation.

All our films exhibit strong room temperature PL resonant to the absorption edge. Since radiative transitions typically originate from the lowest vibrational state of an electronic excited state[26], the absence of a large Stokes shift is indicative of strong exciton vibronic coupling and the existence of delocalized excitons[26]. This is a consequence of the long-range ordering present in our films, that is, the size of crystalline grains, is significantly larger than the typical exciton diffusion lengths[9], such that excitons generated inside a grain recombine before reaching the grain boundary or a defect state.

Figure 3e highlights the most interesting results of the luminescence microscopy experiments. First, the luminescence is linearly polarized perpendicular to the stacking axis, indicating that the room temperature emission is indeed associated with the recombination of the singlet exciton ground state. The π-orbital overlap between adjacent molecules preserves to some extent the monomer selection rules for the ground-state transition dipole at room temperature. In Fig. 3, micro-PL spectra recorded at coordinates $(x,y) = (35,08)$ and $(84,15)$ indicate the luminescence polarization is correlated to the stacking axis orientation, changing sign and magnitude in tandem with the LD. The delocalization of this bandgap singlet exciton state has been described in detailed elsewhere[14]. Second, in addition to the expected singlet exciton recombination feature, the emission spectrum recorded from the grain boundary (for example, $(x,y) = (25,68)$), exhibits an additional strong sharp feature centred at 700 nm. This feature is exclusively observed when the laser excitation beam is focused anywhere along the grain boundary.

To investigate its origins, we placed the sample in a variable temperature liquid He continuous flow cryostat and recorded PL spectra from grain boundaries at various temperatures ranging from 5 to 300 K (Fig. 4). The sharp PL feature redshifts and decreases in intensity, reminiscent of the behaviour of typical emission from localized states in semiconductors. Its polarization

at low temperatures (Fig. 4, inset) was opposite to that of the bandgap exciton recombination. Following this observation, we conducted a systematic PL/LD study of grain boundaries in films deposited under different conditions, surveying ~20 grain boundaries. Images from three representative grain boundaries and the luminescence spectra associated with them are shown in Fig. 5. The first observation is that the relative intensity of the grain boundary luminescence that we define as $I_{relative} = I_{monomer}/I_{total}$ varies greatly from one-grain boundary to another and remains approximately constant along the grain boundary. The second observation is the existence of a correlation between this relative intensity and the relative orientation of adjacent grains. This correlation is readily apparent from the data summarized in Table 1, which indicates that the monomer-like emission intensity increases with increasing boundary angle, reaching a maximum in the more disordered samples where we can no longer speak of a boundary between two crystalline grains but rather of disordered areas with sizes comparable to the crystalline regions (Fig. 5c). In our experiments, these disordered areas are unambiguously distinguished from crystalline grains oriented at 22.5° because they exhibit LD = 0 regardless of the incident light polarization

Discussion

In spite of our study being focused on a specific soluble derivative of phthalocyanine, it is very useful to consider the results of our LD imaging and spatial- and polarization-resolved PL studies in light of the knowledge they bring on the correlations between electronic states and molecular ordering for small molecules systems in general. In recent years, numerous fundamental and applied studies of electron mobilities and exciton diffusion on small-molecule organic semiconductor thin films such as pentacene, rubrene and even phthalocyanines report ever increasing record mobilities and diffusion lengths, far exceeding early predictions that follow from a simple 'oriented gas' model[17]. In all cases, the key factor seems to be the reduction and control of grain boundaries such that continuous improvement in the thin film quality should reveal exciting new physics and bring a better understanding of electronic phenomena in these materials, in the same way that molecular beam epitaxial growth led to very high mobilities and excitons diffusion lengths in inorganic semiconductors.

These materials seem to belong to an intermediate regime where the delocalized carriers and excitons are neither completely localized like the Frenkel excitons nor delocalized like the Wannier excitons in inorganic semiconductors[14,28]. In this context, optical spectroscopy techniques are ideal because they probe the longer-range interactions and excitations while providing spatial resolution on the micron range, which is now sufficient given the increasingly large crystalline grain size. The scanning narrowband laser microscopy we employed here is a powerful technique that has been used to explore optical properties in a variety of systems. Its different flavours (Kerr Rotation, Faraday rotation, MCD, or LD) are employed for studying a vast array of phenomena ranging from spin-polarized electron transport in GaAs to grain boundaries in halide systems[29,30].

Phthalocyanines are the ideal test system for such investigations because they are resistant to oxidation and retain a high yield of radiative recombination in crystalline films. Most of them are readily available and, as mentioned earlier, very well studied as far as the molecular properties are concerned. Moreover, the success we had in depositing thin films with macroscopic order (Fig. 1) proved the robustness of this technique that relies on nucleation and surface tension control, a technique originally

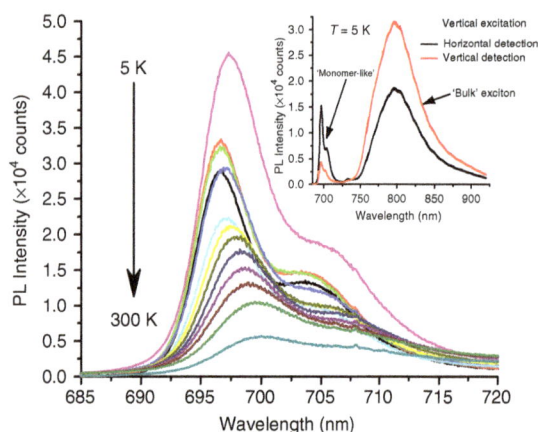

Figure 4 | Temperature evolution of luminescence from grain boundaries. PL spectra from the grain boundary in Fig. 3d recorded at various temperatures between 5 and 300 K. The sharp, 'monomer-like' feature exhibits the typical evolution of radiative emission from highly localized electronic states: the spectrum redshifts and broadens with increasing temperature. The inset displays both x- and y-polarized components of the luminescence spectrum indicating the localized, grain boundary emission and the 'bulk' photoluminescence have opposite polarizations.

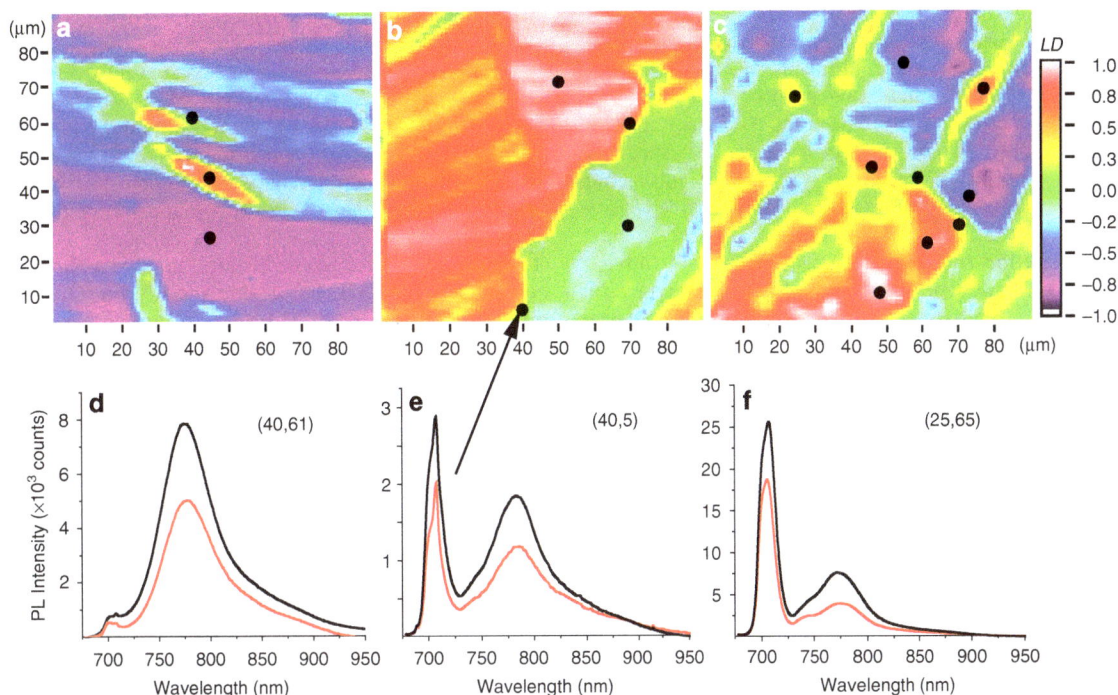

Figure 5 | LD/PL microscopy survey of grain boundaries. (a–c) LD microscopy images of individual grain boundaries from samples fabricated at different concentrations and writing speeds. The linear dichroism is normalized to the absorbance, for a direct comparison between samples (a,b) 0.5%, 30 µm s^{-1} and (c) 0.5%, 20 µm s^{-1}. (d–f) Polarization and spatial resolved photoluminescence spectra recorded at locations marked with black dots in (a–c), respectively. The beam locations are identified through their (x,y) coordinates. The correlation between relative grain orientation and localized emission intensity is summarized in Table 1.

Table 1 | Summary of grain boundary luminescence survey.

$\Delta\varphi$ (°)	$I_{relative} = I_{monomer}/I_{total}$
<5	0
11	0.07
14	0.33
22	0.63
27	0.35
Disordered film	0.8

Measurements of grain boundary luminescence intensity as a function of grains orientation reveal the 'monomer-like' emission is only present at large angle $\Delta\varphi > 5°$ grain boundaries. The largest intensities are measured in samples where crystalline grains are separated by disordered regions. The survey covered twenty distinct grain boundaries. Three of them were highlighted in Fig. 5.

designed and optimized for depositing pentacene thin films[13]. (Our group ultimately obtained macroscopic grain sizes for two distinct substituted metal phthalocyanines as well as two additional molecules in the porphyrin family: tetra-phenyl porphyrin (TPP) and naphthalocyanine (NPc). Optical studies on these species that investigated different aspects of longer-range interactions are beyond the scope of the present paper and will be reported elsewhere).

The necessity of an optical spectroscopy spatially resolved study stems from the progress made in fabricating OFET-like structures that investigate the influence of single-grain boundaries on electron mobilities[6,11,12,31]. These studies looked into how the nature and number of grain boundaries limit the ultimate mobility through the presence of localized states (traps) at the boundary. They also established the nature of grain boundaries in the thin films deposited through all flavours of solution-processing techniques is very different from that of grain boundaries in vapour-deposited thin films, a difference that

originates in the different molecular ordering along the high-mobility axis (π-stacking versus herringbone). Specifically, transport measurements indicated the grains are separated by disordered regions of finite thickness (of the order of 10 nm) rather than molecularly abrupt boundaries[11]. Very recent spatially resolved transient absorption measurements on soluble pentacene thin films also indicate the boundaries should be viewed as disordered regions[32]. In this context, the observation of radiative emission localized at the grain boundary (Figs 3 and 5) is somewhat expected and must be associated with these disordered regions. The unexpected element is the wavelength (or energy) associated with this emission, 700 nm (or 1.77 eV), significantly larger than the bulk luminescence located at 785 nm ($\sim$1.58 eV). In a simplistic approach, we can interpret the 200 meV difference as a quantitative estimate for an energy barrier the excitons/carriers need to overcome to reach the adjacent grain boundary. Scattering from this 200 meV barrier will represent one of the mechanisms limiting the diffusion length/mobility, in addition to trap states present at the boundaries. Imaging of different films and grain boundaries such as the ones presented in Fig. 5 allow us to hypothesize on the nature of electronic states at the boundary and refine this energy barrier concept.

For most films, the grain boundary is certainly narrower than our laser beam diameter ($\sim$5 µm), that explains the bulk exciton contribution to the PL spectra. In the more disordered regions such as the one probed in Fig. 5c, the grain boundaries seem to have evolved into macroscopic disordered regions (LD = 0) that are certainly in the micron range. These are the very same regions where the sharp 700 nm PL feature becomes the dominant feature in the spectrum (Fig. 5c). In addition, our grain boundary survey summarized in Table 1 indicates that the relative intensity of this boundary feature increases with increasing angle $\Delta\varphi$ between stacking axes of adjacent grains. It is also important to note that there are preferred recurrent relative angle orientations, listed in Table 1. In spite of surveying numerous grain boundaries, these were the only orientations we found within an uncertainty of $\sim 2°$.

On the basis of these results, we propose that the grain boundaries are indeed disordered regions (as previously inferred by transport studies) where we observe a PL spectrum reminiscent of a solution spectrum or weakly interacting molecules with reduced orbital overlap. The relatively large polarization observed at low temperatures (Fig. 4, inset) is not a signature of a selection rule at the grain boundary but rather a consequence of the absorption and emission selection rules for the bulk exciton originating from the crystalline regions immediately adjacent to the grain boundary[14].

Finally, we point out the sharp luminescence feature completely disappears in the highly ordered films such as the ones in Fig. 1g,h. We expect the nature of grain boundaries to be entirely different in these films because they are deposited in what has been identified as the convective regime[12], that is, the liquid transforms into a solid film at a well-defined boundary that is <3-μm thick. Studies that investigated correlations between grain boundaries and electronic mobilities[29] indicate that not only is the mobility greatly enhanced when the number of grain boundaries the electron cross is reduced, but more importantly, the mobility is significantly increased in films deposited in the convective regime. Our optical spectroscopy results now provide a quantitative feedback on the bandgap electronic states at the grain boundaries. On the basis of the observations of localized emission, we propose that the electron mobility as well as exciton diffusion will also be reduced in these systems because of the lack of ordering and poor orbital overlap at the grain boundary, even at high excitation rates where trap states are saturated. While lower energy trap states are not luminescent, their presence can be inferred from the evolution of the PL intensity with excitation power shown in the Supplementary Fig. 4b. At low power, the PL intensity is proportional to $N^{0.8}$, where N represents the number of incident excitation photons, suggesting the presence of a non-radiative relaxation mechanism for excited carriers that may involve low-energy traps. This mechanism is most likely the Shockley–Read–Hall recombination, previously identified in solution-processed thin films[33]. For excitation densities larger than 10^{25} photons per second per centimetre^3, the emission intensity increases as the square root of excitation power, a possible signature of the Auger exciton–exciton annihilation at large excitation powers[34]. In these films, the observation of the sharp luminescence feature is independent of the nature of non-radiative traps or the complex nature of non-radiative quenching mechanisms, because the spectral shape and relative PL intensity does not change with excitation power as evidenced by Supplementary Fig. 4a. Moreover, absorbance spectra measurements, PL power studies or PL lifetime measurements (Supplementary Figs 2, 4 and 5) indicate the two transitions originate from entirely spatially separated states that do not interact with each other through energy transfer mechanisms. PL decays for both features are characterized by a single exponential and the same lifetime, in contrast to systems where the presence of energy transfer shortens the higher energy feature lifetime[35]. Parameters such as film thickness or excitation power only seem to affect the overall PL intensity for both features.

In conclusion, the adaptation of polarization-resolved laser scanning techniques to crystalline organic semiconductor thin films have the potential for revealing a wealth of information in regards to the nature of excited states in these systems, with emphasis on longer-range phenomena. Using the LD/PL scanning microscopy in tandem with novel thin film deposition techniques such as the hollow capillary pen-writing technique that relies on careful control of nucleation, we were able to

unveil the nature of excited states at a single-grain boundary in phthalocyanine thin films. Specifically, we concluded that scattering from an energy barrier created in a disordered region is one of the mechanisms limiting exciton diffusion length in the presence of grain boundaries, in addition to the low-energy traps. In a broader sense, the macroscopic grain sizes open up the organic semiconductor systems for the first time to the wealth of bandgap spectroscopy experiments that can investigate the nature of excited states and the correlations between the crystalline structure and these excited states in a sample free of grain boundaries or defects. For example, our group is making strides towards understanding the bandgap luminescence and the spatial extent of the bulk exciton wavefunction in these films some of which were recently reported elsewhere[14].

Methods

Materials and thin film fabrication. The dye molecules of 2,3,9,10,16,17,23, 24-Octakis(octyloxy)-29H,31H-phthalocyanine (H_2Pc-OC_8) were purchased from Sigma Aldrich. The materials were purified by column chromatography using chloroform as an eluent. Thin films were deposited on substrates by a solution processing pen-writing technique at room temperature using either toluene or chloroform as the solvent[13]. This method employs a hollow borosilicate glass capillary pen and solutions with concentration of 0.1–1 wt% are held in the pen by capillary forces. Film deposition is accomplished by allowing droplets of solution at the end of the pen to make contact with the substrate and then laterally translating the pen at a controlled speed ranging between 0.02 and 2 mm s^{-1}. The deposition, conducted at room temperature, produces highly ordered films with uniform thicknesses ranging between 20 nm and 1 μm depending on the deposition parameters. The substrates were thoroughly cleaned and treated with methanol to improve the wettability and to assist with forming uniform films.

LD and PL experiments. The absorption and LD spectra were measured at room temperature in the range of 400–1,100 nm using a quasi-monochromatic (1-nm bandwidth), tuneable, incoherent, tungsten halogen monochromator light source. A continuous wave (CW)-focused HeNe laser was employed as probe beam for LD microscopy measurements. The same laser served as quasi-resonant excitation for spatially resolved PL measurements (Fig. 2a). To record LD images of individual crystalline grains, we adapted a polarization-resolved microscopy technique formerly employed to image spin drift and diffusion in GaAs[29]. A piezoelastic modulator was placed in the beam path to modulate the light polarization from x to y polarized at a frequency of 100 kHz. The laser beam was focused to a $\sim$5-μm spot using a $\times 10$ objective lens mounted on a piezo nanopositioning stage. A long working distance (15 mm) objective lens was chosen to insure that the experiment can be performed at temperatures as low as 5 K in an existing Oxford cryostat. These low-temperature experiments were reported elsewhere[14]. The 2D LD images were obtained by raster scanning the focusing objective lens. Spatially resolved PL was simultaneously collected with the same lens in backscattering geometry and spectrally resolved using an Acton spectrometer coupled with a Roper Scientific liquid nitrogen cooled charge-coupled device. PL decay times presented in the Supplementary Information section were measured using the 2 ps, 86 MHz, 400 nm frequency doubled output of a Ti-Sapphire oscillator as excitation and a PicoQuant Time Correlated Single Photon Counting System with an Instrument Response Function (IRF) of 120ps connected to a PMT detector mounted on the side port of the Acton Spectrometer. All PL/LD experiments were carried out at low excitation powers (<100 μW) to avoid exciton–exciton annihilation or exciton–polaron interactions (Supplementary Fig. 4). For the PL decay measurements presented in Supplementary Fig. 5, we estimated the number of photons per pulse absorbed by the films is $\sim 3 \times 10^6$, while the number of molecules illuminated, estimated from the film crystal structure and the beam size, is a thousand times larger. All spectra were corrected for the charge-coupled device response and grating reflectivity across the spectral range of interest. To compensate for the difference in grating reflectivity for s- and p-polarized light, a quarter waveplate was inserted between the PL polarization analyser and the spectrometer slit. A correction factor was also introduced to account for the differences in the s and p reflectivity and transmittance of the beam splitter used in these experiments. Temperature-dependent PL studies were conducted in a continuous flow Oxford Instruments optical cryostat where the sample temperature was continuously varied from 5 to 300 K.

X-ray diffraction. We independently confirmed the 'edge-on' stacking through out-of-plane X-ray scattering studies performed at the NSLS-I facility of the Brookhaven National Laboratory. The 2D scattering map for sample A, shown in Supplementary Fig. 1, reveals clear peaks at $2\theta = 3°$, $6°$ and $9°$, corresponding to the (100) orthorhombic reflection with a lattice parameter $\mathbf{a} = 23.7$ Å. This number matches the lattice constant of α-phase H_2Pc and confirms the in-plane orientation of the stacking axis. The $2\theta = 3°$ region of the scattering map is re-plotted in Supplementary Fig. 1b on a logarithmic scale to highlight the arc-like diffuse scattering, which indicates the presence of some disorder in the film. A complete 3D structural determination is not yet available for $H_2Pc\text{-}OC_8$.

References

1. Kippelen, B. & Bredas, J. L. Organic photovoltaics. *Energy Environ. Sci.* **2**, 251–261 (2009).
2. Chidichimo, G. & Filippelli, L. Organic solar cells: problems and perspectives. *Int. J. Photoenergy* **2010**, 123534 (2010).
3. Payne, M. M., Parkin, S. R., Anthony, J. E., Kuo, C. C. & Jackson, T. N. Organic field-effect transistors from solution-deposited functionalized acenes with mobilities as high as 1 cm²/Vs. *J. Am. Chem. Soc.* **127**, 4986–4987 (2005).
4. Tang, Q., Li, H., Liu, Y. & Hu, W. High-performance air-stable n-type transistors with an asymmetrical device configuration based on organic single-crystalline submicrometer/nanometer ribbons. *J. Am. Chem. Soc.* **128**, 14634–14639 (2006).
5. Diao, Y. *et al.* Solution coating of large-area organic semiconductor thin films with aligned single-crystalline domains. *Nat. Mater.* **12**, 665–671 (2013).
6. Liu, S. H., Wang, W. C. M., Briseno, A. L., Mannsfeld, S. C. E. & Bao, Z. N. Controlled deposition of crystalline organic semiconductors for field-effect-transistor applications. *Adv. Mater.* **21**, 1217–1232 (2009).
7. Sergeyev, S., Pisula, W. & Geerts, Y. H. Discotic liquid crystals: A new generation of organic semiconductors. *Chem. Soc. Rev.* **36**, 1902–1929 (2007).
8. Forrest, S. R. The path to ubiquitous and low-cost organic electronic appliances on plastic. *Nature* **428**, 911–918 (2004).
9. Najafov, H., Lee, B., Zhou, Q., Feldman, L. C. & Podzorov, V. Observation of long-range exciton diffusion in highly ordered organic semiconductors. *Nat. Mater.* **9**, 938–943 (2010).
10. Holt, J., Singh, S., Drori, T., Ye, Z. & Vardeny, Z. V. Optical probes of π-conjugated polymer blends with strong acceptor molecules. *Phys. Rev. B* **79**, 195210 (2009).
11. Wo, S. T., Headrick, R. L. & Anthony, J. E. Fabrication and characterization of controllable grain boundary arrays in solution-processed small molecule organic semiconductor films. *J. Appl. Phys.* **111**, 073716 (2012).
12. Cour, I. *et al.* Origin of stress and enhanced carrier transport in solution-cast organic semiconductor films. *J. Appl. Phys.* **114**, 093501 (2013).
13. Headrick, R. L., Wo, S., Sansoz, F. & Anthony, J. E. Anisotropic mobility in large grain size solution processed organic semiconductor thin films. *Appl. Phys. Lett.* **92**, 063302 (2008).
14. Rawat, N. *et al.* Macroscopic molecular ordering and exciton delocalization in crystalline phthalocyanine thin films. *J. Phys. Chem. Lett.* **6**, 1834–1840 (2015).
15. Kobayashi, N., Ogata, H., Nonaka, N. & Luk'yanets, E. A. Effect of peripheral substitution on the electronic absorption and fluorescence spectra of metal-free and zinc phthalocyanines. *Chemistry* **9**, 5123–5134 (2003).
16. Orti, E., Bredas, J. L. & Clarisse, C. Electronic-structure of phthalocyanines—theoretical investigation of the optical-properties of phthalocyanine monomers, dimers, and crystals. *J. Chem. Phys.* **92**, 1228–1235 (1990).
17. Davydov, A. S. *Theory of Molecular Excitons* (Plenum, 1971).
18. Andzelm, J., Rawlett, A. M., Orlicki, J. A. & Snyder, J. F. Optical properties of phthalocyanine and naphthalocyanine compounds. *J. Chem. Theory Comput.* **3**, 870–877 (2007).
19. Djurisic, A. B. *et al.* Spectroscopic ellipsometry of metal phthalocyanine thin films. *Appl. Opt.* **42**, 6382–6387 (2003).
20. Garab, G. & van Amerongen, H. Linear dichroism and circular dichroism in photosynthesis research. *Photosynth. Res.* **101**, 135–146 (2009).
21. Flora, W. H., Mendes, S. B., Doherty, 3rd W. J., Saavedra, S. S. & Armstrong, N. R. Determination of molecular anisotropy in thin films of discotic assemblies using attenuated total reflectance uv-visible spectroscopy. *Langmuir.* **21**, 360–368 (2005).
22. Verma, S., Ghosh, A., Das, A. & Ghosh, H. N. Ultrafast exciton dynamics of J- and H-aggregates of the porphyrin-catechol in aqueous solution. *J. Phys. Chem. B* **114**, 8327–8334 (2010).
23. Alfredsson, Y. *et al.* Phase and molecular orientation in metal-free phthalocyanine films on conducting glass: characterization of two deposition methods. *Thin Solid Films* **493**, 13–19 (2005).
24. Muccini, M., Murgia, M., Taliani, C., Degli Esposti, A. & Zamboni, R. Optical properties and the photoluminescence quantum yield of organic molecular materials. *J. Opt. A Pure Appl. Opt.* **2**, 577–583 (2000).
25. Freyer, W. F., Neacsu, C. C. & Raschke, M. B. Absorption, luminescence, and raman spectroscopic properties of thin films of benzo-annelated metal-free porphyrazines. *J. Lumin.* **128**, 661–672 (2008).
26. Kasha, M., Rawls, H. R. & El-Bayoumi, M. A. The exciton model in molecular spectroscopy. *Pure Appl. Chem.* **11**, 371–392 (1965).
27. Tempelaar, R., Stradomska, A., Knoester, J. & Spano, F. C. Anatomy of an exciton: Vibrational distortion and exciton coherence in h- and j-aggregates. *J. Phys. Chem. B* **117**, 457–466 (2013).
28. Bardeen, C. J. The structure and dynamics of molecular excitons. *Annu. Rev. Phys. Chem.* **65**, 127–148 (2014).
29. Crooker, S. A. *et al.* Imaging spin transport in lateral ferromagnet/semiconductor structures. *Science* **309**, 2191–2195 (2005).
30. Kaminsky, W., Claborn, K. & Kahr, B. Polarimetric imaging of crystals. *Chem. Soc. Rev.* **33**, 514–525 (2004).
31. Rivnay, J. *et al.* Large modulation of carrier transport by grain-boundary molecular packing and microstructure in organic thin films. *Nat. Mater.* **8**, 952–958 (2009).
32. Wong, C. Y., Cotts, B. L., Wu, H. & Ginsberg, N. S. Exciton dynamics reveal aggregates with intermolecular order at hidden interfaces in solution-cast organic semiconducting films. *Nat. Commun.* **6**, 5946 (2015).
33. Kyaw, A. K. K. *et al.* Intensity dependence of current-voltage characteristics and recombination in high-efficiency solution-processed small-molecule solar cells. *Acs Nano* **7**, 4569–4577 (2013).
34. List, E. J. W. *et al.* Interaction of singlet excitons with polarons in wide band-gap organic semiconductors: a quantitative study. *Phys. Rev. B* **64**, 155204 (2001).
35. Furis, M., Hollingsworth, J. A., Klimov, V. I. & Crooker, S. A. Time- and polarization-resolved optical spectroscopy of colloidal cdse nanocrystal quantum dots in high magnetic fields. *J. Phys. Chem. B* **109**, 15332–15338 (2005).

Acknowledgements

We thank Dr Rory Waterman from the Chemistry Department at the University of Vermont for invaluable help with column chromatography purification procedures and former UVM graduate students Lan Zhou and Yingping Wang for assistance with X-ray diffraction measurements. Funding for this research was provided by Furis' NSF CAREER Award DMR-1056589. We also acknowledge partial support from R Headrick's NSF award number DMR-1307017. Partial funding for the equipment used in this research was provided by NSF–MRI program under award DMR-0821268

Author contributions

Z.P., N.R. and L.M. performed the experiments, R.L.H. performed the X-ray scattering experiments, Z.P. and M.F. analysed data; N.R. and I.C. purified the samples and fabricated the thin films; and M.F., N.R. and Z.P. wrote the paper.

Additional information

18

Structural identification of electron transfer dissociation products in mass spectrometry using infrared ion spectroscopy

Jonathan Martens[1], Josipa Grzetic[1], Giel Berden[1] & Jos Oomens[1,2]

Tandem mass spectrometry occupies a principle place among modern analytical methods and drives many developments in the 'omics' sciences. Electron attachment induced dissociation methods, as alternatives for collision-induced dissociation have profoundly influenced the field of proteomics, enabling among others the top-down sequencing of entire proteins and the analysis of post-translational modifications. The technique, however, produces more complex mass spectra and its radical-driven reaction mechanisms remain incompletely understood. Here we demonstrate the facile structural characterization of electron transfer dissociation generated peptide fragments by infrared ion spectroscopy using the tunable free-electron laser FELIX, aiding the elucidation of the underlying dissociation mechanisms. We apply this method to verify and revise previously proposed product ion structures for an often studied model tryptic peptide, $[AlaAlaHisAlaArg + 2H]^{2+}$. Comparing experiment with theory reveals that structures that would be assigned using only theoretical thermodynamic considerations often do not correspond to the experimentally sampled species.

[1] Radboud University, Institute for Molecules and Materials, FELIX Laboratory, Toernooiveld 7c, 6525ED Nijmegen, The Netherlands. [2] Van 't Hoff Institute for Molecular Sciences, University of Amsterdam, Science Park 908, 1098XH Amsterdam, The Netherlands. Correspondence and requests for materials should be addressed to J.O. (email: j.oomens@science.ru.nl).

M ass spectrometry-based analysis of peptides and proteins in the bioanalytical and clinical sciences relies on the gas-phase dissociation of their molecular ions to give sequence fragments from which the original primary structure can be inferred. Sequencing by the increasingly popular electron-induced dissociation methods (ExD, such as electron capture and transfer dissociation, ECD and ETD) has recently seen rapid development and widespread application. ECD in Fourier transform ion cyclotron resonance (FT-ICR) mass spectrometry (MS)[1–3] and more recently ETD in a much broader range of mass spectrometers[4–7] constitute the two primary variations of this method. These methods have shown impressive improvements over collision-induced dissociation tandem MS, primarily in the sense that labile post-translational modifications are not detached during activation revealing their position along the backbone, and that sequence coverage is increased, making the sequencing of intact proteins in top-down strategies possible[8–12].

Electron attachment to multiply protonated peptides or proteins leads to extensive backbone fragmentation, which most often takes the form of backbone $N-C_\alpha$ bond cleavages to provide the c- and z-type ion series[1,2,13,14]. The process of electron-induced $N-C_\alpha$ bond cleavage in peptide cation radicals has been the subject of many recent experimental and theoretical studies[13,15–20]. However, the mechanisms involved in electron attachment to peptide ions, the possible role of excited electronic states, and the structural rearrangements and fragmentation reactions that can follow, remain only partially understood. Furthermore, answers to questions regarding the ratios of fragment ions produced and their structures, especially the odd electron fragments, remain, at least partially, elusive. This is undoubtedly related to the uncertainty in the detailed nature of the (open-shell) dissociation products. For example, ExD product ions are known to undergo hydrogen atom migration reactions that increase/decrease the expected masses of the fragment ions, having direct practical implications for ExD-based sequencing applications[21,22].

Several reaction mechanisms have been proposed for ExD of which the Cornell mechanism from the group of McLafferty[2] and the more recent Utah–Washington (UW) mechanism from the groups of Tureček and Simons[19,23–25] are best known. Frison et al.[26] have discussed the different structures of c-type fragments that would result from different ECD mechanisms and were able to spectroscopically identify an amide c-type ion. While not being able to exclude the possibility that this structure results from a rearrangement after the ECD process, this assignment appears to support the UW mechanism, rather than mechanisms that would directly produce enol-imine c-type ions.

In ExD dissociation, the radical is typically on the z-type fragment, although H-atom migration reactions in the dissociating molecule may transfer the radical to the c-type fragment, resulting in changes in the m/z values of the fragments and complicating the interpretation of ExD MS/MS data. As a result, apparent unit mass shifts of backbone sequence ions in ExD spectra are thus not rare and result in an increased occurrence of ions with overlapping masses. As an illustration, for a peptide with the sequence AlaAlaHisAlaArg, hydrogen atom migration to the open-shell $z_3^{\bullet+}$ fragment would give a closed-shell z_3^+ ion having the same chemical formula as the c_4^+ fragment and makes this mass peak unassignable for sequencing purposes.

Unfortunately, the detailed information about an ion's molecular structure is scarce and often undecipherable from MS/MS data alone. Infrared spectroscopy allows the structural characterization of both the trapped precursor and fragment ions. Infrared ion spectroscopy (IRIS) has been used extensively to

determine the gas-phase structures of molecular ions in MS, and specifically in the effort to elucidate peptide fragmentation mechanisms involved in collisional dissociation[27–34]. Tunable infrared free-electron lasers (FELs), such as FELIX at our institute[35], have played an important role in the recent development of IRIS[36], and are especially useful to obtain fingerprint infrared spectra of ionic species on commercial MS platforms[37]. However, to date, only a single example of infrared spectroscopy on ECD-generated fragments has been reported, in which a small closed-shell c-type ion was examined[26].

Structural characterization of peptidic ExD product ions is something that has been highly sought after for a number of years using a variety of methods, including collisional activation[38,39], ion mobility methods[40] and more recently ultraviolet photo-dissociation studies[41–43]. However, in comparison with infrared spectroscopy these methods provide limited information, regarding the structure and conformation of gas-phase ions. Structural characterization of the dissociation products by infrared spectroscopy and quantum chemistry provides a stringent test to ascertain and confirm uncertain aspects of the fragmentation mechanisms and allows for additional information to be extracted from MS/MS data. Here we present the first direct structural characterization of ETD-generated fragments using IRIS and demonstrate the strengths of this technique for addressing the questions surrounding gas-phase peptide radicals.

Results

ETD MS/MS of $[AAHAR + 2H^+]^{2+}$. The ETD MS/MS spectrum of the $[AAHAR + 2H]^{2+}$ (263 m/z) ion is presented in red in Fig. 1. We have characterized each of the $z^\bullet$-type fragments (depicted in the peptide sequence shown in Fig. 2) from the ETD MS/MS spectrum using infrared spectroscopy, providing a comprehensive identification of their structures and conformations. This detailed characterization forms a basis for analysing the reactions, leading to their formation.

Tryptic peptides of the AAXAR type and their ETD MS/MS behaviour have been extensively studied previously[39,42,44–47]. Here, consistent with previously reported results, we observe that fragment ion intensity is approximately split over the charge reduced ion (ETnoD) and $z^\bullet$-type sequence ions (Supplementary Fig. 1). For AAHAR, having arginine in the C-terminal position, it is not surprising that C-terminal $z^\bullet$-type fragments primarily retain the proton after dissociation and that they are the principle

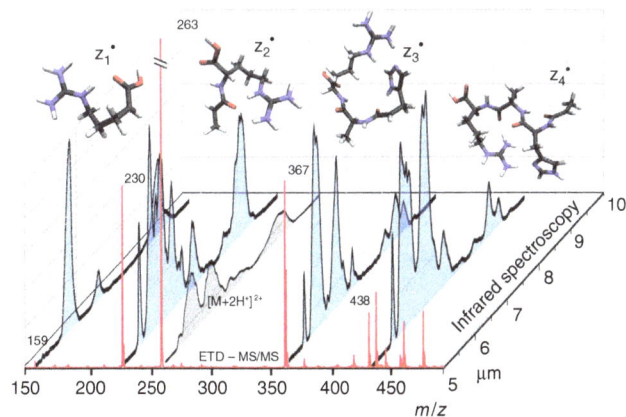

Figure 1 | The ETD MS/MS spectrum of $[AAHAR + 2H]^{2+}$ with the corresponding infrared spectra. The infrared spectra of the ETD-generated fragments are shown in black/blue and that of the precursor peptide in black/grey. Supplementary Fig. 1 contains the comprehensive ETD MS/MS results.

Figure 2 | Dissociation scheme and notation used for product ions. c- and z-type peptide fragments typically result from ETD MS/MS. Here we label only the discussed sequence ions from ETD of $[AAHAR+2H]^{2+}$. Note that fragments carrying a '•' symbol are open-shell radicals and those without are closed shell.

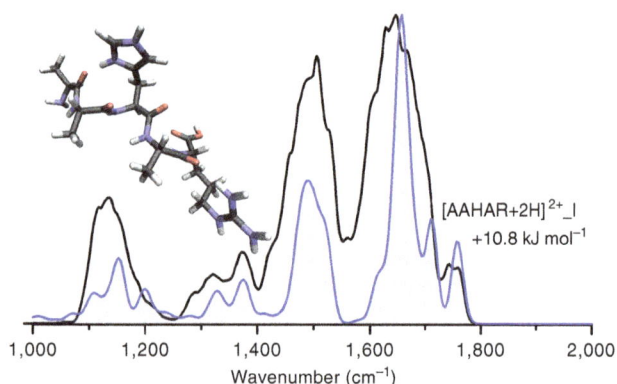

$[AAHAR+2H]^{2+}\_I$
$+10.8$ kJ mol^{-1}

Figure 3 | The infrared spectrum of $[AAHAR+2H]^{2+}$ 263 m/z. The experimental spectrum is presented in black and the spectrum of the assigned calculated structure from this study is shown in blue along with the structure and relative free energy at 298 K.

Table 1 | Summary of structural properties and relative free energies for selected calculated structures.

	m/z	His tautomer	Radical	Rel. ΔG (kJ mol^{-1})
$[AAHAR+2H^+]^{2+}$	263			
I		—	—	$+10.8^*$
II		—	—	0.0
$z_1^{\bullet+}$	159			
I		—	α	$+5.4^*$
II		—	δ (Arg)	0.0
III		—	β (Arg)	$+10.1$
IV		—	γ (Arg)	$+14.7$
$z_2^{\bullet+}$	230			
I		—	α	0.0^*
II		—	α	$+10.2$
III		—	δ (Arg)	$+26.5$
$z_3^{\bullet+}$	367			
I		N3	α	$+22.4^*$
II		N1	α	0.0
III		N3	β (His)	0.0
$z_4^{\bullet+}$	438			
I		N3	α	0.0^*
II		N3	α	$+62.5$
III		N3	—	$+54.4$
IV		N3	β (His)	$+5.7$
z_3^+	368			
I		N1	—	$+28.5^*$
c_4_I		—	—	—
II		N3	—	0.0

Rel., relative.
An asterisk (*) indicates the structure assigned spectroscopically.

fragments in the ETD MS/MS spectrum. The electron attachment process is no doubt affected by the protonation sites and conformation of the parent peptide. In its doubly protonated state, this peptide has been shown to protonate on the imidazole group of the histidine side chain and the guanidine group of the arginine side chain[45], and our spectroscopic data and calculations confirm this.

Structure of $[AAHAR+2H^+]^{2+}$ precursor peptide at m/z 263. Figure 3 presents the infrared spectrum obtained for the doubly protonated peptide $[AAHAR+2H]^{2+}$ at m/z 263. Comparison with the calculated spectrum (blue) allows assignment of the protonation sites (His and Arg sidechains) and the conformation to be made. The band just $<1,800$ cm^{-1} is consistent with the free (or weakly H-bonded) carboxyl group. Both charged sites are hydrogen bound to backbone carbonyl groups. In terms of hydrogen bonding, involving the charged sites and the C and N termini, this assignment confirms a previously proposed structure[44], although it has an overall more extended conformation, easily distinguished by comparison with the infrared spectra (Supplementary Fig. 2).

For singly charged fragments retaining the His residue, the imidazole ring can tautomerize (having the H on either the N1 or N3 position) and we address this issue for the $z_3^{\bullet+}$ and $z_4^{\bullet+}$ fragments. Note that all fragments carrying a '•' symbol are open-shell radicals and those without are closed shell, where H's and electrons are implicit. Backbone sequence fragments studied here with the structural identification based on our spectroscopic results, as detailed in the following are summarized in Table 1.

$z_1^{\bullet+}$ fragment structure. The $z_1^{\bullet+}$ fragment is the smallest C-terminal fragment obtained upon ETD of the precursor peptide. Calculated structure $z_1^{\bullet}\_I$ was identified to match most closely with experiment, as demonstrated in the top panel of Fig. 4. This structure is the lowest-energy calculated structure from our selection of ~30 structures from the molecular dynamics (MD) procedure. In structure $z_1^{\bullet}\_I$, the radical is located at the α-carbon adjacent to the C terminus and the carboxyl C=O stretch is found at 1,635 cm^{-1}. On the basis of this structure, radicals at other positions along the carbon chain of this fragment were defined and optimized, most of which were higher in energy. H-atom migration to the δ-carbon of the Arg side chain gave a structure 5.4 kJ mol^{-1} more stable; however, a calculated barrier of 115.2 kJ mol^{-1} (Supplementary Fig. 3) must be overcome to reach it and it is not consistent with the experimental infrared spectrum. Barriers of >100 kJ mol^{-1} for H-atom migrations have been reported for different ETD fragments[39,48,49]. With the radical in the α-position, conjugation with the carbonyl occurs giving partial double-bond character to the CC-bond, while lowering the bond order of the carbonyl, causing a significant red shift of its stretching mode relative to a carbonyl stretch $>1,700$ cm^{-1} in the β-, γ- and δ-radicals (Supplementary Fig. 4). This is a clear demonstration that the position of the radical can strongly influence the vibrational spectrum, in this case the C=O stretch, highlighting the value of infrared spectroscopy for characterizing open-shell peptide fragments.

$z_2^{\bullet+}$ fragment structure. Figure 5 and Supplementary Fig. 5 present infrared spectra for the $z_2^{\bullet+}$ fragment from ETD of

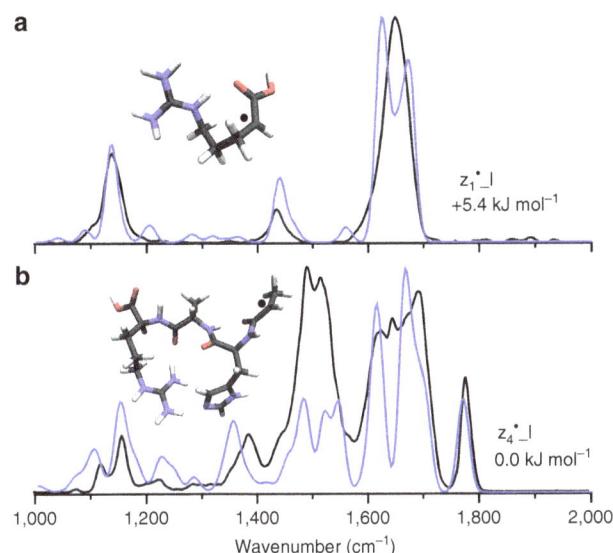

Figure 4 | The infrared spectra of the $z_1^{\bullet+}$ and $z_4^{\bullet+}$ fragments from ETD of [AAHAR + 2H]$^{2+}$. The experimental spectra are presented in black in both cases and assigned calculated structures are shown in blue for (**a**) the $z_1^{\bullet+}$ fragment and (**b**) the $z_4^{\bullet+}$ fragment. The associated relative free energies (298 K) and structures are inlayed with the radical sites labelled by '•'.

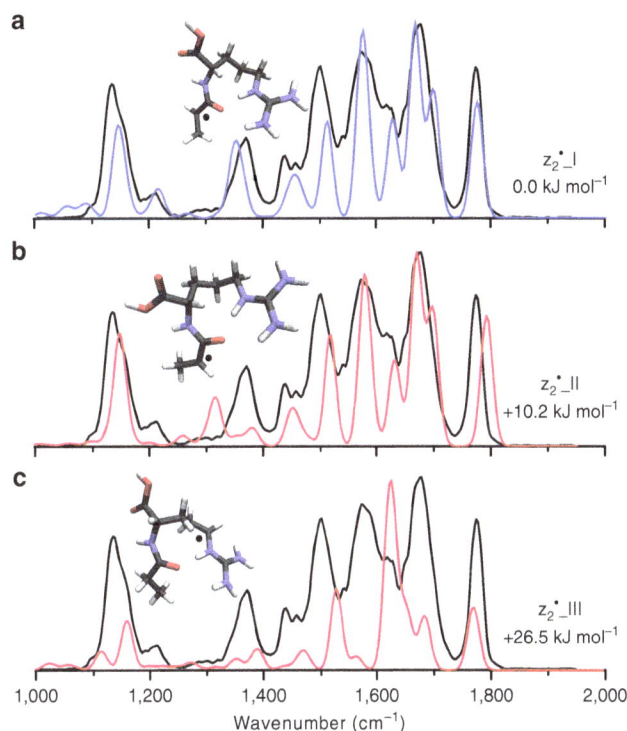

Figure 5 | The infrared spectrum of the $z_2^{\bullet+}$ fragment from ETD of [AAHAR + 2H]$^{2+}$. The experimental spectrum is presented in black and is compared with computed spectra for different low-energy structures. (**a**) The calculated spectrum for the assigned structure is shown in blue. (**b,c**) The calculated spectra for structures disregarded on the basis of spectral mismatch are shown in red. Calculated structures and relative free energies (298 K) are inlayed for each plot.

[AAHAR + 2H$^+$]$^{2+}$. The calculated spectrum of $z_2^{\bullet}$_I, presented in the top panel, matches the experimental spectrum well. This structure is the lowest-energy calculation obtained from our computational procedure. A previously proposed structure[39], labelled here as $z_2^{\bullet}$_II and presented in the centre panel, closely resembles $z_2^{\bullet}$_I. However, the alternate hydrogen bonding orientation of the C terminus is distinguished by comparison with experimental and calculated infrared spectra in the 1,300–1,400 cm^{-1} region and the carbonyl stretching region just <1,800 cm^{-1}. This refinement of the conformation reduces the relative energy by 10.2 kJ mol^{-1}. The bottom panel in the figure shows a comparison with the infrared spectrum calculated for the product of hydrogen migration from the α-carbon of the Ala residue to the δ-carbon of the Arg residue, a species 26.5 kJ mol^{-1} higher in energy and readily distinguishable spectroscopically.

$z_3^{\bullet+}$ fragment structure. Figure 6 presents the infrared spectrum for the $z_3^{\bullet+}$ fragment from ETD of [AAHAR + 2H$^+$]$^{2+}$ with the spectrum of the assigned calculated structure $z_3^{\bullet}$_I in blue and spectra of unassigned alternative structures in red. This assignment was made after considering different conformers, imidazole tautomers and products, resulting from hydrogen atom migration interconverting between the α-radical and the β-radical on the His side chain. A low-energy His N1 tautomer ($z_3^{\bullet}$_II) was identified to be 22.4 kJ mol^{-1} lower in energy than $z_3^{\bullet}$_I, a His N3 tautomer; however, the calculated spectrum of this species does not match as well to the experiment, especially in the 1,200–1,400 cm^{-1} region (and the 3400–3600 cm^{-1} region displayed in Supplementary Fig. 6). Furthermore, a β-radical structure ($z_3^{\bullet}$_III) was found to be 22.4 kJ mol^{-1} more stable, but we do not assign this species on the basis of its spectral mismatch (Fig. 6 and Supplementary Fig. 6). The majority of stable structures we identified for each structure/ tautomer of the $z_3^{\bullet+}$ ion features a stabilizing hydrogen bonding interaction between the imidazole side chain of His and the guanidinium side chain of Arg, leaving little flexibility for the orientation of the carboxyl group and giving a free C–OH group and a

C=O weakly hydrogen bonded to the adjacent amide N–H. Being very sensitive to local environment, the position of the carboxyl C=O stretch just <1,800 cm^{-1} can be used as a diagnostic signature. Structures $z_3^{\bullet}$_II and $z_3^{\bullet}$_III feature hydrogen bonds between the imidazole nitrogen and hydrogens of the two primary nitrogens of the guanidinium group, while in $z_3^{\bullet}$_I the hydrogen bond of the imidazole nitrogen is shared between the hydrogens of the secondary nitrogen and one primary nitrogen.

$z_4^{\bullet+}$ fragment structure. The bottom panel of Fig. 4 presents the infrared spectrum for the $z_4^{\bullet+}$ fragment from ETD of [AAHAR + 2H$^+$]$^{2+}$ and the assigned calculated structure (blue). This is the overall lowest-energy structure obtained after an extensive MD-based search over different conformers for various tautomers and structures obtained by hydrogen migration from the α-radical to the His β-radical (see comparison of $z_4^{\bullet}$_IV in Supplementary Fig. 7). Similar to $z_3^{\bullet}$_I, the assigned structure, $z_4^{\bullet}$_I, has a hydrogen bond between the imidazole side chain and the guanidinium group of the Arg residue. Structure $z_4^{\bullet}$_I offers a refinement over a previously proposed structure[39] (here, re-optimized at the currently applied level of theory), giving both a better spectral match (see $z_4^{\bullet}$_II in Supplementary Fig. 7) and being ~60 kJ mol^{-1} lower in energy.

Fragment ion at m/z 368 can be z_3^+ or c_4^+. While inter- and intramolecular (between c and z fragment pairs) hydrogen migration reactions are commonly observed in ETD MS/MS, their behaviour is still relatively weakly understood. For [AAHAR + 2H$^+$]$^{2+}$, only for the $z_3^{\bullet+}$ cation do we observe an

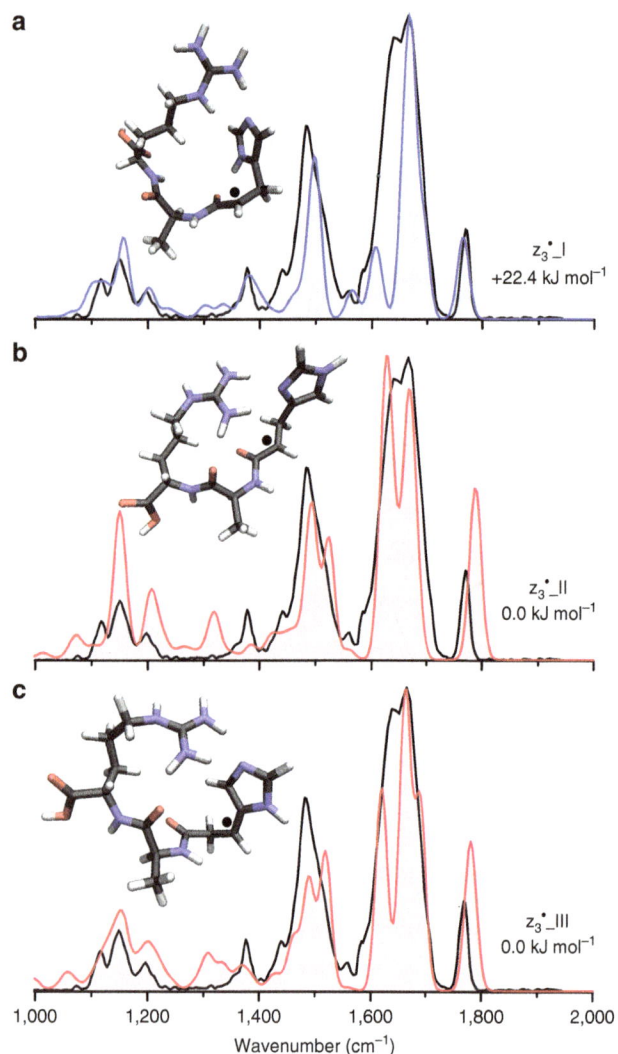

Figure 6 | The infrared spectrum of the $z_3^{\bullet+}$ fragment from ETD of [AAHAR + 2H]$^{2+}$. The experimental spectrum is presented in black and is compared with computed spectra for different low-energy structures. (**a**) The calculated spectrum for the assigned structure is shown in blue. (**b,c**) The calculated spectra for structures disregarded on the basis of spectral mismatch are shown in red. Calculated structures and relative free energies (298 K) are inlayed for each plot.

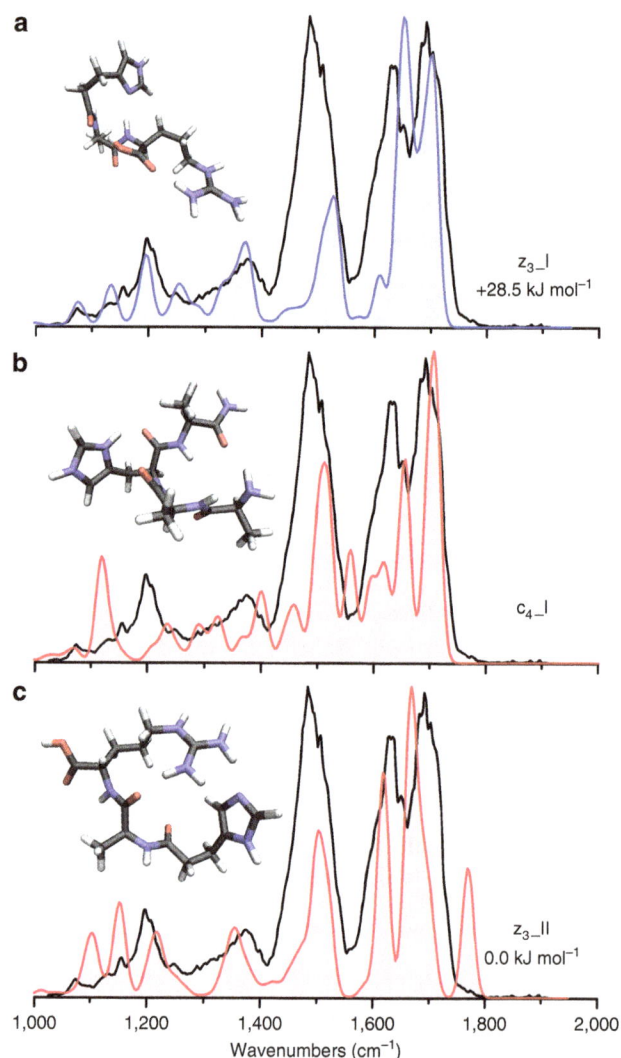

Figure 7 | The infrared spectrum of the *m/z* 368 fragment from ETD of [AAHAR + 2H]$^{2+}$. The experimental spectrum is presented in black and is compared with computed spectra for different low-energy structures. (**a**) The spectrum of the assigned calculated closed-shell z_3^+ structure is shown in blue with the structure and relative free energy (298 K) inlayed. (**b**) c_4_I is a low-energy c_4^+ conformation and its calculated infrared spectrum is presented in red. (**c**) z_3_II is the lowest energy calculated z_3^+ structure identified in this work.

appreciable extent of such a reaction, where we see both the open-shell $z_3^{\bullet+}$ fragment (*m/z* 367) and a closed-shell z_3^+ fragment (*m/z* 368). Highlighting the complications that can arise from hydrogen atom migrations in ETD, the c_4^+ fragment ($C_{15}N_7H_{25}O_4$) has the same chemical formula and overlaps the closed-shell z_3^+ fragment also at *m/z* 368. Identification and consideration of the hydrogen migration products (loss/gain) are important for assigning fragment ions and correct sequencing[38].

In Fig. 7, we identify the *m/z* 368 fragment ion as the z_3^+ species (z_3_I) based on infrared spectral matching. This structure features an alternative hydrogen bonding arrangement in comparison with the open-shell $z_3^{\bullet+}$ and $z_4^{\bullet+}$ fragments described above, most significantly affecting the C terminus (trans configuration) and red shifting the carboxyl C=O stretch for z_3_I away from the position just $<1,800\,cm^{-1}$ for the open-shell $z_3^{\bullet+}$ and $z_4^{\bullet+}$ fragments. A closed-shell equivalent of the geometry of the open-shell $z_3^{\bullet}\_I$ structure is defined as z_3_II and is 28.5 kJ mol^{-1} lower in energy than z_3_I. Calculated spectra

for structure z_3_II and c_4_I, a low-energy c_4^+ conformation, are presented in the bottom two panels of Fig. 7 and support the assignment of z_3_I.

Discussion

These results demonstrate the first use of IRIS to characterize the structures of ETD-generated peptide fragments. Using a model tryptic pentapeptide, precursor ion conformation has been related to the observed ETD fragmentation pattern, and the structures and conformations of the various fragment ions. We show that it is possible to distinguish both conformational details and different radical species, when this approach is combined with routine computational modelling.

We conclude that if structural assignments were made only on the basis of theoretical (thermodynamic) considerations, these assignments would in many cases not match the species observed in experiment—highlighting the potential for IRIS to diagnostically identify gas-phase organic radicals and, more specifically, the mechanisms associated with peptide fragmentation in electron attachment methods. Our results suggest that hydrogen atom transfer necessary for radical migration often does not occur after ETD (without additional activation), leading to the frequent observation of non-equilibrium product ions. Understanding intermolecular hydrogen atom migration is also of practical importance, as it causes shifts in ETD fragment masses and makes sequence ion assignments more complicated.

Methods

Ion spectroscopy in a modified ion trap mass spectrometer. The experiment is based on a commercial quadrupole ion trap mass spectrometer (Bruker, AmaZon Speed ETD) coupled to the infrared beam line of the FELIX FEL. $[M+2H]^{2+}$ peptide ions are generated by electrospray ionization. AAHAR (GeneCust (Luxemburg), 95% purity) solutions of 10^{-5}–$10^{-6}\,mol\,l^{-1}$ (in 50:50 acetonitrile:water, $\sim 1\%$ formic acid) are introduced at $120\,\mu l\,h^{-1}$ flow rates and desolvated by a pressurized nebulizing gas (N_2). The key hardware modifications to the instrument providing optical access to the ion population in the trap were the introduction of a new ring electrode having 3 mm holes at its top and bottom, the installation of mirrors below the trap to direct the beam back out of the instrument and optical windows in the vacuum housing. In ETD experiments, ions were accumulated for 0.1–15 ms in the trap, mass isolated and then reacted with fluoranthene radical anions for $\sim 250\,ms$. A fragment ion of interest was mass isolated in a subsequent MS/MS stage and irradiated with the tunable infrared beam from the FEL. In the experiments reported here, the FELIX FEL was set to produce infrared radiation in the form of 5–10 μs macropulses at 5 or 10 Hz and of 30–60 mJ (bandwidth $\sim 0.4\%$ of the centre frequency). Resonant absorption of infrared radiation leads to an increase in the internal energy of the molecule aided by intramolecular vibrational redistribution of the absorbed energy. When a sufficient number of photons is absorbed (here, typically in a single macropulse), unimolecular dissociation occurs and produces frequency-dependent fragment ion intensities in the mass spectrometer (Supplementary Note 1). Relating the parent and fragment ion intensities in the observed mass spectral data (yield = ΣI(fragment ions)/ΣI(parent + fragment ions)) generates an infrared vibrational spectrum. The yield at each infrared point is obtained from averaged mass spectra and is linearly corrected for laser power; the frequency is calibrated using a grating spectrometer.

Computational chemistry. We have employed a molecular mechanics (MM)/MD approach using AMBER 12 (refs 50,51). Molecular structures manually defined based on chemical intuition where first optimized for each ion at the B3LYP/6-31 + + G(d,p) level in Gaussian09 (ref. 52). Restrained electrostatic potential (RESP) charges from these initial results were used for parameterization of the nonstandard peptide ions in the antechamber program. After minimization within AMBER, a simulated annealing procedure up to 1,000 K was used with a 1 fs step size. Five hundred structures were obtained as snapshots throughout the procedure and after MM minimization were grouped based on structural similarity using prtraj in AMBER. Of these, 30–50 unique structures were then each optimized at the B3LYP/6-31 + + G(d,p) level[51,53,54] and vibrational spectra were calculated within the harmonic oscillator model (vibrational frequencies were scaled by 0.975). This computational approach was applied to all structural isomers considered for each ion, except for the small $z_1^{\bullet+}$ ion, where the MM/MD conformational search was only applied once using the alpha-radical species. Calculated line spectra were broadened using a Gaussian function with a full-width at half-maximum of $25\,cm^{-1}$ to facilitate comparison with experiment. Additional calculations using the LC-BLYP and M06 functionals for a selection of $z_1^{\bullet+}$ and $z_2^{\bullet+}$ structures, and ab initio MP2 calculations for the $z_1^{\bullet+}$ structures were performed to verify the validity of the choice of functional[49,55] and these results are summarized in Supplementary Table 1 and Supplementary Fig. 8. In general, vibrational frequencies were found to be best modelled at the B3LYP/6-31 + + G(d,p) level and calculated free energies are mostly consistent between these levels of theory. Supplementary Data 1–4 contain optimized geometries of the assigned z-type fragments.

References

1. Kruger, N. A., Zubarev, R. A., Horn, D. M. & McLafferty, F. W. Electron capture dissociation of multiply charged peptide cations. *Int. J. Mass. Spectrom.* **185–187**, 787–793 (1999).
2. Zubarev, R. A., Kelleher, N. L. & McLafferty, F. W. Electron capture dissociation of multiply charged protein cations. a nonergodic process. *J. Am. Chem. Soc.* **120**, 3265–3266 (1998).
3. Zubarev, R. A. *et al.* Electron capture dissociation for structural characterization of multiply charged protein cations. *Anal. Chem.* **72**, 563–573 (2000).
4. Syka, J. E. P., Coon, J. J., Schroeder, M. J., Shabanowitz, J. & Hunt, D. F. Peptide and protein sequence analysis by electron transfer dissociation mass spectrometry. *Proc. Natl Acad. Sci. USA* **101**, 9528–9533 (2004).
5. Pitteri, S. J., Chrisman, P. A., Hogan, J. M. & McLuckey, S. A. Electron transfer ion/ion reactions in a three-dimensional quadrupole ion trap: reactions of doubly and triply protonated peptides with SO2•. *Anal. Chem.* **77**, 1831–1839 (2005).
6. Xia, Y. *et al.* Implementation of ion/ion reactions in a quadrupole/time-of-flight tandem mass spectrometer. *Anal. Chem.* **78**, 4146–4154 (2006).
7. Coon, J. J. *et al.* Protein identification using sequential ion/ion reactions and tandem mass spectrometry. *Proc. Natl Acad. Sci. USA* **102**, 9463–9468 (2005).
8. Swaney, D. L., McAlister, G. C. & Coon, J. J. Decision tree-driven tandem mass spectrometry for shotgun proteomics. *Nat. Meth.* **5**, 959–964 (2008).
9. Breuker, K. & McLafferty, F. W. Native electron capture dissociation for the structural characterization of noncovalent interactions in native cytochrome c. *Angew. Chem. Int. Ed.* **42**, 4900–4904 (2003).
10. Breuker, K., Oh, H., Lin, C., Carpenter, B. K. & McLafferty, F. W. Nonergodic and conformational control of the electron capture dissociation of protein cations. *Proc. Natl Acad. Sci. USA* **101**, 14011–14016 (2004).
11. Oh, H. *et al.* Secondary and tertiary structures of gaseous protein ions characterized by electron capture dissociation mass spectrometry and photofragment spectroscopy. *Proc. Natl Acad. Sci. USA* **99**, 15863–15868 (2002).
12. Breuker, K. & McLafferty, F. W. Stepwise evolution of protein native structure with electrospray into the gas phase, 10 − 12 to 102s. *Proc. Natl Acad. Sci. USA* **105**, 18145–18152 (2008).
13. Zubarev, R. A. Reactions of polypeptide ions with electrons in the gas phase. *Mass. Spectrom. Rev.* **22**, 57–77 (2003).
14. Horn, D. M., Zubarev, R. A. & McLafferty, F. W. Automated de novo sequencing of proteins by tandem high-resolution mass spectrometry. *Proc. Natl Acad. Sci. USA* **97**, 10313–10317 (2000).
15. Tureček, F. & Julian, R. R. Peptide radicals and cation radicals in the gas phase. *Chem. Rev.* **113**, 6691–6733 (2013).
16. Simons, J. Mechanisms for S–S and N–Cα bond cleavage in peptide ECD and ETD mass spectrometry. *Chem. Phys. Lett.* **484**, 81–95 (2010).
17. Anusiewicz, I., Skurski, P. & Simons, J. Refinements to the Utah–Washington mechanism of electron capture dissociation. *J. Phys. Chem. B* **118**, 7892–7901 (2014).
18. Li, X., Lin, C., Han, L., Costello, C. E. & O'Connor, P. B. Charge remote fragmentation in electron capture and electron transfer dissociation. *J. Am. Soc. Mass. Spectrom.* **21**, 646–656 (2010).
19. Syrstad, E. A. & Tureček, F. Toward a general mechanism of electron capture dissociation. *J. Am. Soc. Mass. Spectrom.* **16**, 208–224 (2005).
20. Zhurov, K. O., Fornelli, L., Wodrich, M. D., Laskay, U. A. & Tsybin, Y. O. Principles of electron capture and transfer dissociation mass spectrometry applied to peptide and protein structure analysis. *Chem. Soc. Rev.* **42**, 5014–5030 (2013).
21. Savitski, M. M., Kjeldsen, F., Nielsen, M. L. & Zubarev, R. A. Hydrogen rearrangement to and from radical z fragments in electron capture dissociation of peptides. *J. Am. Soc. Mass. Spectrom.* **18**, 113–120 (2007).
22. Liu, J., Liang, X. & McLuckey, S. A. On the value of knowing a z• ion for what it is. *J. Proteome. Res.* **7**, 130–137 (2008).
23. Sobczyk, M. *et al.* Coulomb-assisted dissociative electron attachment: application to a model peptide. *J. Phys. Chem. A* **109**, 250–258 (2004).
24. Anusiewicz, I., Berdys-Kochanska, J. & Simons, J. Electron attachment step in electron capture dissociation (ECD) and electron transfer dissociation (ETD). *J. Phys. Chem. A* **109**, 5801–5813 (2005).
25. Chen, X. & Tureček, F. The arginine anomaly: arginine radicals are poor hydrogen atom donors in electron transfer induced dissociations. *J. Am. Chem. Soc.* **128**, 12520–12530 (2006).
26. Frison, G. *et al.* Structure of electron-capture dissociation fragments from charge-tagged peptides probed by tunable infrared multiple photon dissociation. *J. Am. Chem. Soc.* **130**, 14916–14917 (2008).
27. Lucas, B. *et al.* Investigation of the protonation site in the dialanine peptide by infrared multiphoton dissociation spectroscopy. *Phys. Chem. Chem. Phys.* **6**, 2659–2663 (2004).
28. Erlekam, U. *et al.* Infrared spectroscopy of fragments of protonated peptides: direct evidence for macrocyclic structures of b5 ions. *J. Am. Chem. Soc.* **131**, 11503–11508 (2009).
29. Bythell, B. J., Erlekam, U., Paizs, B. & Maître, P. Infrared spectroscopy of fragments from doubly protonated tryptic peptides. *Chemphyschem* **10**, 883–885 (2009).
30. Polfer, N. C., Oomens, J., Suhai, S. & Paizs, B. Spectroscopic and theoretical evidence for oxazolone ring formation in collision-induced dissociation of peptides. *J. Am. Chem. Soc.* **127**, 17154–17155 (2005).

31. Polfer, N. C., Oomens, J., Suhai, S. & Paizs, B. Infrared spectroscopy and theoretical studies on gas-phase protonated leu-enkephalin and its fragments: direct experimental evidence for the mobile proton. *J. Am. Chem. Soc.* **129**, 5887–5897 (2007).

32. Polfer, N. C. & Oomens, J. Vibrational spectroscopy of bare and solvated ionic complexes of biological relevance. *Mass. Spectrom. Rev.* **28**, 468–494 (2009).

33. Yoon, S. H. *et al.* IRMPD spectroscopy shows that agg forms an oxazolone b2 + ion. *J. Am. Chem. Soc.* **130**, 17644–17645 (2008).

34. Perkins, B. R. *et al.* Evidence of diketopiperazine and oxazolone structures for ha b2 + ion. *J. Am. Chem. Soc.* **131**, 17528–17529 (2009).

35. Oepts, D., van der Meer, A. F. G. & van Amersfoort, P. W. The free-electron-laser user facility FELIX. *Infrared Phys. Technol.* **36**, 297–308 (1995).

36. Oomens, J., Sartakov, B. G., Meijer, G. & Von Helden, G. Gas-phase infrared multiple photon dissociation spectroscopy of mass-selected molecular ions. *Int. J. Mass. Spectrom.* **254**, 1–19 (2006).

37. Bakker, J. M., Besson, T., Lemaire, J., Scuderi, D. & Maître, P. Gas-phase structure of a π-allyl – palladium complex: efficient infrared spectroscopy in a 7T Fourier transform mass spectrometer. *J. Phys. Chem. A* **111**, 13415–13424 (2007).

38. Hamidane, H. B. *et al.* Electron capture and transfer dissociation: peptide structure analysis at different ion internal energy levels. *J. Am. Soc. Mass. Spectrom.* **20**, 567–575 (2009).

39. Ledvina, A., Chung, T., Hui, R., Coon, J. & Tureček, F. Cascade dissociations of peptide cation-radicals. part 2. Infrared multiphoton dissociation and mechanistic studies of z-ions from pentapeptides. *J. Am. Soc. Mass. Spectrom.* **23**, 1351–1363 (2012).

40. Moss, C. L. *et al.* Assigning structures to gas-phase peptide cations and cation-radicals. an infrared multiphoton dissociation, ion mobility, electron transfer, and computational study of a histidine peptide ion. *J. Phys. Chem. B* **116**, 3445–3456 (2012).

41. Shaffer, C. J., Marek, A., Pepin, R., Slovakova, K. & Tureček, F. Combining UV photodissociation with electron transfer for peptide structure analysis. *J. Mass. Spectrom.* **50**, 470–475 (2015).

42. Nguyen, H. T. H., Shaffer, C. J. & Tureček, F. Probing peptide cation–radicals by near-uv photodissociation in the gas phase. structure elucidation of histidine radical chromophores formed by electron transfer reduction. *J. Phys. Chem. B* **119**, 3948–3961 (2015).

43. Nguyen, H. T. H., Shaffer, C. J., Pepin, R. & Tureček, F. U. V. Action spectroscopy of gas-phase peptide radicals. *J. Phys. Chem. Lett.* **6**, 4722–4727 (2015).

44. Tureček, F., Moss, C. L. & Chung, T. W. Correlating ETD fragment ion intensities with peptide ion conformational and electronic structure. *Int. J. Mass. Spectrom.* **330–332**, 207–219 (2012).

45. Tureček, F. *et al.* The histidine effect. electron transfer and capture cause different dissociations and rearrangements of histidine peptide cation-radicals. *J. Am. Chem. Soc.* **132**, 10728–10740 (2010).

46. Zimnicka, M., Moss, C., Chung, T., Hui, R. & Tureček, F. Tunable charge tags for electron-based methods of peptide sequencing: design and applications. *J. Am. Soc. Mass. Spectrom.* **23**, 608–620 (2012).

47. Chung, T., Hui, R., Ledvina, A., Coon, J. & Tureček, F. Cascade dissociations of peptide cation-radicals. part 1. scope and effects of amino acid residues in penta-, nona-, and decapeptides. *J. Am. Soc. Mass. Spectrom.* **23**, 1336–1350 (2012).

48. Chung, T. W. & Tureček, F. Backbone and side-chain specific dissociations of z ions from non-tryptic peptides. *J. Am. Soc. Mass. Spectrom.* **21**, 1279–1295 (2010).

49. Riffet, V., Jacquemin, D. & Frison, G. H-atom loss and migration in hydrogen-rich peptide cation radicals: The role of chemical environment. *Int. J. Mass. Spectrom.* **390**, 28–38 (2015).

50. Case, D.A. *et al.* AMBER 12 (University of California, San Francisco, CA, 2012).

51. Martens, J. K., Grzetic, J., Berden, G. & Oomens, J. Gas-phase conformations of small polyprolines and their fragment ions by IRMPD spectroscopy. *Int. J. Mass. Spectrom.* **377**, 179–187 (2015).

52. Frisch, M. J. *et al.* Gaussian 09 (Gaussian, Inc., Wallingford, CT, 2009).

53. Halls, M. D. & Schlegel, H. B. Comparison of the performance of local, gradient-corrected, and hybrid density functional models in predicting infrared intensities. *J. Chem. Phys.* **109**, 10587–10593 (1998).

54. Kapota, C., Lemaire, J., Maître, P. & Ohanessian, G. Vibrational signature of charge solvation vs salt bridge isomers of sodiated amino acids in the gas phase. *J. Am. Chem. Soc.* **126**, 1836–1842 (2004).

55. Riffet, V., Jacquemin, D., Cauët, E. & Frison, G. Benchmarking DFT and TD-DFT functionals for the ground and excited states of hydrogen-rich peptide radicals. *J. Chem. Theory. Comput.* **10**, 3308–3318 (2014).

Acknowledgements

We gratefully acknowledge the FELIX staff, particularly Dr B. Redlich and Dr A.F.G. van der Meer. In addition, the authors thank Professor F. Tureček and Dr C. Schaffer for discussion. Finally, the authors express their gratitude for the technical assistance from staff at Bruker Daltonics in Bremen, Germany, while implementing this experiment, in particular Dr C. Gebhardt. Financial support for this project was provided by the Chemical Sciences division of the 'Nederlandse organisatie voor Wetenschappelijk Onderzoek' (NWO) under VICI project no. 724.011.002. We also thank NWO Physical Sciences (EW) and the SurfSARA Supercomputer Center for providing the computational resources. This work is part of the research program of FOM, which is financially supported by NWO.

Author contributions

J.M., G.B. and J.O. conceived and designed the experiments. J.M. and J.G. performed the experiments and computations. The manuscript was co-written by all authors.

Additional information

19

Message in a molecule

Tanmay Sarkar[1], Karuthapandi Selvakumar[1], Leila Motiei[1] & David Margulies[1]

Since ancient times, steganography, the art of concealing information, has largely relied on secret inks as a tool for hiding messages. However, as the methods for detecting these inks improved, the use of simple and accessible chemicals as a means to secure communication was practically abolished. Here, we describe a method that enables one to conceal multiple different messages within the emission spectra of a unimolecular fluorescent sensor. Similar to secret inks, this molecular-scale messaging sensor (m-SMS) can be hidden on regular paper and the messages can be encoded or decoded within seconds using common chemicals, including commercial ingredients that can be obtained in grocery stores or pharmacies. Unlike with invisible inks, however, uncovering these messages by an unauthorized user is almost impossible because they are protected by three different defence mechanisms: steganography, cryptography and by entering a password, which are used to hide, encrypt or prevent access to the information, respectively.

[1] Department of Organic Chemistry, Weizmann Institute of Science, Rehovot 7610001, Israel. Correspondence and requests for materials should be addressed to D.M. (email: david.margulies@weizmann.ac.il).

Nowadays, the use of invisible inks to write messages, which can be revealed only when exposed to heat, light or a chemical solution, is mostly associated with children's games. However, only a century ago exceptionally simple chemicals were frequently used in times of war for espionage purposes[1,2]. The main advantage of using these inks was their accessibility to field agents, which enabled straightforward writing and reading of confidential information[3]. However, one drawback of using this technology is the ease by which messages can be exposed, which has led, for example, to the capture of the 'lemon juice spies' in World War I (WWI)[1]. A significant improvement in the ability to secure information by chemical means has been achieved with the development of molecular and biomolecular steganographic systems, in which specific chemical stimuli trigger the appearance of text and images. These data can be created by various sources, such as fluorescent materials[4-12], bacteria[13], antibodies[14], photonic crystals[15], NMR chemical shifts[16] and molecular computing systems[17-20]. Another important advantage of using molecular steganography systems, namely, their small scale, has also been demonstrated by the ability to conceal messages within individual DNA strands[21]. Finally, advances in the area of molecular logic gates[22-26] have resulted in alternative methods of securing information[22,27,28] by using multi-analyte fluorescent molecular sensors that can produce ID-codes[29] or can authorize password entries[30-41].

Herein we present a different approach to molecular information protection, which relies on the ability of a molecular-scale messaging sensor (m-SMS) to convert randomly selected chemical signals into unpredictable emission patterns and, in doing so, communicate short, chemically encoded messages with maximal security. This sensor is the second member of the combinatorial fluorescent molecular sensor family, developed by our group[42], which mimics the function of the olfactory system by integrating several nonspecific signalling receptors on a single molecular platform[43]. Unlike its predecessor[41-43], however, or any other fluorescent probe that responds to several analytes[24,44] or an analyte group[43], m-SMS was designed to operate as a universal sensor that can discriminate among a vast number of distinct chemical species. We show that this property not only distinguishes m-SMS from other types of fluorescent molecular sensors, but also from other chemical security systems[4-22,27-41] by enabling it to function as a molecular cipher device that can convert distinct chemical structures into unique encryption keys. In this way, the system can be used not only to hide the data (steganography), but also to encrypt and decrypt it (cryptography), as well as provide password protection when a higher level of security is needed. Because this system does not depend on using specific chemical inputs, unique instrumentations or complex experimental protocols, it is also very simple to operate. We show that m-SMS and/or the chemical ingredients can be concealed and delivered on plain letter paper and that the messages can be rapidly revealed using a low-cost, handheld spectrometer. This makes the m-SMS technology similar to the ancient technology of invisible inks in terms of simplicity, accessibility and the ease by which different messages can be concealed and exposed using common chemicals from various locations and in a short time.

Results

Design principles. The structure of m-SMS (Fig. 1a) consists of a *cis*-amino proline scaffold that is appended with three spectrally overlapping fluorophores: fluorescein (Flu), sulforhodamine B and nile blue (NB), which serve as a fluorescence resonance energy transfer (FRET) donor1–acceptor1/donor2–acceptor2 system, respectively. In addition, the sensor consists of various recognition elements for binding distinct chemical species. The

boronic acid and dipicolylamine (DPA) groups, for example, provide m-SMS with an affinity towards different saccharides[45] and metal ions[46], respectively. The thiourea and sulfonamide functionalities serve as additional metal ion-binding sites[47-49], as well as anion[50] receptors and hydrogen-bonding motifs[51,52]. Additional binding interactions may involve hydrogen bonding with the amides and carboxylic acid of m-SMS, in addition to hydrophobic interactions and π-stacking with the various aromatic groups. Finally, the Flu structure and protonation state are highly pH dependent[53], whereas solvatochromic NB[54] can interact with DNA and hydrophobic analytes (Fig. 1a). Additional recognition sites could also be formed upon the binding of analytes. DPA–metal ion complexes, for example, are known to interact with anions such as phosphates[55], whereas deprotonation of Flu by a base should enable the phenolic ligand to coordinate with metal ions[56]. This versatility of artificial receptors is counter intuitive to traditional fluorescent molecular sensor design[57], because it aims at creating a sensor that is inherently nonspecific. In this way, the binding of different analytes should induce the formation of distinct emission signatures by affecting FRET, photo-induced electron transfer, dye conjugation or charge transfer processes[57]. For example, the binding of metal ions to DPA could disrupt or enhance photo-induced electron transfer[58], whereas changes in pH or solvents could alter Flu conjugation[53] or intramolecular charge transfer processes within NB. In addition, because the different signalling and recognition elements are integrated on a single molecular platform, the interaction of m-SMS with any chemical species is likely to change the distance between the probes, which would affect the FRET efficiency. This covalent integration of dyes should also facilitate hiding, sending and extracting the molecular device without affecting the molar ratio between them and consequently, without changing the device's photophysical properties.

Multi-analyte identification. The unusual sensing mechanism underlying m-SMS was demonstrated by measuring its response to diverse chemical species (Fig. 1b) including different solvents (top left), metal ions (top right), saccharides (middle left), as well as its response to changing the pH (middle right) or polarity (bottom left) of the solution, and to the presence of complex mixtures such as those that can be found in soft drinks and medications (bottom right). Different emission signatures were also generated in the presence of different sugar phosphates, proteins and by changing analyte concentrations (Supplementary Figs 1 and 2). By analysing these patterns using linear discriminant analysis (LDA), which is an efficient pattern recognition algorithm for classifying unknown samples[59], we could straightforwardly identify 45 representative analytes (Fig. 1c). Thirty-eight unknown samples that were randomly selected from the training set were identified by m-SMS with 97% accuracy.

Molecular cryptography. This ability of m-SMS to produce a wide range of nearly unpredictable emission fingerprints resembles the function of pseudo-random number generators, namely, cipher devices that can effectively encrypt text by associating each letter with an approximate random number. One of the most well-known pseudo-random number generator devices is the Enigma machine[60,61], which was used by the Germans during World War II (WWII) to protect military communication. With the Enigma technology, the sender and receiver possessed identical cipher machines that were used to encrypt and decrypt the text, respectively. In addition, to prevent a third party with an identical machine from spying on these messages, the receiver must also have setup the correct initial state of his

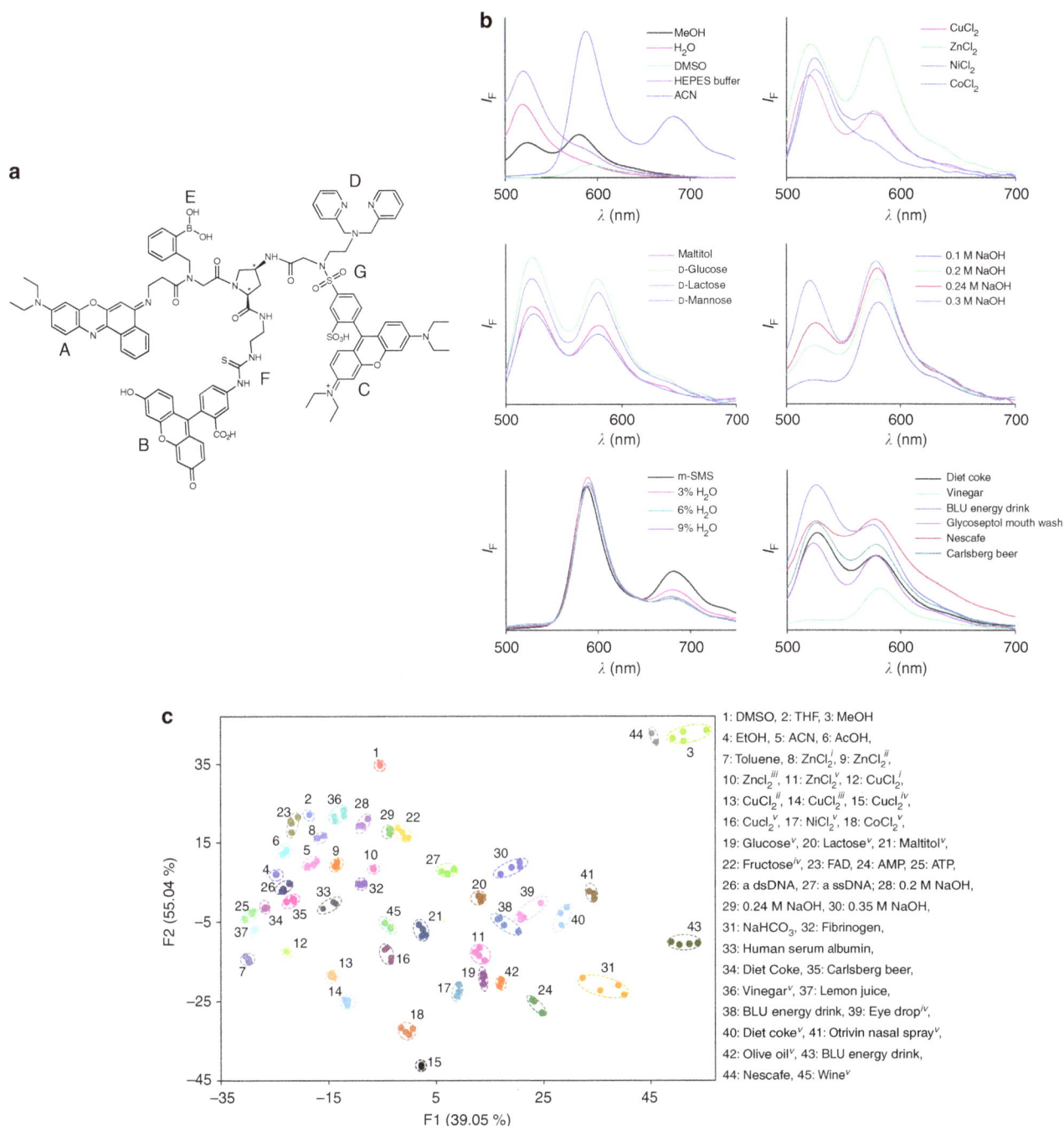

Figure 1 | m-SMS operates as a universal sensor that can discriminate among multiple different analytes. (**a**) The structure of m-SMS integrates three fluorophores: solvatochromic nile blue (A), pH-sensitive fluorescein (B) and sulforhodamine B (C), as well as distinct recognition elements, such as dipicolylamine (D), boronic acid (E), thiourea (F) and sulfonamide (G). (**b**) Representative emission patterns generated by m-SMS in response to different analytes or conditions. The emission was recorded in different solvents (top left) and upon adding 2 μl of an aqueous solution of metal ions* (top right, 300 mM) and saccharides* (middle left, 13 mM) or by changing the pH** (middle right, 0.1–0.3 M NaOH), polarity*** (bottom left, 3–9% H_2O) and upon adding commercial products* (bottom right). Initial conditions: m-SMS in *EtOH-AcOH (10 mM) and NaOH (11 mM), **EtOH-AcOH (10 mM) and ***acetonitrile (ACN). The concentration of m-SMS was 500 nM in all the solutions except for the measurements in ACN, where it was 5 μM. $\lambda_{ex} = 480$ nm. (**c**) Linear discrimination analysis (LDA) of 45 representative patterns generated by different analytes under diverse conditions. Initial conditions: m-SMS in EtOH-AcOH (10 mM) and [i]3, [ii]6, [iii]8, [iv]9 and [v]11 mM of NaOH. DMSO, dimethylsulphoxide; dsDNA, double-stranded DNA; ssDNA, single-stranded DNA; THF, tetrahydrofuran.

machine in order to obtain the right message. To elucidate the function of an Enigma-like molecular machine, we first show how m-SMS can be used to encrypt and decrypt a very simple text: 'open sesame' (Fig. 2). Initially, the sender converts the text to numbers using a public alphanumeric code to obtain a numeric

sequence (Fig. 2a). Note that this alphanumeric code does not need to be secure and can be used to write various other messages. In the next step, the sender dissolves m-SMS in a chosen solution (60 μl EtOH) to which 2 μl of a randomly selected chemical input (chemical *x*, 1 M NaHCO₃) is added. A random

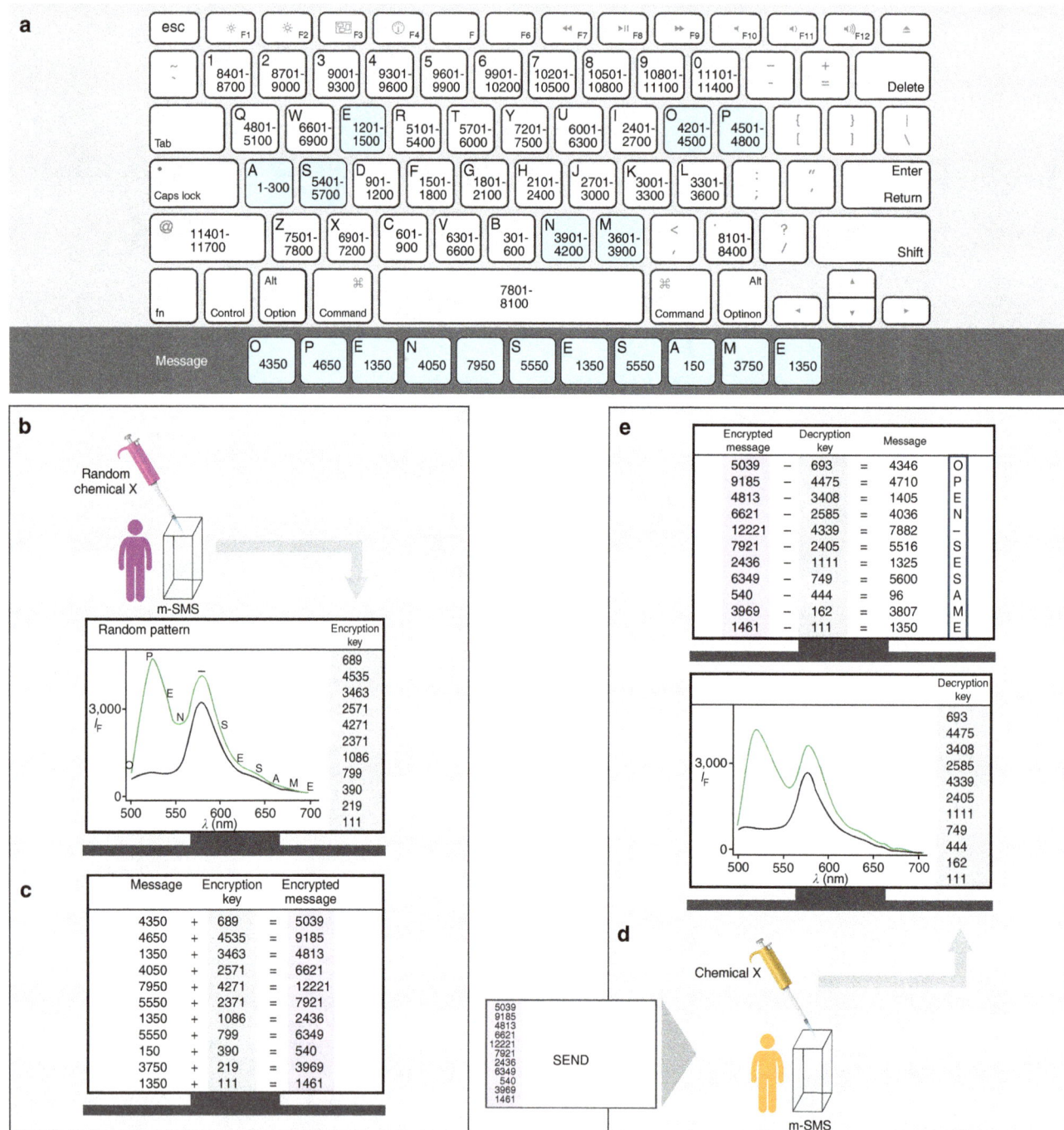

Figure 2 | Cryptographic protection by an Enigma-like molecular cipher device. (**a**) The sender converts his message to numbers by using a public alphanumeric code. (**b**) He then dissolves m-SMS in a chosen solution, verifies the initial emission intensity (black line) and records the emission pattern generated after adding a random chemical input (green line). The resulting intensity values, recorded every 20 nm (denoted in green letters), provide a unique encryption key. (**c**) The sender then encrypts the message by adding the encryption key to the original message and sends the encrypted message (cipher text) to the recipient. (**d**) The recipient, who possesses an identical m-SMS cipher device, repeats this procedure by setting up the correct initial state of the system (for example, solvent, sensor concentration and detector gain) and adding the same chemical x. (**e**) The original message is then revealed by subtracting the resulting values (green line) from the cipher text. Conditions: 500 nM m-SMS in EtOH, chemical x = NaHCO$_3$ (2 μl, 1 M), λ_{ex} = 480 nm. The following illustrations were used under a license from Shutterstock.com: keyboard (credit: Alhovik), pipette (credit: extender_01) and man character (credit: Leremy).

encryption key is then generated by recording the emission every 20 nm and associating each value with the corresponding letter (Fig. 2b). The sender then adds this encryption key to the original message to afford an encrypted message (cipher text; Fig. 2c) that can be safely sent to a recipient with an identical molecular device. To obtain the original message, the receiver simply needs to generate the decryption key by setting up the correct initial state of the system (for example, sensor concentrations, solvents and detector gain), adding the same chemical input (Fig. 2d), and subtracting the resulting values from the cipher text (Fig. 2e).

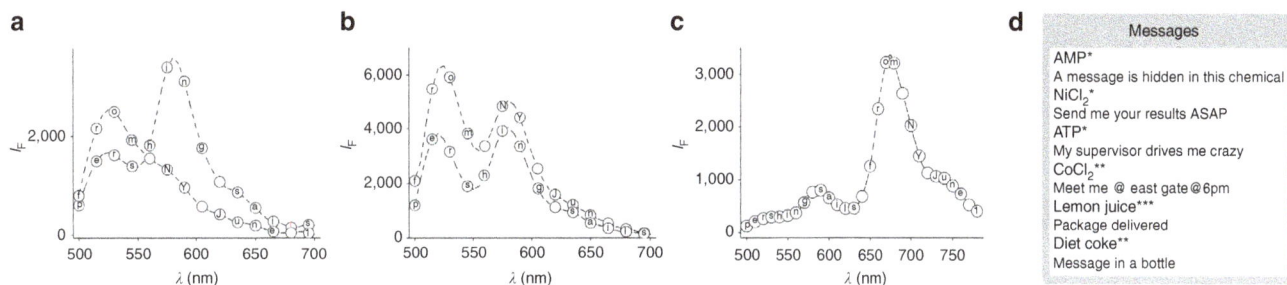

Figure 3 | Encrypting longer messages by sequentially adding chemical inputs. (**a**) Encrypting a message by recording the emission spectra generated after adding NaOH (2 µl, 0.2 M, red letters) and then after adding CuCl$_2$ (2 µl, 0.3 mM, blue letters) to 500 nM SMS in EtOH-AcOH (10 mM). (**b**) Encrypting the same message by recording the emission spectra after adding NaOH (2 µl, 0.35 M, red letters) and then GenTeal eyedrop (2 µl, blue letters) to 500 nM SMS in EtOH-AcOH (10 mM). (**c**) Encrypting the same message by using a single, broad emission spectrum obtained after adding NaOH (0.5 µl, 0.35 M) and CuCl$_2$ (1 µl, 0.3 mM) to 5 µM SMS in acetonitrile. These experiments (**a**-**c**) also demonstrate how the same message can be differently encrypted by changing the chemical inputs (**a** versus **b**) or by changing the initial state of the system (**a** versus **c**). (**d**) Representative messages that were successfully decrypted by untrained, randomly selected users. Initial conditions: m-SMS (500 nM) in *EtOH, **EtOH-AcOH (10 mM) and NaOH (6 mM), and ***EtOH-AcOH (10 mM) and NaOH (10 mM).

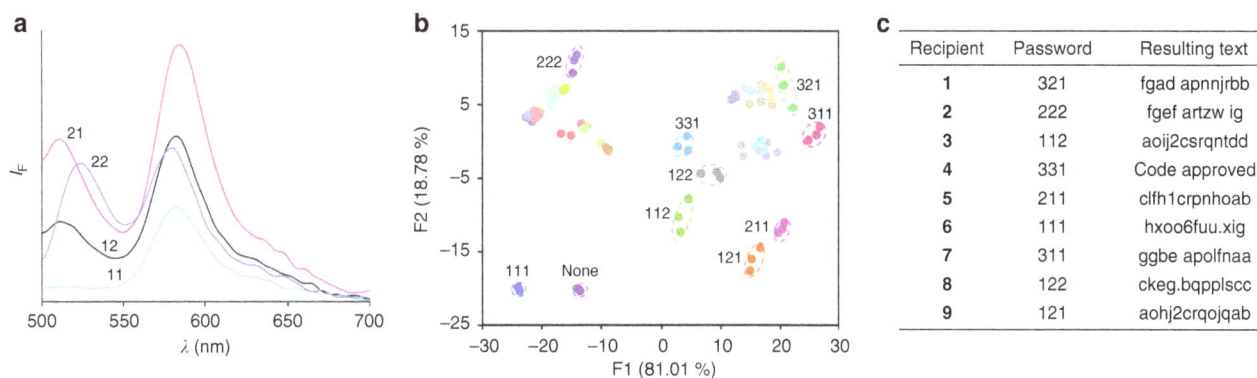

Figure 4 | Password protection by generating sequence-dependent encryption keys. By appropriately choosing chemical inputs, m-SMS can operate as a molecular keypad lock that generates the correct encryption/decryption keys (emission patterns) only when the chemical inputs are introduced in the right order. (**a**) Different encryption keys generated by introducing the four possible combinations of two-digit chemical 'passwords' consisting of ZnCl$_2$ (**1**) and Na$_3$PO$_4$ (**2**) as inputs signals. (**b**) LDA mapping of the encryption keys generated in response to the 27 possible combinations of three-digit chemical passwords, where ZnCl$_2$ (**1**), Na$_3$PO$_4$ (**2**) and NaOH (**3**) serve as input signals. The clusters corresponding to the nine unique encryption keys are denoted in circles. Conditions: each digit corresponds to the addition of 2 µl of **1** (0.08 M), **2** (0.08 M) or **3** (0.1 M) to 60 µl m-SMS (500 nM) in EtOH. (**c**) Text obtained by decrypting the cipher text with the correct password (331) and by the other eight unique combinations.

Figure 3 shows how longer messages can be encrypted by sequentially adding chemical inputs. For clarity, messages encrypted by two inputs are presented. The text 'Pershing sails from NY June 1' was selected for this experiment because, in the context of hidden messages, this is a well-known message that was written by a spy during WWII.[2] Hence, with this message, we intend to highlight the analogy between m-SMS and the simplest stereographic technologies in terms of the ease by which messages can be concealed and exposed by untrained users. In Fig. 3a, the encryption key was generated by first adding NaOH (0.2 M), then CuCl$_2$ (0.3 mM) and recording the emission following each addition. In Fig. 3b, the inputs were changed to NaOH (0.35 M) and eyedrop, which demonstrate the feasibility of encrypting messages with commercially available chemicals. Pharmaceutical liquids are very suitable for this application owing to their high purity and batch-to-batch reproducibility, which enable the sender and receiver to use them as is without performing additional procedures. Figure 3c shows how an entirely different encryption key can be generated with the same inputs used in the first experiment (Fig. 3a, NaOH and CuCl$_2$), but changing the solvent to acetonitrile and the concentrations of the molecular components to 5 µM m-SMS, 0.35 M NaOH and 0.3 M CuCl$_2$. Owing to the stronger intensity of the NB dye under hydrophobic

conditions, the message could be encrypted in a single emission spectrum, which was obtained after the second addition step. This last experiment (Fig. 3c) thus demonstrates the importance of correctly setting up the initial state of the system, which is a fundamental principle underlying the operation of Enigma machines[61]. Following these test cases, 12 different users, including 10 untrained users, were requested to decrypt different messages (2–19 words) by using different chemical inputs (Fig. 3d and Supplementary Table 1). The fact that all messages were successfully decrypted confirmed the simplicity, versatility and reliability of this technique.

Molecular password protection. Despite the fact that cryptography makes m-SMS far more secure than secret inks, there is always the possibility that the enemy would obtain the sensor and the correct chemical inputs, and would attempt to recreate the encryption key using a 'brute force search'[2]. Namely, it would measure the response of m-SMS to different concentrations and combinations of these inputs until meaningful text would result from this screening. Figure 4 shows a means for complicating such efforts by entering a password as an additional layer of defence. This approach exploits the principles of molecular

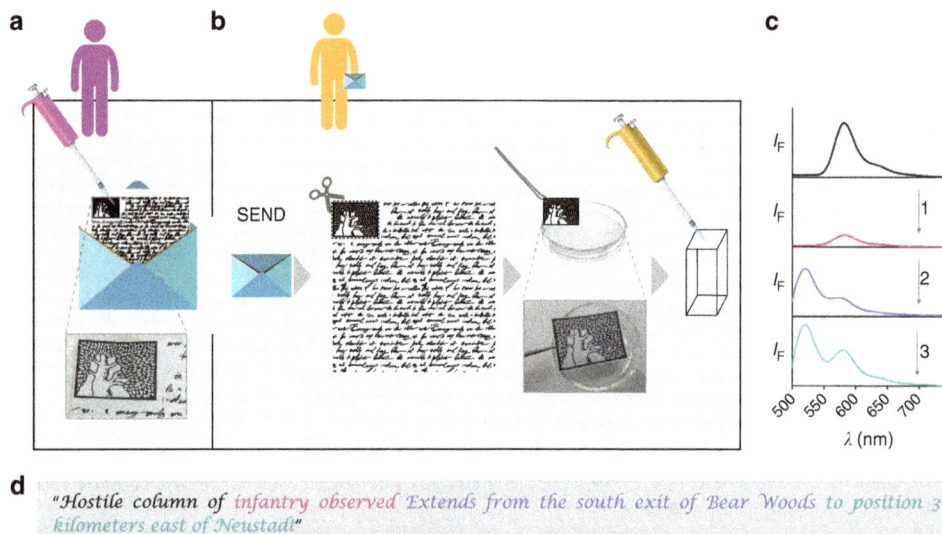

Figure 5 | Steganographic protection by hiding m-SMS on plain letter paper. (**a**) 1.1 µl of m-SMS (440 µM) was hidden on a random spot within the logo of the Weizmann Institute and the letter was sent to a recipient by regular mail. Note that the text within this letter does not contain any valuable information. (**b**) The recipient, who obtained the cipher text and knows the initial conditions, extracts m-SMS from the paper by incubating the logo in 1 ml of EtOH-AcOH (10 mM). (**c**) To uncover the message, the receiver adjusts the correct concentration of m-SMS by calibrating its initial emission intensity (top) and generates the decryption key by recording the emission pattern following the addition of each chemical input (inputs 1–3). (**d**) The resulting text is a message that was encrypted by the Enigma machine. The letter colours correspond to relevant decryption keys shown in **c**. Conditions: 1 µl of (**1**) $NiCl_2$ (0.15 M), (**2**) KOH (2.5 M) and (**3**) Na_4EDTA (0.27 M) were sequentially added to a 60-µl solution of m-SMS (500 nM) in EtOH-AcOH (10 mM). The hand-writing text image (credit: amiloslava) is taken with permission from Shutterstock.com.

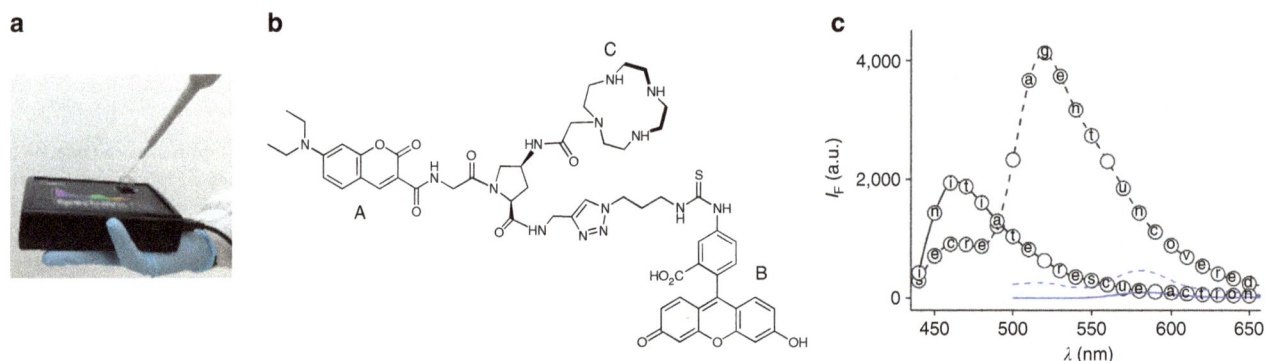

Figure 6 | Versatility of the m-SMS technology. Secret communication was achieved by using (**a**) a hand-held spectrometer, and (**b**) a second molecular cipher device (m-SMS₂) integrating coumarin (A), fluorescein (B) and a cyclen ligand (C). (**c**) Encryption patterns generated by m-SMS (blue lines) or m-SMS₂ (black lines) under the same conditions. The emission of each sensor (250 nM) was recorded in EtOH solution containing NaOAc (1 mM) and $ZnCl_2$ (1.3 mM; dashed line) and after adding AcOH (16 mM; solid line). m-SMS and m-SMS₂ were excited at 480 and 420 nm, respectively.

keypad lock technology[30–41], which largely rely on the tendency of multivalent host–guest complexes and multicomponent assemblies to be entrapped in local minima[41]. We selected $ZnCl_2$ (**1**), Na_3PO_4 (**2**) and NaOH (**3**) as representative entry keys owing to the strong interaction of Zn(II) with DPA ligands[55,56], as well as with NaOH or Na_3PO_4 to yield $Zn(OH)_2$ and zinc phosphate complexes, respectively[62]. Hence, when $ZnCl_2$ is initially added, Zn(II) should readily coordinate to the DPA unit of m-SMS. In contrast, when $ZnCl_2$ is added second, the reaction with an excess of Na_3PO_4 or NaOH in solution should reduce the concentration of free Zn(II) ions and consequently, the amount of the m-SMS-Zn(II) complex. Figure 4a exemplifies how m-SMS can be used to generate four different encryption keys using two-digit chemical passwords: 11, 22, 12 and 21. With three chemical inputs, additional metastable complexes can be formed, which enabled us to identify 9 unique passwords from the 27 possible combinations (Fig. 4b). The relevance of the keypad lock

technique to cryptographic applications was demonstrated by providing nine different recipients with the same chemical inputs (1, 2 and 3), but with distinct individual passwords. As shown in Fig. 4c, only the receiver with the right password could successfully identify the message, whereas the other users only obtained random text.

Molecular steganography. Steganography is the third layer of protection that can be implemented by concealing low quantities of m-SMS on regular paper (Fig. 5). This not only complicates its detection, but also its characterization, which would be needed if an enemy attempts to reproduce the molecular device. Figure 5 depicts a representative experiment in which 1.1 µl of m-SMS was dried on plain letter paper (Fig. 5a) and sent to a second recipient by regular postal services. In this experiment, the letter was printed with a standard printer and the sensor was hidden on a random spot within the logo of the Weizmann Institute (Fig. 5a). To clarify, the text within this letter does not contain any valuable information, but rather, the message is concealed within the emission spectra of m-SMS, which can only be generated by setting up the appropriate conditions. To reveal the message, the receiver merely needs to extract m-SMS from the letter by cutting the logo, incubating it in an appropriate solution, and use this solution to record the fluorescence spectra (Fig. 5b). By setting up the correct initial emission intensity (Fig. 5c, top spectrum) and sequentially adding the right chemical inputs (Fig. 5c, inputs 1–3), the receiver could successfully identify various different messages, such as the one presented in Fig. 5d: 'Hostile column of infantry observed. Extends from the south exit of Bear Woods to position 3 kilometres east of Neustadt', a message that was encrypted by the original Enigma machine.

Versatility of the m-SMS technology. Similar procedures, in which chemical inputs were concealed on letter paper, were also performed (Supplementary Fig. 3), demonstrating an alternative means of hiding and delivering molecular components. In these experiments, chemical inputs with measurable absorption spectra such as $CoCl_2$ (Supplementary Fig. 3b,c) were extracted from the paper and, after determining their concentrations, were added to m-SMS. In addition to commercial chemicals, we also encrypted messages using unique inputs made in our laboratory[63], which shows how messages can be further protected by using synthetic compounds that are difficult to characterize and reproduce (Supplementary Fig. 3a and Supplementary Tables 1 and 2). Finally, to demonstrate that this technology is not limited to particular locations or a specific sensor, we encoded and decoded messages outside the laboratory using a low-cost hand-held spectrofluorometer (Fig. 6a and Supplementary Fig. 4) and we also synthesized a second m-SMS molecule (Fig. 6b, m-SMS$_2$) integrating coumarin and a pH-sensitive Flu probe, as well as a cyclen ligand that can bind various metal ions. Hence, similar to m-SMS (Fig. 1a), m-SMS$_2$ (Fig. 6b) should be able to respond to metal ions, acids and bases. However, it should produce different emission patterns owing to the shorter excitation and emission wavelength of the FRET donor (that is, coumarin), as well as the distinct affinity of cyclen and DPA towards different metal ions. To demonstrate that this new molecular cipher device can generate entirely different encryption keys, the message 'secret agent uncovered initiate rescue action' was encrypted by recording the emission of m-SMS$_2$ before and after adding 16 mM acetic acid. We then attempted to decrypt the resulting cipher text by using both m-SMS$_2$ and the original m-SMS. As shown in Fig. 6c, although the same chemical inputs were used, only the first molecular device successfully decrypted the messages. The second device generated a meaningless text. This last experiment thus shows that even if a third party manages to reproduce m-SMS and spy on the experimental settings, a new cipher device can be readily created by replacing one or several receptors, linkers or dyes.

Discussion
Given recent concerns regarding global electronic surveillance[64], the ability of m-SMS to convert different chemical structures into unique emission patterns demonstrates a potential means to bypass using electronic communication systems and thereby ensure that important messages are secure. Interestingly, even this first prototype provides a very high security level owing to its ability to generate numerous unpredictable encryption keys (cryptography), as well as the difficulty of finding and characterizing the molecular device and/or chemical inputs (steganography), and in particular cases, the order by which the

inputs are introduced (password protection). In addition, as with Enigma cryptographic systems, to break such a defence one also needs to set up the correct initial state of the system, which can be determined by the type of solvents and concentrations used, as well as by the instrumentation setup. We can estimate, for example, the maximal number of patterns that can be generated by using six different concentrations of m-SMS (Supplementary Fig. 2a) at six different pH values (Supplementary Fig. 2b) and upon the addition of six different concentrations of copper ions (Supplementary Fig. 2c). By setting the detector to six different 'gain' values (Supplementary Fig. 2d), even a single chemical input (that is, $CuCl_2$), out of the numerous chemicals that can be discriminated by m-SMS (Fig. 1b), should afford a maximal number of $6^4 = 1,296$ encryption keys. Improving the performance of such systems should be readily achieved by increasing the number of recognition and signalling elements, which would maximize the number of analytes that can be discriminated by a unimolecular cipher device. Other important features of this technology, namely, its versatility and simplicity, have also been demonstrated by creating different m-SMS devices, encrypting messages with a wide range of randomly selected chemicals, as well as by hiding the molecular components on plain paper and sending them by regular mail, akin to invisible inks. Considering the unlimited number of chemical structures that can, in principle, be used as inputs, this work indicates that a unique message could be hidden within each and every molecule around us.

Methods
Synthesis and characterization of m-SMS and m-SMS$_2$. Detailed synthesis and characterization of the m-SMSs are available in the Supplementary Methods.

Multi-analyte sensing. Different analytes and their combinations were identified by adding them to m-SMS (500 nM) in an ethanol solution containing 10 mM of AcOH (EtOH-AcOH). In a typical experiment, a chemical input (2 µl) was added to 60 µl of m-SMS in EtOH-AcOH and the emission pattern was recorded by a BioTek synergy H4 hybrid multi-mode microplate reader (BioTek, Inc.) using black flat-bottom polystyrene 384-well microplates (Corning). This process was performed in four replicates and emission intensity values obtained at 520, 580 and 654 nm were analysed by LDA using XLSTAT version 2014.1.01. LDA reduces the dimensionality of the data into two canonical factors (F1 and F2), which enables classifying unknown samples according to the proximity of the data points (F1, F2) to the clusters obtained by the training set.

Encryption and decryption of messages. Messages were ciphered and deciphered by adding one or several chemical inputs to m-SMS or m-SMS$_2$ and recording the emission spectra with a BioTek synergy H4 hybrid multi-mode microplate reader or by using a portable SpectroVis Plus spectrophotometer (Vernier) connected to a laptop computer equipped with LoggerPro software. The intensity and shape of the spectral patterns, which provide the encryption/decryption keys, were varied by changing the chemical inputs and their concentrations, as well as by altering the initial state of the system. For example, different fluorescence fingerprints were readily obtained by changing the solvent, pH, photomultiplier gain (current amplification), sensor concentration and by combing of these parameters. In a typical experiment, generally, the encryption and decryption keys were generated by dissolving the molecular sensor (500 nM) in 60 µl EtOH or EtOH-AcOH (10 mM), adding 1–2 µl of chemical inputs, and recording the emission intensity values every 4–15 nm. This experiment was performed in triplicate. Steganographic protection was achieved by pipetting 1–2 µl of m-SMS or chemical inputs such as $CoCl_2$ on the Weizmann Institute logo. The logo was printed on plain A4 paper by a standard HP colour LaserJet printer (M651). $CoCl_2$ was extracted from the paper with 300 µl of water and its concentration was determined according to its extinction coefficient ($\varepsilon_{510\,nm} = 4.85\,M^{-1}\,cm^{-1}$).

References
1. Macrakis, K. *Prisoners, Lovers, and Spies: The Story of Invisible Ink from Herodotus to al-Qaeda* (Yale Univ., 2015).
2. Jamil, T. Steganography: the art of hiding information in plain sight. *IEEE Potentials* **18**, 10–12 (1999).

3. Macrakis, K., Bell, E. K., Perry, D. L. & Sweeder, R. D. Invisible ink revealed: concept, context, and chemical principles of 'cold war' writing. *J. Chem. Edu.* **89**, 529–532 (2012).

4. Kishimura, A., Yamashita, T., Yamaguchi, K. & Aida, T. Rewritable phosphorescent paper by the control of competing kinetic and thermodynamic self-assembling events. *Nat. Mater.* **4**, 546–549 (2005).

5. Mutai, T., Satou, H. & Araki, K. Reproducible on-off switching of solid-state luminescence by controlling molecular packing through heat-mode interconversion. *Nat. Mater.* **4**, 685–687 (2005).

6. Perruchas, S. *et al.* Mechanochromic and thermochromic luminescence of a copper iodide cluster. *J. Am. Chem. Soc.* **132**, 10967–10969 (2010).

7. Yoon, S.-J. *et al.* Multistimuli two-color luminescence switching via different slip-stacking of highly fluorescent molecular sheets. *J. Am. Chem. Soc.* **132**, 13675–13683 (2010).

8. Yan, D. *et al.* Reversibly thermochromic, fluorescent ultrathin films with a supramolecular architecture. *Angew. Chem. Int. Ed.* **50**, 720–723 (2011).

9. Li, K. *et al.* Reversible photochromic system based on rhodamine B salicylaldehyde hydrazone metal complex. *J. Am. Chem. Soc.* **136**, 1643–1649 (2014).

10. Sun, H. *et al.* Smart responsive phosphorescent materials for data recording and security protection. *Nat. Commun.* **5**, 3601 (2014).

11. Wu, Y. *et al.* Quantitative photoswitching in bis(dithiazole)ethene enables modulation of light for encoding optical signals. *Angew. Chem. Int. Ed.* **53**, 2090–2094 (2014).

12. Hou, X. *et al.* Tunable solid-state fluorescent materials for supramolecular encryption. *Nat. Commun.* **6**, 6884–6892 (2015).

13. Palacios, M. A. *et al.* InfoBiology by printed arrays of microorganism colonies for timed and on-demand release of messages. *Proc. Natl Acad. Sci. USA* **108**, 16510–16514 (2011).

14. Kim, K.-W., Bocharova, V., Halámek, J., Oh, M.-K. & Katz, E. Steganography and encrypting based on immunochemical systems. *Biotechnol. Bioeng.* **108**, 1100–1107 (2011).

15. Burgess, I. B. *et al.* Encoding complex wettability patterns in chemically functionalized 3D photonic crystals. *J. Am. Chem. Soc.* **133**, 12430–12432 (2011).

16. Ratner, T., Reany, O. & Keinan, E. Encoding and processing of alphanumeric information by chemical mixtures. *ChemPhysChem* **10**, 3303–3309 (2009).

17. Shoshani, S., Piran, R., Arava, Y. & Keinan, E. A molecular cryptosystem for images by DNA computing. *Angew. Chem. Int. Ed.* **51**, 2883–2887 (2012).

18. Poje, J. E. *et al.* Visual displays that directly interface and provide read-outs of molecular states via molecular graphics processing units. *Angew. Chem. Int. Ed.* **53**, 9222–9225 (2014).

19. Ling, J., Naren, G., Kelly, J., Moody, T. S. & de Silva, A. P. Building pH sensors into paper-based small-molecular logic systems for very simple detection of edges of objects. *J. Am. Chem. Soc.* **137**, 3763–3766 (2015).

20. Ling, J., Naren, G., Kelly, J., Fox, D. B. & de Silva, A. P. Small molecular logic systems can draw the outlines of objects via edge visualization. *Chem. Sci.* **6**, 4472–4478 (2015).

21. Clelland, C. T., Risca, V. & Bancroft, C. Hiding messages in DNA microdots. *Nature* **399**, 533–534 (1999).

22. de Silva, A. P. *Molecular Logic-Based Computation* (Royal Society of Chemistry, 2012).

23. Andreasson, J. & Pischel, U. Molecules with a sense of logic: a progress report. *Chem. Soc. Rev.* **44**, 1053–1069 (2015).

24. de Silva, A. P. & Uchiyama, S. Molecular logic and computing. *Nat. Nanotechnol.* **2**, 399–410 (2007).

25. Szaciłowski, K. *Infochemistry* (Wiley, 2013).

26. Baroncini, M., Semeraro, M. & Credi, A. Processing chemical and photonic signals by artificial multicomponent molecular systems. *Isr. J. Chem.* **51**, 23–35 (2011).

27. Credi, A. Molecules that make decisions. *Angew. Chem. Int. Ed.* **46**, 5472–5475 (2007).

28. Strack, G., Luckarift, H. R., Johnson, G. R. & Katz, E. in *Biomolecular Information Processing* 103–116 (Wiley-VCH Verlag GmbH & Co. KGaA, 2012).

29. de Silva, A. P., James, M. R., McKinney, B. O.F., Pears, D. A. & Weir, S. M. Molecular computational elements encode large populations of small objects. *Nat. Mater.* **5**, 787–789 (2006).

30. Margulies, D., Felder, C. E., Melman, G. & Shanzer, A. A molecular keypad lock: a photochemical device capable of authorizing password entries. *J. Am. Chem. Soc.* **129**, 347–354 (2007).

31. Guo, Z., Zhu, W., Shen, L. & Tian, H. A fluorophore capable of crossword puzzles and logic memory. *Angew. Chem. Int. Ed.* **46**, 5549–5553 (2007).

32. Strack, G., Ornatska, M., Pita, M. & Katz, E. Biocomputing security system: concatenated enzyme-based logic gates operating as a biomolecular keypad lock. *J. Am. Chem. Soc.* **130**, 4234–4235 (2008).

33. Sun, W., Xu, C.-H., Zhu, Z., Fang, C.-J. & Yan, C.-H. Chemical-driven reconfigurable arithmetic functionalities within a fluorescent tetrathiafulvalene derivative. *J. Phys. Chem. C* **112**, 16973–16983 (2008).

34. Andréasson, J., Straight, S. D., Moore, T. A., Moore, A. L. & Gust, D. An all-photonic molecular keypad lock. *Chem. Eur. J* **15**, 3936–3939 (2009).

35. Halámek, J., Tam, T. K., Chinnapareddy, S., Bocharova, V. & Katz, E. Keypad lock security system based on immune-affinity recognition integrated with a switchable biofuel cell. *J. Phys.Chem. Lett.* **1**, 973–977 (2010).

36. Andréasson, J. *et al.* All-photonic multifunctional molecular logic device. *J. Am. Chem. Soc.* **133**, 11641–11648 (2011).

37. Liu, Y. *et al.* An aptamer-based keypad lock system. *Chem. Commun.* **48**, 802–804 (2012).

38. Jiang, X.-J. & Ng, D. K.P. Sequential logic operations with a molecular keypad lock with four inputs and dual fluorescence outputs. *Angew. Chem. Int. Ed.* **53**, 10481–10484 (2014).

39. Carvalho, C. P., Dominguez, Z., Da Silva, J. P. & Pischel, U. A supramolecular keypad lock. *Chem. Commun.* **51**, 2698–2701 (2015).

40. Chen, J., Zhou, S. & Wen, J. Concatenated logic circuits based on a three-way DNA junction: a keypad-lock security system with visible readout and an automatic reset function. *Angew. Chem. Int. Ed.* **54**, 446–450 (2015).

41. Rout, B., Milko, P., Iron, M. A., Motiei, L. & Margulies, D. Authorizing multiple chemical passwords by a combinatorial molecular keypad lock. *J. Am. Chem. Soc.* **135**, 15330–15333 (2013).

42. Rout, B., Unger, L., Armony, G., Iron, M. A. & Margulies, D. Medication detection by a combinatorial fluorescent molecular sensor. *Angew. Chem. Int. Ed.* **51**, 12477–12481 (2012).

43. Rout, B., Motiei, L. & Margulies, D. Combinatorial fluorescent molecular sensors: the road to differential sensing at the molecular level. *Synlett* **25**, 1050–1054 (2014).

44. Chen, K., Shu, Q. & Schmittel, M. Design strategies for lab-on-a-molecule probes and orthogonal sensing. *Chem. Soc. Rev.* **44**, 136–160 (2015).

45. Wu, X. *et al.* Selective sensing of saccharides using simple boronic acids and their aggregates. *Chem. Soc. Rev.* **42**, 8032–8048 (2013).

46. Götzke, L. *et al.* Nickel(II) and zinc(II) complexes of N-substituted di(2-picolyl)amine derivatives: Synthetic and structural studies. *Polyhedron* **30**, 708–714 (2011).

47. Abhayawardhana, P. L., Marzilli, P. A., Fronczek, F. R. & Marzilli, L. G. Complexes possessing rare 'tertiary' sulfonamide nitrogen-to-metal bonds of normal length: fac-[Re(CO)₃(N(SO₂R)dien)]PF₆ complexes with hydrophilic sulfonamide ligands. *Inorg. Chem.* **53**, 1144–1155 (2014).

48. Chohan, Z. H. *et al.* Sulfonamide–metal complexes endowed with potent anti-Trypanosoma cruzi activity. *J. Enzyme Inhib. Med. Chem.* **29**, 230–236 (2014).

49. Vonlanthen, M., Connelly, C. M., Deiters, A., Linden, A. & Finney, N. S. Thiourea-based fluorescent chemosensors for aqueous metal ion detection and cellular imaging. *J. Org. Chem.* **79**, 6054–6060 (2014).

50. Beer, P. D. & Gale, P. A. Anion recognition and sensing: the state of the art and future perspectives. *Angew. Chem. Int. Ed.* **40**, 486–516 (2001).

51. Custelcean, R. Crystal engineering with urea and thiourea hydrogen-bonding groups. *Chem. Commun.* **21**, 295–307 (2008).

52. Adsmond, D. A. & Grant, D. J.W. Hydrogen bonding in sulfonamides. *J. Pharm. Sci.* **90**, 2058–2077 (2001).

53. Sjöback, R., Nygren, J. & Kubista, M. Absorption and fluorescence properties of fluorescein. *Spectrochim. Acta A Mol. Biomol. Spectrosc.* **51**, L7–L21 (1995).

54. Jose, J. & Burgess, K. Benzophenoxazine-based fluorescent dyes for labeling biomolecules. *Tetrahedron* **62**, 11021–11037 (2006).

55. Sakamoto, T., Ojida, A. & Hamachi, I. Molecular recognition, fluorescence sensing, and biological assay of phosphate anion derivatives using artificial Zn(ii)-Dpa complexes. *Chem. Commun.* 141–152 (2009).

56. Chang, C. J. *et al.* Bright fluorescent chemosensor platforms for imaging endogenous pools of neuronal zinc. *Chem. Biol.* **11**, 203–210 (2004).

57. de Silva, A. P. *et al.* Signaling recognition events with fluorescent sensors and switches. *Chem. Rev.* **97**, 1515–1566 (1997).

58. de Silva, A. P., Moody, T. S. & Wright, G. D. Fluorescent PET (photoinduced electron transfer) sensors as potent analytical tools. *Analyst* **134**, 2385–2393 (2009).

59. Anslyn, E. V. Supramolecular analytical chemistry. *J. Org. Chem.* **72**, 687–699 (2007).

60. Lloyd, S. Quantum Engima machines. arXiv:1307.0380 (2013).

61. Sebag-Montefiore, H. *Enigma: the Battle for the Code* (Wiley, 2004).

62. Peng, X., Xu, Y., Sun, S., Wu, Y. & Fan, J. A ratiometric fluorescent sensor for phosphates: Zn²⁺-enhanced ICT and ligand competition. *Org. Biomol. Chem.* **5**, 226–228 (2007).

63. Selvakumar, K., Motiei, L. & Margulies, D. Enzyme — artificial enzyme interactions as a means for discriminating among structurally similar isozymes. *J. Am. Chem. Soc.* **137**, 4892–4895 (2015).

64. Macrakis, K. Supervision: an introduction to the surveillance society by John Gilliom and Torin Monahan. *Technol. Cult.* **55**, 515–516 (2014).

Acknowledgements

This research was supported by the European Research Council Starting Grant 338265.

Author contributions

T.S. and K.S. are equally contributed to this work. T.S., K.S., L.M. and D.M. designed the research; T.S. and K.S. synthesized the sensors, performed the experiments and analysed the data. L.M. and D.M. interpreted the data, wrote the manuscript and all the authors read and commented on the paper.

Additional information

Competing financial interests: The authors declare no competing financial interests.

20

Discovery of protein acetylation patterns by deconvolution of peptide isomer mass spectra

Nebiyu Abshiru[1,2,*], Olivier Caron-Lizotte[2,*], Roshan Elizabeth Rajan[2,3,*], Adil Jamai[2], Christelle Pomies[2], Alain Verreault[1,3] & Pierre Thibault[1,2]

Protein post-translational modifications (PTMs) play important roles in the control of various biological processes including protein-protein interactions, epigenetics and cell cycle regulation. Mass spectrometry-based proteomics approaches enable comprehensive identification and quantitation of numerous types of PTMs. However, the analysis of PTMs is complicated by the presence of indistinguishable co-eluting isomeric peptides that result in composite spectra with overlapping features that prevent the identification of individual components. In this study, we present Iso-PeptidAce, a novel software tool that enables deconvolution of composite MS/MS spectra of isomeric peptides based on features associated with their characteristic fragment ion patterns. We benchmark Iso-PeptidAce using dilution series prepared from mixtures of known amounts of synthetic acetylated isomers. We also demonstrate its applicability to different biological problems such as the identification of site-specific acetylation patterns in histones bound to chromatin assembly factor-1 and profiling of histone acetylation in cells treated with different classes of HDAC inhibitors.

[1]Department of Chemistry, Université de Montréal, PO Box 6128, Station centre-ville, Montréal, Québec, Canada H3C 3J7. [2]Institute for Research in Immunology and Cancer, Université de Montréal, C.P. 6128, Succursale centre-ville, Montréal, Québec, Canada H3C 3J7. [3]Molecular Biology Programme, Université de Montréal, PO Box 6128, Station centre-ville, Montréal, Québec, Canada H3C 3J7. *These authors contributed equally to this work. Correspondence and requests for materials should be addressed to A.V. (email: alain.verreault@umontreal.ca) or to P.T. (email: pierre.thibault@umontreal.ca).

Histone post-translational modifications (PTMs) control access to genetic information and, consequently, participate in several important cellular processes such as DNA repair and replication, nucleosome assembly and transcriptional regulation[1,2]. It is now clear that many chromatin-modifying enzymes and enzymes that act on non-histone proteins, recognize 'combinatorial patterns of PTMs', rather than single protein modifications. For instance, ATP-dependent chromatin remodelers often recognize histones that contain more than one acetylated lysine residue that are located in close proximity of each other[3,4]. Although valuable to infer the functions of PTMs, accurately determining patterns and stoichiometries of PTMs located in isomeric and isobaric peptides remains extremely difficult to achieve even by mass spectrometry (MS). This is because isomeric peptides, which are identical except for the position of the PTMs along the peptide chain, often co-elute during liquid chromatography and generate composite tandem mass spectra (MS/MS) containing indistinguishable fragment ions derived from two or more peptide isomers.

A few MS/MS methods were previously developed to quantify co-eluting isobaric histone peptides. Smith et al.[5] were first to use a fragment ion-based approach to determine the fraction of histone H4 molecules acetylated at specific residues based on the normalized intensity ratio of peptide fragment ions containing either protiated or deuterated acetyl groups to distinguish lysine residues that were, respectively, acetylated or free in vivo. However, this approach was limited by the lack of automated data analysis software and the difficulty to deconvolute co-eluting isomers. Recently, Feller et al.[6] described a similar approach in which second generation fragment ions (known as MS^3) were necessary to obtain diagnostic fragment ions that can be used to estimate the abundance of each isomer by solving a set of linear equations. A limitation of this approach is that singly charged MS^2 fragments, and particularly those with a nonmobile proton, fragment poorly and generate MS^3 spectra that are impossible to assign to specific isomers[7,8]. Sidoli et al.[9] designed a software known as isoScale that can be used to determine the relative abundance of isobaric peptides, based on intensity ratios of fragment ions unique to each isomer. isoScale relies on search engine outputs (for example, Mascot.csv result files) to select isomer-specific fragments used in this analysis. The algorithm EpiProfile recently described by Yuan et al.[10] quantifies isomeric histone peptides by solving a series of linear equations derived from peak heights of unique fragments representative of each isomeric species. Several other approaches have also been proposed to deconvolute composite MS/MS spectra of peptide ions with different sequences[11-13], however, their application is limited when analysing MS/MS spectra of co-fragmented isomeric ions that contain the same modification at different sites.

In this study, we present Iso-PeptidAce, a novel software tool that exploits high resolution LC–MS/MS data for deconvolution of composite spectra and quantification of site-specific acetylation. The resolution of composite MS/MS spectra derived from isomeric peptides is based on features associated with their characteristic fragment ion patterns obtained from the corresponding synthetic peptides and, importantly, does not rely on unique fragment ions to distinguish isomers. The application of Iso-PeptidAce is presented here for peptide isomers that contain multiple acetylated lysine residues but can also be applied to other types of modifications such as isomeric phosphopeptides[14]. The use of Iso-PeptidAce is demonstrated by monitoring temporal changes in acetylation patterns of histones H3 and H4 from human erythroleukemic (K562) cells treated with different classes of histone deacetylase inhibitors (HDACi), and by determining acetylation patterns in histones bound to *Saccharomyces cerevisiae* chromatin assembly factor-1 (CAF1).

Results

Deconvolution of mixed MS/MS spectra by Iso-PeptidAce.

We designed a new software tool named Iso-PeptidAce that deconvolutes composite MS/MS spectra of known isomeric peptides. The software takes as input raw MS and MS/MS files of the individual and the mixed isomers, the FASTA sequence file of the protein of interest, and a set of parameters for peptide-spectrum matching (PSM) such as modifications, precursor and fragment ion tolerances (see Data analysis under the Methods section). The raw files are processed to extract precursor intensity, MS1 and MS2 injection times, and MS2 peak lists containing m/z and intensity values (see Supplementary Methods for a detailed section on spectral deconvolution and normalization of MS signal intensities). Iso-PeptidAce computes the proportion of individual isomers in composite MS/MS spectra based on fragment ion patterns that uniquely identify each isomer. In Iso-PeptidAce, fragment ion patterns are reduced to a set of maximum network flow problems, for which a number of efficient algorithms are known[15]. This approach has previously been used in a wide range of complex problems, such as predicting molecular pathways in complex diseases or selecting single nucleotide polymorphisms and their associated alleles in patient and control groups[16-18]. In our study, we implemented the network flow approach to deconvolute composite fragment spectra of acetylated isomers. Although these isomers share multiple fragment ions, they produce distinct fragment ion patterns that can be transformed into a set of network flow problems. Figure 1a shows a schematic overview of the deconvolution conducted by Iso-PeptidAce for a composite MS/MS spectrum generated from two hypothetical co-eluting isomeric peptides labelled X and Y. For every composite spectrum acquired across the elution curve, fragment ion intensities (Fig. 1a, step 1) are modelled into maximum network flow problems (flow capacity shown as empty bar charts, Fig. 1a, step 2). The networks are filled with fragment ion patterns for isomers X and Y (Fig. 1a, step 3) and merged into a single network with excess flow (represented as overrunning colour bars in Fig. 1a, step 4). The resulting network is processed iteratively by a multivariate optimization technique known as Gradient Descent[19] (Fig. 1a, step 5), to remove the excess flow in the network. Each iteration step converges towards the maximum flow, which refers to the optimal ratio of X and Y compatible with the composite spectrum (Fig. 1a, left inner circle). Finally, individual elution curves for each isomer are generated based on the abundance ratios of MS/MS spectra (Fig. 1a, right inner circle) and the peak intensity or peak area of the precursor ions.

Extraction of elution profiles and fragment ion patterns.

We first used Iso-PeptidAce for MS/MS spectra of synthetic peptides from acetylated isomers of histones H3 and H4 (Supplementary Tables 1 and 2). All possible non-acetylated and acetylated forms of H4 peptide 1-SGRGKGGKGLGKGGAKRHR-19 were synthesized, and the purity of each peptide was confirmed by MS analyses. These peptides were propionylated, digested by trypsin and subjected to LC–MS/MS analysis on a Q-Exactive Plus instrument. We obtained three isomeric groups of H4 peptides acetylated at one, two or three lysine residues, and members of each group generally have very narrow retention time differences (Supplementary Fig. 1). To normalize precursor intensity, we determined the signal intensity response for each peptide as a function of the amount injected. We observed that mono- or multiply-acetylated isomers yielded a higher response than their propionylated counterparts (Supplementary Fig. 2A,B). We then prepared equimolar mixtures of peptides for each group of isomers and analysed them by LC–MS/MS. Although propionylated H4 peptides were readily separated into non-acetylated, mono-,

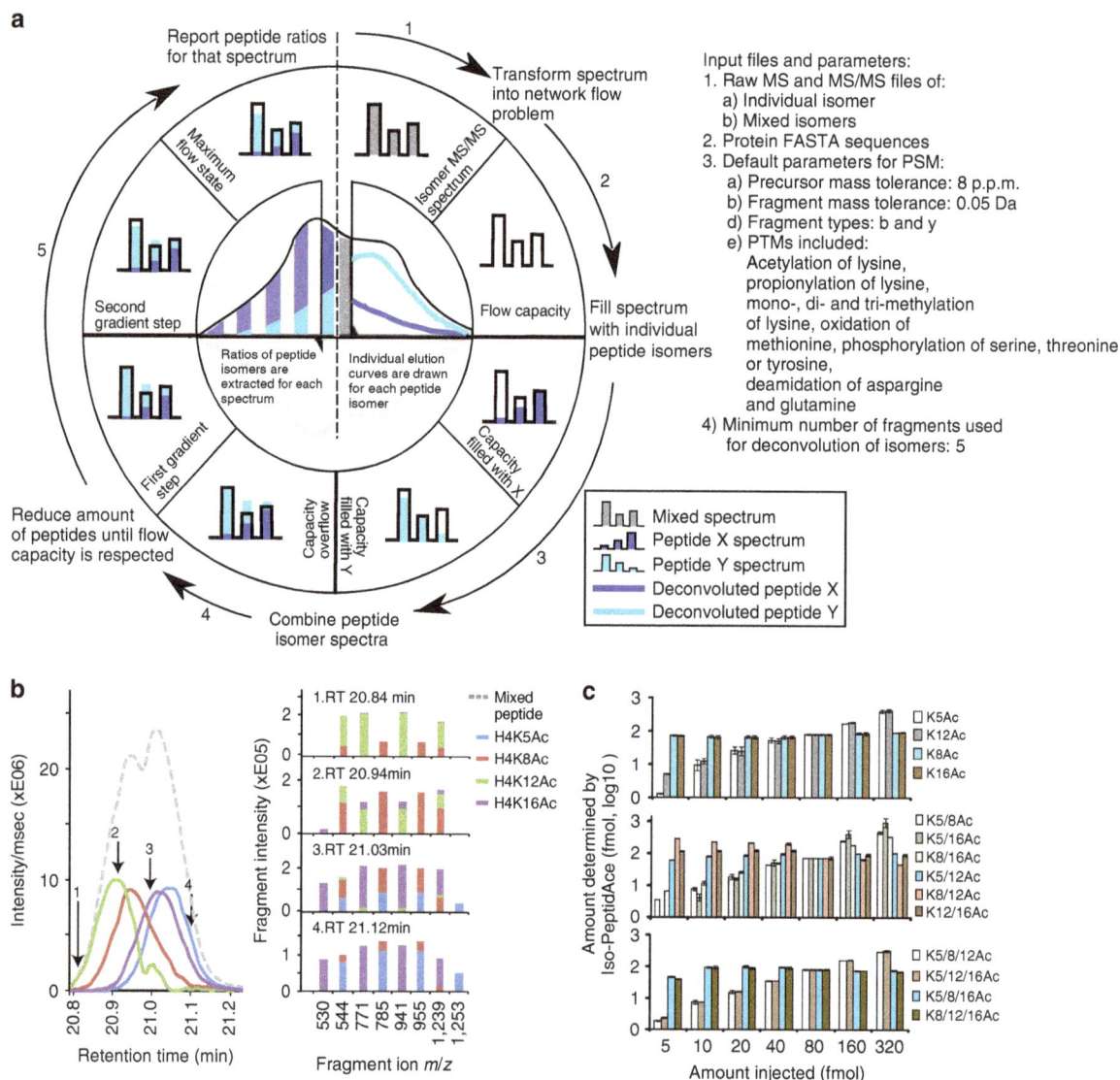

Figure 1 | Deconvolution of co-eluting acetylated isomers by Iso-PeptidAce. (**a**) Schematic showing the process of spectral deconvolution by Iso-PeptidAce. The input files and parameters used for deconvolution of mixtures of isomers are listed on the right side of the panel. The process involves building an individual elution curve for X and for Y based on ratios computed for each MS/MS event. Isomeric peptide ratios are computed in five steps: (1) acquire and normalize fragment intensities of a composite spectrum; (2) model fragment intensities as a network flow with fixed capacities; (3) find the maximum flow for peptide X and for peptide Y; (4) combine both X and Y network flow capacities into the same network flow model; and (5) iteratively reduce the ratio of X and Y until the maximum flow is reached. (**b**) Left panel: elution profiles of a mixture of mono-acetylated isomers before (dashed line) or after (coloured lines) deconvolution. Right panel: the relative intensity of representative fragment ions extracted after isomer deconvolution at four different retention times, RT 20.84, 20.94, 21.03 and 21.12 min. (**c**) Bar graphs showing the amount of each peptide isomer determined by Iso-PeptidAce as a function of the amount of the same peptide injected as part of seven mixtures, each containing known amounts of the mono-, di- or tri-acetylated groups of isomers (from top to bottom). The mean value and the s.d. for each dilution point are shown in Supplementary Table 3A–C. The amount of peptide determined by Iso-PeptidAce in each mixture was normalized to the 80 fmol signal response. The amounts of peptides K5ac and K12ac (upper panel); K5_8ac, K5_16ac and K8_16ac (middle panel); K5_8_12ac and K5_12_16ac (lower panel) were varied from 5 to 320 fmol. Peptides K8ac and K16ac (upper panel); K5_12ac, K8_12ac and K12_16ac (middle panel); K5_8_16ac and K8_12_16ac (lower panel) were added to each of the seven mixtures at a fixed amount of 80 fmol. Each bar graph represents the mean of two technical replicates with error bars showing the relative distance of the maximum and the minimum values from the mean.

di-, tri- and tetra-acetylated peaks (denoted as Ac0, 1, 2, 3 and 4 in Supplementary Fig. 3A, top panel), we observed that peptides within the same isomeric group co-eluted and produced mixed MS/MS spectra (Supplementary Fig. 3A,B). In addition, isomeric peptides shared multiple fragment ions (Supplementary Fig. 4A–E). Supplementary Fig. 4E shows a three-dimensional view of the abundance distribution of fragment ions derived from the four mono-acetylated isomers. H4K5ac and H4K16ac are the only mono-acetylated H4 peptide isomers that produced unique

fragment ions at m/z 1253.7 and 530.3, respectively (Supplementary Fig. 4E, blue and purple bars). The peak intensities of these fragments can be used to infer the relative abundances of H4K5ac and H4K16ac in a given sample. This approach was previously implemented in existing software tools such as IsoScale[9]. Importantly, the two other mono-acetylated isomers, H4K8ac and H4K12ac did not produce any isomer-specific fragment ions (Supplementary Fig. 4E, red and green bars). These peptides share all of their fragments with either H4K5ac or

H4K16ac (Supplementary Fig. 4E, blue and purple bars), and, as a result, fragment ion ratios cannot be used directly to measure the relative proportions of individual peptide isomers. In Iso-PeptidAce, deconvolution of composite MS/MS spectra containing fragments of the four mono-acetylated isomers is achieved by exploiting the distinguishing features associated with isomer-specific fragment ion patterns. Here the fragment ion pattern of a given isomer is defined by the retention time at which the peptide was eluted and selected for fragmentation, the peak height and m/z of each fragment ion, and the sites of modification. For example, the fragment ion pattern for H4K8ac is different from that of H4K12ac by the presence of additional fragmentations at m/z 785.5 and 955.6, and the absence of y-ion fragments at m/z 771.4 and 941.6 (Supplementary Fig. 4C,D) and b-ion at m/z 299.2. In addition, their shared fragments (for example, m/z 242.1, 544.3 and 1239.7) differ slightly in peak intensities (Supplementary Fig. 4C,D). Thus, by exploiting such isomer-specific differences, Iso-PeptidAce deconvoluted the co-eluting mono-acetylated peptides into four distinct elution profiles where the apex of individual isomer peaks were separated by a few seconds (Fig. 1b, left panel, coloured solid lines). The elution curve before the deconvolution is shown in Fig. 1b (grey dashed line). Using Iso-PeptidAce, we were able to determine the relative contribution of each isomer to the abundance of selected fragment ions (Fig. 1b, right panel). Most of these fragments are shared between the different isomers, but their distribution and proportion varies across peptides. Fragmentation patterns representative of H4K8ac and H4K12ac were more prominent at the beginning of elution profiles, whereas those of H4K5ac and H4K16ac were observed at later time points. Similarly, deconvolution of the di- and tri-acetylated isomers by Iso-PeptidAce enabled resolution of each isomeric peptide (Supplementary Fig. 5).

We then assessed the performance of Iso-PeptidAce using seven mixtures of peptides for each group of isomers (mono-, di-, tri-acetylated). Each peptide mixture contained known amounts of propionylated and acetylated isomeric H4 peptides each ranging from 5 to 320 fmols. Within each peptide mixture two to three peptides were selected and added to each mixture at a fixed amount of 80 fmols (Supplementary Table 2A–C). Raw MS and MS/MS files from these samples were submitted to Iso-PeptidAce for deconvolution. The amount of peptide determined by Iso-PeptideAce as a function of amount injected (in fmol) for the mono-, di- and tri-acetylated groups of isomers is shown in Fig. 1c (see Supplementary Table 3A–C for the mean values and error bars). We used the signal intensity observed at 80 fmols for normalization. In the seven mixtures of mono-acetylated peptides we observed an increase in the relative proportion of peptides H4K5ac and H4K12ac whereas peptides H4K8ac and H4K16ac (injected at a constant amount of 80 fmol in each mixture) remained unchanged in each mixture (Fig. 1c, upper panel and Supplementary Table 3A). Similarly, we observed increasing levels of H4K5ac/K8ac, H4K5ac/K16ac and H4K8ac/K16ac among the seven mixtures of di-acetylated peptides (Fig. 1c, middle panel and Supplementary Table 3C) and of H4K5ac/K8ac/K12ac and H4K5ac/K12ac/K16ac in the tri-acetylated mixtures (Fig. 1c, bottom panel and Supplementary Table 3B) in accordance to their expected abundance. Thus, Iso-PeptidAce successfully deconvoluted the composite spectra generated from different co-eluting acetylated isomers present at concentrations ranging over almost two orders of magnitude.

Histone acetylation following HDAC inhibition. We tested our software using tryptic digests of histones isolated from human K562 cells treated with HDAC inhibitors (Fig. 2a). Cells were treated in duplicate for 1, 6 or 24 h with dimethylsulphoxide (DMSO) as solvent control or with the HDAC inhibitors (HDACi) MS-275, SAHA and JNJ-26481585. Histones were isolated by acid extraction and fractionated by reverse phase liquid chromatography (Supplementary Fig. 6). Fractions containing histone H3 or H4 were subjected to propionylation, tryptic digestion and LC–MS/MS analysis before data deconvolution using Iso-PeptidAce (Fig. 2a). Our results show that, 24 h after treatment with SAHA or JNJ-26481585, ~50% of H4 peptides 4-GKGGKGLGKGGAKR-17 were acetylated at the four available lysines, whereas only 20% of H4 molecules were tetra-acetylated in cells treated with MS-275 (Fig. 2b). Using Iso-PeptidAce we then determined acetylation site occupancies at H3K18, H3K23, H4K5, H4K8, H4K12 and H4K16 (Supplementary Table 4A,B). The abundance of H3K23ac is higher than that of H3K18ac in both untreated and HDACi-treated samples (Fig. 2c and Supplementary Table 4A). Although previous studies reported trace amounts of H3K18 methylation in mammalian cells[20], we did not detect this modification in our samples. Cells treated with all three HDACi demonstrated similar increases in H4 acetylation as a function of time. We observed at least an eightfold increase in acetylation at H4K5, H4K8 or H4K12 after a 24 h treatment with MS-275, SAHA or JNJ (Fig. 2d and Supplementary Table 4B). Up to threefold increase in acetylation was observed at H4K16.

Although a number of studies have previously shown that the HDACi MS-275 and SAHA induce global increases in H3 and H4 acetylation, there was no efficient technique to measure site-specific acetylation stoichiometries of lysines in the N-terminal tail of H4 because some of the acetylated isomers cannot be resolved by nano-LC and do not generate isomer-specific fragments. In our study, we observed that both MS-275 and SAHA caused a major increase in acetylation of the four lysines in the N-terminal tail of H4. In addition, for the first time, we report site-specific changes in histone acetylation caused by the more recently characterized pan-HDACi JNJ-26481585 (ref. 21). This drug, also known as Quisinostat, is currently in phase II clinical trial and exhibits antitumour activity in human multiple myeloma and leukaemic cells[22–25], and recent studies have shown that JNJ-26481585 elicits global increases in acetylation of histones H3 and H4 (refs 25,26) However, owing to the inherent limitations of immunoassays, these studies did not report the specific lysine residues affected or the stoichiometries of acetylation. Our approach allowed identification of several H3 and H4 sites that increased in acetylation after JNJ-26481585 treatment. When compared with MS-275 and SAHA, JNJ-26481585 caused a more rapid and higher fold increase in acetylation at all the sites that we investigated (Fig. 2b–d). Consistent with these results, JNJ-26481585 has been reported to be 500-fold more potent than SAHA at inhibiting HDAC1 (ref. 21).

These examples illustrate how Iso-PeptidAce provides an automated and rapid approach to quantify global changes in acetylation site occupancies on histone lysine residues. Our method is implemented as a software tool that deconvolutes composite spectra derived from two or more co-eluting isomeric acetylated peptides. In turn, this enables an accurate assessment of acetylation site occupancies at each of the lysines within tryptic peptides.

Acetylation patterns in CAF1-bound histones. Acetylation of multiple residues in newly synthesized histones has been implicated in nucleosome assembly[27]. However, the molecular function and the biological implications of this acetylation have remained unclear because of redundancy among several acetylation sites in new H3 and H4 molecules. The capability of Iso-PeptidAce to determine acetylation patterns in situations where samples are limiting was evaluated for affinity-purified

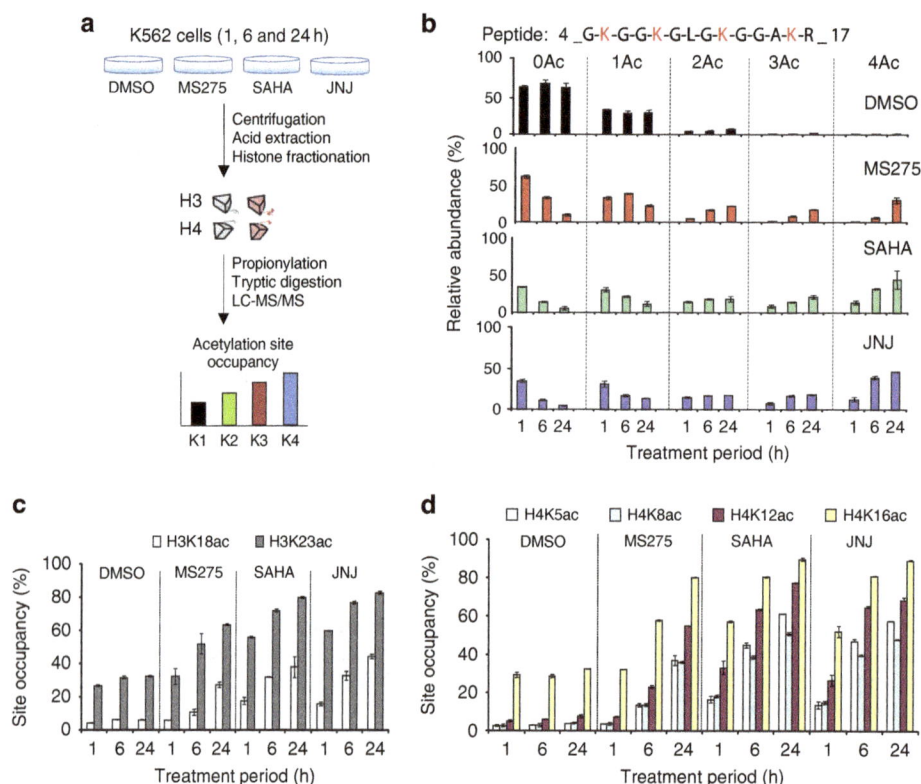

Figure 2 | Analysis of acetylation site occupancies in histones H3 and H4 following HDACi treatment. (**a**) Histone isolation and MS analysis workflow. Total histones were isolated from human erythroleukemic (K562) cells treated for 1, 6 or 24 h with DMSO (control) or with the HDACi MS-275, SAHA and JNJ-26481585. Total histones were fractionated into individual core histones via offline RP-HPLC. Fractions containing histones H3 and H4 were subjected to propionylation, tryptic digestion and LC–MS/MS analysis. Acetylation site occupancy for each lysine residue was determined based on Iso-PeptidAce reported peptide intensities. (**b**) Relative abundance of unmodified (0Ac), mono- (1Ac), di- (2Ac), tri- (3Ac) and tetra-acetylated (Ac) peptide 4-GKGGKGLGKGGAKR-17 before (DMSO) or after treatment with MS-275, SAHA or JNJ-26481585. (**c**) Acetylation site occupancies of histone H3 at positions K18 and K23 and (**d**) histone H4 at positions K5, K8, K12 and K16. Each bar graph represents the mean of two technical replicates with error bars showing the relative distance from the mean of the maximum and minimum values.

histones bound to CAF1, a protein complex that deposits new histone H3/H4 molecules onto nascent DNA during replication[28]. As previously described[28], we purified the CAF1 complex from asynchronously growing yeast cells via a tandem affinity purification (TAP)-tagged Cac2 subunit. We also purified total histones from the same culture to compare their acetylation patterns with those of histones bound to CAF1 (Fig. 3a). The occupancies of H3/H4 acetylation sites were clearly different in CAF1-bound versus total histones (Fig. 3b–d and Supplementary Table 5). We also detected mono-, di- and tri-methylation of H3K36 in the total histones, but these PTMs were absent in CAF1-bound histones (Supplementary Table 5). These results argue that CAF1-bound histones did not dissociate and mix with other histones during our affinity purification procedure. Otherwise, the H3 and H4 acetylation patterns in CAF1-bound and total histones would be identical. In CAF1-bound H3 molecules, we found a high acetylation site occupancy (between 49 and 79%) at each of the five lysine residues previously known to be acetylated in new histones from S. cerevisiae: H3K56, K9, K14, K23 and K27, but not at K18 (Fig. 3b). This high degree of acetylation at multiple sites demonstrates that essentially all the H3 molecules bound to CAF1 are acetylated. In contrast, a large fraction (64%, Fig. 3c) of H4 molecules bound to CAF1 were not acetylated at any of the four lysine residues located in their N-terminal tail. Nonetheless, these results provide a potential explanation to a long-standing paradox. Hat1 is an enzyme that acetylates new H4 molecules at lysines 5 and 12, two sites of acetylation that are conserved from yeast to humans[27]. Given this,

the paradox was that S. cerevisiae cells lacking Hat1 or cells where lysines 5 and 12 were mutated showed essentially no phenotype unless several lysines of H3 were also mutated to block their acetylation[29,30]. The fact that, in S. cerevisiae, the acetylation of new H3 molecules is far more abundant than that of new H4 molecules provides a molecular basis for these genetic results.

Our results also revealed an unexpected acetylation pattern in new H4 molecules where, in addition to lysines 5 and 12, lysine 16 was acetylated (Fig. 3c,d). Until now, the dogma was that new H4 molecules were predominantly acetylated at lysines 5 and 12, but not at lysine 16. This was based on pulse labelling of new histones with [3H]-lysine and Edman sequencing to quantify phenylthiohydantoin derivatives of acetyl-lysine and lysine[31]. The sensitivity of this technique rapidly decreases as a function of distance from the N-terminus and this may account for the fact that lysine 16 acetylation in new H4 molecules was not previously reported. Based on MS data, the presence of K16 acetylation in H4 molecules bound to S. cerevisiae CAF1 was previously reported[28] but its prevalence was unknown. For the first time, our experiments demonstrate that the most abundant form of H4 bound to S. cerevisiae CAF1 consists of molecules that are tri-acetylated at K5, K12 and K16, with significantly lower amounts of mono- and di-acetylated forms that also contain K16 acetylation (Fig. 3c). Consistent with this, the acetylation site occupancies at K5, K12 and K16 were roughly 20% each in CAF1-bound H4, a pattern strikingly different from that observed in total histones (Fig. 3d). In S. cerevisiae, the integrity of heterochromatin depends on removal of H4K16 acetylation by

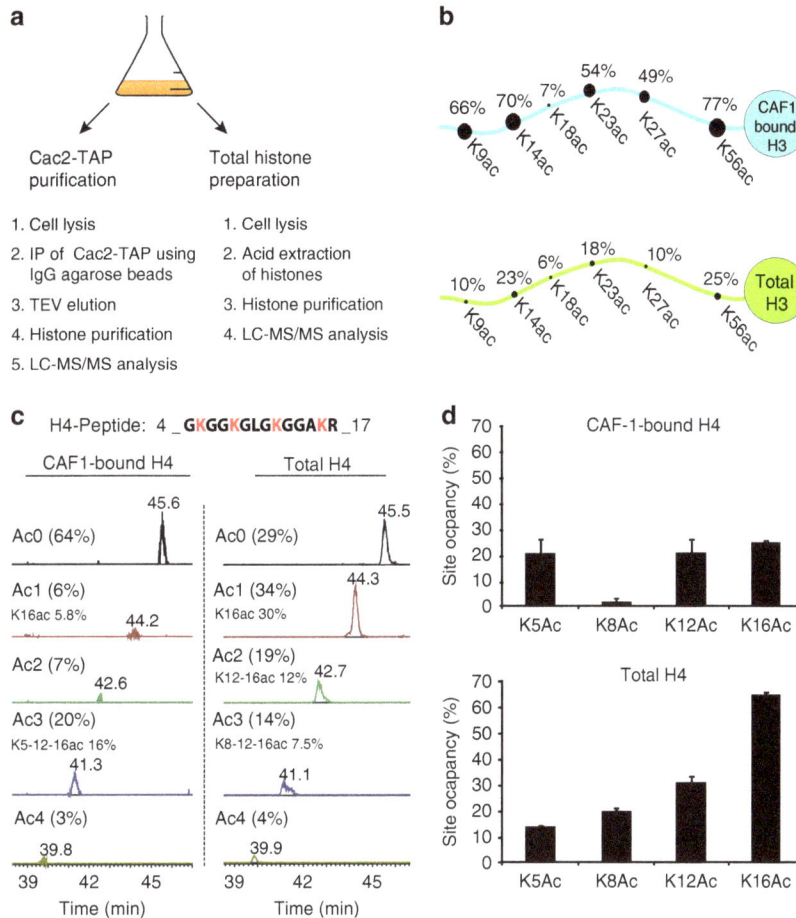

Figure 3 | Analysis of acetylation site occupancies in CAF1 associated histones H3 and H4. (a) Schematic representation of the experimental workflow. Cac2-TAP was purified from 3.5 l of yeast cell culture and the remainder (0.5 l) was processed for total histone preparation. Histones from both samples were fractionated by offline RP-HPLC and analysed by LC–MS/MS. (b) Comparison of lysine acetylation site occupancies in total histone H3 (lower panel) and CAF1-bound H3 (upper panel). (c) Extracted ion chromatograms of histone H4 peptide 4-GKGGKGLGKGGAKR-17 acetylated at zero (Ac0), one (Ac1), two (Ac2), three (Ac3) or four (Ac4) lysine residues in CAF1-bound (left panel) or total histones (right panel). The predominant acetylation pattern in each group of isomers is highlighted in red. (d) Comparison of acetylation site occupancies at positions H4K5, H4K8, H4K12 and H4K16 in CAF1-bound H4 (upper panel) and total histone H4 (lower panel). Each bar graph represents the mean of two technical replicates with error bars showing the relative distance from the mean of the maximum and minimum values.

the deacetylase Sir2 (ref. 32). Our results imply that Sir2-mediated deacetylation of H4K16 in new molecules deposited onto DNA during replication is likely important for propagation of heterochromatin structure in proliferating cells.

Discussion

Iso-PeptidAce was initially created to solve a specific problem that had apparently been neglected, even though it was pointed out by Smith *et al.*[5] more than 12 years ago. The problem of determining the relative abundance of multiple acetylated forms derived from the N-terminal tail of histone H4, which contains four acetylatable lysines, is compounded by the fact that several acetylated isomers co-elute during liquid chromatography and cannot be distinguished from each other based on the masses of MS[2] fragments. The value of determining the relative abundance of acetylated isomers was illustrated by our studies of new histone molecules bound to CAF1. Unexpectedly, the predominant pattern of acetylation that we found consisted of H4 molecules tri-acetylated at K5, K12 and K16. Although further studies are necessary to assess its biological significance, the prevalence of this acetylation pattern has implications for both replication-coupled nucleosome assembly and the roles of histone

deacetylases in the propagation of heterochromatin structures in proliferating cells.

It is known that in addition to lysine acetylation and methylation, histones are also modified by numerous other types of PTMs. Among many others, these include mono- or di-methylation of arginine, and several acylated forms of lysine residues that contain formyl, butyryl, crotonyl, malonyl or succinyl moieties. In the current study we did not determine the abundance of those types of PTMs because our method was developed with synthetic peptides carrying PTMs that were known to occur in yeast. Based on our data, we cannot rule out the possibility that other types of PTMs might also exist on the peptides that we studied. However, when coupled with recently developed multiplexed parallel reaction monitoring approaches[20,33] we anticipate that our method will prove valuable for deconvolution of a large number of isomeric/isobaric peptides bearing numerous patterns of PTMs.

We also expect that our method will prove an asset to investigate a broad range of biological problems related to protein lysine acetylation. For instance, our approach will be valuable to determine the *in vivo* substrate specificity of the multitude of bromodomain (BRD)-containing proteins that exist in model organisms and human cells[3]. Previous *in vitro* studies have clearly shown that BRDs derived from chromatin-modifying enzymes bind with higher affinity to peptides that contain acetylated lysines located in close proximity[3]. Moreover, because large protein complexes contain multiple BRDs located in different subunits of a given complex, deciphering their physiologically relevant acetylated substrates remains a formidable task. Proteomics studies performed in several species have identified a myriad of acetylation sites in proteins with a wide range of biological functions[34–36]. Supplementary Table 6 shows seven examples of known acetylated peptides where isomers could now be distinguished based on their fragmentation patterns. We therefore anticipate that, when combined with techniques to co-purify protein complexes with their acetylated substrates, the method described here will help identify combinatorial patterns of lysine acetylation that are recognized by protein complexes involved in numerous biological processes.

Methods

Synthetic peptide preparations. Synthetic histone H3 and H4 peptides were purchased from GenScript and include four variants of the H3 peptide 17-RKQLATKAAR-26: one with unmodified lysines, two peptides with one acetylated lysine and one peptide with both lysines acetylated (Supplementary Table 1A). A total of 16 H4 peptides were purchased. These included one unmodified, four mono-acetylated, six di-acetylated, four tri-acetylated and one tetra-acetylated form of peptide 1-SGRGKGGKGLGKGGAKRHR-19 (Supplementary Table 1B). The concentration of each peptide was determined based on triplicate ultraviolet absorbance measurements at 205 nm using an Ultraspec 2100 Pro spectrophotometer (GE Healthcare). Each peptide solution was nominally in $1 \mu g \mu l^{-1}$ range. We verified the purity and identity of each peptide by injecting directly 1 pmol of each peptide on a Q-Exactive mass spectrometer (Thermo scientific). All MS files will be publicly available from PeptideAtlas database (data identifier: PASS00658, http://www.peptideatlas.org/PASS/PASS00658) For convenience MS/MS spectra of isomeric peptides are provided as Supplementary Fig. 7.

Synthetic peptide mixtures. Standard curves were prepared for seven mixtures of H4 peptides with amounts of specific peptides ranging from 5 to 320 fmol (Supplementary Table 2A–C). Two or three peptides from each group of isomers were selected and added to the mixtures at a fixed amount of 80 fmols. The peptide mixtures were subjected to propionylation, tryptic digestion and LC–MS/MS analysis on a Q-Exactive mass spectrometer coupled to an EASY nLC II system (Thermo scientific).

Detection efficiency of H3 and H4 acetylated peptides. We prepared five mixtures of H3 or H4 synthetic peptides. The amount of each peptide in mixtures 1, 2, 3, 4 and 5 are 25, 50, 100, 200 and 400 fmols, respectively. These samples were subjected to propionylation, tryptic digestion and LC–MS/MS analysis on a Q-Exactive Plus instrument. The signal intensity of each peptide was plotted against the amount of peptide injected as shown in Supplementary Fig. 2A,B.

Cell treatment with HDAC inhibitors. Suspension cultures of K562 (human erythroleukemic) cells were grown in T-75 flasks at 37 °C and 5% CO_2 in RPMI-1640 medium (Gibco) supplemented with 10% fetal bovine serum (Wisent) and 1% penicillin/streptomycin. Asynchronously growing cells were treated with 1 µM of SAHA, MS-275 or JNJ-26481585 (Selleck Chemicals) for 1, 6 and 24 h. Control cells were treated with DMSO. Two biological replicates were prepared for each condition. After treatment, the cells were collected, centrifuged at 1,000 r.p.m. at room temperature, washed with PBS, and flash frozen in liquid nitrogen. Histones were isolated using acid extraction as previously described[37]. Briefly, nuclei from 10^7 cells were isolated with a hypotonic lysis buffer, followed by histone extraction using 0.2 M H_2SO_4 and TCA precipitation. The concentration of histone samples was determined by a Micro-BCA assay (Pierce). Further purification of core histones was performed by reverse-phase HPLC.

Purification of yeast CAF1. The Cac2-TAP purification was adapted from Zhou et al.[28] Briefly, exponentially growing yeast cells (4 l at ~1.8 × 10[7] cells per ml in YPD medium) were collected by centrifugation and washed twice with ice-cold water. All further steps were carried out on ice. The pellet was resuspended in an equal volume of lysis buffer (25 mM Tris-HCl pH 7.5, 100 mM NaCl, 10 mM $MgCl_2$, 0.1% NP-40, 1 mM EDTA, 10% glycerol, 1 mM DTT and 1 mM PMSF) containing a cocktails of protease inhibitors (5 µM Leupeptin, 25 µM Aprotinin and 5 µM Pepstatin A) and HDAC inhibitors (10 mM nicotinamide, 10 mM sodium butyrate, 10 µM trichostatin A and 100 µM SAHA). Cell suspensions were frozen in liquid nitrogen and lysed in a freezer mill (Spex Certiprep) using two

repeats of four cycles with 2 min of grinding (15 impacts per second) and 2 min of cooling per cycle. The resulting cell lysate powder was thawed out on ice. Ethidium bromide (75 µg ml^{-1}) was added and the lysate was incubated on ice for 1 h with benzonase (25 U ml^{-1}). The extract was clarified by centrifugation at 25,000 r.p.m. for 30 min. The supernatant was then incubated with 200 µl of IgG-Agarose beads (SIGMA) for 2 h at 4 °C. The beads were recovered by centrifugation, extensively washed with lysis buffer followed by several washes with TEV cleavage buffer (10 mM Tris-HCl pH 8, 100 mM NaCl, 1 mM DTT and 0.5 mM EDTA) containing the protease inhibitor and HDAC inhibitor cocktails. To elute Cac2-TAP, the beads were incubated with TEV protease at 4 °C overnight in TEV cleavage buffer. A portion of the TEV eluate (30 µl) was resolved by SDS–PAGE and the presence of Cac2-CBP and histones H3 and H4 was confirmed by immunoblotting. The remainder of the TEV eluate (170 µl) was dried in a Speed-Vac, resuspended in 1 ml H_2O and applied to a C18 column equilibrated with 0.1% trifluoroacetic acid (TFA) in water. The C18 column was washed twice with 1 ml of 0.1% TFA in 5% methanol and eluted with 1 ml of 70% acetonitrile (ACN). The eluate was dried in a Speed-Vac and resuspended in 0.1% TFA for purification of intact histones by offline HPLC.

Fractionation of core histones by RP-HPLC. Separation of intact core histones was achieved using a narrow-bore Zorbax C8 reverse-phase column (2.1 × 150 mm, 5 µm, 300 Å) on an Agilent 1200 HPLC system. Solvent A was aqueous 0.1% TFA (Sigma) and solvent B was 0.1% TFA in 100% acetonitrile (ACN). Approximately 2 µg of acid extracted histones were loaded onto the C8 column at a flow rate of 150 µl min^{-1}. Histones were eluted from the column using a gradient of 5–80% solvent B in 60 min. Fractions were collected in a 96-well plate at a rate of one fraction per minute. Fractions containing histones H3 or H4 were pooled and dried completely in a Speed-vac concentrator.

Propionylation and trypsin digestion of histones. In-solution tryptic digestion of synthetic peptides or core histones was performed as previously described[38]. Briefly, a total of 2 µg of HPLC purified histones H3 and H4 were subjected to propionylation by adding 200 µl of freshly prepared 2:1 (v/v) water: propionic anhydride (Sigma) mixture and vortexing the mixture for 1 h at room temperature. The samples were then dried in a Speed-vac at 4 °C. The dried samples were resuspended in 50 mM ammonium bicarbonate, vortexed for 2 min and subjected to a second round of evaporation at 4 °C. The samples were collected and resuspended in 100 µl of 50 mM ammonium bicarbonate and vortexed for 5 min to re-dissolve the proteins. Our digestion solution was prepared by adding 200 µl of 50 mM ammonium bicarbonate in a vial containing 20 µg of lyophilized trypsin (Promega). Roughly 0.5 µl of this solution was added to each histone sample and digested overnight at 37 °C. After digestion, samples were dried completely in a Speed-vac and then resuspended in 0.2% formic acid before LC–MS/MS analyses.

LC–MS/MS analyses of histone digests. MS data were acquired in duplicate on a Q-Exactive Plus mass spectrometer (Thermo scientific) coupled to an EASY nLC II system (Thermo scientific). A total of 1 µg of histone H3 and H4 digests generated from the control or HDACi-treated cells were first desalted on a Jupiter C18 (3 µm particles, Phenomenex) trap column (4 mm length, 360 µm i.d.) for 5 min at 10 µl min^{-1}, before their elution onto a C18 analytical column (18 cm length, 150 µm i.d.). A linear gradient from 5 to 60% ACN (containing 0.2% formic acid) at 600 nl min^{-1} over 60 min was used for peptide elution. The MS instrument was operated in positive ion mode and capillary voltage of 1.6 kV. MS scans were acquired in the Orbitrap analyser over the range of 300–1,500 m/z at a resolution of 70,000 and automatic gain control target value of 1.0×10^6. An inclusion list containing m/z, charge state and collision energy values of H3 and H4 peptides was used to trigger MS/MS acquisition. Every precursor ion found in the inclusion list was automatically selected for fragmentation in the HCD cell at a normalized collision energy setting of 27. The fragments were analysed in the Orbitrap at a resolution of 35,000 and a target value of 5.5×10^5. The dynamic exclusion setting was disabled to acquire multiple MS/MS spectra per peptide.

Data analysis. Data analysis was performed using the Iso-PeptidAce software. Raw MS and MS/MS files of individual synthetic peptides, mixtures of isomeric peptides or histone digests and Fasta files with H3 and H4 protein sequences were submitted to Iso-PeptidAce. The default settings for deconvolution of composite MS/MS spectra are: precursor mass tolerance: 8 p.p.m., fragment mass tolerance: 0.05 Da, the minimum number of fragment ions (per isomer) considered for deconvolution: 5, types of fragment ions considered: b and y, digestion enzyme: trypsin, missed cleavages: Iso-PeptidAce uses a no-enzyme in silico protein digestion routine to parse the provided FASTA files for potential matches. PTMs included in peptide-spectrum matching: carbamidomethylation of cysteine (C, 57.0215 Da), oxidation of methionine (M, 15.9949), phosphorylation of serine, threonine or tyrosine (S/T/Y, 79.9663), deamidation of asparagine and glutamine (N/Q 0.9840/0.9847), acetylation of lysine (K, 42.0106), propionylation of lysine (K, 56.0262), mono-methylation of lysine (K, 14.0157), di-methylation of lysine (K, 28.0313), tri-methylation of lysine (K, 42.0470), acetylation of protein N-terminus (42.0106). All the PTMs were considered as variable modifications. The output from Iso-PeptidAce is a combined result file (spreadsheet) containing intensity values for all the acetylated and non-acetylated forms of the H3 and H4 peptides obtained after deconvolution. A representative result file for the H4 peptide GKGGKGLGKGGAKR is shown in Supplementary Table 7.

A total of 16 different isoforms of this peptide were detected (shown in each column in Supplementary Table 7). These peptides are divided into five groups (per sample) based on their m/z values (shown in each row in Supplementary Table 7): one un-acetylated (0Ac, m/z 747.94), four mono-acetylated (1Ac, m/z 740.93), six di-acetylated (2Ac, 733.93), four tri-acetylated (3Ac, m/z 726.92) and one tetra-acetylated (4Ac, m/z 719.91). Each intensity value in Supplementary Table 7 is normalized based on the peptide's response factor, which is determined from the slope of the lines shown in Supplementary Fig. 2A,B. The slope of each line was determined from the linear equations that best fit the MS signal responses. Acetylation site occupancy at a specific Lys residue was determined from the ratio of the sum of intensities of peptides bearing the ac-Lys to the sum of intensities of all the 16 peptide isoforms. Iso-PeptidAce can be downloaded from the website: http://proteomics.iric.ca/tools/Iso-PeptidAce. Detailed information on Iso-PeptidAce is also provided as Supplementary Methods.

References

1. Unnikrishnan, A., Gafken, P. R. & Tsukiyama, T. Dynamic changes in histone acetylation regulate origins of DNA replication. *Nat. Struct. Mol. Biol.* **17**, 430–437 (2010).
2. Zentner, G. E. & Henikoff, S. Regulation of nucleosome dynamics by histone modifications. *Nat. Struct. Mol. Biol.* **20**, 259–266 (2013).
3. Filippakopoulos, P. & Knapp, S. The bromodomain interaction module. *FEBS Lett.* **586**, 2692–2704 (2012).
4. Filippakopoulos, P. *et al.* Histone recognition and large-scale structural analysis of the human bromodomain family. *Cell* **149**, 214–231 (2012).
5. Smith, C. M. *et al.* Mass spectrometric quantification of acetylation at specific lysines within the amino-terminal tail of histone H4. *Anal. Biochem.* **316**, 23–33 (2003).
6. Feller, C., Forné, I., Imhof, A. & Becker, P. B. Global and specific responses of the histone acetylome to systematic perturbation. *Mol. Cell* **57**, 559–571 (2015).
7. Huang, Y. *et al.* Statistical characterization of the charge state and residue dependence of low-energy CID peptide dissociation patterns. *Anal. Chem.* **77**, 5800–5813 (2005).
8. Kapp, E. A. *et al.* Mining a tandem mass spectrometry database to determine the trends and global factors influencing peptide fragmentation. *Anal. Chem.* **75**, 6251–6264 (2003).
9. Sidoli, S. *et al.* Middle-down hybrid chromatography/tandem mass spectrometry workflow for characterization of combinatorial post-translational modifications in histones. *Proteomics* **14**, 2200–2211 (2014).
10. Yuan, Z.-F. *et al.* EpiProfile Quantifies histone peptides with modifications by extracting retention time and intensity in high-resolution mass spectra. *Mol. Cell. Proteomics* **14**, 1696–1707 (2015).
11. Bern, M. *et al.* Deconvolution of mixture spectra from ion-trap data-independent-acquisition tandem mass spectrometry. *Anal. Chem.* **82**, 833–841 (2010).
12. Wang, J., Bourne, P. E. & Bandeira, N. MixGF: spectral probabilities for mixture spectra from more than one peptide. *Mol. Cell. Proteomics* **13**, 3688–3697 (2014).
13. Zhang, B., Pirmoradian, M., Chernobrovkin, A. & Zubarev, R. A. DeMix workflow for efficient identification of cofragmented peptides in high resolution data-dependent tandem mass spectrometry. *Mol. Cell. Proteomics* **13**, 3211–3223 (2014).
14. Courcelles, M., Bridon, G., Lemieux, S. & Thibault, P. Occurrence and detection of phosphopeptide isomers in large-scale phosphoproteomics experiments. *J. Proteome Res.* **11**, 3753–3765 (2012).
15. Ahuja, R. K., Magnanti, T. L. & Orlin, J. B. in *Network Flows: Theory, Algorithms, and Applications* (Prentice Hall, Upper Saddle River, 1993).
16. Cho, D.-Y., Kim, Y.-A. & Przytycka, T. M. Chapter 5: network biology approach to complex diseases. *PLoS Comput. Biol.* **8**, e1002820 (2012).
17. Albert, R. Scale-free networks in cell biology. *J. Cell Sci.* **118**, 4947–4957 (2005).
18. Dutt, S., Dai, Y., Ren, H. & Fontanarosa, J. in *Bioinformatics and Computational Biology* Vol. 5462 (ed. Rajasekaran, S.) 211–223 (Springer, 2009).
19. SNYMAN, J. A. in *Practical Mathematical Optimization: an Introduction to Basic Optimization Theory and Classical and New Gradient-Based Algorithms* (Springer Publishing, 2005).
20. Tang, H. *et al.* Multiplexed parallel reaction monitoring targeting histone modifications on the QExactive mass spectrometer. *Anal. Chem.* **86**, 5526–5534 (2014).
21. Arts, J. *et al.* JNJ-26481585, a novel 'Second-Generation' oral histone deacetylase inhibitor, shows broad-spectrum antitumoral activity. *Clin. Cancer Res.* **15**, 6841–6851 (2009).
22. Stühmer, T. *et al.* Preclinical anti-myeloma activity of the novel HDAC-inhibitor JNJ-26481585. *Br. J. Haematol.* **149**, 529–536 (2010).
23. Tong, W.-G. *et al.* Preclinical antileukemia activity of JNJ-26481585, a potent second-generation histone deacetylase inhibitor. *Leuk. Res.* **34**, 221–228 (2010).
24. Deleu, S. *et al.* The effects of JNJ-26481585, a novel hydroxamate-based histone deacetylase inhibitor, on the development of multiple myeloma in the 5T2MM and 5T33MM murine models. *Leukemia* **23**, 1894–1903 (2009).
25. Venugopal, B. *et al.* A phase I study of quisinostat (JNJ-26481585), an oral hydroxamate histone deacetylase inhibitor with evidence of target modulation and antitumor activity, in patients with advanced solid tumors. *Clin. Cancer Res.* **19**, 4262–4272 (2013).
26. Carol, H. *et al.* Initial testing (stage 1) of the histone deacetylase inhibitor, quisinostat (JNJ-26481585), by the Pediatric Preclinical Testing Program. *Pediatr. Blood Cancer* **61**, 245–252 (2014).
27. Li, Q., Burgess, R. & Zhang, Z. All roads lead to chromatin: multiple pathways for histone deposition. *Biochim. Biophys. Acta* **1819**, 238–246 (2013).
28. Zhou, H., Madden, B. J., Muddiman, D. C. & Zhang, Z. Chromatin assembly factor 1 interacts with histone H3 methylated at lysine 79 in the processes of epigenetic silencing and DNA repair. *Biochemistry* **45**, 2852–2861 (2006).
29. Kelly, T. J., Qin, S., Gottschling, D. E. & Parthun, M. R. Type B histone acetyltransferase Hat1p participates in telomeric silencing. *Mol. Cell. Biol.* **20**, 7051–7058 (2000).
30. Qin, S. & Parthun, M. R. Histone H3 and the histone acetyltransferase Hat1p contribute to DNA double-strand break repair. *Mol. Cell. Biol.* **22**, 8353–8365 (2002).
31. Sobel, R. E., Cook, R. G., Perry, C. A., Annunziato, A. T. & Allis, C. D. Conservation of deposition-related acetylation sites in newly synthesized histones H3 and H4. *Proc. Natl Acad. Sci. USA* **92**, 1237–1241 (1995).
32. Rusche, L. N., Kirchmaier, A. L. & Rine, J. The establishment, inheritance, and function of silenced chromatin in Saccharomyces cerevisiae. *Annu. Rev. Biochem.* **72**, 481–516 (2003).
33. Gallien, S. *et al.* Targeted proteomic quantification on quadrupole-orbitrap mass spectrometer. *Mol. Cell. Proteomics* **11**, 1709–1723 (2012).
34. Chen, Y. *et al.* Quantitative acetylome analysis reveals the roles of SIRT1 in regulating diverse substrates and cellular pathways. *Mol. Cell. Proteomics* **11**, 1048–1062 (2012).
35. Downey, M. *et al.* Acetylome profiling reveals overlap in the regulation of diverse processes by sirtuins, gcn5, and esa1. *Mol. Cell. Proteomics* **14**, 162–176 (2015).
36. Weinert, B. T. *et al.* Proteome-wide mapping of the Drosophila acetylome demonstrates a high degree of conservation of lysine acetylation. *Sci. Signal.* **4**, ra48 (2011).
37. Shechter, D., Dormann, H. L., Allis, C. D. & Hake, S. B. Extraction, purification and analysis of histones. *Nat. Protoc.* **2**, 1445–1457 (2007).
38. Abshiru, N. *et al.* Chaperone-mediated acetylation of histones by Rtt109 identified by quantitative proteomics. *Proteomics* **81**, 80–90 (2013).

Acknowledgements

We thank Dr Sébastien Lemieux (UdeM) for his invaluable advice on the use of different bioinformatics tools and approaches. We also thank Éric Bonneil (UdeM) for his assistance with MS experiments. This work was carried out with financial support from the Canadian Institute for Health Research to A.V. (CIHR, FRN 125916) and the Natural Sciences and Engineering Research Council (NSERC 311598) to P.T. The Institute for Research in Immunology and Cancer receives infrastructure support from IRICoR, the Canadian Foundation for Innovation, and the Fonds de Recherche du Québec—Santé.

Author contributions

N.A. designed and carried out the proteomics experiments. O.C.-L. built and tested the Iso-PeptidAce deconvolution software. R.E.R. and C.P. produced HDAC inhibitor-treated cells and performed histone extraction. A.J. and R.E.R. isolated histones bound to yeast CAF1. N.A. analysed data with help from O.C.-L. P.T. and A.V. conceived and led the project. N.A., O.C.-L., P.T. and A.V. wrote the manuscript, with input from all the authors.

Additional information

Accession codes. MS files are available from PeptideAtlas database (data identifier: PASS00658).

21

Programming a topologically constrained DNA nanostructure into a sensor

Meng Liu[1,2], Qiang Zhang[2], Zhongping Li[1], Jimmy Gu[1], John D. Brennan[2] & Yingfu Li[1,2]

Many rationally engineered DNA nanostructures use mechanically interlocked topologies to connect individual DNA components, and their physical connectivity is achieved through the formation of a strong linking duplex. The existence of such a structural element also poses a significant topological constraint on functions of component rings. Herein, we hypothesize and confirm that DNA catenanes with a strong linking duplex prevent component rings from acting as the template for rolling circle amplification (RCA). However, by using an RNA-containing DNA [2] catenane with a strong linking duplex, we show that a stimuli-responsive RNA-cleaving DNAzyme can linearize one component ring, and thus enable RCA, producing an ultra-sensitive biosensing system. As an example, a DNA catenane biosensor is engineered to detect the model bacterial pathogen *Escherichia coli* through binding of a secreted protein, with a detection limit of $10\,\text{cells}\,\text{ml}^{-1}$, thus establishing a new platform for further applications of mechanically interlocked DNA nanostructures.

[1]Department of Biochemistry and Biomedical Sciences, McMaster University, 1280 Main Street West, Hamilton, Ontario, Canada L8S 4K1. [2]Biointerfaces Institute, McMaster University, 1280 Main Street West, Hamilton, Ontario, Canada L8S 4O3. Correspondence and requests for materials should be addressed to J.D.B. (email: brennanj@mcmaster.ca) or to Y.L. (email: liying@mcmaster.ca).

DNA is not only important in biological systems as genetic material, it has also become a key player in synthetic biology. DNA can be engineered into catalysts (DNAzymes) and molecular receptors (DNA aptamers), making DNA a functionally versatile polymer. DNA, as a highly programmable material based on predictable Watson–Crick base-pairing interactions, has become a valuable macromolecule for rational engineering of molecular machines for potential nanotechnological applications.

In recent years, tremendous progress has been made toward building DNA-based nanodevices with increasing structural complexity and functional capabilities[1-12]. One important feature of many reported DNA nanostructures, such as DNA Borromean rings[1] and DNA catenanes[3], is the use of mechanically interlocked topologies to connect individual DNA components. The mechanical interlocking between DNA strands can be easily achieved in the case of DNA through the formation of a linking duplex between partner rings before ring closure. The existence of a linking duplex is not only essential to the creation of a strong connectivity between partner rings but also necessary for the stability of these well-defined structures.

As we will show in this work, the linking-duplex feature also enables the use of topologically interlocked architectures, such as DNA catenanes, for the design of amplified biosensors for bioanalytical applications. The biosensing strategy is based on the following idea: the strong physical engagement of two mechanically interlocked single-stranded DNA rings in a DNA [2] catenane (termed D2C in this study for simplicity) with a strong linking duplex makes the component rings unsuitable as the template for rolling circle amplification (RCA), an isothermal DNA amplification technique[13-15]. However, when one of the component rings is engineered to be a substrate of a stimuli-responsive RNA-cleaving DNAzyme (RCD), the system can be programmed into a biosensor that is capable of reporting a target of interest in three sequential reactions: target-induced RNA cleavage, nucleolytic conversion of the cleavage product into a DNA primer, and DNA amplification via RCA. By this approach, we establish an amplified biosensing system that is capable of achieving ultra-sensitive detection of *Escherichia coli* (*E. coli*) at a concentration as low as 10 cells ml^{-1} without cell culture.

Results

Inability of a D2C with a strong linking duplex to undergo RCA. The conceptual cornerstone of this work is the assumption that the components of a D2C with a strong linking duplex are unable to undergo RCA. To test this idea, we synthesized a D2C (Fig. 1a) consisting of two component single-stranded DNA rings, named ^{C}DNA$_i$ and r^{C}DNA$_{ii}$ (Fig. 1b; r stands for the single ribonucleotide, ribo-A, in the sequence of ^{C}DNA$_{ii}$). Briefly, the linear DNA r^{L}DNA$_{ii}$ was circularized into r^{C}DNA$_{ii}$ using ^{C}DNA$_i$ as the ligation template (sequences of all DNA species are provided in Supplementary Table 1). The resultant D2C, denoted rD2C1, contains a strong linking duplex of 24 bp, which translates into two helical turns (boxed nucleotides in Fig. 1b). The reaction yield of rD2C1 was determined to be 58% (Fig. 1c).

We then performed RCA reactions with gel-purified ^{C}DNA$_i$, r^{C}DNA$_{ii}$ and rD2C1. Agarose gel analysis indicated that RCA products were produced with ^{C}DNA$_i$ (using DP1 as the primer; Fig. 1d) and r^{C}DNA$_{ii}$ (using DP2 as the primer; Fig. 1d). In contrast, no RCA products were observed for rD2C1 using the same set of primers (Fig. 1d). This experiment shows that the topological constraint imposed by a strong linking duplex indeed prevents φ29 DNA polymerase (φ29DP) from replicating interlocked circular templates. In a control experiment, it was found that RCA was not inhibited when the linking duplex of the DNA catenane was made of 9 bp (Supplementary Fig. 1).

Enabling RCA via cleavage of a component ring by a DNAzyme. The inability of φ29DP to carry out RCA with topologically

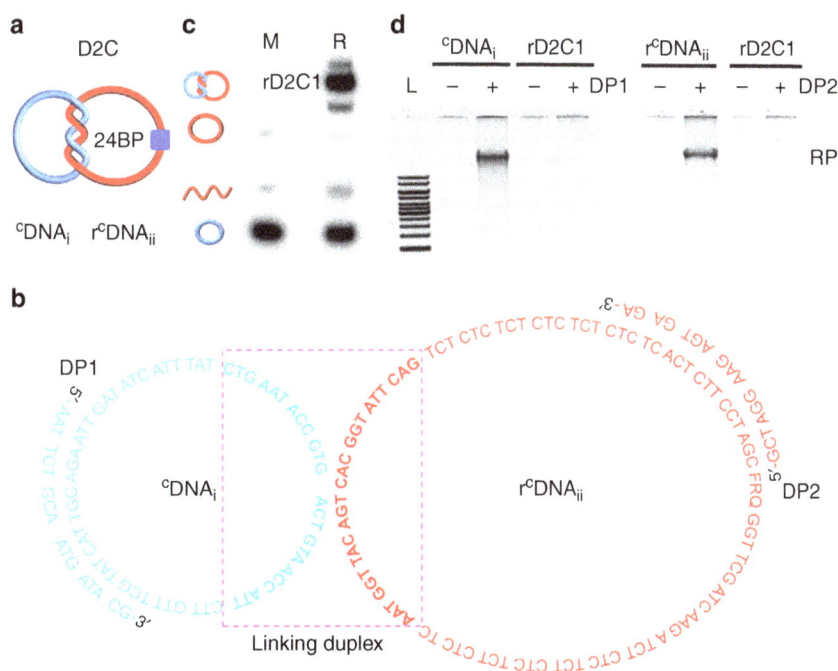

Figure 1 | Inability of a D2C to undergo RCA. (a) Schematic illustration of a D2C made of ^{C}DNA$_i$ and r^{C}DNA$_{ii}$, with a linking duplex of 24 bp. **(b)** Sequences of rD2C1, DP1 and DP2. Boxed nucleotides represent the 24-bp linking duplex. F: fluorescein-dT; R, adenosine ribonucleotide; Q, dabcyl-dT. **(c)** Synthesis of rD2C1 by circularizing linear DNA$_{ii}$ over ^{C}DNA$_i$ as the template. Lane M, markers made of ^{L}DNA$_{ii}$, ^{C}DNA$_i$ and r^{C}DNA$_{ii}$. Lane R: circularization mixture. **(d)** RCA reactions with gel-purified ^{C}DNA$_i$, r^{C}DNA$_{ii}$ and rD2C1 using DP1 and DP2 as primers. Lane L, DNA ladders ranging from 1 to 10 kbp; RP, RCA product.

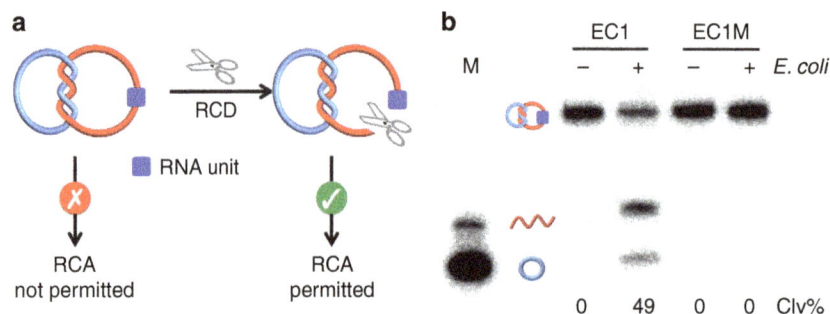

Figure 2 | Cleavage of an RNA-containing D2C by an RCD. (**a**) Restoration of RCA compatibility of an rD2C using an RCD. (**b**) Cleavage of rD2C1 by EC1, an E. coli-responsive DNAzyme. Concentration of E. coli: 10^5 cells ml^{-1}. Reaction mixtures were analyzed by 10% denaturing PAGE. EC1M: a mutant EC1 that cannot be activated by E. coli. Both rCDNA$_{ii}$ and CDNA$_i$ in rD2C1 were radioactively labelled with ^{32}P to facilitate DNA visualization on the gel. Clv%: per cent cleavage.

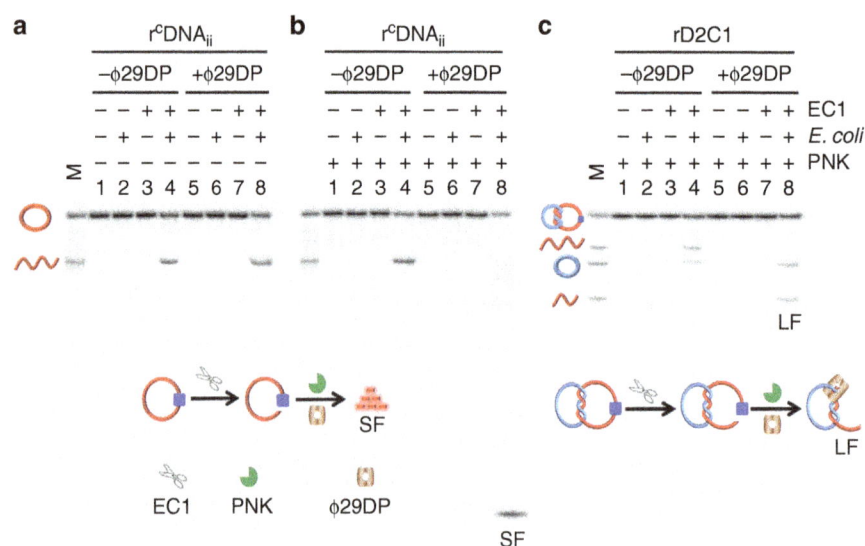

Figure 3 | 3′–5′ exonucleolytic activity of ϕ29DP on rCDNA$_{ii}$ and rD2C1. Degradation of EC1-mediated cleavage product of rCDNA$_{ii}$ (**a,b**) and rD2C1 (**c**) by ϕ29DP and PNK. Concentration of E. coli: 10^5 cells ml^{-1}. Reaction mixtures were analyzed by 20% denaturing PAGE. LF, large DNA fragment; SF, small DNA fragment. M lanes contain various DNA markers as indicated. rCDNA$_{ii}$, both rCDNA$_{ii}$ and CDNA$_i$ in rD2C1 were radioactively labelled with ^{32}P to facilitate DNA visualization on the gel.

constrained DNA catenanes provides a novel avenue to explore these intricate DNA assemblies for practical applications. Given our interest in DNAzymes and biosensing, we turned our attention next to the engineering of a highly unique biosensing system that takes advantage of topologically constrained DNA catenanes, DNAzymes and RCA.

The working principle of our biosensing system is shown in Fig. 2a. It uses an RCD to cleave the embedded RNA linkage within rD2C1 (note that the rCDNA$_{ii}$ was designed to contain a single ribonucleotide; Fig. 1b). It is expected that the cleavage and linearization of rCDNA$_{ii}$ by the RCD will release the topological constraint on the DNA assembly, which converts CDNA$_i$ into a suitable template for RCA.

In theory, our approach should be compatible with any RCD, making it a platform for detection of any species recognized by an allosteric DNAzyme. For this work, we employed EC1, which was previously isolated by us from a random-sequence DNA pool using in vitro selection for specific detection of E. coli, a model bacterial pathogen[16,17]. EC1 was found to be activated by a protein molecule secreted specifically by E. coli cells (illustrated in Supplementary Fig. 2). Therefore, the use of EC1 enables the detection of this pathogen. As illustrated in Fig. 2b, EC1 was indeed able to cleave the rCDNA$_{ii}$ present in rD2C1 in an

E. coli-dependent manner, resulting in CDNA$_i$ and a linear DNA$_{ii}$. An inactive DNAzyme mutant, EC1M, was also tested as a control. No cleavage product was observed for EC1M, indicating that the cleavage reaction is highly dependent on the DNAzyme sequence. We also examined the cleavage activity at different reaction times and found that the cleavage activity reached a plateau in 1 h (Supplementary Fig. 3). Thus, this reaction time was used for the remaining experiments.

Upon demonstrating EC1-mediated cleavage of rCDNA$_{ii}$ in rD2C1, we examined the use of the cleavage reaction mixture for initiating the RCA reaction with ϕ29DP. Because ϕ29DP has 3′–5′ exonucleolytic activity that can degrade single-stranded DNA from the 3′-end but does not digest double-stranded DNA[18,19], we speculated that the system should not require an external primer, as ϕ29DP should be able to convert the linearized DNA$_{ii}$ into a primer for RCA. To evaluate this hypothesis, we determined whether ϕ29DP could digest EC1-linerized DNA$_{ii}$. From the data presented in Fig. 3a, it is clear that ϕ29DP could not digest linearized DNA$_{ii}$ (comparing lanes 4 and 8).

There are two possible reasons for this finding. The first possibility is that ϕ29DP is incapable of digesting an RNA-terminated DNA molecule. However, testing with an

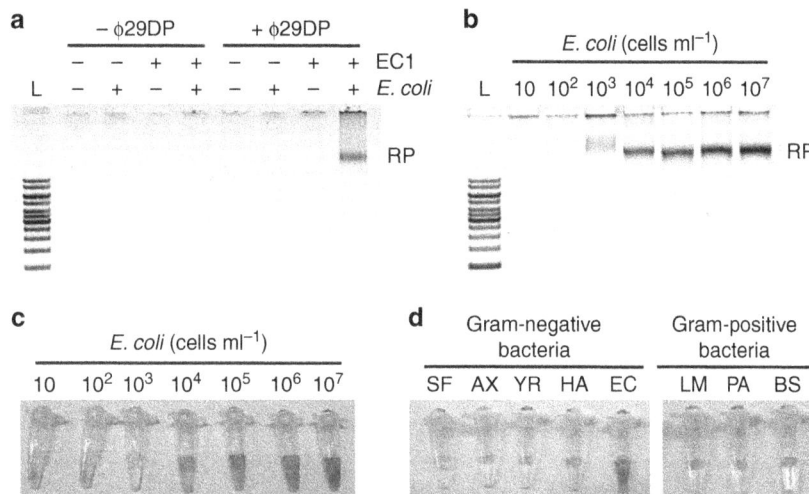

Figure 4 | _E. coli_-dependent RCA reaction. (**a**) RCA reactions of rD2C1 in the presence of _E. coli_ (10^5 cells ml^{-1}) analyzed using 0.6% agarose gel electrophoresis. Note every reaction also contained PNK and dNTPs. L, DNA ladders ranging from 1 to 10 kbp; RP, RCA product. (**b**) Determination of detection sensitivity through analysis of RP using 0.6% agarose gel electrophoresis. (**c**) Determination of detection sensitivity via the colourimetric assay enabled by PW17 peroxidase DNAzyme. (**d**) Analysis of assay specificity using the colourimetric assay. The gram-negative bacteria used were _Serratia fonticola_ (SF), _Achromobacter xylosoxidans_ (AX), _Yersinia ruckeri_ (YR) and _Hafnia alvei_ (HA). The gram-positive bacteria used were _Leuconostoc mesenteroides_ (LM) and _Pediococcus acidilactici_ (PA).

RNA-terminated oligonucleotide ruled out this possibility (Supplementary Fig. 4). The second scenario is that φ29DP is not able to digest an RNA-terminated DNA molecule containing a 2′,3′-cyclic phosphate on the RNA moiety, which is a common product of RNA cleavage[20]. To test this, we treated the reaction mixture with T4 polynucleotide kinase (PNK), which is known to be capable of removing the terminal 2′,3′-cyclic phosphate in RNA[21]. As shown in Fig. 3b, treatment with PNK indeed facilitated the digestion of EC1-linearized rDNA$_{ii}$, as evidenced by the accumulation of small cleavage fragments (labelled SF in lane 8).

We next investigated the combined action of φ29DP and PNK on EC1-linearized rDNA$_{ii}$ within rD2C1. As shown in Fig. 3c, φ29DP degraded complexed DNA$_{ii}$ to a product of ∼60 nt (labelled LF, representing long fragment; lane 8). We also conducted the digestion assays using different incubation times (Supplementary Fig. 5). The progressive accumulation of LF and disappearance of the cleaved rDNA$_{ii}$ was observed. These experiments allow us to conclude that the combination of φ29DP and PNK can remove the single-stranded fragment at the 3′-end of the EC1-linearized DNA$_{ii}$ from rD2C1. It is expected that the trimmed DNA$_{ii}$ can now function as a primer to initiate RCA over the complexed ^{C}DNA$_i$ template.

To verify the point above, we carried out the RCA reaction with the rD2C1 assembly. The reaction was performed in two sequential steps: activation of EC1 by _E. coli_ and PNK treatment, followed by the addition of φ29DP and dNTPs. As expected, RCA products were indeed observed following this procedure (the last lane of Fig. 4a; the other lanes represent various controls). The RCA products were further analyzed through partial digestion with _Eco_RV (Supplementary Fig. 6) as the sequence of ^{C}DNA$_i$ was designed to contain a recognition sequence for this restriction enzyme. The appearance of the expected characteristic DNA banding pattern on the gel, which consists of monomeric, dimeric and other higher-ordered DNA amplicons, verified that the RCA products indeed contained the correct repetitive sequences.

Quantitative detection of _E. coli_ using the DNA catenane sensor. We then investigated the feasibility of performing quantitative analysis using the DNA catenane sensor. Samples containing at _E. coli_ 10–10^7 cells ml^{-1} were assessed for RCA amplified detection using a gel-staining method. By this method, we were able to detect as low as 10^3 cells ml^{-1} (Fig. 4b). Although gel-based RP analysis can perform quantitative detection of _E. coli_, the procedure is extremely inconvenient. To overcome this issue, we developed a colorimetric assay by modifying the sequence of ^{C}DNA$_i$ (the new sequence is named ^{C}DNA$_i$CD) such that the RCA product contained a repetitive sequence of PW17, a peroxidase-like DNAzyme capable of generating a colourimetric signal[22–26]. In the presence of hemin, PW17 catalyzes the H_2O_2-mediated oxidation of 2,2′-azino-bis(3-ethylbenzthia-zoline-6-sulphonic acid) (ABTS) into a coloured product. As shown in Fig. 4c, this colourimetric method was indeed able to detect _E. coli_ in a concentration-dependent manner and registered a detection sensitivity of 10^3 cells ml^{-1}, similar to what was observed with the gel-based method.

We also evaluated the bacterial detection specificity using the colourimetric assay. We selected four other Gram-negative and three Gram-positive bacteria that were previously tested for EC1-based detection. It was observed that none of these bacteria were able to produce a positive signal, indicating that the rD2C1/EC1 system retained the high recognition specificity for _E. coli_ (Fig. 4d). To further evaluate the specificity, we checked the potential influence of small RNAs on _E. coli_ detection due to the fact that small RNAs (for example, microRNA) are suitable primers for RCA. For this experiment, we used the total small RNAs extracted from breast cancer cell line MCF-7. Agarose gel and colourimetric results indicated that the small RNAs were not able to induce the RCA reaction (Supplementary Fig. 7). This high specificity is attributed to the unique topologically constrained structure of the DNA catenane.

Enhancing detection sensitivity using hyperbranched RCA. Finally we examined the possibility of performing a double-primed hyperbranched RCA (HRCA)[27] reaction with the rD2C1/

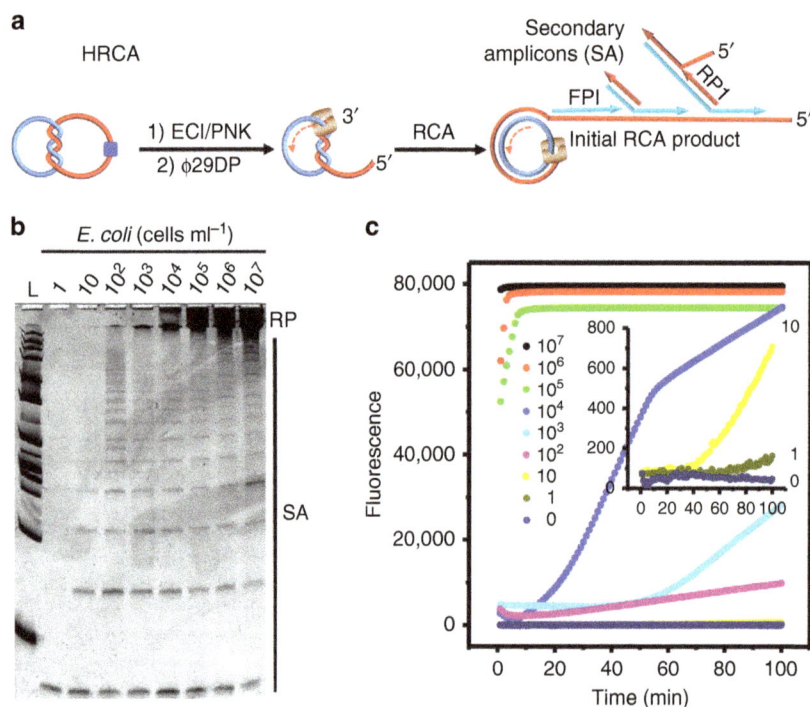

Figure 5 | *E. coli*-dependent HRCA reaction. (**a**) Schematic illustration of HRCA. FP1, forward primer; RP1, reverse primer. (**b**) Denaturing PAGE analysis of HRCA products. RP, RCA products; SA, secondary amplicons produced from the initial RCA products. (**c**) Real-time monitoring of HRCA reactions at various *E. coli* concentrations (cells ml^{-1}).

EC1 system to further increase the detection sensitivity. In HRCA (Fig. 5a), multiple priming events can be continuously initiated by a forward primer (FP1) and a reverse primer (RP1) as the original RCA product strand elongates, resulting in an exponential amplification[28].

We found that HRCA was indeed functional with the rD2C1/EC1 system: as shown in Fig. 5b, in addition to the observation of the RCA products on denaturing polyacrylamide gel electrophoresis (PAGE), a series of shorter DNA molecules were also produced, representing various secondary amplicons produced from the primary amplicons (that is, initial RCA products). The HRCA reactions, in response to varying concentrations of *E. coli*, were also monitored in real time through the use of EvaGreen, a DNA-binding dye (Fig. 5c). We found that this method exhibited much enhanced detection sensitivity, as it was able to detect *E. coli* at a concentration of as low as 10 cells ml^{-1} without cell culture.

It is interesting to note that RCA reactions with ϕ29DP remain functional even in the presence of 50% human blood (Supplementary Fig. 8). In comparison, PCR with Taq DNA polymerase was completely inhibited by less than 0.2% human blood[29-31]. To demonstrate the performance of the assay when using more complex samples, we spiked whole blood with *E. coli* and demonstrated that under these conditions the DNA catenane sensor was still able to detect *E. coli* at a concentration of 10 cells ml^{-1} (Supplementary Fig. 9). This observation is consistent with a recent report where EC1 was used to detect *E. coli* cells in human blood[32].

The ability of the DNA catenane sensor to distinguish between live and dead *E. coli* cells was also investigated. For this experiment, we used lysozyme to kill *E. coli* (10^5 cells ml^{-1}) and compared the signal responses of the catenane sensor in the presence of live and dead bacteria using HRCA. Negligible signal was observed with dead cells while a high activity was seen with live *E. coli* cells (Supplementary Fig. 10).

We also performed an *E. coli* detection experiment with an enzyme-linked immunosorbent assay (ELISA) using a commercial ELISA kit designed to detect *E. coli* host cell protein (Supplementary Fig. 11). It was found that the ELISA method was able to detect *E. coli* at a concentration of 10^3 cells ml^{-1}. Therefore, our amplified DNA catenane sensor offers a detection sensitivity that is 100 times better than the ELISA method, while the test time for both methods are similar (~3 h; see Supplementary Table 2 for additional information).

Discussion

We have shown for the first time that mechanically interlocked DNA [2] catenanes with a strong linking duplex impose a significant topological constraint on their component DNA rings, making them unsuitable as the template for RCA. We have further demonstrated that such DNA nanostructures can be uniquely exploited for the design of a biosensing system where the elimination of the topological engagement, achieved simply through the cleavage of one interlocked DNA ring in an analyte-dependent manner, frees up the other ring for the RCA reaction. As an example, we have produced one of the two interlocked DNA rings to contain a RNA linkage so that an RCD can be used to cleave one interlocked ring. Through the use of an RCD whose activity is specifically triggered by a secreted protein in *E. coli*, we have shown that the featured biosensing system is capable of achieving ultra-sensitive detection of this bacterial pathogen.

The biosensing system featured here offers some distinct advantages over existing detection methods for *E. coli*, such as cell culturing, PCR and ELISA (see Supplementary Table 2 for additional information). The use of RCA and HRCA for signal amplification makes this system extremely sensitive for bacterial detection, which can achieve the detection of as low as 10 cells ml^{-1} without a cell-culturing step. The assay is also more compatible with point-of-care or field applications because

RCA is an isothermal process and there is no need for DNA extraction (as in the case of PCR). In addition, the system functions well with biological samples (no interferences from small RNAs and compatibility with blood samples).

It is conceivable that the same design can be extended to other RCDs, DNA-cleaving DNAzymes, as well as ribozymes and protein enzymes that have DNA or RNA-cleaving activities. Although rolling circle amplification was exploited to achieve signal amplification in this study, it should be feasible to take advantage of other signal amplification strategies, such as the DNAzyme cross-amplification system developed by Levy and Ellington that does not need a DNA polymerase[33]. We envision that the concept presented in this study should open up new opportunities for exploring mechanically interlocked DNA architectures for many potential applications in chemical biology, medical diagnostics, and environmental monitoring.

Methods

Preparation of rD2C1. A total of 100 pmol of $r^L DNA_{ii}$ was first labelled with γ-[^{32}P]ATP at the 5′ end using T4 PNK according to the manufacturer's protocol. To ensure that all DNA molecules contain the 5′ phosphate required for the subsequent ligation reaction, PNK-mediated end-labelling solution containing 5′-^{32}P labelled $r^L DNA_{ii}$ was further incubated with 2 mM non-radioactive ATP at 37 °C for 30 min. Then 120 pmol of $^C DNA_i$, prepared using the procedure detailed in Supplementary Methods, was added and heated to 90 °C for 30 s. After cooling to room temperature and leaving the solution for 15 min, 10 µl of 10 × T4 DNA ligase buffer and 10 U T4 DNA ligase were added (total 100 µl) and incubated at room temperature for 2 h. The obtained rD2C1 molecules were concentrated by standard ethanol precipitation and purified by 10% dPAGE.

E. coli-dependent RCA reaction. A single colony of *E. coli* K12 freshly grown on Luria broth (LB) agar plate was taken and used to inoculate 2 ml of LB. After shaking at 37 °C for 14 h at 250 r.p.m., the bacterial culture was serially diluted in 10-fold intervals. Overall, 100 µl of each diluted solution was plated onto LB agar plate (done in triplicate) and cultured at 37 °C for 15 h to obtain the cell counts. Colonies in each plate were counted; the average number of colonies from the three plates was taken as the number of *E. coli* cells for this dilution. This number was then used to calculate the number of cells for the other dilutions. A total of 500 µl of each dilution was centrifuged at 13,000g for 20 min at 4 °C and re-suspended in 100 µl of 1 × RB (50 mM HEPES, 150 mM NaCl, 15 mM MgCl₂, pH 7.5). After being frozen at −20 °C, *E. coli* cells were sonicated for 1 min and put on the ice for 5 min. This process was repeated three times. Then the cell suspension containing different numbers of *E. coli* cells were centrifuged at 13,000g for 10 min at 4 °C. The obtained crude intracellular mixture produced by *E. coli* cells (CIM-EC) in the supernatant was used for the following experiment.

A cleavage reaction mixture containing 5 µl of CIM-EC, 1 µl of rD2C1 (5 µM), 4 µl of EC1 (50 µM) and 10 µl of 2 × RB was incubated at room temperature for 60 min. Then 1 µl of PNK (10 U µl^{-1}) was added and incubated at 37 °C for 30 min. The RCA reaction was initiated by the addition of 1 µl of φ29DP (10 U µl^{-1}), 1 µl of dNTPs (50 mM), 5 µl of 10 × RCA reaction buffer and 22 µl of water. The reaction mixtures were incubated at 30 °C for 60 min before heating at 90 °C for 5 min. The resultant RCA products were analyzed by 0.6% agarose gel electrophoresis.

Colourimetric detection. rD2C1 used for the colourimetric detection of *E. coli* was made of $r^C DNA_{ii}$ and $^C DNA_i CD$, prepared using the procedure described in Supplementary Methods. After the cleavage reaction described above, 1 µl of PNK (10 U µl^{-1}) was added and incubated at 37 °C for 30 min. The RCA reaction was then initiated by the addition of 1 µl of φ29DP (10 U µl^{-1}), 1 µl of dNTPs (50 mM), 2 µl of hemin (100 µM), 5 µl of 10 × RCA reaction buffer and 20 µl of water. The reaction mixtures were incubated at 30 °C for 60 min before heating at 65 °C for 20 min. After cooling to room temperature, 2 µl of ABTS (50 mM) and 1 µl of H₂O₂ (8.8 mM) were added, and the colourimetric result was recorded immediately using a digital camera.

HRCA reaction. Following the cleavage reaction, 1 µl of PNK (10 U µl^{-1}) was added and incubated at 37 °C for 30 min. The HRCA reaction was then initiated by the addition of 1 µl of φ29DP (10 U µl^{-1}), 1 µl of dNTPs (50 mM), 1 µl of FP1 (50 µM), 1 µl of RP1 (50 µM), 5 µl of 10 × RCA reaction buffer, 2.5 µl of 20 × EvaGreen and 17.5 µl of water. These reactions were carried out in BioRad CFX96 qPCR system set to a constant temperature of 30 °C, and the fluorescence intensity was recorded in 1 min intervals.

Other experiments. Details for the following experiments are provided in Supplementary Methods: preparation of $r^L DNA'_{ii}$ and $r^C DNA'_{ii}$, and rD2C1′; comparison of the cleavage activity of EC1 and EC1M in the presence of *E. coli*; degradation of rD2C1 by φ29DP in the presence of EC1, *E. coli* and PNK; cell culture and miRNA extraction; restriction digestion of RCA products; effect of blood on RCA reactions; detection of *E. coli* in blood samples; and detection of *E. coli* using an ELISA kit.

References

1. Mao, C., Sun, W. & Seeman, N. C. Assembly of Borromean rings from DNA. *Nature* **386**, 137–138 (1997).
2. Wang, H., Du, S. M. & Seeman, N. C. Tight single-stranded DNA knots. *J. Biomol. Struct. Dyn.* **10**, 853–863 (1993).
3. Schmidt, T. L. & Heckel, A. Construction of a structurally defined double-stranded DNA catenane. *Nano Lett.* **11**, 1739–1742 (2011).
4. Ackermann, D. *et al.* A double-stranded DNA rotaxane. *Nat. Nanotechnol.* **5**, 436–442 (2010).
5. Zhang, F., Nangreave, J., Liu, Y. & Yan, H. Structural DNA nanotechnology: state of the art and future perspective. *J. Am. Chem. Soc.* **136**, 11198–11211 (2014).
6. Chen, Y. J., Groves, B., Muscat, R. A. & Seelig, G. DNA nanotechnology from the test tube to the cell. *Nat. Nanotechnol.* **10**, 748–760 (2015).
7. Pinheiro, A. V., Han, D., Shih, W. M. & Yan, H. Challenges and opportunities for structural DNA nanotechnology. *Nat. Nanotechnol.* **6**, 763–772 (2011).
8. Elbaz, J., Cecconello, A., Fan, Z., Govorov, A. O. & Willner, I. Powering the programmed nanostructure and function of gold nanoparticles with catenated DNA machines. *Nat. Commun.* **4**, 2000 (2013).
9. Liu, X. Q., Lu, C. H. & Willner, I. Switchable reconfiguration of nucleic acid nanostructures by stimuli-responsive DNA machines. *Acc. Chem. Res.* **47**, 1673–1680 (2014).
10. Li, T., Lohmann, F. & Famulok, M. Interlocked DNA nanostructures controlled by a reversible logic circuit. *Nat. Commun.* **5**, 4940 (2013).
11. Zhang, D. Y. & Seelig, G. Dynamic DNA nanotechnology using strand-displacement reactions. *Nat. Chem.* **3**, 103–113 (2011).
12. Wu, Z. S., Shen, Z., Tram, K. & Li, Y. Engineering interlocking DNA rings with weak physical interactions. *Nat. Commun.* **5**, 4279 (2014).
13. Fire, A. & Xu, S. Q. Rolling replication of short DNA circles. *Proc. Natl Acad. Sci. USA* **92**, 4641–4645 (1995).
14. Liu, D., Daubendiek, S. L., Zillman, M. A., Ryan, K. & Kool, E. T. Rolling circle DNA synthesis: small circular oligonucleotides as efficient templates for DNA polymerases. *J. Am. Chem. Soc.* **118**, 1587–1594 (1996).
15. Zhao, W., Ali, M. M., Brook, M. A. & Li, Y. Rolling circle amplification: applications in nanotechnology and biodetection with functional nucleic acids. *Angew. Chem. Int. Ed.* **47**, 6330–6337 (2008).
16. Ali, M. M., Aguirre, S. D., Lazim, H. & Li, Y. Fluorogenic DNAzyme probes as bacterial indicators. *Angew. Chem. Int. Ed.* **50**, 3751–3754 (2011).
17. Aguirre, S. D., Ali, M. M., Salena, B. J. & Li, Y. A sensitive DNA enzyme-based fluorescent assay for bacterial detection. *Biomolecules* **3**, 563–577 (2013).
18. Liu, M., Zhang, W., Zhang, Q., Brennan, J. D. & Li, Y. Biosensing by tandem reactions of structure switching, nucleolytic digestion, and DNA amplification of a DNA assembly. *Angew. Chem. Int. Ed.* **54**, 9637–9641 (2015).
19. Blanco, L. & Salas, M. Characterization of a 3′→5′ exonuclease activity in the phage φ29-encoded DNA polymerase. *Nucleic Acids Res.* **13**, 1239–1249 (1985).
20. Silverman, S. K. *In vitro* selection, characterization, and application of deoxyribozymes that cleave RNA. *Nucleic Acids Res.* **33**, 6151–6163 (2015).
21. Schurer, H., Lang, K., Schuster, J. & Morl, M. A universal method to produce *in vitro* transcripts with homogeneous 3′ ends. *Nucleic Acids Res.* **30**, e56 (2002).
22. Li, Y. & Sen, D. A catalytic DNA for porphyrin metallation. *Nat. Struct. Biol* **3**, 743–747 (1996).
23. Travascio, P., Li, Y. & Sen, D. DNA-enhanced peroxidase activity of a DNA aptamer-hemin complex. *Chem. Biol.* **5**, 505–517 (1998).
24. Travascio, P., Witting, P. K., Mauk, A. G. & Sen, D. The peroxidase activity of a hemin-DNA oligonucleotide complex: free radical damage to specific guanine bases of the DNA. *J. Am. Chem. Soc.* **123**, 1337–1348 (2001).
25. Cheglakov, Z., Weizmann, Y., Basnar, B. & Willner, I. Diagnosing viruses by the rolling circle amplified synthesis of DNAzymes. *Org. Biomol. Chem.* **5**, 223–225 (2007).
26. Tian, Y., He, Y. & Mao, C. Cascade signal amplification for DNA detection. *ChemBioChem.* **7**, 1862–1864 (2006).
27. Zhang, D., Brandwein, M., Hsuih, T. & Li, H. Amplification of target-specific, ligation-dependent circular probe. *Gene* **211**, 277–285 (1998).

28. Lizardi, P. M. *et al.* Mutation detection and single-molecule counting using isothermal rolling-circle amplification. *Nat. Genet.* **19**, 225–232 (1998).

29. Al-Soud, W. A. & Radstrom, P. Effect of amplification facilitators on diagnostic PCR in the presence of blood, feces and meat. *J. Clin. Microbiol.* **38**, 4463–4470 (2000).

30. Al-Soud, A. W. & Radstrom, P. Capacity of nine thermo-stable DNA polymerases to mediate DNA amplification in the presence of PCR-inhibiting samples. *Appl. Environ. Microbiol.* **64**, 3748–3753 (1998).

31. Schrader, C., Schielke, A., Ellerbroek, L. & Johne, R. PCR inhibitors— occurrence, properties and removal. *J. Appl. Microbiol.* **113**, 1014–1026 (2012).

32. Kang, D. K. *et al.* Rapid detection of single bacteria in unprocessed blood using integrated comprehensive droplet digital detection. *Nat. Commun.* **5**, 5427 (2014).

33. Levy, M. & Ellington, A. D. Exponential growth by cross-catalytic cleavage of deoxyribozymogens. *Proc. Natl Acad. Sci. USA* **100**, 6416–6421 (2003).

Acknowledgements

Funding for this work was provided by Natural Sciences and Engineering Research Council of Canada (NSERC) Discovery Grants (Y.L. and J.D.B.), Pro-Lab Developments Inc., Collaborative Health Research Projects (CHRP) Initiative from Canadian Institutes of Health Research (CIHR), the Canada Foundation for Innovation (CFI) and the Ministry for Research and Innovation (MRI) of Ontario. Part of the work was conducted at the McMaster Biointerfaces Institute. J.D.B. holds the Canada Research Chair in Bioanalytical Chemistry and Biointerfaces.

Author contributions

M.L., Q.Z., J.D.B. and Y.L. designed the research project; M.L. and Q.Z. performed the experiments and discussed the experimental results; Z.L. and J.G. performed some experiments; M.L., J.D.B., and Y.L. analyzed the data and wrote the paper.

Additional information

22

Ultrafast multidimensional Laplace NMR for a rapid and sensitive chemical analysis

Susanna Ahola[1], Vladimir V. Zhivonitko[2,3], Otto Mankinen[1], Guannan Zhang[4], Anu M. Kantola[1], Hsueh-Ying Chen[4], Christian Hilty[4], Igor V. Koptyug[2,3] & Ville-Veikko Telkki[1]

Traditional nuclear magnetic resonance (NMR) spectroscopy relies on the versatile chemical information conveyed by spectra. To complement conventional NMR, Laplace NMR explores diffusion and relaxation phenomena to reveal details on molecular motions. Under a broad concept of ultrafast multidimensional Laplace NMR, here we introduce an ultrafast diffusion-relaxation correlation experiment enhancing the resolution and information content of corresponding 1D experiments as well as reducing the experiment time by one to two orders of magnitude or more as compared with its conventional 2D counterpart. We demonstrate that the method allows one to distinguish identical molecules in different physical environments and provides chemical resolution missing in NMR spectra. Although the sensitivity of the new method is reduced due to spatial encoding, the single-scan approach enables one to use hyperpolarized substances to boost the sensitivity by several orders of magnitude, significantly enhancing the overall sensitivity of multidimensional Laplace NMR.

[1] Faculty of Science, NMR Research Group, University of Oulu, PO Box 3000, 90014 Oulu, Finland. [2] Laboratory of Magnetic Resonance Microimaging, International Tomography Center SB RAS, Instututskaya Street 3A, 630090 Novosibirsk, Russia. [3] Department of Natural Sciences, Novosibirsk State University, Pirogova Street 2, 630090 Novosibirsk, Russia. [4] Department of Chemistry, Texas A&M University, 3255 TAMU, College Station, Texas 77843-3255, USA. Correspondence and requests for materials should be addressed to V.-V.T. (email: ville-veikko.telkki@oulu.fi).

Nuclear magnetic resonance (NMR) spectroscopy is one of the most powerful analytical techniques in chemical sciences[1]. NMR is also one of the very few methods for measuring molecular self-diffusion without an invasive tracer[2], and the relaxation experiments reveal the effect of random molecular motion on the recovery of initially perturbed nuclear magnetization[3]. Although the frequency content of conventional, oscillating NMR signal is analysed by a Fourier transform, the relaxation and diffusion data consist of exponentially decaying components, and the distribution of diffusion coefficients or relaxation times can be extracted from the experimental data by an inverse Laplace transform[2]. Consequently, these methods can be referred to as Laplace NMR (LNMR).

Like in traditional NMR spectroscopy, the resolution and information content of LNMR can be increased by a multidimensional approach[2,4,5]. The approach has entered routine use only in recent years, after the development of a sufficiently reliable and robust multidimensional Laplace inversion algorithm[6–8]. However, the conventional scheme of multidimensional NMR based on the repetition of the experiment with varying evolution delay leads to a long experiment time, restricting the investigation of fast processes. The need for multiple repetitions also practically prevents the utilization of significant sensitivity gain provided by nuclear spin hyperpolarization methods, such as dynamic nuclear polarization (DNP)[9], parahydrogen-induced polarization (PHIP)[10] and spin-exchange optical pumping[11].

As a solution to these problems, we introduce a concept of ultrafast multidimensional LNMR, which is based on continuous spatial encoding. Recently, these principles have been successfully applied in ultrafast multidimensional NMR spectroscopy[12–14] (also with hyperpolarization[15]), in one-dimensional (1D) LNMR experiments[16,17], as well as for correlating diffusion and spectroscopic data[18]. Here, we describe an ultrafast diffusion–T_2 relaxation (D–T_2) correlation experiment. Supported by our recent work[19], in which we proposed an ultrafast T_1–T_2 relaxation correlation experiment, it proves that the principles are applicable to a broad range of multidimensional LNMR experiments and can be used to efficiently correlate relaxation times and diffusion coefficients as well as to investigate chemical exchange phenomena.

Results

Ultrafast D–T_2 correlation LNMR experiment.

The ultrafast D–T_2 correlation experiment (Fig. 1a) begins with spatial encoding of diffusion data along the longitudinal (z) axis of a sample tube, similar to single-scan diffusion-ordered spectroscopy[17]. After the first $\pi/2$ pulse, spins at different z positions experience the frequency-swept refocusing π-pulse at different times because of the simultaneously applied magnetic field gradient pulse. Consequently, the value of the wave vector q (proportional to the strength of the gradient)[2] becomes linearly dependent on the position, being zero at the top and maximum at the bottom (Fig. 1b). Subsequently, the magnetization is stored along the longitudinal direction for the mixing (diffusion) period, followed by the second pair of the radio frequency (RF) frequency-swept and gradient pulses. Because of diffusion, the resulting stimulated echo is the most intense at the top and weakest at the bottom (Fig. 1c). The final T_2-encoding part comprises a Carr-Purcell-Meiboom-Gill (CPMG) loop[20], and the magnetization profile along the z direction is imaged at each CPMG echo point similar to multiple-echo magnetic resonance imaging (MRI)[21]. After Fourier transform in the spatial frequency (k) dimension, the resulting data set is analogous to that obtained from the conventional D–T_2 correlation experiment[22] comprising

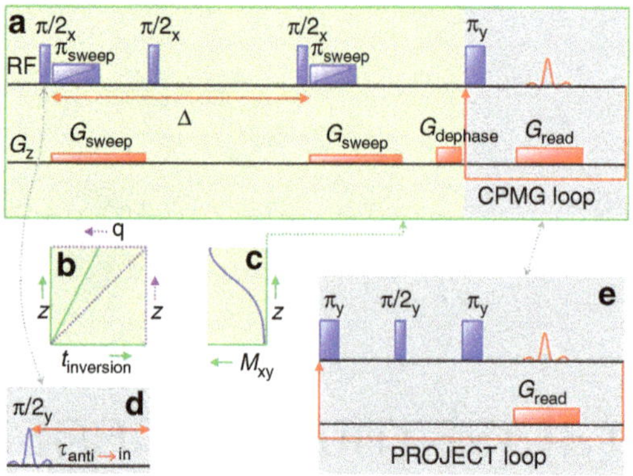

Figure 1 | Ultrafast D–T_2 correlation LNMR experiment. (**a**) Pulse sequence. (**b**) Spatial dependence of the inversion time $t_{inversion}$ and the value of wave vector q due to the frequency-swept radio frequency and gradient pulse pair. (**c**) Transverse magnetization profile after the diffusion encoding. (**d**) A soft pulse and a delay replacing the first $\pi/2$ pulse to convert the antiphase signal to in-phase in PHIP experiments. (**e**) PROJECT loop replacing the CPMG loop in order to eliminate J modulation.

pulsed-field-gradient stimulated-echo[23] and CPMG blocks. The ultrafast experiment, however, is measured in a single scan. The number of points in the indirect (diffusion) dimension typically varies from tens to hundreds, which are collected in a repetitive manner in the conventional experiments. Therefore, the experiment time in the ultrafast version can be one to two orders of magnitude (or more) shorter. The price to pay is the reduced sensitivity because of the spatial encoding[14]. However, if the concentration of the sample is high, the sensitivity is not an issue in high-field NMR spectrometers, and the sensitivity losses can be alleviated by a moderate amount of averaging or overcompensated through the use of hyperpolarized substances, as we show below. Spatial encoding also requires a homogeneous sample, although slight inhomogeneity can be compensated in post-processing (see below).

Resolving different physical environments of molecules.

In the first experimental demonstration, we show that, contrary to a ^{1}H NMR spectrum, the ultrafast D–T_2 correlation experiment resolves differing physical environments of water molecules in a sample consisting of water and silica gel 60 porous powder with an average pore diameter of 6 nm and particle size of 60–200 μm (Fig. 2a). The experimental data after Fourier transform in the spatial frequency dimension are shown in Fig. 2b. The observed diffusion-encoded magnetization profile along the z direction is weighted by the excitation-detection profile of the coil and the slight sample heterogeneity. To eliminate this weighting, we measured in a separate experiment the coil excitation-detection profile, that is, the 1D MRI of the sample along the z axis (Fig. 2b), with the same imaging parameters as in the CPMG loop of the ultrafast D–T_2 experiment. Each row in the D–T_2 data set was then divided along the z direction by this profile. Before the two-dimensional (2D) Laplace inversion, the data outside the region affected by the frequency-swept inversion pulse were also removed, and the z axis was converted into a q axis using the linear relationship between these two quantities. The resulting D–T_2 map includes two dominant peaks: one, with smaller D and shorter T_2, arising from water in the pores, and the other from the bulk water in the spaces between the particles of the porous material. There are also some additional minor peaks that arise from imperfectly compensated sample heterogeneity and noise. The largest artefact has an amplitude of about 28% of the highest peak. The ultrafast experiment is more sensitive to local field

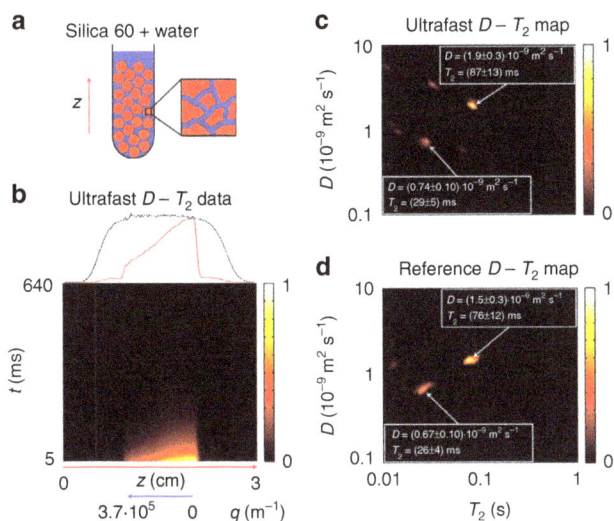

Figure 2 | Resolving different physical environments of molecules.
(**a**) Schematics of the sample consisting of porous silica gel 60 powder immersed in water (1% H_2O in D_2O). (**b**) Experimental ultrafast D–T_2 data after the Fourier transform in the spatial frequency dimension. The first row (red) is shown on the top along with the coil sensitivity profile (black). (**c**) Ultrafast D–T_2 map including one peak arising from water in the pores ($D = 0.74 \cdot 10^{-9}$ m^2 s^{-1}, $T_2 = 29$ ms) and the other from water between the particles ($D = 1.9 \cdot 10^{-9}$ m^2 s^{-1}, $T_2 = 87$ ms). The map is the result of a 2D Laplace inversion of experimental data corrected using the coil sensitivity profile in the region affected by the frequency-swept pulse ($z = 0.96$–1.99 cm). (**d**) Corresponding reference map obtained in the conventional D–T_2 correlation experiment. The experiments were carried out at 300 MHz ^{1}H frequency. The total time in the conventional experiment was 46 min, using eight scans per increment, while only 2 min 30 s were required with a total of 32 scans in the ultrafast experiment.

Figure 3 | Improved chemical resolution. ^{1}H NMR spectra and ultrafast (PROJECT-based) D–T_2 maps of 1.65 M hexane, 0.79 M pentadecane and a mixture of 1.36 M hexane and 0.65 M pentadecane in CCl$_4$. Although the compounds are not resolved in the spectrum of the mixture, they are resolved in the D–T_2 map. The experiments were carried out on a 600-MHz NMR spectrometer.

inhomogeneities along the sample axis than the conventional one, because the various evolution delays are encoded in the layers of the sample, whereas in the conventional experiment the signal corresponding to a single evolution time is measured from the entire sample volume inside the NMR coil. Artefacts due to background gradients in a heterogeneous sample could be removed by using a bipolar gradient[24] in the ultrafast D–T_2 experiment. In this alternative implementation, the G_{sweep} gradient would be replaced with a pair of gradients with opposite amplitudes. The π_{sweep} pulses associated with the gradients would have the same sweep direction. We also carried out a conventional D–T_2 correlation experiment to serve as a reference. The resulting D–T_2 map (Fig. 2d) shows the same dominant peaks as the ultrafast map and, within the error limits, results in the same D and T_2 values, proving that the ultrafast method works. It is notable that, despite the fourfold number of scans, the measurement time for the ultrafast experiment was 18 times shorter than for the conventional experiment.

Improved chemical resolution. In the following, we will demonstrate that chemical resolution lacking in the NMR spectrum of hydrocarbons can be revealed by the ultrafast D–T_2 experiment. However, homonuclear scalar coupling present in these molecules modulates the CPMG echo amplitudes and severely complicates the T_2 data. This problem can be solved by replacing the CPMG block with a PROJECT (Periodic Refocusing of J Evolution by Coherence Transfer)[25] block (Fig. 1e). The PROJECT is a cyclic analogue of the perfect echo experiment[26], in which a $\pi/2$ pulse at the midpoint of a double spin echo refocuses the J modulation.

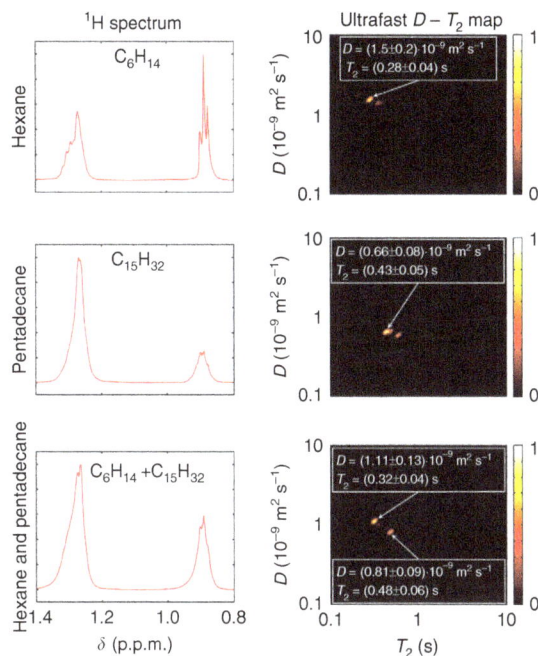

The ^{1}H NMR spectrum of hexane includes two peaks, one arising from hydrogens in methyl (CH$_3$-) groups at 0.9 p.p.m. and the other from methylene (-CH$_2$-) groups at 1.3 p.p.m., and the same groups resonate at the same frequencies also in the case of pentadecane. Consequently, these two chemicals are not resolved in the ^{1}H spectrum of their mixture (Fig. 3). However, the ultrafast D–T_2 experiments (experimental data in Supplementary Fig. 1) result in maps that are unique for each compound, with larger D and shorter T_2 for hexane than for pentadecane, and the compounds are resolved also in the map of the mixture (Fig. 3). The amplitudes of the two signals are roughly equal, as expected based on the concentrations used. Importantly, the D values obtained in the ultrafast measurements performed on individual solutions are in good agreement with the values measured by the standard pulsed-field-gradient stimulated-echo NMR (hexane: $1.62 \cdot 10^{-9}$ m^2 s^{-1}, pentadecane: $0.66 \cdot 10^{-9}$ m^2 s^{-1}), confirming the reliability of the ultrafast experiments. At the same time, for the mixed hexane-pentadecane sample, the observed D value of hexane in the mixture, $(1.11 \pm 0.13) \cdot 10^{-9}$ m^2 s^{-1}, is smaller than that of hexane in the reference individual sample, which is physically reasonable according to the scaling laws for diffusion coefficients in mixtures of alkanes (large pentadecane molecules hinder the diffusion of hexane)[27]. As for T_2, the imaging magnetic field gradients in the PROJECT block of the ultrafast experiments make the observed values significantly shorter than in the standard PROJECT experiment without a gradient because of two factors well-known from, for example, T_2 maps measured with MRI or in T_2 measurements performed using single-sided NMR with an inhomogeneous magnetic field: diffusion in inhomogeneous field speeds up the echo attenuation[2] and the gradients make the experiment more sensitive to B_1 inhomogeneity[28]. Although in standard PROJECT experiments of the hexane and pentadecane samples the observed T_2 value of hexane (3.0 s) was higher than that of pentadecane (1.2 s), consistently with the scaling law for relaxation times[29], in the ultrafast experiment it is the other way around because of above-mentioned reasons. We note that the T_2 shortening is minor in the water/silica sample (Fig. 2) because strong local field inhomogeneities caused by the porous material are present both in the ultrafast and reference experiments.

It is also worth to note that the two compounds are not resolved in the 1D T_2 and D distributions (Supplementary Fig. 1e and f) obtained by the Laplace inversion of the first row or column of the 2D ultrafast data, indicating that the 2D approach increased the resolution of the experiment.

Boosting sensitivity by hyperpolarization. In principle, the sensitivity of LNMR can be increased by several orders of magnitude by means of nuclear spin hyperpolarization[9–11], broadening the applicability of the method to low concentration samples. However, the conventional multidimensional LNMR approach practically prevents the use of hyperpolarized substances, because the experiment has to be repeated multiple times with varying evolution time. Hyperpolarization should be regenerated before each repetition, which is extremely laborious and time consuming, and it may even take hours in the case of some implementations of DNP[9]. Furthermore, the polarization level might vary among repetitions, causing serious artefacts in the experimental data. The ultrafast multidimensional approach, realized in a single-scan manner, can overcome these problems.

First, we produced hyperpolarization using the PHIP method by bubbling a mixture of propyne and parahydrogen (prepared by cooling H_2 to 77 K in the presence of a paramagnetic material[30], see the Supplementary Methods) through the solution of hydrogenation catalyst in deuterated acetone. The resulting hydrogenation reaction produced hyperpolarized propene (Fig. 4a). Based on the hyperpolarized and thermally polarized ^1H spectra shown in Fig. 4b, the sensitivity enhancement factor given by PHIP was estimated to be about 500. The ultrafast $D–T_2$ experiment was modified by replacing the hard $\pi/2$ excitation pulse with a selective excitation of the methylene resonance of propene, followed by a delay for converting the antiphase PHIP signal into an in-phase signal (Fig. 1d). This was done to ensure that the opposite components of the antiphase multiplet would not cancel each other when reading the magnetization profile in the PROJECT loop. The single-scan $D–T_2$ map of hyperpolarized propene is shown in Fig. 4c (experimental data in Supplementary Fig. 2). The map expectedly contains a single peak with a realistic D value, although the concentration of propene was only about 40 mM, and the experiment time was notably reduced to 0.5 s. In this experiment, the observed T_2 (0.17 s) was exceptionally much shortened as compared with the thermally polarized reference value (9.5 s) because of the effect of very strong read gradients used in the experiment and the large diffusion coefficient of propene. The value would approach the reference value by decreasing the read gradient amplitude.

In the second demonstration, the ultrafast $D–T_2$ measurement was applied to a hyperpolarized spin system prepared via dissolution DNP[9]. A 5-μl sample of dimethyl sulfoxide (DMSO) in D_2O (v/v 18:7) with 15 mM of 4-hydroxy-2,2,6,6-tetramethylpiperidine 1-oxyl radicals was first hyperpolarized by microwave irradiation of 94.005 GHz at 1.4 K in a field of 3.35 T for about 30 min. Subsequently, the sample was dissolved with superheated water. This sample solution was rapidly transferred to an injection loop, and driven into a flow cell in a 400-MHz NMR magnet using water from a high-pressure pump[31]. ^1H NMR spectra of hyperpolarized DMSO in H_2O, both with and without solvent suppression, are shown for reference in Fig. 4d. The final concentration of DMSO was determined

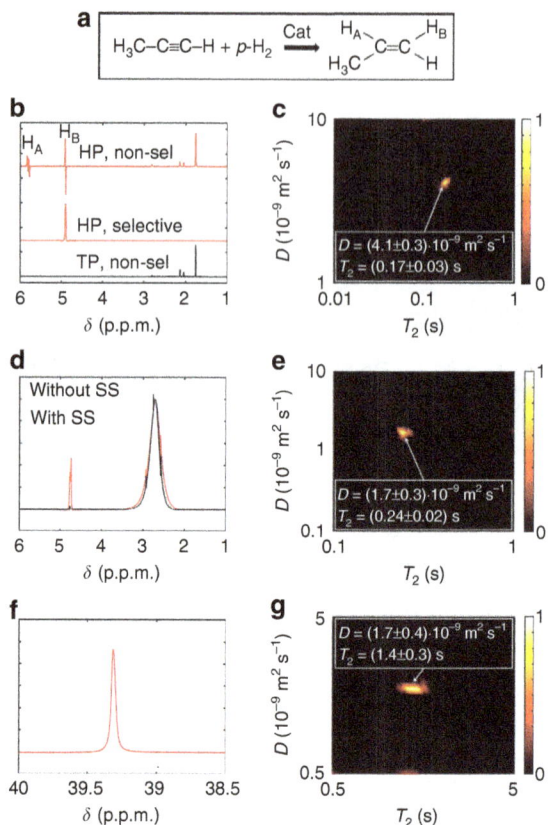

Figure 4 | Boosting sensitivity by hyperpolarization. (**a**) Hydrogenation of propyne into propene with parahydrogen (p-H_2) to produce PHIP. Red symbols indicate the hyperpolarized hydrogens. (**b**) Top: Hyperpolarized (HP) ^1H NMR spectrum measured right after bubbling a mixture of p-H_2 and propyne through the solution of [Rh(COD)(DPPB)]BF$_4$ catalyst in deuterated acetone. The antiphase propene multiplets at 4.9 and 5.8 p.p.m. indicate a strong PHIP effect. Middle: the corresponding spectrum recorded after a selective excitation of the methylene (4.9 p.p.m.) signal followed by a delay converting the antiphase signal into an in-phase signal. Bottom: The spectrum measured after the decay of PHIP hyperpolarization due to relaxation. (**c**) Single-scan $D–T_2$ map of hyperpolarized propene using the selective excitation of the 4.9 p.p.m. signal. The experiment time was only 0.5 s. The PHIP experiments were carried out at 600 MHz ^1H frequency. (**d**) ^1H NMR spectra of DNP hyperpolarized DMSO in H_2O, both with and without solvent suppression (SS). (**e**) Single-scan $D–T_2$ map of hyperpolarized DMSO measured after solvent suppression. (**f**) ^{13}C NMR spectrum of DNP hyperpolarized DMSO in H_2O. (**g**) Corresponding single-scan $D–T_2$ map. The DNP NMR measurements were carried out at 400 MHz for ^1H and 100 MHz for ^{13}C.

to be 34 mM. The broadening of the DMSO signal in the spectra is due to radiation damping[32], which arises because of the strong hyperpolarized signal. Figure 4e shows the $D–T_2$ map of DMSO, measured in a single scan 3 s after stopping the in- and out-flow of the sample by switching an injection valve. These data were acquired with solvent suppression. It is important to assure that the sample convection after the transport does not disturb the spatial encoding. In these experiments the sample convection was significantly reduced because of the use of flow cell with liquid-driven injection[31], and single-scan $D–T_2$ test experiments showed that the convection was insignificant after the 3 s stabilization delay. In the homogeneous sample, the observed T_2 value is shortened (T_2 in the thermally polarized reference experiment was 3.0 s), because of the effects described above. The D value, which is strongly dependent on DMSO concentration[33], was found to be $(1.7 \pm 0.3) \cdot 10^{-9} \, m^2 \, s^{-1}$. This value is in agreement with a reference measurement using a conventional stimulated echo pulse sequence on a stationary, non-hyperpolarized sample at $T = 300 \, K$, which yielded $D = 1.3 \cdot 10^{-9} \, m^2 \, s^{-1}$. The agreement of the measured diffusion coefficients indicates that the sample solution was very nearly stationary during the measurement, in agreement with our previous pulsed field gradient-based characterization of high-pressure liquid-driven injection into a flow cell[31]. This demonstration shows that the challenges related to fast transport of hyperpolarized substances and their stabilization before an ultrafast LNMR experiment can be overcome.

Figure 4g shows D–T_2 map of the corresponding experiment, in which ^{13}C nuclei, instead of ^{1}H, were hyperpolarized and detected in the ultrafast D–T_2 experiment. In this experiment, a final concentration of DMSO of 288 mM was used. The signal-to-noise ratio (SNR) in the D–T_2 experiment was rather low (about 24) because of small gyromagnetic ratio of [13]C (one-fourth of that of ^{1}H) and low natural abundance of ^{13}C isotope (1%), but, still, the 2D Laplace inversion resulted in the expected single component D–T_2 map with a D value in good agreement with the ultrafast ^{1}H and reference experiments. Again, the observed T_2 value (1.4 s) was shorter than in the thermally polarized reference experiment (2.8 s) because of the use of the read gradients. We stress that this kind of multidimensional LNMR experiment with a low-sensitivity heteronucleus is absolutely unattainable with the conventional technique relying on thermal polarization. Here, the ^{13}C signal enhancement was estimated to be about 3,200 by comparing the DMSO signal to the signal obtained from a sample of known concentration in a separate DNP experiment. Consequently, the ultrafast method opens unprecedented prospects of applications of multidimensional LNMR, as DNP is the most universal hyperpolarization method applicable to any NMR active nuclei with sufficiently long relaxation time.

Discussion

When dealing with LNMR, one has to keep in mind that the inverse Laplace transform of noisy data is an ill-posed problem, meaning that there exist an infinite number of relaxation time or diffusion coefficient distributions consistent with the experimental data[5]. The inversion becomes feasible with the use of regulator smoothing the distributions[34,35]. However, the inversion algorithms tend to broaden intrinsically narrow peaks, lowering the resolution, but also to split up intrinsically broad peaks into a series of narrow peaks (the so-called uniform-penalty smoothing, however, provides a satisfactory solution to these problems)[35]. Consequently, it is advisable to confirm the reliability of the results by reference experiments (described above) and simulations. We have physical reasons to assume that both the silica/water sample and the mixed hexane–pentadecane sample produce two rather sharp D and T_2 peaks: one from water in the small pores and the other from bulk water in between the particles in the former case, and one from each compound in the latter case. We simulated the D–T_2 correlation LNMR data for two components with a zero peak width, varying the peak amplitudes and separation as well as the SNR, and performed 1D and 2D Laplace inversions of the data (see Supplementary Figs 3 and 4 as well as Supplementary Discussion). The results confirmed that the resolution in the LNMR distributions can be improved by increasing SNR, as has been shown earlier[35], and that it is rather high when SNR is high: if SNR is 1,000, peaks of equal amplitude with relaxation times and/or diffusion coefficients differing by a factor of 1.5 can be resolved. If SNR is 10,000, this factor can be as low as 1.25 (Supplementary Fig. 5). These SNR values are achievable with high field NMR

instruments, if the sample concentration is high enough or if signal is amplified by hyperpolarization. On the other hand, systematic errors due to hardware imperfections become more significant relative to the noise variation with increasing SNR, and may become a limiting factor of the accuracy of the method. In any case, the simulations affirm that the compounds of the hexane/pentadecane mixture can be reliably separated in the ultrafast D–T_2 experiment, because SNR in the experiment was 8,000 (Supplementary Fig. 1).

As already stated, the acceleration of multidimensional LNMR by the ultrafast method based on spatial encoding leads to sensitivity reduction, which depends on numerous experimental parameters. To obtain an estimate, we carried out both ultrafast and conventional D–T_2 correlation experiments on a doped water sample using the same number of scans and almost identical q and t values. The SNR of the ultrafast experiment determined from the echo amplitude data, $E(q,t)$, was about four times lower than in the reference experiment. However, remarkably, the SNR per unit time was slightly better (by the factor of 1.8) in the ultrafast experiment, as 64-fold number of scans were accumulated in the same experiment time. Pathan et al.[36] have reported a similar sensitivity gain per unit time provided by ultrafast 2D NMR spectroscopy. It is worth noting that the single-scan sensitivity loss factor (4) is much smaller than a typical sensitivity gain of several orders of magnitude given by hyperpolarization methods. Consequently, by combining hyperpolarization with ultrafast techniques, one can increase significantly the overall sensitivity of multidimensional LNMR.

An important question related to sensitivity is: what is the limiting concentration in order to obtain reliable results with ultrafast D–T_2 LNMR. The lower concentration limit depends on many factors, including the desired D and T_2 resolution (Supplementary Fig. 5), the strength of the magnetic field of the NMR spectrometer, the number of observed nuclei in the molecule and the gyromagnetic ratios. The SNR from the hexane–pentadecane mixture was about 8,000, obtained with 128 scans (Supplementary Fig. 1). The overall concentration was about 2 M. As the SNR in the LNMR experiment should be at least 100 in order to resolve two components (Supplementary Fig. 5), we can estimate that under these conditions the minimum concentration for a single-scan experiment is about 0.3 M. The limit can be decreased moderately by accumulation and substantially (by orders of magnitude) by hyperpolarization.

One limitation of the ultrafast D–T_2 experiment is that the frequency-swept refocusing π-pulse does not work perfectly immediately after it has been switched on[14]. This limits the measurement of the echo amplitudes corresponding to very small and very large q values simultaneously in a single experiment, and consequently impacts the range of the diffusion coefficients that can be detected. Another limitation is that, although it is necessary to measure only a single data point at each echo maximum in the CPMG loop in the conventional method, the ultrafast method requires collecting a sufficient number of points in the presence of the imaging gradient in order to obtain the magnetization profile along the sample axis. This requirement limits the shortest possible echo time in the CPMG loop, and consequently the shortest observable T_2 value. However, the T_2 limit can be decreased by increasing the imaging gradient strength. The experimental results reported in this manuscript show that, regardless of these limitations, the typical moderate range of D and T_2 in liquids falls well within the capacity of the experiment.

The additional elements of the ultrafast D–T_2 experiment as compared with the conventional method are the frequency-swept π-pulse and the imaging gradients of the CPMG (or PROJECT) loop. Most modern NMR spectrometers are able to generate shaped pulses and typically provide at least the z gradient; hence, the ultrafast D–T_2 experiment can be carried out with standard NMR instrumentation. Setting up the experimental parameters can be somewhat more complicated than in the conventional experiments: first, one has to set the length and amplitude of the diffusion gradient (G_{sweep}) to correspond to the highest desired q value, then the bandwidth of the frequency-swept pulse has to be adjusted to match the Larmor frequencies (linearly dependent on the position due to the presence of G_{sweep}) of the nuclei in the sensitive region of the radio frequency coil and finally the amplitude of the read gradient, the dwell time and the number of collected points has to be set so that the 1D image has a proper field-of-view and resolution. It is also advisable to include a correction for the coil sensitivity profile, as described above. However, it is possible to automate the measurement process so that, after the desired D and T_2 ranges are specified, the measurement programme calculates all the pulse programme parameters and applies the sensitivity profile correction to the measured data.

We believe that the results presented above, along with our recent publication introducing the T_1–T_2 correlation experiment[19], open a broad field of ultrafast multidimensional NMR, because, although not trivial, the basic blocks of the experiments can be used to develop multiple 2D and even three-dimensional ultrafast LNMR experiments for correlating relaxation times and diffusion coefficients as well as for investigating chemical exchange phenomena via relaxation and diffusion data, as we will show elsewhere. The data can be efficiently combined with spectral resolution by frequency-selective pulses (shown above), Hadamard-encoding[37] or echo planar spectroscopic imaging type detection[38]. Moreover, we envision exploiting long-lived singlet-states[39] in the ultrafast LNMR experiments in the investigation of slow diffusion and chemical exchange phenomena. The methods offer unprecedented opportunities to study fast processes in real-time, and to use various nuclear spin hyperpolarization techniques[9–11] to increase the experimental sensitivity by several orders of magnitude. As demonstrated in this paper, the methods provide information about chemical structure and physical environments of molecules that is not available in the conventional NMR spectra, offering extremely interesting prospects, for example, for investigations of polymerization, gel formation, phases of ionic liquids, phase transitions, transport of substances into cells, metabolism and protein folding.

References

1. Keeler, J. *Understanding NMR spectroscopy* (Wiley, 2010).
2. Callaghan, P. T. *Translational Dynamics and Magnetic Resonance: Principles of Pulsed Gradient Spin Echo NMR* (Oxford Univ., 2011).
3. Kowalewski, J. & Maler, L. *Nuclear Spin Relaxation in Liquids* (CRP, 2006).
4. Song, Y.-Q. Magnetic resonance of porous media (MRPM): A perspective. *J. Magn. Reson.* **229**, 12–24 (2013).
5. Bernin, D. & Topgaard, D. NMR diffusion and relaxation correlation methods: New insights in heterogeneous materials. *Curr. Opin. Colloind In* **18**, 166–172 (2013).
6. Venkataramanan, L., Song, Y.-Q. & Hürlimann, M. D. Solving Fredholm integrals of the first kind with tensor product structure in 2 and 2.5 dimensions. *IEEE Trans. Signal. Proc.* **50**, 1017–1026 (2002).
7. Song, Y.-Q. *et al.* T_1–T_2 correlation spectra obtained using a fast two-dimensional Laplace inversion. *J. Magn. Reson.* **154**, 261–268 (2002).
8. Granwehr, J. & Roberts, P. J. Inverse laplace transform of multidimensional relaxation data without non-negativity constraint. *J. Chem. Theory Comput.* **8**, 3473–3482 (2012).
9. Ardenkjær-Larsen, J. H. *et al.* Increase in signal-to-noise ratio of >10,000 times in liquid-state NMR. *Proc. Natl Acad. Sci. USA* **100**, 10158–10163 (2003).
10. Bowers, S. R. . in *Encyclopedia of Nuclear Magnetic Resonance* vol. 9 (eds Gant, D. M. & Harris, R. K.) 750–769 (Wiley, 2002).
11. Goodson, B. M. Nuclear magnetic resonance of laser-polarized noble gases in molecules, materials, and organisms. *J. Magn. Reson.* **155**, 157–216 (2002).
12. Frydman, L., Scherf, T. & Lupulescu, A. The acquisition of multidimensional NMR spectra within a single scan. *Proc. Natl Acad. Sci. USA* **99**, 15858–15862 (2002).
13. Pelupessy, P. Adiabatic single scan two-dimensional NMR spectrocopy. *J. Am. Chem. Soc.* **125**, 12345–12350 (2003).
14. Tal, A. & Frydman, L. Single-scan multidimensional magnetic resonance. *Prog. Nucl. Mag. Res. Sp* **57**, 241–292 (2010).
15. Frydman, L. & Blazina, D. Ultrafast two-dimensional nuclear magnetic resonance spectroscopy of hyperpolarized solutions. *Nat. Phys.* **3**, 415–419 (2007).
16. Loening, N. M., Thrippleton, M. J., Keeler, J. & Griffin, R. G. Single-scan longitudinal relaxation measurements in high-resolution NMR spectroscopy. *J. Magn. Reson.* **164**, 321–328 (2003).
17. Thrippleton, M. J., Loening, N. M. & Keeler, J. A fast method for the measurement of diffusion coefficients: one-dimensional DOSY. *Magn. Reson. Chem.* **41**, 441–447 (2003).
18. Smith, P. E. S., Donovan, K. J., Szekely, O., Baias, M. & Frydman, L. Ultrafast NMR T_1 relaxation measurements: probing molecular properties in real time. *Chemphyschem* **14**, 3138–3145 (2013).
19. Ahola, S. & Telkki, V.-V. Ultrafast two-dimensional NMR relaxometry for investigating molecular processes in real-time. *Chemphyschem* **15**, 1687–1692 (2014).
20. Meiboom, S. & Gill, D. Modified spin-echo method for measuring nuclear relaxation times. *Rev. Sci. Instrum* **29**, 688–691 (1958).
21. Haacke, E. M., Brown, R. W., Thompson, M. R. & Venkatesan, R. *Magnetic Resonance Imaging: Physical Principles and Sequence Design* (Wiley, 1999).
22. Godefroy, S. & Callaghan, P. T. 2D relaxation/diffusion correlations in porous media. *Magn. Reson. Imaging* **21**, 381–383 (2003).
23. Tanner, J. E. Use of the stimulated echo in NMR diffusion studies. *J. Chem. Phys.* **52**, 2523–2526 (1970).
24. Cotts, R. M., Hoch, M. J. R., Sun, T. & Markert, J. T. Pulsed field gradient stimulated echo methods for improved NMR diffusion measurements in heterogeneous systems. *J. Magn. Reson.* **83**, 252–266 (1989).
25. Aguilar, J. A., Nilsson, M., Bodenhausen, G. & Morris, G. A. Spin echo NMR spectra without J modulation. *Chem. Commun.* **48**, 811–813 (2012).
26. Takegoshi, K., Ogura, K. & Hikichi, K. A perfect spin echo in a weakly homonuclear J-coupled two spin-1/2 system. *J. Magn. Reson.* **84**, 611–615 (1989).
27. Freed, D. E., Burcaw, L. & Song, Y.-Q. Scaling laws for diffusion coefficients in mixtures of alkanes. *Phys. Rev. Lett.* **94**, 067602 (2005).
28. Majumdar, S., Orphanoudakis, S. C., Gmitro, A., O'Donnell, M. & Gore, J. C. Errors in measurements of T_2 using multiple pulse MRI techniques. *Magn. Reson. Med.* **3**, 397–417 (1986).
29. Freed, D. E. Dependence on chain length of NMR relaxation times in mixtures of alkanes. *J. Chem. Phys.* **126**, 174502 (2007).
30. Telkki, V.-V. *et al.* Microfluidic gas flow imaging utilizing parahydrogen-induced polarization and remote detection NMR. *Angew. Chem. Int. Ed.* **49**, 8363–8366 (2010).
31. Chen, H. Y. & Hilty, C. Implementation and characterization of flow injection in dissolution DNP-NMR. *Chemphyschem* **16**, 2646–2652 (2015).
32. Krishnan, V. V. & Murali, N. Radiation damping in modern NMR experiments: Progress and challenges. *Prog. Nucl. Mag. Res. Sp* **68**, 41–57 (2013).
33. Packer, K. J. & Tomlinson, D. J. Nuclear spin relaxation and self-diffusion in the binary system, dimethyl sulfoxide (dmso) + water. *Trans. Faraday Soc* **67**, 1302–1314 (1971).
34. Provencher, S. W. A constrained regularization method for inverting data represented by linear algebraic or integral equations. *Comput. Phys. Commun.* **27**, 213 (1982).
35. Borgia, G. C., Brown, R. J. S. & Fantazzini, P. Uniform-penalty inversion of multiexponential decay data. *J. Magn. Reson.* **132**, 65–77 (1998).
36. Pathan, M., Akoka, S., Tea, I., Charrier, B. & Giraudeau, P. "Multi-scan single shot" quantitative 2D NMR: a valuable alternative to fast conventional quantitative 2D NMR. *Analyst* **136**, 3157–3163 (2011).
37. Kupče, E., Nishida, T. & Freeman, R. Hadamard NMR spectroscopy. *Prog. Nucl. Mag. Res. Sp* **42**, 95–122 (2003).
38. Mansfield, P. Spatial mapping of the chemical shift in NMR. *Magn. Reson. Med.* **1**, 370–386 (1984).
39. Levitt, M. H. Singlet nuclear magnetic resonance. *Annu. Rev. Phys. Chem.* **63**, 89–105 (2012).

Acknowledgements

The Laplace inversion programme was provided by the late Professor P. Callaghan. The financial support by grants from the Academy of Finland (#279721, #289649 and #294027), the Emil Aaltonen foundation, the Council on Grants of the President of the Russian Federation (MK-1329.2014.3) and the Welch Foundation (Grant A-1658) is gratefully acknowledged.

Author contributions

S.A.: implementation of the ultrafast D–T_2 experiment, carrying out the experiments with porous material, hydrocarbons, PHIP and DNP, analysis of the results; V.V.Z: implementation of the ultrafast D–T_2 experiment, planning and carrying out the experiments with hydrocarbons and PHIP; O.M.: carrying out the experiments with porous materials and hydrocarbons; G.Z. and H.-Y.C: carrying out the DNP experiments; A.M.K.: implementation of the ultrafast D–T_2 experiment; C.H.: planning and carrying out the DNP experiments; I.V.K.: planning the PHIP experiments; V.-V.T.: corresponding author, planning the ultrafast D–T_2 experiment, carrying out the experiments and simulations, analysis of the results, writing the first version of the manuscript. All authors contributed to the revision of the manuscript.

Additional information

Competing financial interests: The authors declare no competing financial interests.

23

Gelation process visualized by aggregation-induced emission fluorogens

Zhengke Wang[1,2,*], Jingyi Nie[1,2,*], Wei Qin[3], Qiaoling Hu[1,2] & Ben Zhong Tang[3]

Alkaline-urea aqueous solvent system provides a novel and important approach for the utilization of polysaccharide. As one of the most important polysaccharide, chitosan can be well dissolved in this solvent system, and the resultant hydrogel material possesses unique and excellent properties. Thus the sound understanding of the gelation process is fundamentally important. However, current study of the gelation process is still limited due to the absence of direct observation and the lack of attention on the entire process. Here we show the entire gelation process of chitosan LiOH-urea aqueous system by aggregation-induced emission fluorescent imaging. Accompanied by other pseudo *in situ* investigations, we propose the mechanism of gelation process, focusing on the formation of junction points including hydrogen bonds and crystalline.

[1] MOE Key Laboratory of Macromolecular Synthesis and Functionalization, Department of Polymer Science and Engineering, Zhejiang University, Hangzhou 310027, China. [2] Key Laboratory of Adsorption and Separation Materials and Technologies of Zhejiang Province, Hangzhou 310027, China. [3] Department of Chemistry, Hong Kong Branch of Chinese National Engineering Research Center for Tissue Restoration and Reconstruction, Hong Kong University of Science and Technology, Clear Water Bay, Hong Kong 999077, China. * These authors contributed equally to this work. Correspondence and requests for materials should be addressed to Z.W. (email: wangzk@zju.edu.cn) or to Q.H. (email: huql@zju.edu.cn) or to B.Z.T. (email: tangbenz@ust.hk).

Polysaccharide has gained tremendous attention as a useful and renewable resource[1]. As one of the most important polysaccharide, chitosan (CS) has generated a great deal of interest and provides a wide range of applications[2-8]. In the utilization of CS material, hydrogel is a major and vital branch[9]. Numerous methods have been proposed for the fabrication of CS based hydrogels. Generally speaking, CS based hydrogel can be fabricated via covalent/ionic cross-linking, complexation with another polymer or aggregation after CS grafting and so on[2,10,11]. Solubilization of CS in an acidic aqueous medium is the simplest and conventional way to prepare a CS hydrogel. However, the lack of mechanical strength of CS hydrogels via acidic solvent has been a serious impediment. Several methods such as irradiation and reinforcement have been used to enhance their mechanical strength. But these methods are not very desirable due to the sacrifice of intrinsic properties of CS[12]. Fortunately, a class of novel solvent system, alkali-urea aqueous solutions, had been introduced[13,14]. The resultant CS hydrogel showed significant improvement in hardness, strength and toughness without any cross-linking agents[13]. Furthermore, the distinct gelation behaviour offers innovative hydrogel design strategy[15]. Thus, the study of gelation mechanism shows great industrial and academic value.

Although hypothesis of gelation mechanism had been discussed, understanding of the gelation process is still limited[12-14]. First, in previous studies only the thermal gelation of CS alkali-urea aqueous solution was in the centre of focus. However, a rinse procedure is required after thermal gelation to remove LiOH and urea. The rinse procedure is not only a prerequisite of any application, but also a crucial stage in the formation of robust CS hydrogel according to our research. Second, although visualization is the most powerful way in the investigation of gelation process, there are several defects in the current studies. First, for observation like scanning electron microscope or transmission electron microscope, the investigation suffers from a disadvantage because the native stage of hydrogel is characterized by the presence of water and the need to remove water before examinations inevitably affects the morphology of the hydrogel[16]. Second, due to the existence of large amount of LiOH and urea, the formation of salt crystals could hardly be avoided, which is interference in the observation. Third, the transformation during gelation process had not been studied. In summary, it is necessary to monitor the CS LiOH-urea system in situ and in real time, with the entire gelation process involved.

Fluorescent imaging is an ideal tool for such research purpose. Nevertheless, the research on the gelation process is internally related to the evolution of aggregation state for CS LiOH-urea system[17]. For conventional fluorescent agents, aggregation leads to weakened emission due to the notorious aggregation-caused quenching effect[18]. Opposite to the common aggregation-caused quenching effect, a unique photophysical process termed aggregation-induced emission (AIE) had been observed[19]. The fluorescence of AIE fluorogens is boosted by the restriction of intramolecular rotation[18]. So an AIE fluorogen becomes highly emissive in the aggregation state[20].

Tetraphenylethene is an archetypical AIE fluorogen. In our previous work, we synthesized a novel AIE fluorogenic probe, that is, tetraphenylethene-labelled chitosan (TPE-CS)[21,22]. TPE-CS is a perfect candidate for the investigation on gelation process. Unlike conventional fluorescent agents such as fluorescein isothiocyanate, tetraphenylethene is photostable in the strong alkaline medium[23]. Aggregates formation in the system renders stronger photoluminescence rather than causing quenching. Besides fluorescent images, TPE-CS also provides fluorescent spectroscopy information, which is correlated to the aggregation state of system. In the present work, AIE fluorogens are employed to realize the visualization of the gelation process of CS LiOH-urea solution. The visual characterization is supported by other in situ or pseudo in situ investigations, focusing on the formation of junction points. Finally, an understanding of gelation mechanism is proposed, covering the unique and intriguing transition from solution to hydrogel.

Results

Fluorescent TPE-CS hydrogel via LiOH-urea solvent. Three TPE-CS samples, molecular structure of TPE-CS shown in Fig. 1a, were prepared with different degree of labelling (DL). ^{1}H NMR spectra of TPE-CS showed that labelling the CS with tetraphenylethene fluorogens had been successfully done (Supplementary Fig. 1). The DL of CS was 0.93, 1.54 and 3.56 mol% for the three samples respectively. These samples were denoted as A, B and C with the increase of DL. The powders of TPE-CS samples were light yellow in appearance under normal room lighting and emitted a strong blue light on ultraviolet illumination (Fig. 1b). These TPE-CS samples showed different solubility in LiOH-urea aqueous solvent (Fig. 1c). Sample A and B formed transparent solution, while sample C only existed as sediment. The results showed that the solubility of TPE-CS in LiOH-urea aqueous solution was reduced by the increase of DL, owing to the hydrophobic tetraphenylethene fluorogens. However, the fluorescence of the TPE-CS is intensified with the increase of DL, which was shown in the ultraviolet photographs (Fig. 1c) and validated by the photoluminescence (PL) spectra (Fig. 1d).

Thus the DL should be controlled by balancing two opposite factors. On one hand, heavily labelled sample was favoured for strong photoluminescence signal. On the other hand, lower DL was favoured to minimize interference in the behaviour of CS. Taking both factors into consideration, sample B was optimum and was utilized as the raw material in the present work.

Typical preparation procedures of CS hydrogel via LiOH-urea solvent were demonstrated in Supplementary Fig. 2, including thermal gelation and the succedent rinse procedure. TPE-CS can be fabricated into hydrogel by the same preparation procedures (Fig. 1e). In addition, the gelation behaviour of CS and TPE-CS was tested by rheological method (Supplementary Fig. 3)[24]. The well-matched gelation time of CS and TPE-CS indicated that, with proper DL, the introduction of tetraphenylethene fluorogens did not exert much interference on the gelation process of CS.

Visualization of gelation process. CS gelation process was monitored by confocal laser scanning fluorescence microscope (Fig. 2). In previous studies, extremely dilute CS solution (10^{-6} g ml^{-1}) was used for imaging, with drying procedure involved[12]. However, for the preparation of hydrogel with high mechanical performance, the concentration of CS was in the range of 10^{-2} g ml^{-1}. So the in situ visualization within this concentration range would provide more direct and effective information on the gelation process. The fluorescent image of solution showed no specific patterns (Fig. 2a). However, some bright areas started to appear after heat absorption started, corresponding to the thermal gelation stage (Supplementary Fig. 4). As time goes on, the bright areas kept developing. Then the development slowed down, and finally the pattern became stable (Fig. 2b). This indicated that the thermal gelation of the system possessed certain terminal point. However, after LiOH and urea were completely removed the structure of gel further developed (Fig. 2c). This indicated that the terminal point of thermal gelation was not the termination of gelation process. After rinse stage, the bright areas subdivided and contracted, forming dark areas. Eventually, CS hydrogel came into being with

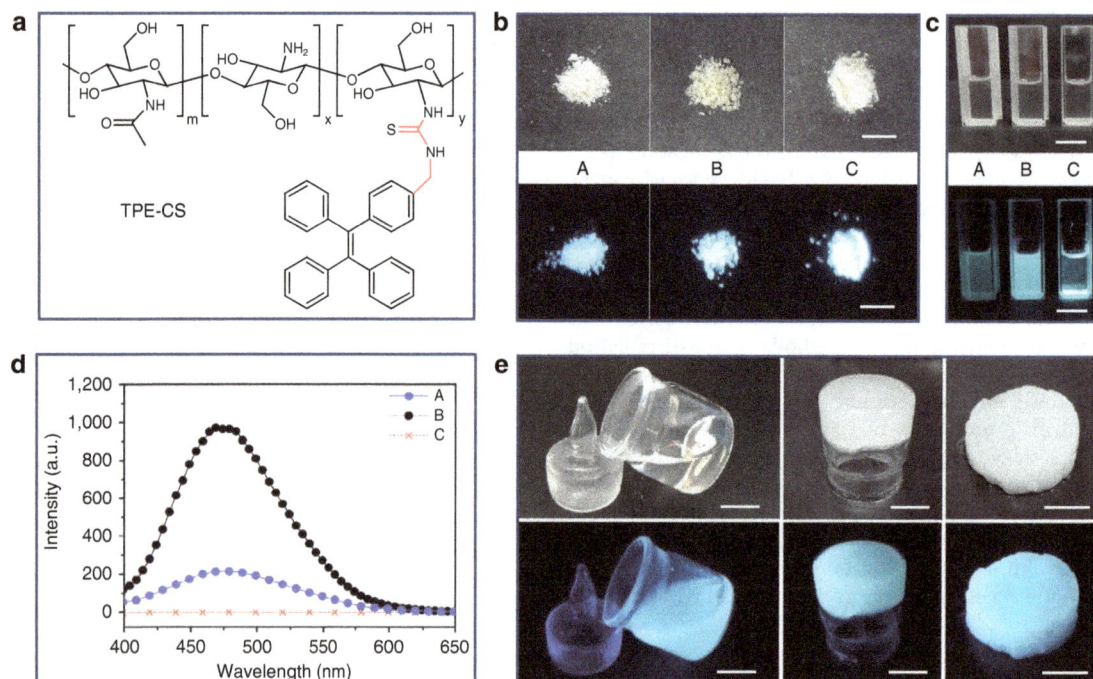

Figure 1 | TPE-CS samples in the form of powder solution and hydrogel. (**a**) Molecular structure of TPE-CS. (**b**) Digital images of powder samples; Scale bar, 1.0 cm. (**c**) TPE-CS samples after treated by freeze-thaw dissolving procedure in LiOH-urea aqueous solution; Scale bar, 1.0 cm. (**d**) Photoluminescence spectra of TPE-CS solution, the intensity corresponding to sample C was denoted as zero, since sample C cannot be dissolved in the LiOH-urea solvent. (**e**) Digital images of formation of hydrogel with TPE-CS: solution of TPE-CS, TPE-CS gel after thermal gelation, and TPE-CS hydrogel after complete removal of LiOH and urea; Scale bar, 5.0 mm. The fluorescent photographs were taken under ultraviolet illumination (365 nm).

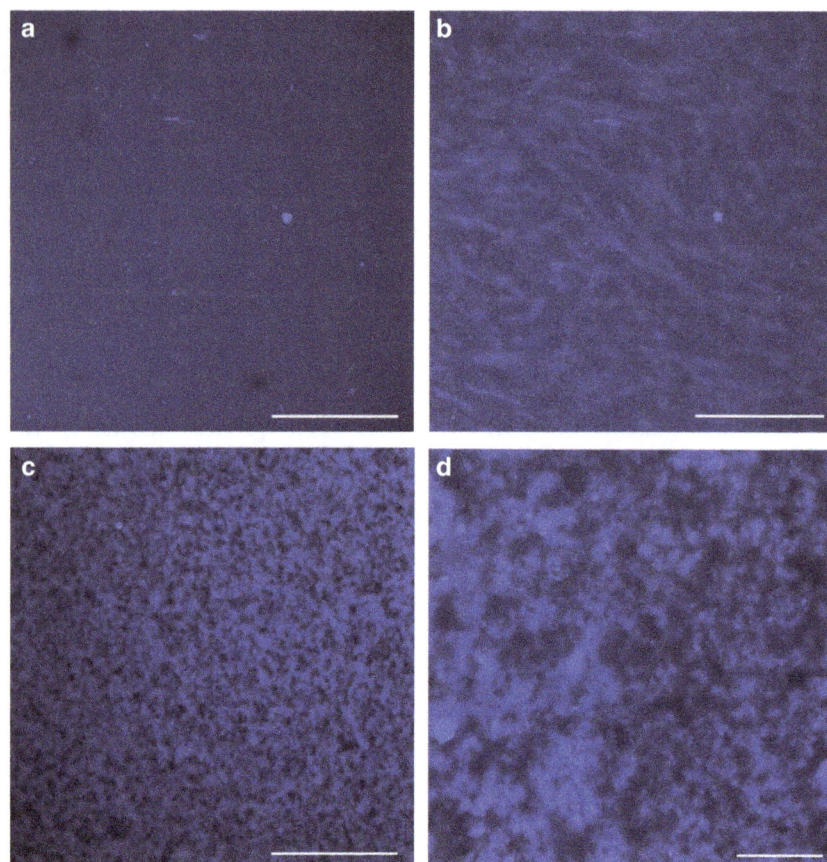

Figure 2 | Confocal laser scanning fluorescence microscope images of the gelation process of TPE-CS. (**a**) solution; (**b**) gel after thermal gelation; (**c,d**) hydrogel after rinse procedure; Scale bar, 250 μm (**a–c**); 25 μm (**d**).

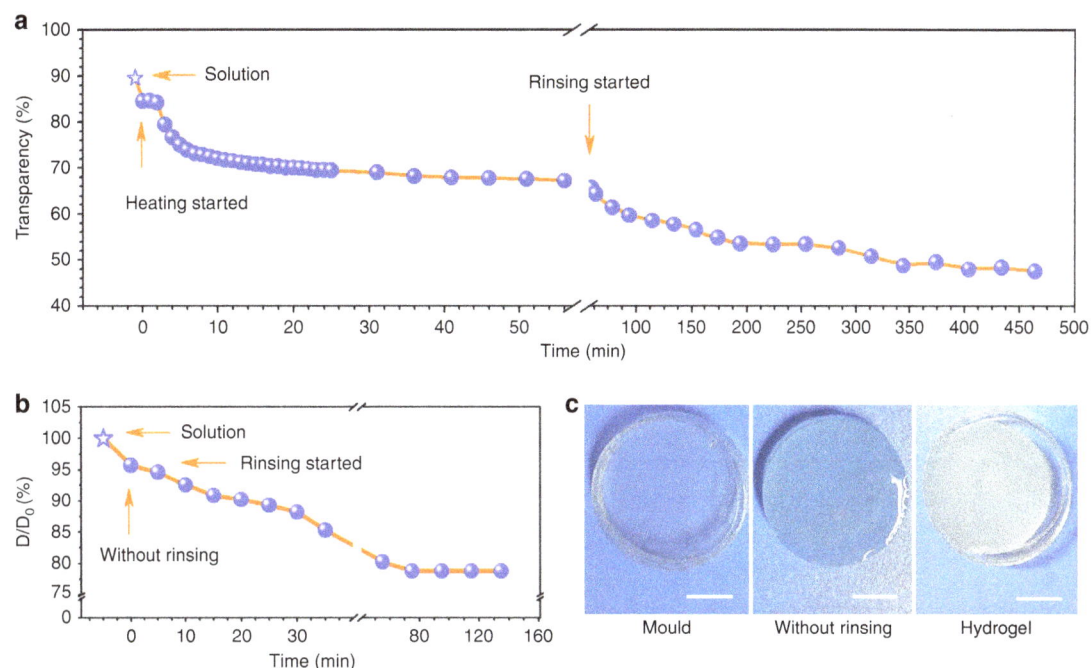

Figure 3 | Change of macroscopic properties of CS LiOH-urea system in the gelation process. (**a**) Evolution of transparency of CS LiOH-urea system. (**b**) Dimension shrinkage rate of CS gel, D_0 is the diameter of mould and D is the diameter of gel. (**c**) Images of CS hydrogel and the mould used in preparation; Scale bar, 1.0 cm.

a reticular structure (Fig. 2d and Supplementary Fig. 5). In addition to fluorescent images, TPE-CS also provides fluorescent spectroscopy information, which was in good accordance with the aggregation state of system (Supplementary Figs 6,7 and Supplementary Discussion). The results above indicated that gelation initiated after heat absorption, however, the rinse stage was a crucial stage in the formation of CS hydrogel structure.

In addition to the fluorescent study, macroscopic property evolution of CS LiOH-urea system was studied by pseudo *in situ* investigations to further understand the gelation process. Optical transparency reflected the homogeneity of materials. In the gelation process, the solution was transparent originally, however, certain structure arose due to gelation, thus increased the light scattering and decreased the transparency of system[25]. Figure 3a showed the transparency as a function of time in the entire gelation process. First, the transparency maintained for a short period after contacted with heat source. Then the transparency gradually decreased and finally reached a constant value, corresponding to the terminal point of thermal gelation stage. The decrease continued in the rinse stage, indicating the further development of structure in this stage. The shrinking behaviour in the gelation process is an important physical phenomenon of hydrogel, which is closely related to the solvophobic and solvophilic interactions. The shrinking behaviour of CS LiOH-urea system was followed in the entire process (Fig. 3b). The volume of gel slightly decreased after thermal gelation stage. When the gel was immersed in water after thermal gelation, it showed evident shrinkage rather than swelling, and the diameter of CS hydrogel was 78% of the value of mould (Fig. 3c). Moreover, the volume change was not a simple shrinkage caused by the loss of components. In addition, the removal of OH$^-$ is closely related to the volume change of CS gel (Supplementary Figs 8,9). The results mentioned above strongly indicated that the shrinkage was due to the further development of gel structure. The evolution of macroscopic property during the gelation process was in good accordance with the information provided by fluorescent images. In addition to thermal gelation,

the CS LiOH-urea system went through crucial transformation in the rinse stage.

Formation and development of junction points. To further understand the formation and development of junction points during the two stages, other *in situ* or pseudo *in situ* investigations were employed.

The thermal gelation process of CS solutions was studied by dynamic viscoelastic method (Fig. 4a and Supplementary Fig. 10). Storage modulus G' represented the elastic behaviour of the system, which was essentially related to the formation of junction points in the system. So the evolution of G' was followed with time when the system was cured at a predetermined temperature. Taking the G' curves of 40 °C as an example (Supplementary Fig. 10c). First, the heat absorption led to the rise of system temperature and consequently a minor decrease in G'. As system temperature continuously increased, a sharp increase in G' was observed. In this period, the storage modulus increased faster than loss modulus (G''), thus a crossover of G' and G'' curves occurred, which was usually determined as the gelation point[26,27]. In previous work, little attention was paid to the behaviour after this point. However, in fact G' continued increasing after the crossover, indicating that more junction points were formed. The increase of G' gradually slowed down. This was because the motion of CS chains was restricted, and the formation of more junction points became more difficult. Finally, the G' curve presented a plateau. In the thermal gelation stage, elevated temperature is the driving force of gelation. Different temperatures were used in the isothermal curing tests. When the predetermined temperature was 25 °C, the increase of G' could hardly be observed, while a minor increase was detected at 32 °C. The G' curve reached plateau more quickly when temperature was set at elevated value (60 and 80 °C). Although occurrence of plateau happened at different times, the evolution of G' curves presented similar pattern at all temperatures.

The plateau of G' indicated the end of thermal gelation. However, it did not mean the increase of toughness terminated.

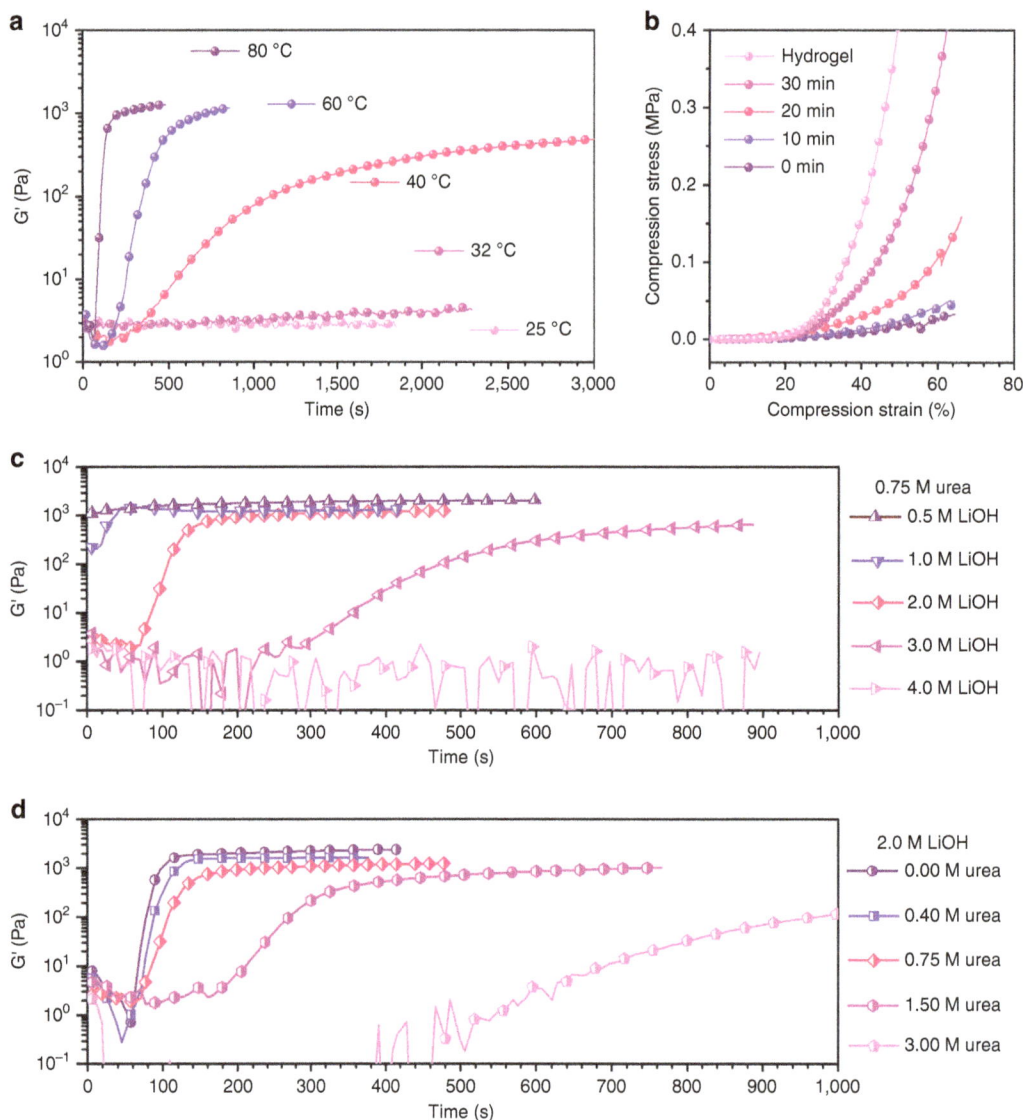

Figure 4 | Strength evolution of CS LiOH-urea system in the entire gelation process. (**a**) Evolution of storage modulus (G') of CS LiOH-urea solution with time; the solutions were cured at marked temperatures. (**b**) Evolution of compression stress of CS LiOH-urea gel with different rinse time. (**c,d**) Relation between gelation behaviour and the concentration of LiOH (**c**) and the concentration of urea (**d**).

The variation of toughness was characterized by compression stress in the rinse stage. As shown in Fig. 4b, if not rinsed at all after thermal gelation, the CS gel proved to be brittle, which failed at a compression stress ≤ 0.05 MPa and strain $\leq 70\%$. During the rinse stage, the failure-compressive stress increased gradually. After LiOH and urea were completely removed from the system, CS hydrogel was formed. The hydrogel possessed failure-compressive stress 15 times higher than gel without rinsing, and reached a failure-compressive strain higher than 90%. This indicated that the system could absorb much more energy before fracture. The strength of CS hydrogel was mainly gained in the rinse stage.

To explore the driving force of strength increase, a serial of experiments were performed as control (Supplementary Fig. 9). The results showed that: (1) experimental groups with OH^-, with or without urea, all proved to be brittle and had low failure strain; (2) experimental groups without OH^-, even treated with urea or other inorganic salt (LiCl or NaCl), proved to be tough and had high failure-compressive strain. The results validated that, in the rinse stage, the removal of OH^- is the primary contribution to the increase of toughness.

The results above also indicated that LiOH and urea are not equal in the interaction with CS, so the further investigations were performed to clarify the different roles that LiOH and urea played during sol-gel transition (Fig. 4c,d and Supplementary Fig. 11). The occurrence of plateau was accelerated by the decrease of both the concentration of LiOH and urea, and was delayed contrariwise. However, the influence on the evolution of G' curves was quite different in the two cases. The c(LiOH) in the system determined the primary G' value (Fig. 4c). With lower c(LiOH), the primary G' value got closer to the value of plateau. However, the change of c(urea) did not affect the primary G' value nor the process of evolution (Fig. 4d). This also indicated that urea played a subordinate role of dissolution and gelation in the system.

Crystalline is a very important form of physical junction in hydrogel[28]. CS is a well-known semi-crystalline polymer in the solid state. So it is possible that the polymer chains would rearrange and form crystallites. The evolution of crystalline in the whole gelation process was studied by X-ray diffraction (Fig. 5). The broad peak belonged to the X-ray diffraction profile of liquid water (Supplementary Fig. 12). LiOH and urea in the solution did not show any peaks since they were at amorphous state. There was no characteristic peak of CS in the profile of CS LiOH-urea solution, this indicated that there was no crystalline in the solution before thermal gelation. With the initiation of thermal gelation, a small peak appeared at $2\theta = 20°$. This peak could be attributed to the form II crystals of CS[29]. This demonstrated the formation of low amount of small crystalline. The crystalline was already formed in the elevated temperature and the amount of crystalline did not increase when the temperature decreased (Supplementary Fig. 13). Intensity of this peak increased with gelation time, indicating the formation of more crystalline. However, the intensity of peak did not increase in the rinse stage. The results indicated that the formation of crystalline mainly happened in the thermal gelation stage. The formation of crystalline was not affected by the introduction of tetraphenylethene (Supplementary Fig. 14).

Discussion

We proposed the hypothesis on the evolution of CS LiOH-urea aqueous system during the gelation process.

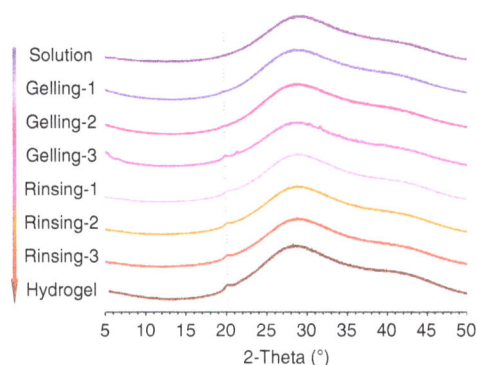

Figure 5 | Evolution of crystalline during gelation process of CS LiOH-urea system. XRD patterns of CS LiOH-urea system in the entire gelation process.

There existed dynamic intermolecular interactions among LiOH, urea and polysaccharide in the alkali/urea/cellulose solution[30]. On the molecular level, the oxygen atom in OH^- is electron-richer due to the extra electron, and consequently very competitive in the formation of hydrogen bonds with $-NH_2$ and $-OH$ on the CS chain. Several possible forms of hydrogen bonds between CS and OH^- were shown in Supplementary Fig. 15a. Due to the saturability of hydrogen bonds, these $-NH_2$ and $-OH$ can no longer form inter/intramolecular hydrogen bonds. When large number of $-NH_2$ and $-OH$ groups were attached by OH^-, CS became soluble in this aqueous system (Fig. 6a). On the other hand, the role of urea in the intermolecular interaction is not decisive. If the OH^- detached from $-NH_2$ and $-OH$ on the CS chains, the oxygen atom in urea could serve as the hydrogen-bonding receptor for hydrogen atoms in these groups. Moreover, the hydrogen atoms in urea serve as the hydrogen-bonding donor for the electron-rich atoms in these groups (Supplementary Fig. 15b). Although the interactions are in a dynamic detached re-attached pattern, the addition of urea decelerates the formation of inter/intramolecular hydrogen bonds among CS.

The existence of terminal point in thermal gelation stage indicated the equilibrium of intermolecular interactions. The intermolecular interactions in the system can be described as the overall reaction of two competitive interactions. Reaction 1 is the formation of H-bonds between OH^- and CS, and reaction 2 is the formation of inter/intramolecular H-bonds among CS. The thermal gelation stage can be interpreted as the shift of equilibrium of the overall reaction at elevated temperature. The frequency of detach and re-attach of OH^- with CS increases with the rise of system temperature. Due to the restriction of macromolecule, reaction 2 is less susceptible to heat than reaction 1. So the equilibrium shifts to direction of forming more inter/intramolecular H-bonds among CS and even enable formation of crystalline (Fig. 6b,c). The termination of thermal gelation stage corresponds to the establishment of new equilibrium at the elevated temperature.

The rinse stage could also be interpreted with the shift of equilibrium. In this stage, the driving force is the concentration change of reactants. The removal of OH^- largely promotes the

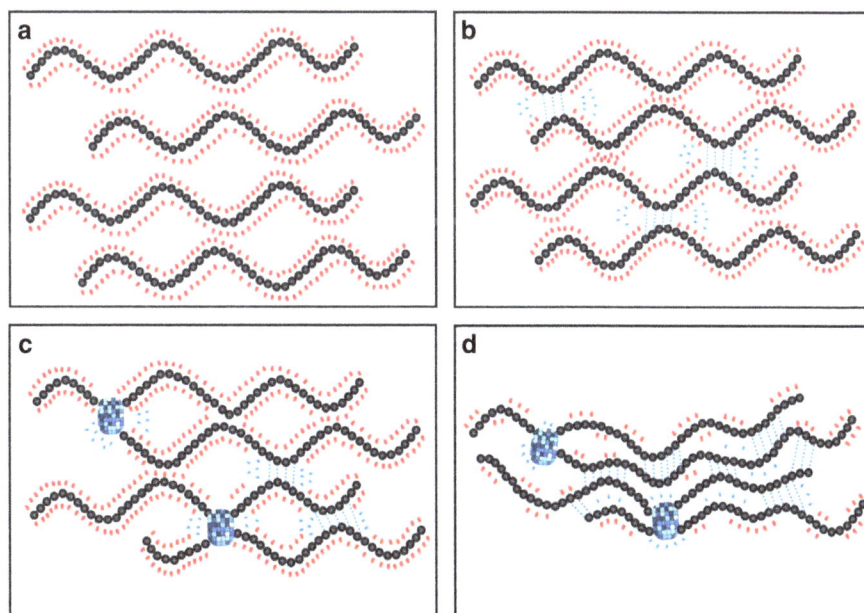

Figure 6 | Schematic illustration of the formation of junction points of CS hydrogel in LiOH-urea solution. (**a**) Well-dissolved solution; (**b**) formation of inter/intramolecular hydrogen bonds of CS, induced by heating; (**c**) formation of crystalline in CS gel; (**d**) further formation of hydrogen bonds of CS, induced by the removal of LiOH and urea. The partial CS chains in the figure may belong to the same CS macromolecule or different CS macromolecules.

reverse course of reaction 1, and greatly promotes the formation of inter/intramolecular H-bonds among CS (Fig. 6d). So this stage is main stage for the increase of toughness and the volume shrinkage.

During the entire gelation process, the detachment of OH$^-$ from CS chains render the possibility of crystallization. However, in the rinse stage, CS chains were not rearranged into the lattice. This could be explained that CS chains still had mobility in the early stage of gelation, and was able to reform and get into the lattice. While in the rinse stage, chains were bounded by inter/intramolecular hydrogen bonds formed in the thermal gelation. This indicated that the formation of H-bonds was more localized in the rinse stage.

The gelation process of CS LiOH-urea solution comprised two distinct stages. In the thermal gelation stage, the system showed similarity to typical thermogels (for example, PLGA-PEG-PLGA). However, there also exists difference in the gelation mechanism of the two-gelation system. The thermosensitivity of PLGA-PEG-PLGA originates from the balanced structure of hydrophilicity and hydrophobicity of the macromolecules[31–33], while the thermosensitivity of CS LiOH-urea solution originates from the change of intermolecular interaction discussed above. The rinse stage can be considered as the removal of additives. The nature of this stage is transforming the medium to poor solvent by decreasing c(LiOH) and c(urea). Unlike random precipitation caused by introducing poor solvent, the structure of CS hydrogel further develops based on the embryonic structure formed in thermal gelation stage. Essentially, the two stages have something in common, which is the decrease of interactions between solvent and macromolecules and the increase of inter/intramolecular interactions of CS.

In this work, we visualized the entire gelation process of CS LiOH-urea aqueous system by AIE fluorescent imaging for the first time. Accompanied by other pseudo *in situ* investigations, we proposed the mechanism of gelation process. The formation of CS hydrogel via CS LiOH-urea aqueous system is unique. The entire process has two distinct but integrated stages, that is, the thermal gelation stage and rinse stage. Thermal gelation stage is driven by the elevated temperature, forming the embryonic structure and crystalline. While the rinse stage is driven by the continuous change of system components, greatly increasing interintramolecular hydrogen bonds. The two distinct stages can be interpreted under a unified understanding, which is the intermolecular interaction of solvent and polymer. The present work provided further understanding of the gelation process, and facilitated the utilization of polysaccharide. Furthermore, with the boom of AIE filed, this work could be a starting point, various AIE fluorogens would be available for the visualization of various gelation systems.

Methods

Materials and reagents. CS was prepared in our laboratory by heterogeneous N-deacetylation from commercial chitin. (Zhejiang Gold Shell Pharmaceutical Co., Ltd.) The average viscosity molecular weight (Mη) of CS was 1.12×10^6 Da and degree of deacetylation (DD) was 94.4%. Mη, degree of deacetylation were determined following the process reported in ref. 34. Lithium hydroxide monohydrate (LiOH · H$_2$O), urea and acetic acid were purchased from Sinopharm Chemical Reagent Co., Ltd. 1-[4-(isothiocyanatomethyl)phenyl]-1, 2, 2-triphenylethene was synthesized, and labelled onto CS chain to fabricate TPE-CS according to the procedures described in our previous work[21].

Preparation of TPE-CS solution and hydrogel. LiOH · H$_2$O and urea were dissolved in deionized water to form a transparent solution with weight percentage of 4.8 and 8.0 wt.%, respectively. The resultant aqueous solution was used as solvent in this work. A certain amount of TPE-CS powder was mixed in the solvent. The mixture was treated by a freeze (-20 °C)-thawing (20 °C) process three times to get transparent TPE-CS LiOH-urea solution. For the study of solubility, 20 mg TPE-CS powder was mixed in 2 g solvent and then went through the dissolving

procedure mentioned above. Fluorescence (FL) spectra of the TPE-CS LiOH-urea solution were measured on a Perkin-Elmer LS 55 spectrofluorometer. For the preparation of hydrogel, the concentration of TPE-CS was 2.0 wt.%. The solution was held for 30 min in a 60 °C thermostatic water bath, and turned to TPE-CS gel containing LiOH and urea. The gel was then fully washed in deionized water to remove LiOH and urea after the mould was unloaded.

Fluorescent images. The fluorescent images of gelation process were obtained by confocal laser scanning fluorescence microscope (Leica TCS SP5 CLSM). A TPE-CS LiOH-urea aqueous solution was prepared first, and the concentration of TPE-CS was 2 wt.%. The solution was then filled in a quartz culture dish and the original temperature of solution was -4 °C. In thermal gelation stage, the heat was provided by a heat source with constant temperature (60 °C). The timing started once the sample contacted with the heat source. The structure of TPE-CS hydrogel was observed after the LiOH and urea was completely removed.

Transparency. The light transmission of the CS solution, gel and hydrogel was examined at room temperature using a WGT-S Haze and Luminous Transmittance Instrument. The solution was first filled in a cuvette with thickness of 10 mm and side length of 40 mm. The predetermined temperature of heat source was 60 °C in the thermal gelation stage. In the rinse stage, the CS gel was unloaded from the cuvette and supported by a sample holder during tests. The gel sample was rinsed in 500 ml deionized water before every test.

Shrinking behaviours of the hydrogel. The test of volume shrinking behaviours of CS LiOH-urea system were carried out by measuring the diameter change of hydrogel discs[35]. The mould was unloaded after enough thermal gelation time, and then the gel was rinsed for 5 min in 100 ml deionized water before every measurement.

Rheological tests. Rheological tests were performed on AR-G2 rheometer (TA Co., USA). The freshly prepared and degassed CS solution were used as samples, with c(CS) = 2 wt.%. Silicon oil was placed around the rim to prevent water evaporation during the measurement. The temperature was set to 4 °C first to prevent gelation of sample, and then maintained at a predetermined temperature as soon as the test started. The change in dynamic modulus was monitored as a function of time during the gelation process. All the rheological experiments were performed within the linear viscoelastic region.

Mechanical tests. The mechanical property of CS hydrogels were obtained on a universal materials testing machine (Instron, 5543A) at a strain rate of 2% min^{-1} for compression tests at room temperature. Samples were prepared to be cylinder. The mould used in this section had a diameter of 33 mm and a depth of 4 mm. Several identical samples were prepared and went through thermal gelation stage. Then the samples were rinsed in 200 ml deionized water for desired time, respectively.

X-ray measurements. Wide-angle X-ray diffraction profiles were collected with an Ultima IV analytical instrumentation (Rigaku Corporation) using Ni-filtered Cu Kα radiation scanned at 2 deg (2θ) s^{-1} in the 2θ range 5–60°. The solution was first filled in a tailor-made specimen holder with thickness of 2 mm. The predetermined temperature of heat source was 60 °C in the thermal gelation stage. The sample was heated by the source for 5 min before every measurement. The treatments were denoted as gelling 1, gelling 2 and gelling 3. After thermal gelation the gel was rinsed for 30 min in 500 ml deionized water before every measurement. The treatments were denoted as rinsing 1, rinsing 2 and rinsing 3.

References

1. Gandini, A. Polymers from renewable resources: a challenge for the future of macromolecular materials. *Macromolecules* **41**, 9491–9504 (2008).
2. Berger, J., Reist, M., Mayer, J. M., Felt, O. & Gurny, R. Structure and interactions in chitosan hydrogels formed by complexation or aggregation for biomedical applications. *Eur. J. Pharm. Biopharm.* **57**, 35–52 (2004).
3. Baldrick, P. The safety of chitosan as a pharmaceutical excipient. *Regul. Toxicol. Pharmacol.* **56**, 290–299 (2010).
4. Wang, Z., Hu, Q. & Wang, Y. Preparation of chitosan rods with excellent mechanical properties: one candidate for bone fracture internal fixation. *Sci. China-Chem.* **54**, 380–384 (2011).
5. Nie, J., Wang, Z., Zhang, K. & Hu, Q. Biomimetic multi-layered hollow chitosan-tripolyphosphate rod with excellent mechanical performance. *Rsc. Advances* **5**, 37346–37352 (2015).
6. Casettari, L. & Illum, L. Chitosan in nasal delivery systems for therapeutic drugs. *J.Control. Release* **190**, 189–200 (2014).

7. Aider, M. Chitosan application for active bio-based films production and potential in the food industry: Review. *Lwt-Food Sci. Technol.* **43**, 837–842 (2010).
8. Miretzky, P. & Fernandez Cirelli, A. Hg(II) removal from water by chitosan and chitosan derivatives: A review. *J. Hazard. Mater.* **167**, 10–23 (2009).
9. Ladet, S., David, L. & Domard, A. Multi-membrane hydrogels. *Nature* **452**, 76–79 (2008).
10. Berger, J. *et al.* Structure and interactions in covalently and ionically crosslinked chitosan hydrogels for biomedical applications. *Eur. J. Pharm. Biopharm.* **57**, 19–34 (2004).
11. Kozicki, M. *et al.* Hydrogels made from chitosan and silver nitrate. *Carbohydr. Polym.* **140**, 74–87 (2016).
12. Duan, J., Liang, X., Cao, Y., Wang, S. & Zhang, L. High strength chitosan hydrogels with biocompatibility via new avenue based on constructing nanofibrous architecture. *Macromolecules* **48**, 2706–2714 (2015).
13. Sun, Y. F., Li, Y. L., Nie, J. Y., Wang, Z. K. & Hu, Q. L. High-strength Chitosan Hydrogels Prepared from LiOH/Urea Solvent System. *Chem. Lett.* **42**, 838–840 (2013).
14. Li, C., Han, Q. Y., Guan, Y. & Zhang, Y. J. Thermal gelation of chitosan in an aqueous alkali-urea solution. *Soft Matter* **10**, 8245–8253 (2014).
15. Nie, J. *et al.* Orientation in multi-layer chitosan hydrogel: morphology, mechanism, and design principle. *Scient. Rep.* **5**, 7635 (2015).
16. Fergg, F., Keil, F. J. & Quader, H. Investigations of the microscopic structure of poly(vinyl alcohol) hydrogels by confocal laser scanning microscopy. *Colloid. Polym. Sci.* **279**, 61–67 (2001).
17. Weng, L. H., Zhang, L. N., Ruan, D., Shi, L. H. & Xu, J. Thermal gelation of cellulose in a NaOH/thiourea aqueous solution. *Langmuir* **20**, 2086–2093 (2004).
18. Hong, Y., Lam, J. W. Y. & Tang, B. Z. Aggregation-induced emission: phenomenon, mechanism and applications. *Chem. Commun.* **12**, 4332–4353 (2009).
19. Luo, J. D. *et al.* Aggregation-induced emission of 1-methyl-1, 2, 3, 4, 5-pentaphenylsilole. *Chem. Commun.* 1740–1741 (2001).
20. Hong, Y., Lam, J. W. Y. & Tang, B. Z. Aggregation-induced emission. *Chem. Soc. Rev.* **40**, 5361–5388 (2011).
21. Wang, Z. *et al.* Long-term fluorescent cellular tracing by the aggregates of AIE bioconjugates. *J. Am. Chem. Soc.* **135**, 8238–8245 (2013).
22. Jia, J. *et al.* Monitoring layer-by-layer self-assembly process of natural polyelectrolytes by fluorescent bioconjugate with aggregation-induced emission characteristic. *J. Mater. Chem. B* **2**, 8406–8411 (2014).
23. Xu, J. *et al.* FITC and Ru(phen)(3)(2 +) co-doped silica particles as visualized ratiometric pH indicator. *Nanoscale Res. Lett.* **6**, 561 (2011).
24. Chen, L., Ci, T., Li, T., Yu, L. & Ding, J. Effects of molecular weight distribution of amphiphilic block copolymers on their solubility, micellization, and temperature-induced sol gel transition in water. *Macromolecules* **47**, 5895–5903 (2014).
25. Suzuki, A., Yamazaki, M., Kobiki, Y. & Suzuki, H. Surface domains and roughness of polymer gels observed by atomic force microscopy. *Macromolecules* **30**, 2350–2354 (1997).
26. Cai, J. & Zhang, L. Unique gelation behavior of cellulose in NaOH/Urea aqueous solution. *Biomacromolecules* **7**, 183–189 (2006).
27. Li, C., Han, Q., Guan, Y. & Zhang, Y. Thermal gelation of chitosan in an aqueous alkali-urea solution. *Soft Matter* **10**, 8245–8253 (2014).
28. Ricciardi, R., Auriemma, F., De Rosa, C. & Laupretre, F. X-ray diffraction analysis of poly(vinyl alcohol) hydrogels, obtained by freezing and thawing techniques. *Macromolecules* **37**, 1921–1927 (2004).
29. Samuels, R. J. Solid-state characterization of the structure of chitosan films S. *J. Polym. Sci. Part B-Polym. Phys.* **19**, 1081–1105 (1981).
30. Jiang, Z. *et al.* Intermolecular interactions and 3D structure in cellulose-NaOH-urea aqueous system. *J.Phys. Chem. B* **118**, 10250–10257 (2014).
31. Yu, L., Chang, G., Zhang, H. & Ding, J. Temperature-induced spontaneous sol–gel transitions of poly(D,L-lactic acid-co-glycolic acid)-b-poly(ethylene glycol)-b-poly(D,L-lactic acid-co-glycolic acid) triblock copolymers and their end-capped derivatives in water. *J. Polym. Sci. Part A: Polym. Chem.* **45**, 1122–1133 (2007).
32. Yu, L., Zhang, H. & Ding, J. D. A subtle end-group effect on macroscopic physical gelation of triblock copolymer aqueous solutions. *Angew. Chem. Int. Ed. Engl.* **45**, 2232–2235 (2006).
33. Park, M. H., Joo, M. K., Choi, B. G. & Jeong, B. Biodegradable Thermogels. *Acc. Chem. Res.* **45**, 424–433 (2012).
34. Fan, M., Hu, Q. & Shen, K. Preparation and structure of chitosan soluble in wide pH range. *Carbohydr. Polym.* **78**, 66–71 (2009).
35. Mi, P. *et al.* A novel stimuli-responsive hydrogel for K + -induced controlled-release. *Polymer* **51**, 1648–1653 (2010).

Acknowledgements

This work was financially supported by the Key Science Technology Innovation Team of Zhejiang Province (No. 2013TD02), the National Natural Science Foundation of China (No. 21104067, 21274127, 21374099 and 51473144), the University Grants Committee of Hong Kong (AoE/P-03/08), the Research Grants Council of Hong Kong (16305015, 16301614 and N_HKUST 604/14) and the Innovation and Technology Commission (ITC-CNERC14SC01). We thank Prof. Shen Jiacong and Dr. Zhang Ling for their help in this research.

Author contributions

Z.W. and J.N. designed and conducted the experiments, analysed the experimental data and wrote the paper; W.Q. synthesized 1-[4-(isothiocyanatomethyl)phenyl]-1, 2, 2-triphenylethene; Z.W., Q.H. and B.Z.T. supervised the project.

Additional information

24

Ultrahigh-throughput discovery of promiscuous enzymes by picodroplet functional metagenomics

Pierre-Yves Colin[1], Balint Kintses[1,†], Fabrice Gielen[1], Charlotte M. Miton[1], Gerhard Fischer[1], Mark F. Mohamed[1], Marko Hyvönen[1], Diego P. Morgavi[2,3], Dick B. Janssen[4] & Florian Hollfelder[1]

Unculturable bacterial communities provide a rich source of biocatalysts, but their experimental discovery by functional metagenomics is difficult, because the odds are stacked against the experimentor. Here we demonstrate functional screening of a million-membered metagenomic library in microfluidic picolitre droplet compartments. Using bait substrates, new hydrolases for sulfate monoesters and phosphotriesters were identified, mostly based on promiscuous activities presumed not to be under selection pressure. Spanning three protein superfamilies, these break new ground in sequence space: promiscuity now connects enzymes with only distantly related sequences. Most hits could not have been predicted by sequence analysis, because the desired activities have never been ascribed to similar sequences, showing how this approach complements bioinformatic harvesting of metagenomic sequencing data. Functional screening of a library of unprecedented size with excellent assay sensitivity has been instrumental in identifying rare genes constituting catalytically versatile hubs in sequence space as potential starting points for the acquisition of new functions.

[1] Department of Biochemistry, University of Cambridge, 80 Tennis Court Road, Cambridge CB2 1GA, UK. [2] INRA, UMR1213 Herbivores, F-63122 Saint-Genès-Champanelle, France. [3] Clermont Université, VetAgro Sup, UMR Herbivores, BP 10448, F-63000 Clermont-Ferrand, France. [4] Groningen Biomolecular Science and Biotechnology Institute, University of Groningen, Nijenborgh 4, 9747 AG, Groningen, The Netherlands. † Present address: Synthetic and Systems Biology Unit, Institute of Biochemistry, Biological Research Centre of the Hungarian Academy of Sciences, Szeged 6726, Hungary. Correspondence and requests for materials should be addressed to F.H. (email: fh111@cam.ac.uk).

A much broader range of enzyme catalysts than currently known is needed to address a number of challenges—from the implementation of green, sustainable processes in white biotechnology[1] to a fundamental understanding of the evolutionary origin and mechanistic basis of biocatalysis. Microbial ecosystems are viewed as enormous reservoirs of genetic diversity[2], but 99% of environmental microorganisms are understood to be unculturable as yet[3]. Extensive sequencing efforts using DNA that was directly extracted from environmental samples (eDNA) have already given unprecedented insight into the make-up and genetic diversity of such ecosystems[4]. However, extrapolation from sequence to protein function is not trivial and often gives only a rough idea of functional assignments[5]. As a consequence, annotations frequently prove to be incorrect when tested experimentally[6]. Large-scale experimental characterization would be the preferred basis of functional annotation of proteins, but is currently time-consuming and expensive: which is why sequencing dominates metagenomic explorations. Furthermore, it is especially difficult to predict entirely new enzymes without precedent: genes without significant homology, encoding catalysts for unfamiliar reactions that have not yet been comprehensively assigned are likely to be overlooked (or remain unannotated). As a consequence, the percentage of genes with unknown functions in newly sequenced genomes has remained constant over the last decade (at ~30–40%)[7].

Functional annotation is further complicated by increasing evidence for catalytic promiscuity (that is, the ability of enzymes to process more than one type of substrate). Promiscuous side activities assist evolutionary adaptation by providing a head start activity after gene duplication, enabling a smoother route from one enzyme function to another by avoiding loss of function during their interconversion[8,9]. Enzyme promiscuity is even harder (and often impossible) to predict by sequence analysis than primary activities[10], thus adding multiple dimensions to the challenge of experimental identification of function. As a consequence, the functional potential of this 'underground network'[11] remains largely unexplored. Functional characterizations of protein families suggest however that promiscuity is an intrinsic enzyme property[12,13], a potential marker for evolutionary related proteins[14] and a rich source of new functions[15]. Apart from chance observations no comprehensive system for detection of promiscuity exists. Discovery of promiscuous activities will be valuable: by analogy to natural neofunctionalization based on promiscuity, the identification of promiscuous activities similarly provides starting points for enzyme repurposing by directed evolution and can ultimately yield useful catalysts of practical utility.

Experimental screening of eDNA, where the corresponding proteins are heterologously expressed in a surrogate host, is a powerful method to functionally identify and annotate novel enzymes without relying on homology[16]. However, success depends on the efficiency of the heterologous expression[17] and hit rates are typically extremely low (estimated at one hit per 10^4–10^5 variants, depending on the target activity)[18]. Efficient ultrahigh-throughput systems are therefore essential to beat the odds[19]. Apart from costly robotic screening of individual metagenomic library members, requiring expensive liquid handling systems[20] and labour-intensive procedures, no direct screening system for catalytic product formation (and able to screen large libraries to yield large numbers of hits) has been used to isolate novel catalysts in functional metagenomics. Apart from providing ultrahigh throughput, an experimental system for identifying promiscuous activities in metagenomes must be sensitive enough to pick up potentially weak side activities, as their rates are often several orders of magnitude slower than native activities[15].

To address the challenge of performing sensitive functional screening with ultrahigh throughput, we use water-in-oil droplets, a biomimetic compartment acting as a genotype–phenotype linkage[21,22], in analogy to cells. Monodisperse droplets are produced in microfluidic chips so that quantitative and sensitive measurements can be performed[23] and ultrahigh-throughput sorting[24] enables selection of droplets according to their fluorescence. Using this format, up to ~10^8 biochemical reactions can be performed per day, typically in pico- to femtolitre volumes, resulting in dramatically lower screening costs compared with robotic screening[25]. Microfluidic droplets have been successfully used for directed evolution[25–27] strain selections[28,29] or bioprospecting[30]. We now employ such droplets as vessels to miniaturize cell lysate assays[27]—the most frequently used screening formats in metagenomics[19,31]—to the single-cell level. In this experimental set-up, each library member is represented by a single cell, statistically compartmentalized into one droplet and assayed for catalytic activity after cell lysis. A microfluidic screening platform is used to recover and identify various hydrolases from metagenomic sources by screening for the hydrolysis of two substrates: sulfate monoesters, containing one of the most unreactive functional groups in biology[32], and phosphate triesters. The latter are non-natural environmental pollutants historically used as pesticides[33] that we now use to probe the catalytic potential of microbial communities for degradation of xenobiotics. The high sensitivity of our platform enabled us to isolate enzymes with strong and weak activities, including promiscuous side activities that have not been selected in the course of evolution and could not be predicted by sequence homology.

Results

Screening for new hydrolases in microfluidic droplets.
A metagenomic library of 1,250,000 variants pooled from a variety of sources (combining libraries derived from soil, degraded plant material and cow rumen samples; 'SCV', Supplementary Table 1) was screened in droplets following the strategy illustrated in Fig. 1. Library members were expressed in *Escherichia coli* and single cells were compartmentalized (with a Poisson distribution that gives an average of ~0.8 cells per droplet) into monodisperse 2 pl droplets (a volume found to allow suitably sensitive screening of metagenomic libraries—Fig. 2a and Supplementary Figs 1 and 2). Bacteria were co-compartmentalized with lysis agents and the respective fluorogenic substrates (sulfate monoester **1d** or phosphate triester **2d**; Fig. 1) in two different screening experiments. After a two-day off-chip incubation, the emulsions were re-injected into the sorting chip, where ~20 million droplets (covering the library size more than 15 times and found to be sufficient to recover all hits in a model library; Supplementary Figs 3 and 4) were analysed in ~2 h and the brightest 0.01% (sulfatases) or 0.01 and 0.001% (phosphotriesterases (PTEs), in two sorts performed in parallel and later combined) of the droplets selected (Fig. 2b). This relatively tolerant detection threshold was chosen to avoid loss of potential hits. The selected droplets were subsequently de-emulsified and plasmid DNA was recovered. The high-copy plasmid DNA allowed its direct re-transformation into *E. coli*[27]. The final number of transformants exceeded 10^4 (one droplet gave on average approximately five transforming clones depending on the plasmid copy number found to vary as a function of the insert size; Supplementary Fig. 6). To enrich the hits further, a second microfluidic screening cycle (Fig. 1) was performed with a lower cell/droplet ratio (0.1 cells per droplet), reducing the number of droplets containing multiple cells. Out of ~10 million screened, ~500 droplets were collected as

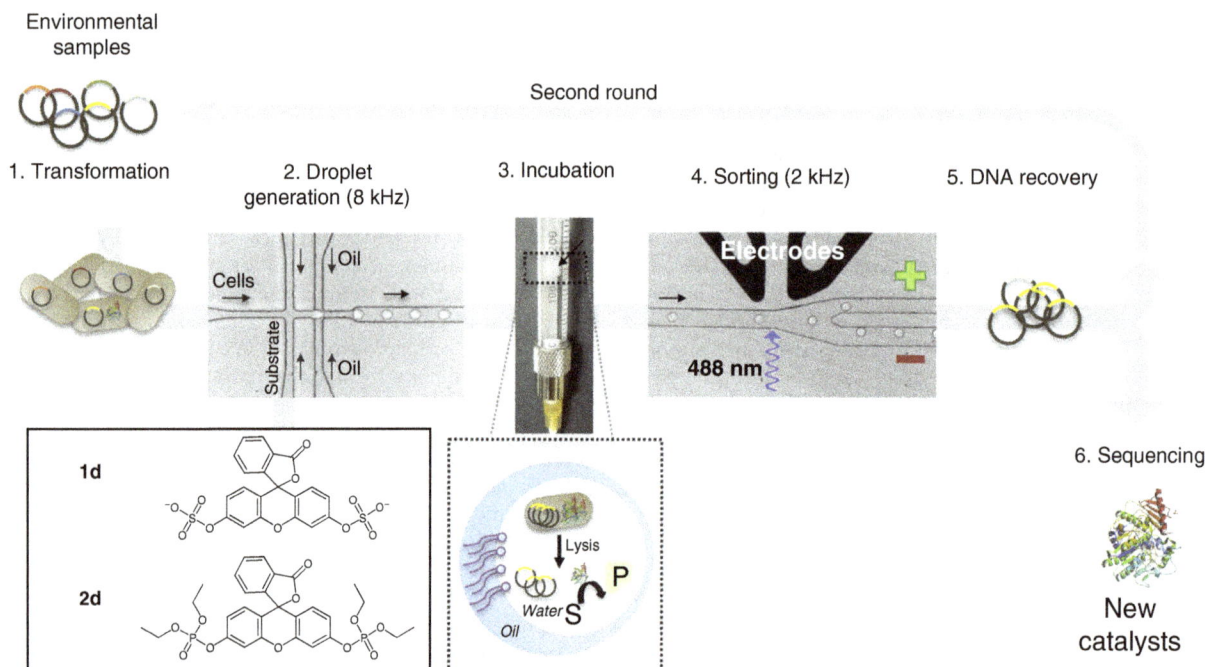

Figure 1 | Functional metagenomic using microfluidic droplets. General procedure. (1) Environmental DNA (eDNA) was cloned into a high-copy plasmid and transformed into *E. coli*. (2) Single bacteria were encapsulated into water-in-oil droplets together with substrate (**1d** or **2d**) and lysis agents. (3) Emulsion droplets were incubated off-chip; after single cell lysis, cytoplasmically expressed protein catalysts were able to turn over substrate. The arrow designates the droplets ($\sim 3 \times 10^7$) at the interface between fluorous oil and mineral oil (on top). (4) Emulsion droplets were re-injected (Supplementary Movie 1) into a sorting chip and strongly fluorescent droplets ('+' channel) were separated from those with fluorescence below the threshold ('−' channel) by dielectrophoresis[24] (Supplementary Movie 2). (5) Selected droplets were de-emulsified and high-copy plasmid DNA was recovered following by re-transformation into *E. coli*. For further enrichment, iterative selections could be performed. (6) Plasmids containing eDNA coding for active catalysts were sequenced. The microfluidic devices are shown in Supplementary Fig. 5.

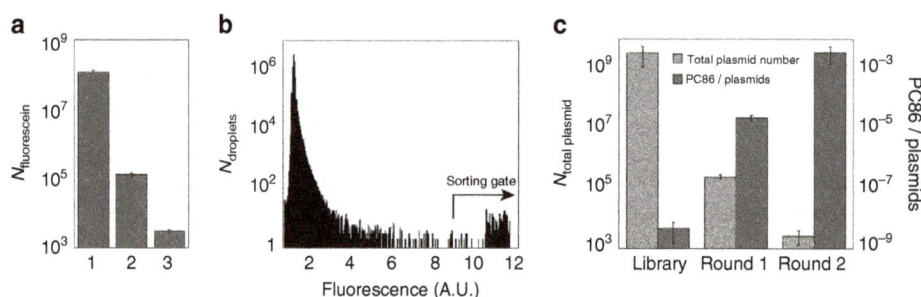

Figure 2 | High sensitivity of the microfluidic platform allows selection and enrichment of active metagenomic variants. (**a**) Minimal number of fluorescein molecules detected. (1) In a 200 µl well of a microtiter plate (conventional technology), (2) in 20 pl droplets[27] and (3) in 2 pl droplets (according to respective calibration curves; Supplementary Fig. 2). Errors bars are set to 10% of the calculated value, corresponding to an estimation of the uncertainty of measurement. (**b**) Fluorescence signal distribution of 10^7 droplets containing metagenomic cell lysate after 2 days of incubation at 22 °C. The sorting gate was set up such that droplets with two- to fivefold increased fluorescence over the population average were selected (sorting gate). All histograms corresponding to the two screening campaigns are shown in Supplementary Fig. 7. (Note: as the photomultiplier tube saturated at around 10 AU, values shown here as $\sim$10 AU may be higher). (**c**) The selection stringency was tested by monitoring the enrichment of a PTE variant PC86 over multiple rounds of microfluidic sorting. Three samples—metagenomic library before selection; DNA recovered from round 1; DNA recovered from round 2—were analysed for (i) their total plasmid content and (ii) the number of plasmids encoding PC86 by quantitative PCR. The proportion of PC86 in each DNA sample (library before selection, round 1 and round 2) was calculated by dividing the number of plasmid PC86 by the total number of plasmids. (Detailed data are shown in Supplementary Fig. 10). Error bars, s.d. from triplicates. AU, arbitrary units.

arylsulfatase hits and $\sim$300 as PTEs (corresponding to $\sim$0.007% or $\sim$0.003% of droplets for sulfatases and PTEs, respectively; Supplementary Figs 7 and 8). The resulting transformants were screened on Petri dishes or in 96-well plates for hydrolysis of sulfate monoester **1c** or phosphate triester **2d**, respectively. In 10% of the variants hydrolytic turnover was confirmed

(Supplementary Fig. 9). Sequencing the plasmids of the confirmed positive hits revealed six and eight unique sequences for sulfatases and PTEs, respectively (Fig. 3). The low positive/negative ratio observed at the end of the screening campaign can be explained mainly by the deliberately tolerant sorting gate chosen to avoid loss of relevant clones at the cost of

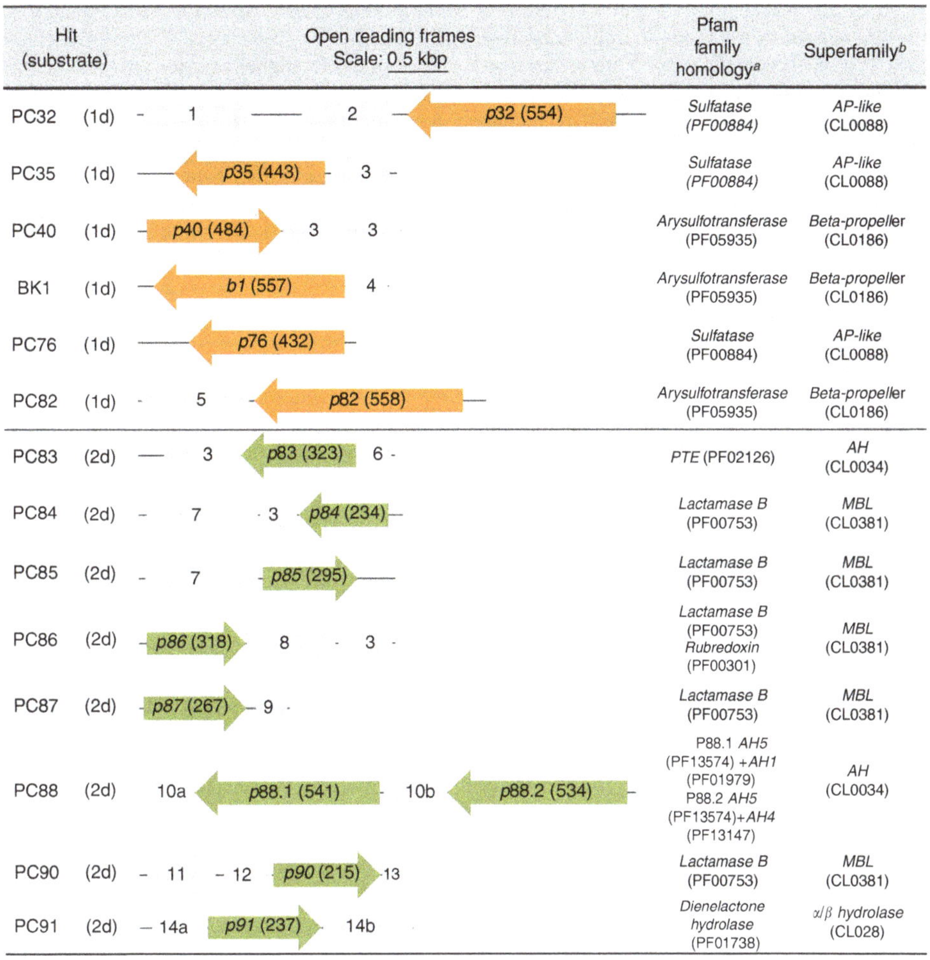

Figure 3 | Unique hits isolated from metagenomic libraries. Open reading frames (ORFs) encoding hits are highlighted in orange or green (sulfate hydrolases/transferases or PTEs, respectively). Numbers between brackets indicate the number of amino acids in the protein sequence. Grey arrows represent other ORFs isolated together with the hit sequence. As it was less obvious which of the p88.1 or p88.2 gene was most likely to code for the active triesterase, both were recloned. Selections were carried out with either sulfate ester **1d** or phosphate triester **2d**. (a,b) Data extracted from the Pfam database[36]. MBL, metallo β-lactamase, also called metallo-hydrolase/oxidoreductase; AH, amidohydrolase. The other genes isolated (grey arrows) have their closest homologues in the NCBI non-redundant database predicted as: (1) transcription regulator; (2) formylglycine generating enzyme; (3) ABC transporter or membrane protein; (4) TonB-dependent receptor; (5) succinate-semialdehyde dehydrogenase; (6) carnitine dehydratase; (7) penicillin-binding protein; (8) radical SAM protein; (9) K$^+$/H$^+$ antiporter; (10a,b) cobalamin biosynthesis protein; (11) YKuD transpeptidase; (12) peptidase M15; (13) aminotransferase; (14a,b) permease (see Supplementary Note 1).

collecting false positives: droplets with a two- to fivefold higher fluorescence than background were selected (Fig. 2b and Supplementary Fig. 7).

Assessing enrichment over microfluidic rounds. Despite the permissive sorting gate applied (Fig. 2b and Supplementary Fig. 7), we sought to quantify the enrichment obtained in our screening campaign. To this end, the abundance of one PTE hit in the library (PC86; Fig. 3) was analysed using quantitative PCR before and after the two rounds of sorting. Figure 2c shows that the plasmid content of this hit in the library increased >1,000-fold during the first round and >100-fold during the second round of sorting, corresponding to an overall enrichment of >10^5 (from 10^{-6}% to nearly 0.5%; Supplementary Fig. 10). Such enrichment capacity betters previously reported hit rates in functional metagenomic selections (estimated to be between 1/10,000 and 1/100,000)[18], demonstrating the utility of the microdroplet format to recover extremely rare hits. An enrichment of more than 1,000-fold in one round of screening

(despite a permissive sorting gate) is similar to or exceeds other studies using microfluidic droplets[24,27,34]. Sorting with greater stringency could be applied to achieve a similar enrichment in fewer rounds[35]. The objective of extending coverage to include hits even with weak activity, however, was made possible by a low detection threshold (<3,000 fluorescein molecules being detected; Fig. 2a and Supplementary Fig. 2) and exemplified the ability to detect low activities (Supplementary Fig. 11).

Characterization of metagenomic hits. (a) *Sulfatase hits*. Our screening workflow yielded six unique variants (Fig. 3) that turned over sulfate ester **1d** (Fig. 1). Three hits possessed genes predicted to be part of the sulfatase family (see Fig. 3 for the assignments by Pfam, the protein families database[36]). One gene (p35) was recloned, the protein P35 was purified and shown to hydrolyse sulfate ester **1a** (Supplementary Fig. 12) with high efficiency ($k_{cat}/K_M = 1.7 \times 10^5$ s^{-1} M^{-1}; Supplementary Table 3) and high proficiency ($k_{cat}/K_M/k_{uncat} = 1.5 \times 10^{14}$ M^{-1}; Supplementary Table 4), demonstrating the ability to recover an enzyme with high specific activities from our metagenomic libraries. Three hits contained genes that were assigned to the Pfam arylsulfotransferase family (*p40*, *b1* and *p82*). We experimentally verified this assignment after recloning and purification of the protein P40 (Fig. 3). P40 was shown to catalyse sulfate transfer from sulfate ester **1a** to phenolate following a ping-pong mechanism (Supplementary Fig. 13) and can therefore be considered as a new adenosine 3-phosphate-5-phosphosulfate-independent sulfotransferase[37]. Reactions in droplets were followed by the release of the fluorescent leaving group of **1d**. Transferases able to transfer sulfate groups to acceptors present in the cell lysate were therefore also selected, indicating that multiple types of related reactions can be recovered by the fluorogenic miniaturized cell lysate assay.

(b) PTE hits. Applying the screening workflow to PTEs (using triester **2d** as substrate) led to the isolation of eight unique and novel sequences, mostly originating from unknown microorganisms (Supplementary Table 2). Only one of these genes (*p83*) was predicted as a potential PTE by Pfam domain recognition (Fig. 3). To confirm PTE activity, we identified the genes most likely to encode triesterase enzymes and cloned them into an expression vector. Nine genes in total (*p83*, *p84*, *p85*, *p86*, *p87*, *p88.1*, *p88.2*, *p90* and *p91*; highlighted in green in Fig. 3) were recloned and their encoded proteins purified. All enzymes were active towards the fluorogenic triester **2d** (Fig. 1) used for screening, confirming the isolation of genuine PTEs (Table 1). When tested with the pesticides paraoxon **2a** or parathion **2b** (Supplementary Fig. 12), 5- to 100-fold slower rates than for the fluorogenic triester **2d** were observed ($8\,M^{-1}\,s^{-1}$ for P84 to $3 \times 10^3\,M^{-1}\,s^{-1}$ for P83; Table 1). All hits degraded the fluorogenic screening substrate and the pesticide, suggesting that triester **2d** is a suitable bait for identifying paraoxon hydrolases. Despite rather low second-order rates (8–$3 \times 10^3\,M^{-1}\,s^{-1}$), most of these enzymes substantially accelerated hydrolysis of **2a** with rate enhancements up to 10^{12} ($k_{cat}/K_M/k_w$), when compared with the spontaneous hydrolysis in water (Supplementary Table 4).

Access to unexplored sequence space. Sequence-similarity networks[38] (SSNs) (Fig. 4 and Supplementary Fig. 15) were constructed to map hit sequences: instead of being clustered together they were spread over three superfamilies (defined by the Pfam database[36]; Fig. 3) and covered a large sequence diversity.

The only predicted triesterase (*p83*) located within the PTE/PTE-like lactonase cluster of the amidohydrolase superfamily (AH; Fig. 4a) and its higher activity towards lactone **8c** (Supplementary Fig. 12 and $k_{cat}/K_M = 5 \times 10^4$ $s^{-1}M^{-1}$; Table 1) were verified experimentally, indicating that P83 is a new lactonase endowed with phosphotriesterase activity.

The remaining hits defied prediction and were scattered over three superfamilies. (i) Two were also AH superfamily members (P88.1 and P88.2) but were located in a completely different region of the SSN (Fig. 4a) and shared very little sequence identity ($<15\%$) with other PTEs from the same superfamily (Supplementary Fig. 16). No other native function was predicted, making them completely unassigned. (ii) All hits assigned to the metallo-β-lactamase (MBL) superfamily (P84, P85, P86, P87 and P90) have close homologues in the NCBI non-redundant database predicted as putative β-lactamases. However, none of the five hits was able to degrade the chromogenic β-lactam **10** (Supplementary Fig. 12), a frequently used probe of β-lactamase activity[14,39]. In the SSNs (Fig. 4b), the genes *p85*, *p86* and *p87* located in proximity to sequences coding for PTEs and lactonases. However, no lactonase activity was detected in any of these hits (with lactones **8a**, **8b**, **8c** and **8d$_{1-2}$**; Supplementary Fig. 12), confirming that sequence similarity is not a good predictor for activity. P84 and P90 are positioned away from other known triesterases and seem more related to glyoxalase II (Fig. 4b), a family of thioesterases from the glyoxalase system that catalyses the hydrolysis of *S*-lactoylglutathione into glutathione and lactate[40]. Experimental verification (Table 1) of their ability to hydrolyse *S*-lactoylglutathione **7** (Supplementary Fig. 12) suggested that these enzymes were a subgroup of the glyoxalase II family. (iii) One hit, P91, is a member of the α/β-hydrolase (α/β) superfamily related to dienelactone hydrolases (DLHs; Fig. 3). Members of this superfamily possess a catalytic triad[41], whereas most bacterial PTEs described to date use catalytic divalent ions to activate a water nucleophile[42]. The X-ray structure of the protein was determined at a resolution of 1.7 Å (Supplementary Fig. 18) and structural alignment with a previously characterized DLH[43] (also identified in the SSNs, shown in Supplementary Fig. 15) confirmed an α/β hydrolase fold (Fig. 5a), and highlighted a homologous catalytic triad in the active site of P91 (C118, D167 and H199; Fig. 5b). Site-directed knockdown mutagenesis of C118, D167 and H199 led to enzyme inactivation,

Table 1 | Catalytic efficiencies of the metagenomic hits towards the tested substrates.

	Sulfate monoester	Phosphate triester		Phosphate monoester	Phosphate diester	Phosphonate monoester	Ester	Thioester	Lactone*
	1a	2a	2d	3	4	5	6	7	8
					k_{cat}/K_M $(s^{-1}M^{-1})$				
P35	2×10^5	—†	ND	2×10^{-1}	9×10^{-3}	3×10^1	—	ND	ND
P83	—	3×10^3	9×10^5	—	—	—	1×10^2	ND	5×10^4
P84	—	7.6	52	—	4×10^{-2}	3×10^{-1}	5×10^2	2×10^3	2×10^3
P85	—	9×10^1	9×10^2	—	5×10^{-3}	2×10^{-1}	4×10^1	ND	—
P86	—	2×10^2	1×10^3	—	9×10^{-3}	1×10^{-1}	3×10^1	ND	—
P87	—	1×10^2	4×10^2‡	—	7×10^{-2}	1×10^{-1}	4×10^1	ND	—
P88.1	—	4×10^2	3×10^3	—	—	—	6×10^2	ND	—
P88.2	—	2×10^2	1×10^3	—	—	1×10^{-1}	1×10^1	ND	—
P90	—	5×10^1	2×10^2	—	—	8×10^{-1}	1×10^2	2×10^3	6×10^3
P91	—	3×10^2	9×10^3	—	—	1×10^{-1}	1×10^4	ND	4×10^3

ND, not determined.

The substrates tested were suggested by previous observations of promiscuity[56] and the superfamily context (Fig. 4). Michaelis–Menten plots and all k_{cat} and K_M values are shown in Supplementary Fig. 14 and Supplementary Table 3, respectively.

*Activity measured using different lactones: **8a** (P84, P90 and P91); **8c** (P83)—no activity detected for **8b** and **8d$_{1-2}$** (see different lactones; Supplementary Fig. 12).

†(−) No activity was detected with high enzyme (1 µM) and substrate (1 mM) concentrations after incubation for 3 h.

‡The magnitude of the fluorescence signal change suggests that only one of the two phosphate groups of triester **2d** was cleaved.

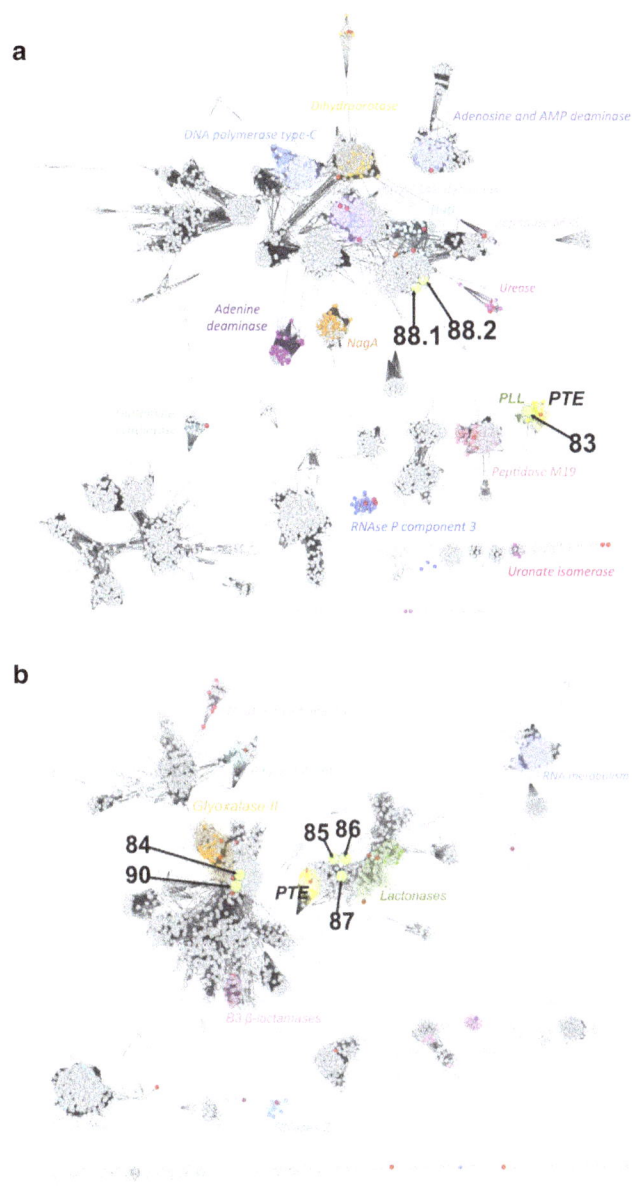

Figure 4 | SSNs highlight the novelty of the triesterases hits. Hits are spread over (**a**) the AH clan (Pfam number: CL0034) and (**b**) the MBL superfamily (Pfam family PF00753). Bright green nodes represent the sequences of metagenomic hits identified in this work. Annotations retrieved from the Uniprot database were used to putatively characterize sequence clusters (represented by coloured nodes). To confirm these annotations, experimentally characterized proteins (red nodes) were added into each network. Previously described OP-degrading enzymes (PTE, phosphotriesterase) are highlighted in yellow; PLL (PTE-like lactonases) are reported in the AH superfamily. 5,042 and 2,984 sequences define the AH and the MBL networks, respectively. Edges lengths represent protein sequence similarity. Only edges corresponding to similarity scores below E-values of $1 \times e^{-9}$ (AH) and $1 \times e^{-14}$ (MBL) are considered; the worst edges displayed correspond to a median 26.5% (or 27.9%) identity over an alignment length of 210 (or 218) residues for the AH (and MBL) networks, respectively. See also the position of additional hits in the α/β superfamily (Supplementary Fig. 15).

consistent with catalysis by the cysteine-based catalytic triad in P91 (Fig. 5c), which was expected to provide nucleophilic and charge relay catalysis. We observed electron density consistent with two orientations of C118. Figure 5b highlights potential

'protecting' residues (E37 and H141) that may play a role in stabilizing an inactive conformation of the nucleophile cysteine, as previously invoked[44]. The lack of electron density in the vicinity of the active site supported the absence of a metal-binding site, underlining a role for the triad instead of a metal cofactor in catalysis. Functional tests further ruled out P91 as a metallo-enzyme: treatment of P91 with chelating agents did not affect its activity towards triester **2a**, whereas a reduction (or complete abolition) of activity was observed in all other PTE hits tested (Supplementary Fig. 17).

Acetylcholine esterases are the biological targets of organophosphate (OP) pesticides, acting as suicide substrates leading to acetylcholine esterase inactivation and belong to the α/β superfamily (see SSNs of the superfamily in Supplementary Fig. 15). It has been reported that α/β hydrolases have evolved in insects (for example, *Lucilla cuprina*) to slowly degrade OPs[45], conferring insecticide resistance in the absence of a metal cofactor. However, P91 is the first metal-free bacterial (or archaeal) triesterase to be described, unrelated to insect carboxylesterases (Supplementary Fig. 15). Most of these carboxylesterases have evolved ageing-resistant variants[46] and are characterized by low turnover rates combined with tight binding[45]. In contrast, P91 is a weak paraoxon binder ($K_M \sim mM$) but with relatively high turnover rates ($\sim 1 s^{-1}$), suggesting that P91 does not suffer enzyme ageing, allowing up to four orders of magnitude higher k_{cat} (ref. 45).

Selection of catalysts via a promiscuous function. The comparison of the Michaelis–Menten parameters (k_{cat}/K_M and K_M) of our triesterase hits with those of (i) OP hydrolases isolated from pesticide-contaminated soils and (ii) enzymes with known promiscuous triesterase activity (Fig. 6a and Supplementary Table 5) suggests that paraoxon **2a** (or parathion **2b** with lower or similar catalytic rates measured; Supplementary Table 3) and, by extension, the synthetic triester **2d** are not the physiological substrates of our hits. This conclusion was also supported by the experimental verification of higher activities in P83, P84, P90 and P91 towards lactone **8c**, thioester **7** and ester **6**, respectively (Table 1 and Supplementary Table 3).

This observation provides further evidence for the widespread existence of latent enzymatic activities, enabling biocatalysts to promote 'unnatural reactions' besides their native function. Most PTEs known to date were isolated from OP-degrading bacteria, providing a selective advantage in highly OP contaminated soils[47]. As OP-based pesticides were only introduced into the environment in the 1940s (ref. 42), these enzymes are examples of rapid evolutionary adaptation. Our screens suggest the availability of additional points for 'head starts' in adaptive evolution, spread over three superfamilies. Previously, only ~ 20 experimentally characterized PTEs were reported in the literature (Supplementary Table 6). We now have increased this number by a third and doubled the number of enzymes with promiscuous triesterase activity.

Many PTEs seem to have evolved from lactonases and retain some degree of lactonase activity[14,48,49]. However, most of our hits were not active (Table 1) towards the natural lactones tested (**8b**, **8c** and **8d$_{1-2}$**), suggesting that they have different evolutionary ancestors than other known PTEs (also supported by their position in the SSNs in Fig. 4 and Supplementary Fig. 15) and implying that diverse native activities can accommodate promiscuous triesterase function. As the hits locate mostly in sparsely annotated regions in their respective SSNs (Fig. 4) and because of the relatively short insert sizes, only limited additional information is provided by the genome context (Fig. 3 and Supplementary Note 1), the native activities and physiological roles of the hits remain unknown.

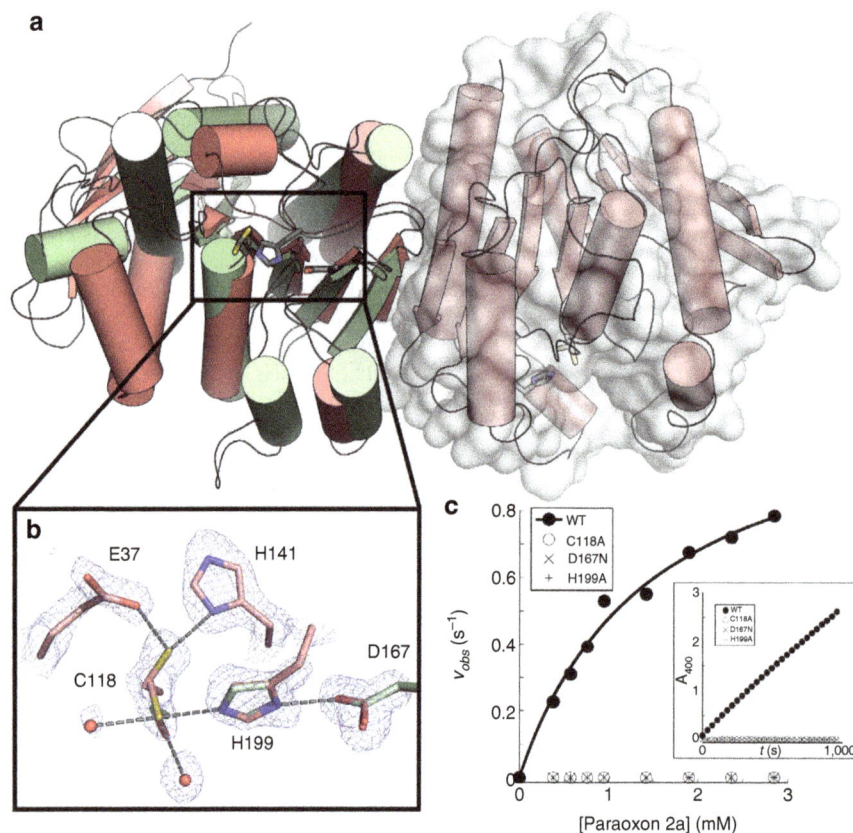

Figure 5 | A triad is the catalytic feature of the α/β-hydrolase fold of the triesterase P91. (**a**) The P91 structure (red) is aligned with DLH (green), a well-characterized α/β hydrolase fold (PDB ID: 1ZIC[43]). (**b**) Active site superposition with DLH reveals the catalytic triad of P91 consisting of C118, D167 and H199. E37 and H141 stabilize an alternative cysteine conformation. (**c**) Site-directed knockdown mutagenesis of the residues in the triad compromises P91's PTE activity ($v_{obs} = v/[E_0]$) (shown by Michaelis-Menten plot and time-course measurements using ~1 mM of substrate (framed)). Conditions: activity towards paraoxon **2a** measured at 25 °C (100 mM Tris-HCl, 150 mM NaCl pH 8.0).

The majority of the triesterase hits constitute first examples of enzymes endowed with triesterase activity within their respective superfamily subgroup underlining that a different region of sequence space is accessible by droplet screening compared with metagenome mining by bioinformatics. Accessing such new isolated but functionally characterized points in sequence space makes them hubs for exploration of close homologous sequences that may share similar activities[50].

Novel entry points to promiscuous networks. The two bait reactions (that is, phosphotriester hydrolysis and arylsulfate hydrolysis) unearthed six different activities in total: sulfate monoesterase, phosphate monoesterase, phosphate diesterase, PTE, phosphonate monoesterase and (thio)-esterase activities (Fig. 6b). Promiscuity emerged as a feature of all hits, with more than two chemically distinct substrates turned over besides the mostly unknown native ones, confirming the insight that promiscuity is ubiquitous[15] and worthwhile harvesting. Many reactions were common among the hits despite their extreme sequence divergence (Supplementary Fig. 16) and their different catalytic machineries (Fig. 5b and Supplementary Fig. 17). A comparison of the observed promiscuity suggests trends that reflect the respective bait: sulfatases favour charged substrates, triesterases neutral ones. Phosphonate hydrolase and phosphodiesterase activities are common to the two groups (Fig. 6b), but triesterases support these activities only at a low level (Table 1). Testing a range of bait substrates in future

experiments may lead to empirical cross-reactivity maps (similar to Fig. 6b) that describe active site recognition features (that is, geometric or electrostatic complementarity, or the availability of reactive active site groups, for example, nucleophiles).

Discussion

One of the main limitations in functional metagenomics is the difficulty of discovering rare genes. Indeed, much work in this field has focused on widely occuring enzymes such as lipases, esterases and carbohydrate-converting enzymes[18]. To identify potentially less widespread activities, effective screening technologies are key[31] to covering larger metagenomic sequence space[21] and beating the odds. In this work, we establish droplet screening as a general method to find new, rare enzymes (6 and 8 hits out of 1,250,000 clones for sulfatase and PTE, respectively) in metagenomics using hydrolytic turnover of two different baits as initial target activities. Previous attempts at harvesting environmental samples by droplet screening have not successfully identified DNA sequences that correspond to the selected phenotype[30]. By contrast, this present work has yielded functional information on new metagenomic sequences from droplet experiments, for the first time. More generally, no metagenomic campaign had so far identified triesterases or sulfatases, possibly because they are much less abundant than the above-mentioned enzymes (and certainly with fewer known examples in the literature). Our screening capacity of $>5 \times 10^6$ assays per hour enables $\sim 10^8$ assays per day so that libraries of

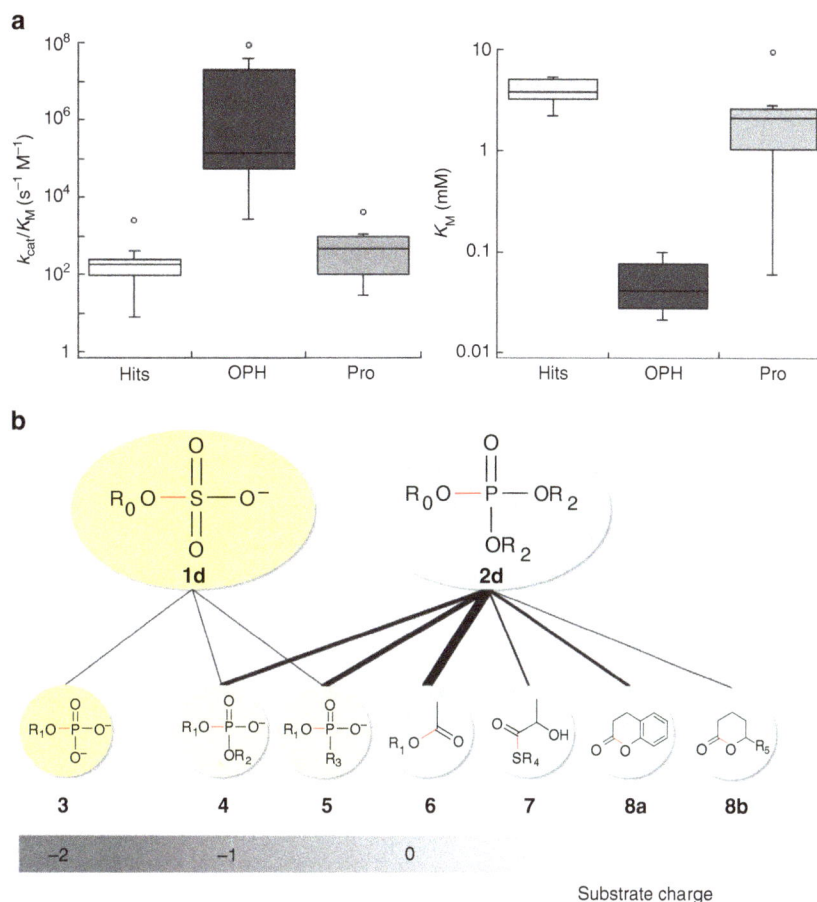

Figure 6 | Promiscuity is a general feature of the selected hits. (**a**) Comparison of the data (catalytic efficiencies k_{cat}/K_M (left) and K_M values (right)) for our hits with known enzymes catalysing phosphotriester hydrolysis suggests that our screening has identified promiscuous activities. 'OPH' summarizes PTE recovered from bacteria isolated from a pesticide-polluted environment and therefore assigned as enzymes that evolved specifically for triester hydrolysis. 'Pro' designates enzymes for which PTE activity was shown to be a promiscuous side activity. The rates reported here are towards paraoxon **2a** (for metagenomic hits) or—in the case of known PTEs—towards paraoxon **2a** or methylparathion **2b**. Full names (Supplementary Table 6) and reported catalytic efficiencies (Supplementary Table 7) of known PTEs are listed. (**b**) The catalytic promiscuity among the triesterase hits reflects bait substrate charge attributes. Substrates **1d** and **2d** are the bait substrates and substrates **3**–**8c** were used for catalytic profiling. The likely bonds cleaved are represented in red. Black lines indicate that the hits share these two activities. Their width represents how often this type of enzyme promiscuity was observed. The x axis groups substrates by their charges and suggests that the bait activities select hits with consistent promiscuous activity patterns in which substrate charge is a key determinant. R_0 = fluorescein; R_1 = 4-nitrophenyl; R_2 = ethyl; R_3 = methyl; R_4 = glutathione (which has two additional negative charges remote from the reaction centre) R_5 = C$_6$H$_{13}$. See Supplementary Fig. 12 for structures of all substrates.

$> 10^6$ variants are easily accessible at low cost[25] without relying on survival-based assays[51] or substrate-induced regulators (for example, in SIGEX)[52]. The high sensitivity of the miniaturized cell lysate assay allowed the isolation of enzymes with catalytic efficiencies (k_{cat}/K_M) ranging from 54 M^{-1} s^{-1} (P84) to 9×10^5 M^{-1} s^{-1} (P83; Table 1), providing evidence that this microfluidic platform is able to identify slow and fast catalysts, with a wide dynamic range extending over four orders of magnitude. Our protocol is straightforward as the readout is directly reporting on enzymatic activity, so that promiscuous activities are accessible. The versatility of microfluidic workflows[53] makes our platform adaptable for different challenges: to further increase our assay sensitivity (or diminish incubation time) bacterial growth in droplets could be implemented after cell encapsulation, by implementation of an additional microfluidic injection (or droplet fusion) step to supply lysis agents and substrates. Furthermore, the microfluidic platform is readily applicable to other enzymatic activities, as long as a fluorogenic assay can be implemented.

Most (if not all) triesterase hits identified in this work must be presumed not to have evolved in response to an environmental OP triester challenge. Promiscuous activities can be recruited during evolution to give rise to new protein functions[8,9,15]. Our hits may therefore constitute alternative starting points for adaptive evolution in OP-contaminated environments giving rise to different OP-degrading enzymes than currently known. The systematic detection of such catalytic starting points can form the basis of evolutionary models that account for the adaptive potential of a microbial community to degrade exogenous compounds, such as pesticides and antibiotics. Likewise, directed evolution in the laboratory[10] can use hits with low promiscuous activities as starting points that can be enhanced to generate new efficient catalysts.

In contrast to methods for active site mapping that rely on single turnover suicide inhibition[54], the droplet format enables detection of multiple turnovers and indeed all catalysts selected show $> 10^3$ turnovers per enzyme molecule. As demonstrated here, weak or promiscuous activities are directly accessible by droplet screening. This is a powerful method to identify new enzymes for further improvement by protein engineering[55] and also new entry points into networks of promiscuous activities[56], for which there is currently no systematic prediction tool in bioinformatics. The choice of the bait substrate alone determines which catalytic processes are identified. The ultrahigh throughput and the high sensitivity of our method make microfluidic droplets a powerful format to explore this 'promiscuome' (i.e. the collection of promiscuous activities from complex meta-proteomes), where other indirect methods fail. Additional optical detection modes and systematic exploration of libraries with assays for new substrates will be useful to exhaustively harvest many different classes of enzymes. Sensitive, fast and affordable microfluidic droplet screening is now validated as an alternative technology for enzyme discovery that can yield information not accessible by other approaches.

Methods

Metagenomic material. Environmental libraries from various soils[57] and vanilla pods were constructed[58] starting with shearing eDNA using a nebulizer and blunting it using Klenow fragment. Next, the resulting 3–5 kbp fragments were cloned into the EcoRV restriction site in the high-copy plasmid pZero-2 (Invitrogen). DNA from cow rumen was partially digested with the restriction enzyme MluCI. DNA fragments with sizes around 3 kbp were isolated by gel electrophoresis and cloned into the EcoRI site of the same high-copy plasmid pZero-2. The resulting ten libraries (Supplementary Table 1) were pooled together, constituting a library of about 1,250,000 variants, called the 'SCV library'.

Chip design and microfluidic device fabrication. The designs for the poly(dimethyl)siloxane (PDMS) chip devices were prepared using Autocad CAD software and designs are shown in Supplementary Fig. 4. The corresponding CAD files can be downloaded from http://openwetware.org/wiki/DropBase (a database of microfluidic device designs). The devices were fabricated with standard soft lithographic procedures[59]. The photoresist material SU-8 2015 was used to obtain a 15 μm channel height. PDMS monomer and curing agent were mixed at a ratio 10:1 and then poured onto the lithographic plate before degassing. After PDMS solidification (65 °C, 4 h), PDMS was activated by exposure to an oxygen plasma and devices were sealed onto a microscope glass slide (or cover slip (thickness: 0.13 mm) for the sorting chip). Hydrophobic modification of the channels surface was achieved by injecting a solution of 1% (v/v) trichloro(1H,1H,2H,2H-perfluorooctyl)silane (Sigma) in HFE-7500 oil into the channels. Electrodes for the sorting devices were prepared using low melting point indium composite solder (51 In/32.5 Bi/16.5 Sn, Indium Corporation).

Preparation of bacterial suspensions for droplet encapsulation. Transformation of ∼10 ng of the metagenomic library SCV into E. coli (E. cloni 10G Elite, Lucigen) yielded ∼10[8] variants on agar plate (containing 40 μg ml[−1] kanamycin) covering the library size ∼100 times. Bacteria were grown overnight at 37 °C, then incubated at 22 °C for 2 days. Colonies were subsequently scraped from the agar plates, washed, filtered using a 5-μm filter (Sartorius) and resuspended in MOPS (100 mM, pH 8.0) containing NaCl (115 mM), kanamycin (40 μg ml[−1]), complete EDTA-free protease inhibitor (one tablet per 50 ml; Roche) and Percoll (25% v/v; Sigma). The cell density was adjusted by dilution to obtain the required cell/droplet ratio after compartmentalization. Assuming a Poisson distribution for bacterial encapsulation[60], a cell density $OD_{600\,nm} \sim 1$ should result in ∼35% of droplets with single cells and ∼20% of droplets with higher occupancy (for ∅ 15 μm droplets).

Generation of water-in-oil picolitre droplets. Water-in-oil droplets (volume ∼2 pl, ∅: 15 μm) were generated using a flow-focusing device (Fig. 1) (dimensions width × height of 16 × 15 μm) bearing three inlets. Two inlets carry aqueous solutions prepared in MOPS buffer (100 mM, pH 8.0), NaCl (115 mM), kanamycin (40 μg ml[−1]), EDTA-free protease inhibitor (one tablet per 50 ml; Roche). The streams from these inlets supplied (i) a cell suspension ($OD_{600\,nm}$ depending on cell occupancy (i.e. number of cell per droplets) required) in Percoll (25%, v/v, Sigma) and (ii) a mixture of the cell lysis reagents BugBuster (20% v/v, Novagen) and lysozyme (30 kU ml[−1]; Novagen), as well as the respective substrate (10 μM of sulfate monoester **1d** or phosphate triester **2d**). From the third inlet, fluorinated oil HFE-7500 (3 M) containing EA surfactant (1%, w/w, RainDance Technologies) was injected. Aqueous solutions and the oil phase were injected using plastic syringes (BD; 1 or 3 ml) at a rate of 50 and 500 μl h[−1], respectively, with PHD 2000 Harvard Apparatus pumps.

Droplet storage and incubation. Droplets were stored in a glass syringe (500 μl, SGE) in HFE-7500 (3 M) with EA surfactant (1%, w/w), covered with mineral oil (Sigma). The fluorous phase containing droplets is not miscible with mineral oil and droplets remain at the interface between HFE-7500 and mineral oil. Droplets were incubated from 1 to up to 3 days in the syringe at room temperature (∼22 °C).

Droplet sorting and electronics. After mounting the incubation syringe vertically on a syringe pump (PHD 2000 Harvard Apparatus), water-in-oil droplets were re-injected into the sorting device at a rate of 30 μl h[−1]. The distance between droplets was increased to facilitate the sorting of single droplets by electric pulses. To this end, additional fluorous oil (HFE-7500 (3 M) with EA surfactant (0.5%, w/w) was injected at 300 μl h[−1] in the sorting chip. This set-up with a parallel channel design (with a width ratio of 1.3 between waste and positive channels, Supplementary Fig. 4) resulted in a droplet sorting rate of 2–2.5 kHz, without the need for pressure equilibration between the two outlets as previously required[27]. A 488-nm laser was focused 180 μm upstream of the sorting junction through a × 40 microscope objective (UPlanFLN, Olympus) for fluorophore excitation and the emitted fluorescent light was collected and amplified using photomultiplier tubes (H8249, Hamamatsu Photonics). The amplified fluorescence signal was processed by a data acquisition card operating at 38 kHz (National Instruments, USB-6009) that was linked to a peak detection algorithm, which recorded fluorescence distributions (LabView 8.2, National Instrument). Hardware triggering was implemented via a voltage comparator (LM339N, Texas Instruments), which compared the voltage readout by the photomultiplier tube with a user-defined arbitrary voltage generated via the acquisition card and doubled using an operational amplifier (LM358N, STMicroelectronics), to generate voltages between 0 and 10 V. A pull-up resistor (1 kΩ) was used to force the logical high state of the comparator to 5 V and send the trigger signal to the pulse generator. Whenever the fluorescence peak reached a user-defined voltage threshold (typically corresponding to two- to fivefold increase over the average fluorescence of droplets not containing active enzymes), a pulse generator triggered a single square pulse of 500 μs length and an amplitude of 0.6–0.8 V_p. This pulse was amplified 1,000-fold by a high voltage amplifier (610E, Trek) and applied to the electrodes on the sorting device. With the current electronics implementation, 500 μs was found to be the minimal pulse length able to trigger the fast camera used as a control, to monitor the success of sorting. The electric pulse applied between the two electrodes dielectrophoretically attracted the highly fluorescent droplet towards the narrower channel (Fig. 1 and Supplementary Fig. 5). The sorting events were recorded with a fast camera (Phantom V7.2) that was triggered by the voltage comparator, to allow analysis of whether the desired droplets with increased fluorescence were selected. Optical inspection of the movies thus recorded monitored that only single droplets were selected for each pulse.

Detection of useful hits in this format was only possible after improvements to the sensitivity of the previously described miniaturized cell lysate assay[27]. This was achieved by adapting our microfluidic platform to generate and sort 2 pl droplet compartments (∅ 15 μm). To test whether smaller droplets led to higher sensitivity, we measured the accumulation of fluorescein over time in two droplet populations with different volumes (2 and 8 pl) containing single bacteria transformed with a metagenomic variant active towards sulfate monoester **1d** (Fig. 1). Fluorescence signal change (average fluorescence of droplets containing bacteria divided by average fluorescence of empty droplets) increased approximately twice as fast in 2 pl droplets than in 8 pl droplets (Supplementary Fig. 1), confirming that sensitivity was improved by scaling down the droplet size.

DNA recovery. Plasmid DNA from fluorescent droplets was recovered by de-emulsification using 1H,1H,2H,2H-perfluorooctanol[27]. The use of a high copy number plasmid (pZero-2; > 1,000 copies per cell) in an E. coli endonuclease knockout strain (E. cloni 10G, Lucigen, bearing the mutation *endA1*) allowed plasmid DNA recovery from droplets even after 2 days of incubation, without additional PCR amplification.

Quantification of enrichment of hits by quantitative PCR. To quantify the total number of plasmids from the library, a set of primers annealing to the vector was used:

   5′-TTTCTGCGGACTGGCTTT-3′ (qPZeroFwd)
   5′-ACAGGATTAGCAGAGCGAGG-3′ (qPZeroRev).

To quantify plasmids containing PC86 (as a representative metagenomic hit), primers annealing to the inserted PC86 sequence were designed:

   5′-ATACCGACGAAGCCCTGT-3′ (qPC86Fwd)
   5′-TCGGCAGGGTCATACACATA-3′ (qPC86Rev).

All primers were supplied by Invitrogen Life Technologies. Quantitative real-time PCR experiments (see also Supplementary Fig. 10) were performed in triplicate using Sensimix SYBR green (Bioline) in the Rotor-Gene 6000 (Corbett Life Sciences). The PCR conditions were as follows: initial DNA denaturation at 95 °C for 10 min, 40 cycles (95 °C for 10 s; 52 °C for 15 s; 72 °C for 20 s) and a temperature gradient enabling determination of the DNA melting temperature (between 72 °C and 95 °C). Reference curves using both sets of primers were obtained with correlation coefficients $R^2 > 0.99$.

Metagenomic screening on plates. Metagenomic libraries (ENR-S, ENR-G, ENR-M and ENR-L; Supplementary Table 1) were transformed into electrocompetent E. coli (E. cloni 10G Elite, Lucigen) and ~100,000 clones were plated on 10 different ∅ 14 cm Petri dishes with Luria Bertani (LB) Agar (1.5%) containing 40 µg ml^{-1} kanamycin. After overnight growth at 37 °C, plates were incubated at 22 °C for 24 h. Colonies were then transferred onto nitrocellulose membranes (Pall Corporation) and lysed by three cycles of 10 min incubation at − 20 °C and 37 °C before being overlaid with 100 mM Tris-HCl pH 8.0, 0.5% agarose (w/v) containing 135 µM of sulfate monoester **1c** (Supplementary Fig. 12). Colonies that turned blue after ~15 h of incubation at 25 °C were isolated and their plasmid DNA extracted using a miniprep kit (Zymo Research) and sequenced.

Fidelity of droplet sorting. We probed whether our miniaturized cell lysate screen in droplets was able to recover hits from metagenomic libraries. Thus, a subset of the library (~ 100,000 clones) (Supplementary Table 1) was screened for sulfate hydrolases in two experiments: (i) in microfluidic droplets using sulfate monoester **1d** (> 10^7 droplets in 100 µl, protocol shown in Fig. 1) and (ii) in a classic colony screening procedure[61] (400,000 colonies using 40 Petri dishes) using sulfate monoester **1c** as a substrate. When the two screens were compared, all hits found on plates were recovered within 0.5 h by droplet screening (Supplementary Figs 3 and 4), except one that was later isolated when the ENR-G library was screened alone. The ability of our system to detect and sort hits was further addressed by analysing the relationship between hit rate and sample size. When covering the library size with a variety of oversampling ratios (by screening a number of droplets corresponding to 2 ×, 10 ×, 20 × and 25 × the library size) we observed an increasing number of recovered model hits (Supplementary Fig. 4), suggesting that the increase in screening capacity indeed leads to more hits (assuming a library contains them). We found that a > 10-fold oversampling of the library is sufficient to recover every hit confidently, thereby achieving perfect coverage (Supplementary Fig. 4).

Protein production and purification. Open reading frames coding for sulfatase P35 and the sulfotransferase P40 were recloned into a modified expression vector pRSFDuet (Novagen) using the restriction sites NdeI–XhoI (P35) and NcoI–XhoI (P40). Recombinant plasmids pRSFDuetP35 and pRSFDuetP40 were transformed into E. coli BL21(DE3). For protein expression transformants were grown in 750 ml LB broth (containing 40 µg ml^{-1} kanamycin) at 37 °C until an OD$_{600\,nm}$ ~ 0.5 was reached. At this stage, expression was induced with isopropyl-β-D-thiogalactoside (1 mM) for 15 h at 25 °C. Cells were harvested by centrifugation, resuspended in LB medium (30 ml) and lysed by sonication. Cell lysate was obtained by centrifugation (30,000g, 1 h, 4 °C) and diluted in a 1:1 ratio with Tris-HCl (50 mM, pH 8.0). All subsequent steps were carried out in this buffer, unless stated otherwise. The desired enzyme was purified from diluted cell lysates using a sequence of three columns as follows: (i) anion exchange (Q-sepharose Fast Flow; GE Healthcare) using NaCl gradients 0-400 mM (P35) or 0-1 M (P40), (ii) affinity chromatography (P-sepharose Fast Flow; GE Healthcare) using (NH$_4$)$_2$SO$_4$ gradients 1-0 M (P35) or 500-0 M (P40), and (iii) gel filtration (Superdex 200) running in a Tris-HCl buffer (20 mM, pH 8.0). Chromatographic steps of the purification were carried out in a AKTA FPLC apparatus (GE Healthcare).

Open reading frames coding for PTEs were recloned into the pASKIBA5plus plasmid (Iba Life) using the BsaI restriction site (P83, P84, P85, P87, P881, P90 and P91) or EcoRI–NcoI restriction sites (P86 and P882). Recombinant plasmids were transformed by electroporation into E. coli BL21(DE3). Cells were grown at 37 °C in LB containing 100 µg ml^{-1} ampicillin (750 ml) until OD$_{600\,nm}$ ~ 0.5. The expression of the amino-terminal Strep-tagged proteins was induced with anhydrotetracyclin (200 µg l^{-1}) at 25 °C for 15 h. Cells were harvested by centrifugation and resuspended in Tris-HCl (30 ml of a 100 mM solution, pH 8.0) containing NaCl (150 mM) before cell lysis by sonication. The lysate was centrifuged (30,000g, 1 h, 4 °C) and the extract was directly loaded onto Strep-Tactin Superflow resin (Iba Life) that was previously equilibrated with Tris-HCl (100 mM, pH 8.0, containing 150 mM NaCl). Washing steps were performed using Tris-HCl (100 mM, pH 8.0), NaCl (150 mM) and Strep-tagged proteins were eluted in Tris-HCl buffer (100 mM, pH 8.0, containing 150 mM NaCl and 2.5 mM d-desthiobiotin (d-biotin)). Columns were regenerated using Tris-HCl buffer (100 mM, pH 8.0; containing 150 mM NaCl, 1 mM EDTA and 1 mM 2-(4-hydroxyphenyl-azo)benzoic acid). Eluted proteins were concentrated to a final volume of 2 ml and further purified by gel filtration (Superdex 200, in 20 mM Tris-HCl pH 8.0).

Mutant P91 C118 was purified in the same way as wild-type P91, but mutants P91 D167N and H199A were prepared using Strep-Tactin spin columns (Iba Life). Mutant-encoding plasmids were transformed by electroporation into E. coli BL21(DE3). Small volume cultures (15 ml) were grown under the same conditions as described above. Cells were pelleted and lysed using 500 µl of a solution containing 1 × Bugbuster (Novagen) and 0.001 × Lysonase (Novagen) in MilliQ water (Millipore) before addition of Tris-HCl buffer (1 ml, 100 mM, pH 8.0; containing 150 mM NaCl) and centrifugation to remove cell debris. After equilibration using Tris-HCl buffer, Strep-Tactin columns were loaded with the cleared cell lysates and centrifuged for 30 s at 700 g (at 4 °C). After a washing step to remove weakly bound proteins, P91 mutants were eluted in Tris-HCl (150 µl of 100 mM, pH 8.0), containing 150 mM NaCl and 2 mM d-biotin. To remove

d-biotin, the elutions were buffer exchanged by successive concentration–dilution cycles using 1 ml Amicon 10 k concentrator columns.

Substrates. Sulfate monoester **1d**, phosphotriester **2d**, phosphate diester **4** and phosphonate **5** were synthetized from the respective chlorides or chloridates[62–64]. Sulfate monoester **1c** was purchased from Fluka. Sulfate monoester **1a**, phosphate monoester **3**, acetate ester **6**, thioester **7**, lactones **8a**, **8b**, **8c**, **8d$_1$**, **8d$_2$** and acetamide **9** were purchased from Sigma-Aldrich. Chromogenic β-lactamase substrate **10** (CENTA) was purchased from Merck Millipore.

Enzyme Assays. All enzymatic assays were conducted at 25 °C in a final volume of 200 µl, in the activity buffer (100 mM MOPS-NaOH pH 8.0, containing 150 mM NaCl) that was used throughout the experiments, unless otherwise stated. Measurements were performed in 96-well plate format and product formation followed in SpectraMax M5 or 190 microplate readers. Hydrolysis of substrates with fluorescein leaving groups (**1d** and **2d**) was monitored at the following wavelengths: $\lambda_{excitation} = 488$ nm and $\lambda_{emission} = 525$ nm. A calibration curve (linearly fit to the equation: fluorescence = 1.2×10^{10} M^{-1} [fluorescein]) was used to quantify fluorescein release. Hydrolysis of esters **1**, **2a**, **3**, **4**, **5**, **6** and acetamide **9** was monitored by measuring absorbance of p-nitrophenolate at 400 nm ($\varepsilon \approx 17,700$ M^{-1} cm^{-1}). Hydrolysis of thioester **7** was measured in the presence of the indicator 5,5'-dithiobis-(2-nitrobenzoic acid) (500 µM); the increase of 2-nitro-5-thiobenzoate was monitored at 412 nm ($\varepsilon \approx 13,100$ M^{-1} cm^{-1})[39]. Hydrolysis of lactones **8b**, **8c**, **8d$_1$**, **8d$_2$** was measured using a pH-shift assay in 2.5 mM Bicine, 200 mM NaCl, 0.3 mM cresol purple pH 8.3 (ref. 65). The pH drop was monitored by the decrease in the absorbance of the indicator dye cresol purple at 577 nm ($\varepsilon \approx 4,000$ M^{-1} cm^{-1}). Hydrolysis of lactone **8a** was measured at 270 nm in an ultraviolet-transparent 96-well plate. Hydrolysis of β-lactam **10** was measured at 405 nm in phosphate buffer (100 mM, pH 8.0). All kinetic parameters were calculated by fitting initial rates v_o to the Michaelis–Menten equations (or, if saturation could not be reached due to solubility limits, to linear correlations) using Kaleidagraph. Example data and details of equations used for fitting are shown in Supplementary Fig. 14.

Sequence similarity networks. All protein sequences from superfamilies (AH: Pfam CL0034; MBLs: Pfam PF00753; α/β: Pfam PF07859, PF00135 and PF01738) were retrieved from the EMBL-EBI Pfam database[36]. To limit the number of sequences and to reduce eventual bias that could induce too many homologous sequences in a network, a cutoff on the sequence identity was applied using CD-HIT[66] (40% for AH and MBL, and 30% for α/β). Our metagenomic hits protein sequences and several known characterized members of each protein families were subsequently added to the protein lists to help network functional annotations. The data sets were composed of 5,042 (AH), 2,984 (MBL) and 1,345 (α/β) sequences. Each data set was subjected to an all-versus-all BLAST (National Center for Biotechnology Information, version 2.2.29 +), considering sequence similarity only when alignment score was below an appropriate threshold (E-values: $1 \times e^{-9}$ (AH), $1 \times e^{-14}$ (MBL) and $1 \times e^{-19}$ (α/β)). The sequence similarity scores (E-values) were then imported into Cytoscape (version 3.0.2) and networks were constructed using the organic layout (in which the lengths of the edges correlate with the dissimilarity of the connected sequences (represented by nodes)). Functional annotations were then imported from the Uniprot database and mapped onto the networks. Recently, the process of generating SSNs from any Pfam families has been made much more convenient by the Enzyme Function Initiative: (http://efi.igb.illinois.edu/efi-est/)[67].

Metal removal. P83, P85, P86, P87, P88.1, P88.2, P90 and P91 were incubated at 4 °C in MOPS (100 mM, pH 8) containing NaCl (150 mM) and the chelators EDTA, (25 mM), pyridine-2,6-dicarboxylic acid (25 mM) and phenanthroline (2 mM). After 4 days of incubation, chelating agents were removed by successive centrifugations using centrifugation filter tubes (Amicon Ultra-2 ml, 10 kDa; Millipore) and washed with MOPS buffer (as above). To restore divalent metals, the samples were incubated at 25 °C with MnCl$_2$ (200 µM) and activity towards phosphate triester **2a** (800 µM) was measured by monitoring the increase of 4-nitrophenolate absorbance at 400 nm. Samples that were subjected to the metal removal procedure, without restoring the metal, were used as controls to assay any residual activity in the apo-enzyme.

Crystallization and structure determination. Crystallization conditions were screened in 96-well plates using the sitting-drop vapour diffusion method at 292 K. Crystals of P91 appeared after 36 h in a condition of 0.2 µl P91 (10 mg ml^{-1}) and 0.2 µl reservoir solution (0.1 M Tris pH 8.5, 0.2 M MgCl$_2$ and 20% (w/v) PEG 8,000). Diffraction data were collected from a single crystal at 100 K at the Swiss Light Source (beamline X06DA) at 0.9188 Å. Data were processed using autoPROC/XDS/AIMLESS software[68]. As the crystal suffered significant radiation damage, affected images were removed from processing, leading to a slightly reduced completeness of 87%—the observed electron density however is of good quality and continuous. Phases were obtained through molecular replacement using BALBES/Phaser[69], which used a putative DLH from Klebsiella pneumoniae

(PDB ID 3F67) as the search model. Two molecules of P91 were found in the asymmetric unit. Iterative structure refinement was performed using Refmac5 (ref. 70) from the CCP4 suite[71] and Coot[72]. 97.7% of residues are found within the Ramachandran-favoured, 2.3% in the allowed and none in the disallowed regions. Loop 74–80 (in both monomers) shows higher flexibility than the rest of the protein and multiple conformations that could not be modelled accurately. Full data collection and refinement statistics are shown in Supplementary Table 8. The structure has been deposited with PDB code 4ZI5.

Site-directed mutagenesis. P91 nucleophile mutant C118A was constructed by overlap extension PCR and cloned into the pASKIBA5plus (Iba Life) using the BsaI restriction site. P91 mutants D167N and H199A were constructed using the QuikChange protocol (Stratagene) with pASKIBA5plus-P91 as DNA template.

References

1. Bornscheuer, U. T. *et al.* Engineering the third wave of biocatalysis. *Nature* **485**, 185–194 (2012).
2. Daniel, R. The metagenomics of soil. *Nat. Rev. Microbiol.* **3**, 470–478 (2005).
3. Amann, R. I., Ludwig, W. & Schleifer, K.-H. Phylogenetic identification and in situ detection of individual microbial cells without cultivation. *Microbiol. Rev.* **59**, 143–169 (1995).
4. Venter, J. C. *et al.* Environmental genome shotgun sequencing of the Sargasso Sea. *Science* **304**, 66–74 (2004).
5. Harrington, E. D. *et al.* Quantitative assessment of protein function prediction from metagenomics shotgun sequences. *Proc. Natl Acad. Sci. USA* **104**, 13913–13918 (2007).
6. Schnoes, A. M., Brown, S. D., Dodevski, I. & Babbitt, P. C. Annotation error in public databases: misannotation of molecular function in enzyme superfamilies. *PLoS Comput. Biol.* **5**, e1000605 (2009).
7. Galperin, M. Y. & Koonin, E. V. From complete genome sequence to 'complete' understanding? *Trends Biotechnol.* **28**, 398–406 (2010).
8. Jensen, R. A. Enzyme recruitment in evolution of new function. *Annu. Rev. Microbiol.* **30**, 409–425 (1976).
9. O'Brien, P. J. & Herschlag, D. Catalytic promiscuity and the evolution of new enzymatic activities. *Chem. Biol.* **6**, 91–105 (1999).
10. Nobeli, I., Favia, A. D. & Thornton, J. M. Protein promiscuity and its implications for biotechnology. *Nat. Biotechnol.* **27**, 157–167 (2009).
11. Notebaart, R. A. *et al.* Network-level architecture and the evolutionary potential of underground metabolism. *Proc. Natl Acad. Sci. USA* **111**, 11762–11767 (2014).
12. Bastard, K. *et al.* Revealing the hidden functional diversity of an enzyme family. *Nat. Chem. Biol.* **10**, 42–49 (2014).
13. Huang, H. *et al.* Panoramic view of a superfamily of phosphatases through substrate profiling. *Proc. Natl Acad. Sci. USA* **112**, E1974–E1983 (2015).
14. Baier, F. & Tokuriki, N. Connectivity between catalytic landscapes of the metallo-beta-lactamase superfamily. *J. Mol. Biol.* **426**, 2442–2456 (2014).
15. Khersonsky, O. & Tawfik, D. S. Enzyme promiscuity: a mechanistic and evolutionary perspective. *Annu. Rev. Biochem.* **79**, 471–505 (2010).
16. Handelsman, J. Metagenomics: application of genomics to uncultured microorganisms. *Microbiol. Mol. Biol. Rev.* **68**, 669–685 (2004).
17. Gabor, E. M., Alkema, W. B. & Janssen, D. B. Quantifying the accessibility of the metagenome by random expression cloning techniques. *Environ. Microbiol.* **6**, 879–886 (2004).
18. Lorenz, P. & Eck, J. Metagenomics and industrial applications. *Nat. Rev. Microbiol.* **3**, 510–516 (2005).
19. Uchiyama, T. & Miyazaki, K. Functional metagenomics for enzyme discovery: challenges to efficient screening. *Curr. Opin. Biotechnol.* **20**, 616–622 (2009).
20. Lafferty, M. & Dycaico, M. J. GigaMatrix™: an ultra high-throughput tool for accessing biodiversity. *J. Lab. Autom.* **9**, 200–208 (2004).
21. Ferrer, M. *et al.* Interplay of metagenomics and in vitro compartmentalization. *Microb. Biotechnol.* **2**, 31–39 (2009).
22. Schaerli, Y. & Hollfelder, F. The potential of microfluidic water-in-oil droplets in experimental biology. *Mol. Biosyst.* **5**, 1392–1404 (2009).
23. Theberge, A. B. *et al.* Microdroplets in microfluidics: an evolving platform for discoveries in chemistry and biology. *Angew. Chem. Int. Ed. Engl.* **49**, 5846–5868 (2010).
24. Baret, J. C. *et al.* Fluorescence-activated droplet sorting (FADS): efficient microfluidic cell sorting based on enzymatic activity. *Lab. Chip.* **9**, 1850–1858 (2009).
25. Agresti, J. J. *et al.* Ultrahigh-throughput screening in drop-based microfluidics for directed evolution. *Proc. Natl Acad. Sci. USA* **107**, 4004–4009 (2010).
26. Colin, P. Y., Zinchenko, A. & Hollfelder, F. Enzyme engineering in biomimetic compartments. *Curr. Opin. Struct. Biol.* **33**, 42–51 (2015).
27. Kintses, B. *et al.* Picoliter cell lysate assays in microfluidic droplet compartments for directed enzyme evolution. *Chem. Biol.* **19**, 1001–1009 (2012).
28. Sjostrom, S. L. *et al.* High-throughput screening for industrial enzyme production hosts by droplet microfluidics. *Lab. Chip.* **14**, 806–813 (2014).
29. Wang, B. L. *et al.* Microfluidic high-throughput culturing of single cells for selection based on extracellular metabolite production or consumption. *Nat. Biotechnol.* **32**, 473–478 (2014).
30. Najah, M. *et al.* Droplet-based microfluidics platform for ultra-high-throughput bioprospecting of cellulolytic microorganisms. *Chem. Biol.* **21**, 1722–1732 (2014).
31. Taupp, M., Mewis, K. & Hallam, S. J. The art and design of functional metagenomic screens. *Curr. Opin. Biotechnol.* **22**, 465–472 (2011).
32. Edwards, D. R., Lohman, D. C. & Wolfenden, R. Catalytic proficiency: the extreme case of S-O cleaving sulfatases. *J. Am. Chem. Soc.* **134**, 525–531 (2012).
33. Hassall, K. A. (ed.) *The Biochemistry and Uses of Pesticides* (VCH, Weinheim, 1990).
34. Scanlon, T. C., Dostal, S. M. & Griswold, K. E. A high-throughput screen for antibiotic drug discovery. *Biotechnol. Bioeng.* **111**, 232–243 (2014).
35. Zinchenko, A. *et al.* One in a million: flow cytometric sorting of single cell-lysate assays in monodisperse picolitre double emulsion droplets for directed evolution. *Anal. Chem.* **86**, 2526–2533 (2014).
36. Finn, R. D. *et al.* Pfam: the protein families database. *Nucleic Acids Res.* **42**, D222–D230 (2014).
37. Malojcic, G. *et al.* A structural and biochemical basis for PAPS-independent sulfuryl transfer by aryl sulfotransferase from uropathogenic *Escherichia coli*. *Proc. Natl Acad. Sci. USA* **105**, 19217–19222 (2008).
38. Atkinson, H. J., Morris, J. H., Ferrin, T. E. & Babbitt, P. C. Using sequence similarity networks for visualization of relationships across diverse protein superfamilies. *PLoS ONE* **4**, e4345 (2009).
39. Stamp, A. L. *et al.* Structural and functional characterization of *Salmonella enterica* serovar Typhimurium YcbL: an unusual type II glyoxalase. *Protein Sci.* **19**, 1897–1905 (2010).
40. Daiyasu, H., Osaka, K., Ishino, Y. & Toh, H. Expansion of the zinc metallo-hydrolase family of the β-lactamase fold. *FEBS Lett.* **503**, 1–6 (2001).
41. Heikinheimo, P., Goldman, A., Jeffries, C. & Ollis, D. L. Of barn owls and bankers: a lush variety of a/b hydrolases. *Structure* **7**, R141–146 (1999).
42. Bigley, A. N. & Raushel, F. M. Catalytic mechanisms for phosphotriesterases. *Biochim. Biophys. Acta* **1834**, 443–453 (2013).
43. Kim, H. K., Liu, J. W., Carr, P. D. & Ollis, D. L. Following directed evolution with crystallography: structural changes observed in changing the substrate specificity of dienelactone hydrolase. *Acta Crystallogr. Sect. D Biol. Crystallogr.* **61**, 920–931 (2005).
44. Cheah, E., Austin, C., Ashley, G. W. & Ollis, D. L. Substrate-induced activation of dienelactone hydrolase: an enzyme with a naturally occurring Cys-His—Asp triad. *Protein Eng.* **6**, 575–583 (1993).
45. Russell, R. J. *et al.* The evolution of new enzyme function: lessons from xenobiotic metabolizing bacteria versus insecticide-resistant insects. *Evol. Appl.* **4**, 225–248 (2011).
46. Jackson, C. J. *et al.* Structure and function of an insect alpha-carboxylesterase (alphaEsterase7) associated with insecticide resistance. *Proc. Natl Acad. Sci. USA* **110**, 10177–10182 (2013).
47. Singh, B. K. Organophosphorus-degrading bacteria: ecology and industrial applications. *Nat. Rev. Microbiol.* **7**, 156–164 (2009).
48. Afriat, L., Roodveldt, C., Manco, G. & Tawfik, D. S. The latent promiscuity of newly identified microbial lactonases is linked to a recently diverged phosphotriesterase. *Biochemistry* **45**, 13677–13686 (2006).
49. Elias, M. & Tawfik, D. S. Divergence and convergence in enzyme evolution: parallel evolution of paraoxonases from quorum-quenching lactonases. *J. Biol. Chem.* **287**, 11–20 (2012).
50. Gabor, E., Niehaus, F., Aehle, W. & Eck, J. Zooming in on metagenomics: molecular microdiversity of Subtilisin Carlsberg in soil. *J. Mol. Biol.* **418**, 16–20 (2012).
51. Sommer, M. O., Dantas, G. & Church, G. M. Functional characterization of the antibiotic resistance reservoir in the human microflora. *Science* **325**, 1128–1131 (2009).
52. Uchiyama, T., Abe, T., Ikemura, T. & Watanabe, K. Substrate-induced gene-expression screening of environmental metagenome libraries for isolation of catabolic genes. *Nat. Biotechnol.* **23**, 88–93 (2005).
53. Kintses, B., van Vliet, L. D., Devenish, S. R. & Hollfelder, F. Microfluidic droplets: new integrated workflows for biological experiments. *Curr. Opin. Chem. Biol.* **14**, 548–555 (2010).
54. Adam, G. C., Burbaum, J., Kozarich, J. W., Patricelli, M. P. & Cravatt, B. J. Mapping enzyme active sites in complex proteomes. *J. Am. Chem. Soc.* **126**, 1363–1368 (2004).
55. Schulenburg, C. & Miller, B. G. Enzyme recruitment and its role in metabolic expansion. *Biochemistry* **53**, 836–845 (2014).
56. Mohamed, M. F. & Hollfelder, F. Efficient, crosswise catalytic promiscuity among enzymes that catalyze phosphoryl transfer. *Biochim. Biophys. Acta* **1834**, 417–424 (2013).
57. Gabor, E. M., de Vries, E. J. & Janssen, D. B. Construction, characterization, and use of small-insert gene banks of DNA isolated from soil and enrichment

cultures for the recovery of novel amidases. *Environ. Microbiol.* **6**, 948–958 (2004).

58. Gabor, E. M., Vries, E. J. & Janssen, D. B. Efficient recovery of environmental DNA for expression cloning by indirect extraction methods. *FEMS Microbiol. Ecol.* **44**, 153–163 (2003).

59. McDonald, J. C. *et al.* Fabrication of microfluidic systems in poly(dimethylsiloxane). *Electrophoresis* **21**, 27–40 (2000).

60. Koster, S. *et al.* Drop-based microfluidic devices for encapsulation of single cells. *Lab. Chip.* **8**, 1110–1115 (2008).

61. Kaltenbach, M., Jackson, C. J., Campbell, E. C., Hollfelder, F. & Tokuriki, N. Reverse evolution leads to genotypic incompatibility despite functional and active site convergence. *eLife* **4**, e06492 (2015).

62. Fischlechner, M. *et al.* Evolution of enzyme catalysts caged in biomimetic gel-shed beads. *Nat. Chem.* **6**, 791–796 (2014).

63. Hendry, P. & Sargeson, A. M. Metal ion promoted phosphate ester hydrolysis. Intramolecular attack of coordinated hydroxide ion. *J. Am. Chem. Soc.* **111**, 2521–2527 (1989).

64. Scheigetz, J., Gilbert, M. & Zamboni, R. Synthesis of fluorescein phosphates and sulfates. *Org. Prep. Proc. Int.* **29**, 561–568 (1997).

65. Chapman, E. & Wong, C.-H. A pH sensitive colorimetric assay for the high-throughput screening of enzyme inhibitors and substrates: a case study using kinases. *Bioorg. Med. Chem.* **10**, 551–555 (2002).

66. Huang, Y., Niu, B., Gao, Y., Fu, L. & Li, W. CD-HIT Suite: a web server for clustering and comparing biological sequences. *Bioinformatics* **26**, 680–682 (2010).

67. Gerlt, J. A. *et al.* Enzyme Function Initiative-Enzyme Similarity Tool (EFI-EST): A web tool for generating protein sequence similarity networks. *Biochim. Biophys. Acta* **854**, 1019–1037 (2015).

68. Vonrhein, C. *et al.* Data processing and analysis with the autoPROC toolbox. *Acta Crystallogr. Sect. D Biol. Crystallogr.* **67**, 293–302 (2011).

69. McCoy, A. J. *et al.* Phaser crystallographic software. *J. Appl. Crystallogr.* **40**, 658–674 (2007).

70. Murshudov, G. N., Vagin, A. A. & Dodson, E. J. Refinement of macromolecular structures by the maximum-likelihood metho. *Acta Crystallogr. Sect. D Biol. Crystallogr.* **53**, 240–255 (1997).

71. CCP4. The CCP4 suite: programs for protein crystallography. *Acta Crystallogr. Sect. D Biol. Crystallogr.* **50**, 760–763 (1994).

72. Emsley, P. & Cowtan, K. Coot: model-building tools for molecular graphics. *Acta Crystallogr. Sect. D Biol. Crystallogr.* **60**, 2126–2132 (2004).

Acknowledgements

This research was funded by the Engineering and Physical Sciences Research Council (EPSRC) and the Biological and Biotechnological Research Council (BBSRC). F.H. is an ERC Starting Investigator. P.-Y.C. holds a fellowship of the EU ITN PhosChemRec, C.M.M. of the ITN ProSA and ENEFP. B.K. and M.F.M. were supported by postdoctoral Marie-Curie fellowships. We thank Sean Devenish and Nobuhiko Tokuriki for useful comments on the manuscript, and Gabrielle Potocki-Veronese and Stephane Emond for help with the design of the library strategy. We thank Raindance for the gift of EA surfactant. We thank the Paul Scherrer Institut, Villigen, Switzerland, for the provision of beamtime at beamline X06SA at the Swiss Light Source for crystallographic data collection.

Author contributions

P.-Y.C. conducted and evaluated metagenomic screenings, characterized the hits, measured kinetics and carried out initial crystal screens. B.K. initiated this work, set-up microfluidic droplet sorting, designed, conducted and analysed enrichment experiments and initial sulfatase screenings. F.G. implemented the voltage comparator for sorting, wrote the Labview interface and, with P.-Y.C., redesigned the droplet-sorting module. B.K. and C.M.M. constructed metagenomic libraries. P.-Y.C. and C.M.M. constructed sequence similarity networks. G.F. and M.H. determined and analysed the crystal structure of P91. M.F.M. synthesized enzyme substrates. D.J. and D.P.M. contributed metagenomic libraries. P.-Y.C., B.K., F.G. and F.H. wrote the manuscript incorporating comments from all other authors. FH directed the research.

Additional information

Accession codes: The sequences for all hits were submitted to the NCBI GenBank and can be found with the following accession numbers: KP212134 (*p32*), KP212135 (*p35*), KP212136 (*p40*), KP212137 (*bk1*), KP212138 (*p76*), KP212139 (*p82*), KP212140 (*p83*), KP212141 (*p84*), KP212142 (*p85*), KP212143 (*p86*), KP212144 (*p87*), KP212145 (*p88.1*), KP212146 (*p88.2*), KP212147 (*p90*), KP212148 (*p91*). The structure of P91 is available from the Protein Data Bank (ID: 4ZI5).

Permissions

List of Contributors

Randy P. Carney and Francesco Stellacci
Department of Materials Science and Engineering, Massachusetts Institute of Technology, Cambridge, Massachusetts 02139, USA
Institute of Materials, École Polytechnique Fédérale de Lausanne, Station 12, 1015 Lausanne, Switzerland

Jin Young Kim
Department of Materials Science and Engineering, Massachusetts Institute of Technology, Cambridge, Massachusetts 02139, USA

Huifeng Qian and Rongchao Jin
Department of Chemistry, Carnegie Melon University, Pittsburgh, Pennsylvania 15213, USA

Hakim Mehenni and Osman M. Bakr
Center for Solar and Alternative Energy Science and Engineering, King Abdullah University of Science and Technology, Thuwal 23955-6900, Saudi Arabia
Physical Sciences and Engineering Division, King Abdullah University of Science and Technology, Thuwal 23955-6900, Saudi Arabia

Anton Tadich
Australian Synchrotron, 800 Blackburn Road, Clayton, Victoria 3168, Australia
Department of Physics, La Trobe University, Bundoora, Victoria, Australia. w Present addresses: Center for Nanotechnology & Advanced Materials, Bar Ilan University, Ramat Gan, Israel (K.J.R.); School of Physics, Monash University, Clayton, Victoria, Australia (M.T.E.); School of Physics and Advanced Materials, University of Technology, Sydney, New South Wales, Australia (O.S.)

Nikolai Dontschuk, Alastair Stacey, Steven Prawer, Jiri Cervenka, Steven Prawer, Jiri Cervenka and Olga Shimoni
The School of Physics, The University of Melbourne, Melbourne, Victoria 3010, Australia

Kevin J. Rietwyk, Alex Schenk, Mark T. Edmonds and Chris I. Pakes
Department of Physics, La Trobe University, Bundoora, Victoria, Australia. Present addresses: Center for Nanotechnology & Advanced Materials, Bar Ilan University, Ramat Gan, Israel (K.J.R.); School of Physics, Monash University, Clayton, Victoria, Australia (M.T.E.); School of Physics and Advanced Materials, University of Technology, Sydney, New South Wales, Australia (O.S.)

Jin-Young Lee, Jae Gyeong Lee, Minjee Seo, Lilin Piao, Je Hyun Bae, Sung Yul Lim and Taek Dong Chung
Department of Chemistry, Seoul National University, Seoul 151-747, Korea

Seok-Ha Lee and Young June Park
Department of Electrical and Computer Engineering, Seoul National University, Seoul 151-744, Korea

Chao Zhao and Daniel H. Appella
Laboratory of Bioorganic Chemistry, NIDDK, NIH, DHHS, 9000 Rockville Pike, Bethesda, Maryland 20892, USA

Travis Hoppe
Laboratory of Biochemistry and Genetics, NIDDK, NIH, DHHS, 9000 Rockville Pike, Bethesda, Maryland 20892, USA

Mohan Kumar Haleyur Giri Setty and Indira Hewlett
Laboratory of Molecular Virology, Center for Biologics Evaluation and Research, FDA, 9000 Rockville Pike, Bethesda, Maryland 20892, USA

Danielle Murray and Tae-Wook Chun
Laboratory of Immunoregulation, NIAID, NIH, DHHS, 9000 Rockville Pike, Bethesda, Maryland 20892, USA

Slavomír Nemšák, Arunothai Rattanachata, Catherine S. Conlon, Armela Keqi and Charles S. Fadley
Department of Physics, University of California Davis, Davis, California 95616, USA
Materials Sciences Division, Lawrence Berkeley National Laboratory, Berkeley, California 94720, USA

Andrey Shavorskiy, Osman Karslioglu, Ioannis Zegkinoglou, , Hendrik Bluhm3
Chemical Sciences Division, Lawrence Berkeley National Laboratory, Berkeley, California 94720, USA

Peter K. Greene, Kai Liu and Edward C. Burks
Department of Physics, University of California Davis, Davis, California 95616, USA

Farhad Salmassi and Eric M. Gullikson
Materials Sciences Division, Lawrence Berkeley National Laboratory, Berkeley, California 94720, USA

See-Hun Yang
IBM Almaden Research Center, San Jose, California 95120, USA

Xing-You Lang, Hong-Ying Fu, Chao Hou, Gao-Feng Han, Yong-Bing Liu and Qing Jiang
Key Laboratory of Automobile Materials, Ministry of Education, and School of Materials Science and Engineering, Jilin University, Changchun 130022, China

Ping Yang
Cardiovascular medicine, Sino-Japan Friendship Hospital, Jilin University, Changchun 130033, China

Stefan Harmsen and Ruimin Huang
Department of Radiology, Memorial Sloan Kettering Cancer Center, 1275 York Avenue, New York, New York 10065, USA

Matthew A. Bedics and Michael R. Detty
Department of Chemistry, University at Buffalo, The State University of New York, Buffalo, New York 14260-3000, USA

Matthew A. Wall
Department of Radiology, Memorial Sloan Kettering Cancer Center, 1275 York Avenue, New York, New York 10065, USA
Department of Chemistry, Hunter College of the City University of New York, 695 Park Avenue, New York, New York 10065, USA

Moritz F. Kircher
Department of Radiology, Memorial Sloan Kettering Cancer Center, 1275 York Avenue, New York, New York 10065, USA
Center for Molecular Imaging and Nanotechnology (CMINT), Memorial Sloan

Kettering Cancer Center, 1275 York Avenue, New York, New York 10065, USA
Department of Radiology, Weill Cornell Medical College, 445 East 69th Street, New York, New York 10021, USA

Maged F. Serag, Maram Abadi and Satoshi Habuchi
Biological and Environmental Sciences and Engineering Division, King Abdullah University of Science and Technology (KAUST), Thuwal 23955-6900, Saudi Arabia

Justin D. Besant and Wenhan Liu
Institute for Biomaterials and Biomedical Engineering, University of Toronto, Toronto, Canada M5S 3G9

Jagotamoy Das and Ian B. Burgess
Department of Pharmaceutical Science, Leslie Dan Faculty of Pharmacy, University of Toronto, Toronto, Canada M5S 3M2

Edward H. Sargent
Department of Electrical and Computer Engineering, Faculty of Engineering, University of Toronto, Toronto, Canada M5S 3G4

Shana O. Kelley
Institute for Biomaterials and Biomedical Engineering, University of Toronto, Toronto, Canada M5S 3G9
Department of Pharmaceutical Science, Leslie Dan Faculty of Pharmacy, University of Toronto, Toronto, Canada M5S 3M2
Department of Biochemistry, Faculty of Medicine, University of Toronto, Toronto, Canada M5S 1A8

Yiqing Lu, Jie Lu, Jiangbo Zhao, James A. Piper and Ewa M. Goldys
Advanced Cytometry Laboratories, ARC Centre of Excellence for Nanoscale BioPhotonics (CNBP), Macquarie University, Sydney, New South Wales 2109, Australia

Janet Cusido and Francisco M. Raymo
Laboratory for Molecular Photonics, University of Miami, 1301 Memorial Drive, Coral Gables, Florida 33146-0431, USA

Jingli Yuan
State Key Laboratory of Fine Chemicals, School of Chemistry, Dalian University of Technology, Dalian 116024, China

Sean Yang and Robert C. Leif
Newport Instruments, 3345 Hopi Place, San Diego, California 92117-3516, USA

Yujing Huo
Department of Electronic Engineering, Tsinghua University, Beijing 100084, China

J. Paul Robinson
Purdue University Cytometry Laboratories, Bindley Bioscience Center, Purdue University, West Lafayette, Indiana 47907, USA.

Dayong Jin
Advanced Cytometry Laboratories, ARC Centre of Excellence for Nanoscale BioPhotonics (CNBP), Macquarie University, Sydney, New South Wales 2109, Australia
Purdue University Cytometry Laboratories, Bindley Bioscience Center, Purdue University, West Lafayette, Indiana 47907, USA

Alice Mattiuzzi and Ivan Jabin
Laboratoire de Chimie Organique, Université Libre de Bruxelles (U.L.B.), CP 160/06, 50 avenue F.D. Roosevelt, 1050 Brussels, Belgium

Claire Mangeney
ITODYS, Université Paris Diderot-Paris 7 and CNRS, UMR n°7086, 15 rue Jean de Baïf, 75013 Paris, France

Clément Roux
Department of Chemistry, University of Canterbury, MacDiarmid Institute for advanced materials and nanotechnology, Christchurch, New Zealand

Olivia Reinaud
Laboratoire de Chimie et Biochimie Pharmacologiques et Toxicologiques, PRES Sorbonne Paris Cité, Université Paris Descartes and CNRS, UMR n°8601, 45 rue des Saints-Pères, 75006 Paris, France

Luis Santos, Jean-François Bergamini, Philippe Hapiot and Corinne Lagrost
Sciences Chimiques de Rennes, Equipe MaCSE, Université de Rennes 1 and CNRS, UMR n° 6226, Campus de Beaulieu, 35042 Rennes cedex, France

Hideaki Ogata, Tobias Krämer, Maurice van Gastel, Frank Neese and Wolfgang Lubitz
Max Planck Institute for Chemical Energy Conversion, Mülheim an der Ruhr 45470, Germany

Hongxin Wang and Stephen P. Cramer
Department of Chemistry, University of California, Davis, California 95616, USA
Division of Physical Biosciences, Lawrence Berkeley National Laboratory, Berkeley, California 94720, USA

David Schilter and Thomas B. Rauchfuss
Department of Chemistry, University of Illinois, Urbana, Illinois 61801, USA

Vladimir Pelmenschikov
Institut für Chemie, Technische Universität Berlin, Berlin 10623, Germany

Leland B. Gee and Aubrey D. Scott
Department of Chemistry, University of California, Davis, California 95616, USA

Yoshitaka Yoda
Division of Research and Utilization, SPring-8/JASRI, Hyogo 679-5198, Japan

Yoshihito Tanaka
Materials Dynamics Laboratory, RIKEN SPring-8, Hyogo 679-5148, Japan

Amir Lichtenstein, Ehud Havivi, Ronen Shacham and Igor Presman
Tracense Ltd., Hanadiv 71 Street, Hertzelia 46485, Israel

Ehud Hahamy and Ronit Leibovich
Tracense Ltd., Hanadiv 71 Street, Hertzelia 46485, Israel
School of Mathematics, The Raymond and Beverly Sackler Faculty of Exact Sciences Tel-Aviv University, Tel Aviv 69978, Israel

Alexander Pevzner, Vadim Krivitsky, Guy Davivi, Roey Elnathan and Yoni Engel
School of Chemistry, The Raymond and Beverly Sackler Faculty of Exact Sciences Tel-Aviv University, Tel Aviv 69978, Israel

Eli Flaxer
School of Chemistry, The Raymond and Beverly Sackler Faculty of Exact Sciences Tel-Aviv University, Tel Aviv 69978, Israel
Afeka Tel Aviv Academic College of Engineering, Tel Aviv 69978, Israel

Fernando Patolsky
School of Chemistry, The Raymond and Beverly Sackler Faculty of Exact Sciences Tel-Aviv University, Tel Aviv 69978, Israel
The Center for Nanoscience and Nanotechnology, Tel-Aviv University, Tel Aviv 69978, Israel
Department of Materials Science and Engineering, The Iby and Aladar Fleischman Faculty of Engineering, Tel-Aviv University, Tel Aviv 69978, Israel

Jung Lee, Hyun Taek Chang, Hyosung An, Sora Ahn and Jina Shim
Department of Chemical Engineering, Hanyang University, Seoul 133-791, Korea

Jong-Man Kim
Department of Chemical Engineering, Hanyang University, Seoul 133-791, Korea. Institute of Nano Science and Technology, Hanyang University, Seoul 133-791, Korea

Yasufumi Takahashi and Tomokazu Matsue
WPI-Advanced Institute for Materials Research, Tohoku University, Sendai 980-8577, Japan
Graduate School of Environmental Studies, Tohoku University, Sendai 980-8579, Japan

Patrick R. Unwin
Department of Chemistry, University of Warwick, Coventry CV4 7AL, UK

Akichika Kumatani
WPI-Advanced Institute for Materials Research, Tohoku University, Sendai 980 8577, Japan

Hirokazu Munakata and Kiyoshi Kanamura
Graduate School of Urban Environmental Sciences, Tokyo Metropolitan University, Tokyo 192-0397, Japan

Hirotaka Inomata, Komachi Ito, Kosuke Ino and Hitoshi Shiku
Graduate School of Environmental Studies, Tohoku University, Sendai 980-8579, Japan

Yuri E. Korchev
Division of Medicine, Imperial College London, London W12 0NN, UK

Jiun-Yi Shen, Wei-Chih Chao, Chun Liu, Hsiao-An Pan, Steven Chun-Wei Chou, Kuo-Chun Tang and Pi-Tai Chou
Department of Chemistry, Center for Emerging Material and Advanced Devices, National Taiwan University, Taipei 10617, Taiwan

Hsiao-Ching Yang and Yi-Kang Lan
Department of Chemistry, Fu-Jen Catholic University, New Taipei City 24205, Taiwan

Chi-Lin Chen
Department of Chemistry, Center for Emerging Material and Advanced Devices, National Taiwan University, Taipei 10617, Taiwan
Department of Chemistry, Fu-Jen Catholic University, New Taipei City 24205, Taiwan

Li-Ju Lin, Jinn-Shyan Wang and Jyh-Feng Lu
School of Medicine, Fu-Jen Catholic University, New Taipei City 24205, Taiwan

Z. Pan, N. Rawat, I. Cour, L. Manning, R.L. Headrick and M. Furis
Department of Physics, Materials Science Program, University of Vermont, Burlington, Vermont 05405, USA. w Present Addresses: CGG, Houston, Texas 77072, USA (Z.P.); Intel Corporation, Hillsboro, Oregon 97124, USA (I.C.)

Jonathan Martens, Josipa Grzetic and Giel Berden
Radboud University, Institute for Molecules and Materials, FELIX Laboratory, Toernooiveld 7c, 6525ED Nijmegen, The Netherlands

Jos Oomens
Radboud University, Institute for Molecules and Materials, FELIX Laboratory, Toernooiveld 7c, 6525ED Nijmegen, The Netherland
Van 't Hoff Institute for Molecular Sciences, University of Amsterdam, Science Park 908, 1098XH Amsterdam, The Netherlands

Tanmay Sarkar, Karuthapandi Selvakumar, Leila Motiei and David Margulies
Department of Organic Chemistry,Weizmann Institute of Science, Rehovot 7610001, Israel

Nebiyu Abshiru and Pierre Thibault
Department of Chemistry, Université de Montréal, PO Box 6128, Station centre-ville, Montréal, Québec, Canada H3C 3J7
Institute for Research in Immunology and Cancer, Université de Montréal, C.P. 6128, Succursale centre-ville, Montréal, Québec, Canada H3C 3J7

Roshan Elizabeth Rajan
Institute for Research in Immunology and Cancer, Université de Montréal, C.P. 6128, Succursale centre-ville, Montréal, Québec, Canada H3C 3J7
Molecular Biology Programme, Université de Montréal, PO Box 6128, Station centre-ville, Montréal, Québec, Canada H3C 3J7

Alain Verreault
Department of Chemistry, Université de Montréal, PO Box 6128, Station centre-ville, Montréal, Québec, Canada H3C 3J7
Molecular Biology Programme, Université de Montréal, PO Box 6128, Station centre-ville, Montréal, Québec, Canada H3C 3J7

Olivier Caron-Lizotte, Adil Jamai and Christelle Pomies
Institute for Research in Immunology and Cancer, Université de Montréal, C.P. 6128, Succursale centre-ville, Montréal, Québec, Canada H3C 3J7

Meng Liu and Yingfu Li
Department of Biochemistry and Biomedical Sciences, McMaster University, 1280 Main Street West, Hamilton, Ontario, Canada L8S 4K1
Biointerfaces Institute, McMaster University, 1280 Main Street West, Hamilton, Ontario, Canada L8S 4O3

Qiang Zhang and John D. Brennan
Biointerfaces Institute, McMaster University, 1280 Main Street West, Hamilton, Ontario, Canada L8S 4O3

Zhongping Li and Jimmy Gu
Department of Biochemistry and Biomedical Sciences, McMaster University, 1280 Main Street West, Hamilton, Ontario, Canada L8S 4K1

Susanna Ahola, Anu M. Kantola, Ville-Veikko Telkki and Otto Mankinen
Faculty of Science, NMR Research Group, University of Oulu, PO Box 3000, 90014 Oulu, Finland

Vladimir V. Zhivonitko and Igor V. Koptyug
Laboratory of Magnetic Resonance Microimaging, International Tomography Center SB RAS, Instututskaya Street 3A, 630090 Novosibirsk, Russia Department of Natural Sciences, Novosibirsk State University, Pirogova Street 2, 630090 Novosibirsk, Russia

Guannan Zhang, Hsueh-Ying Chen and Christian Hilty
Department of Chemistry, Texas A&M University, 3255 TAMU, College Station, Texas 77843-3255, USA

Zhengke Wang, Jingyi Nie and Qiaoling Hu
MOE Key Laboratory of Macromolecular Synthesis and Functionalization, Department of Polymer Science and Engineering, Zhejiang University, Hangzhou 310027, China
Key Laboratory of Adsorption and Separation Materials and Technologies of Zhejiang Province, Hangzhou 310027, China

Wei Qin and Ben Zhong Tang
Department of Chemistry, Hong Kong Branch of Chinese National Engineering Research Center for Tissue Restoration and Reconstruction, Hong Kong University of Science and Technology, Clear Water Bay, Hong Kong 999077, China

Pierre-Yves Colin, Balint Kintses, Fabrice Gielen, Charlotte M. Miton, Gerhard Fischer, Mark F. Mohamed, Marko Hyvönen and Florian Hollfelder
Department of Biochemistry, University of Cambridge, 80 Tennis Court Road, Cambridge CB2 1GA, UK

Diego P. Morgavi
INRA, UMR1213 Herbivores, F-63122 Saint-Genès-Champanelle, France. 3 Clermont Université, VetAgro Sup, UMR Herbivores, BP 10448, F-63000 Clermont-Ferrand, France

Dick B. Janssen
Groningen Biomolecular Science and Biotechnology Institute, University of Groningen, Nijenborgh 4, 9747 AG, Groningen, The Netherlands

Index